Pierre Meystre   Murray Sargent III

Elements of Quantum Optics

Second Edition

With 123 Figures

Springer-Verlag
Berlin Heidelberg New York
London Paris Tokyo
Hong Kong Barcelona
Budapest

Professor Dr. Dr. Rer. Nat. Habil. Pierre Meystre
Professor Murray Sargent III, Ph. D.
Optical Sciences Center, University of Arizona, Tucson, AZ 85721, USA

ISBN 3-540-54190-X Springer-Verlag Berlin Heidelberg New York
ISBN 0-387-54190-X Springer-Verlag New York Berlin Heidelberg

ISBN 3-540-52160-7 1. Auflage Springer-Verlag Berlin Heidelberg New York
ISBN 0-387-52160-7 1st edition Springer-Verlag New York Berlin Heidelberg

Library of Congress Cataloging-in-Publication Data
Meystre, Pierre. Elements of quantum optics / Pierre Meystre, Murray Sargent III. – 2nd ed. p.cm. Includes bibliographical references and index.
ISBN 3-540-54190-X (Berlin). – ISBN 0-387-54190-X (U.S.)
1. Quantum optics. I. Sargent, Murray. II. Title. QC446.2.M48 1991 535–dc20 91-21949

Typesetting: Camera ready by author
This text was prepared using the PS™ Technical Word Processor
54/3140-543210 – Printed on acid-free paper

To

Renée and Pierre-André
Helga, Nicole, and Tina

Preface

This book grew out of a 2-semester graduate course in laser physics and quantum optics. It requires a solid understanding of elementary electromagnetism as well as at least one, but preferably two, semesters of quantum mechanics. Its present form resulted from many years of teaching and research at the University of Arizona, the Max-Planck-Institut für Quantenoptik, and the University of Munich. The contents have evolved significantly over the years, due to the fact that quantum optics is a rapidly changing field. Because the amount of material that can be covered in two semesters is finite, a number of topics had to be left out or shortened when new material was added. Important omissions include the manipulation of atomic trajectories by light, superradiance, and descriptions of experiments.

Rather than treating any given topic in great depth, this book aims to give a broad coverage of the basic elements that we consider necessary to carry out research in quantum optics. We have attempted to present a variety of theoretical tools, so that after completion of the course students should be able to understand specialized research literature and to produce original research of their own. In doing so, we have always sacrificed rigor to physical insight and have used the concept of "simplest nontrivial example" to illustrate techniques or results that can be generalized to more complicated situations. In the same spirit, we have not attempted to give exhaustive lists of references, but rather have limited ourselves to those papers and books that we found particularly useful.

The book is divided into three parts. Chapters 1 through 3 review various aspects of electromagnetic theory and of quantum mechanics. The material of these chapters, especially Chaps. 1 and 3, represents the minimum knowledge required to follow the rest of the course. Chapter 2 introduces many nonlinear optics phenomena by using a classical nonlinear oscillator model, and is usefully referred to in later chapters. Depending on the level at which the course is taught, one can skip Chaps. 1 through 3 totally or, at the other extreme, give them considerable emphasis.

Chapters 4 through 11 treat semi-classical light-matter interactions. They contain more material than we have typically been able to teach in a one-semester course. Especially if much time is spent on the Chaps. 1 - 3, some of Chaps. 4 - 11 must be skipped. However, Chap. 4 on the density matrix, Chap. 5 on the interaction between matter and cw fields, Chap. 6 on semi-classical laser theory, and to some extent Chap. 8 on nonlinear spectroscopy are central to the book and cannot be ignored. In contrast one could omit Chap. 7 on optical bistability, Chap. 9 on phase conjugation, Chap. 10 on optical instabilities, or Chap. 11 on coherent transients.

Chapters 12 through 17 discuss aspects of light-matter interaction that require the quantization of the electromagnetic field. They are tightly knit together and it is difficult to imagine skipping one of them in a one-semester course. Chapter 12 draws an analogy between electromagnetic field modes and harmonic oscillators to quantize the field in a simple way. Chapter 13 discusses simple aspects of the interaction between a single mode of the field and a two-level atom. Chapter 14 on reservoir theory is essential for the discussion of resonance fluorescence (Chap. 15) and squeezing (Chap. 16). These chapters are strongly connected to the nonlinear spectroscopy discussion of Chap. 8. In resonance fluorescence and in squeezing the quantum nature of the field appears mostly in the form of noise. We conclude in Chap. 17 by giving elements of the quantum theory of the laser, which requires a proper treatment of quantum fields to all orders.

In addition to being a textbook, this book contains many important formulas in quantum optics that are not found elsewhere except in the original literature or in specialized monographs. As such, and certainly for our own research, this book is a very valuable reference. One particularly gratifying feature of the book is that it reveals the close connection between many seemingly unrelated or only distantly related topics, such as probe absorption, four-wave mixing, optical instabilities, resonance fluorescence, and squeezing.

We are also indebted to the many people who have made important contributions to this book: they include first of all our students, who had to suffer through several not-so-debugged versions of the book and have helped with their corrections and suggestions. Special thanks to S. An, B. Capron, T. Carty, P. Dobiasch, J. Grantham, A. Guzman, D. Holm, J. Lehan, R. Morgan, M. Pereira, G. Reiner, E. Schumacher, J. Watanabe, and M. Watson. We are also very grateful to many colleagues for their encouragements and suggestions. Herbert Walther deserves more thanks than anybody else: this book would not have been started or completed without his constant encouragement and support. Thanks are due especially to the late Fred Hopf as well as to J. H. Eberly, H. M. Gibbs, J. Javanainen, S. W. Koch, W. E. Lamb, Jr., H. Pilloff, C. M. Savage, M. O. Scully, D. F. Walls, K. Wodkiewicz, and E. M. Wright. We are also indebted to the Max-Planck-Institut für Quantenoptik and to the U.S. Office of Naval Research for direct or indirect financial support of this work.

The text and figures of this book were prepared and printed using the Scroll Systems PS™ Technical Word Processor.

Tucson, August 1989

Pierre Meystre

Murray Sargent III

Preface to the Second Edition

This edition contains a significant number of changes designed to improve clarity. We have also added a new section on the theory of resonant light pressure and the manipulation of atomic trajectories by light. This topic is of considerable present interest and has applications both in high resolution spectroscopy and in the emerging field of atom optics. Smaller changes include a reformulation of the photon-echo problem in a way that reveals its relationship to four-wave mixing, as well as a discussion of the quantization of standing-waves versus running-waves of the electromagnetic field. Finally, we have also improved a number of figures and have added some new ones.

We thank the readers who have taken the time to point out to us a number of misprints. Special thanks are due to Z. Bialynicka-Birula, S. Haroche, K. Just, S. LaRochelle, E. Schumacher, and M. Wilkens.

Tucson, February 1991

P. M. M. S. III

Readers interested in the latest electronic version of this book can contact Scroll Systems, Inc., 5530 N. Camino Escuela, Tucson, AZ 85718.

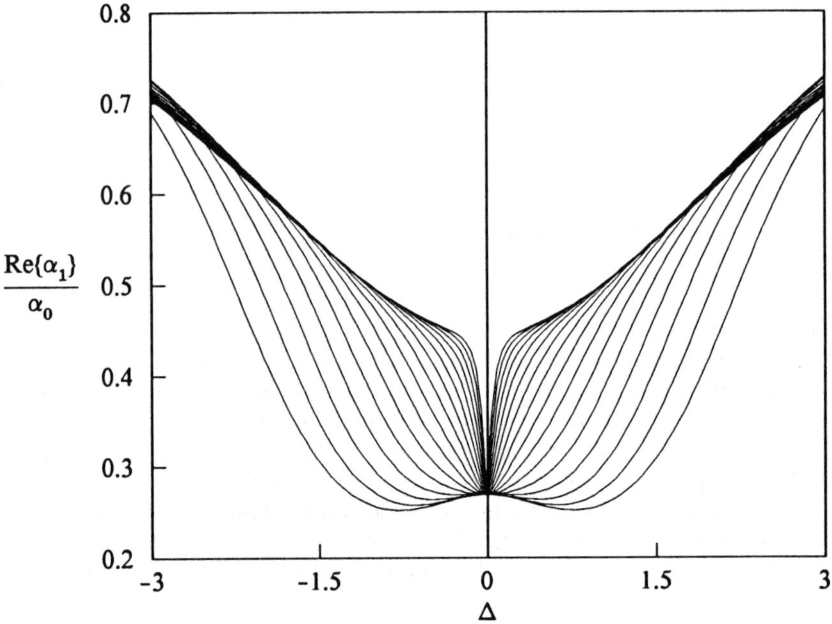

Probe absorption in an inhomogeneously broadened two-level medium versus pump-probe beat frequency Δ for the dipole decay time $T_2 = 1$, the population difference decay time $T_1 = T_2, \ldots, 64T_2$, and dimensionless pump intensity $I_2 = 4$. The physics illustrated in this figure is dicussed in Sec. 8-3. α_1 is given by Eq. (8.28) as approximated by Eqs. (8.48) and (5.52). The reduction of absorption for $|\Delta|$ < the resonant Rabi frequency $\mathcal{R}_0$ (= 2 for $T_1 = T_2 = 1$) occurs due to the addition of curves like those in Figs. 8-5 and 8-7, which acquire appreciable values only at and beyond the generalized Rabi frequency $\mathcal{R}$.

CONTENTS

Chapter 1
CLASSICAL ELECTROMAGNETIC FIELDS

In this book we present the basic ideas needed to understand how laser light interacts with various kinds of matter. Among the important consequences is an understanding of the laser itself. The present chapter summarizes classical electromagnetic fields, which describe laser light remarkably well. The chapter also discusses the interaction of these fields with a medium consisting of classical simple harmonic oscillators. It is surprising how well this simple model describes linear absorption, a point discussed from a quantum mechanical point of view in Sec. 3-3. The rest of the book is concerned with *nonlinear* interactions of radiation with matter. Chapter 2 generalizes the classical oscillator to treat simple kinds of nonlinear mechanisms, and shows us a number of phenomena in a relatively simple context. Starting with Chap. 3, we treat the medium quantum mechanically. The combination of a classical description of light and a quantum mechanical description of matter is called the *semiclassical* approximation. This approximation is not always justified (see Chaps. 12 through 18), but there are remarkably few cases in quantum optics where we need to quantize the field.

In the present chapter, we limit ourselves both to *classical* electromagnetic fields and to *classical* media. Section 1-1 briefly reviews Maxwell's equations in a vacuum. We derive the wave equation, and introduce the slowly-varying amplitude and phase approximation for the electromagnetic field. Section 1-2 recalls Maxwell's equations in a medium. We then show the roles of the in-phase and in-quadrature parts of the polarization of the medium through which the light propagates, and give a brief discussion of Beer's law of light absorption. Section 1-3 discusses the classical dipole oscillator. We introduce the concept of the self field and show how it leads to radiative damping. Then we consider the classical Rabi problem, which allows us to introduce the classical analog of the optical Bloch equations. The derivations in Sec. 1-1 to 1-3 are not necessarily the simplest ones, but they correspond as closely as possible to their quantum mechanical counterparts that appear later in the book.

Section 1-4 is concerned with the coherence of the electromagnetic field. We review the Young and Hanbury Brown-Twiss experiments. We introduce the notion of *n*th order coherence. We conclude this section by a brief comment on antibunching, which provides us with a powerful test of the quantum nature of light.

With knowledge of Secs. 1-1 through 1-4, we have all the elements needed to understand an elementary treatment of the free-electron laser (FEL), which is presented in Sec. 1-5. The FEL is in some ways the simplest laser to understand, since it can largely be described classically, i.e., there is no need to quantize the matter.

1-1. Maxwell's Equations in a Vacuum

In the absence of charges and currents, Maxwell's equations are given by

$$\nabla \cdot \mathbf{B} = 0 \tag{1}$$

$$\nabla \cdot \mathbf{E} = 0 \tag{2}$$

$$\nabla \times \mathbf{E} = -\frac{\partial \mathbf{B}}{\partial t} \tag{3}$$

$$\nabla \times \mathbf{B} = \mu_0 \epsilon_0 \frac{\partial \mathbf{E}}{\partial t}, \tag{4}$$

where **E** is the electric field, **B** is the magnetic field, μ_0 is the permeability of free space, and ϵ_0 is the permittivity of free space (in this book we use MKS units throughout). Alternatively it is useful to write c^2 for $1/\mu_0\epsilon_0$, where c is the speed of light in the vacuum. Taking the curl of Eq. (3) and substituting the time rate of change of Eq. (4), we find

$$\nabla \times \nabla \times \mathbf{E} = -\frac{1}{c^2}\frac{\partial^2 \mathbf{E}}{\partial t^2}. \tag{5}$$

This equation can be simplified by noting that $\nabla\times\nabla = \nabla(\nabla\cdot\ ) - \nabla^2$ and using Eq. (2). We find the wave equation

$$\nabla^2 \mathbf{E} - \frac{1}{c^2}\frac{\partial^2 \mathbf{E}}{\partial t^2} = 0. \tag{6}$$

This tells us how an electromagnetic wave propagates in a vacuum. By direct substitution, we can show that

$$\mathbf{E}(\mathbf{r},t) = \mathbf{E}_0\, f(\mathbf{K}\cdot\mathbf{r} - \nu t) \tag{7}$$

is a solution of Eq. (6) where f is an arbitrary function, $\mathbf{E}_0$ is a constant, ν is an oscillation frequency in radians/second ($2\pi\times Hz$), $\mathbf{K}$ is a constant vector in the direction of propagation of the field, and having the magnitude $K \equiv |\mathbf{K}| = \nu/c$. This solution represents a transverse plane wave propagating along the direction of $\mathbf{K}$ with speed $c = \nu/K$.

A property of the wave equation (6) is that if $\mathbf{E}_1(\mathbf{r},t)$ and $\mathbf{E}_2(\mathbf{r},t)$ are solutions, then the superposition $a_1\mathbf{E}_1(\mathbf{r},t) + a_2\mathbf{E}_2(\mathbf{r},t)$ is also a solution, where a_1 and a_2 are any two constants. This is called the principle of superposition. It is a direct consequence of the fact that differentiation is a linear operation. In particular, the superposition

$$\mathbf{E}(\mathbf{r},t) = \sum_k \mathbf{E}_k\, f(\mathbf{K}_k\cdot\mathbf{r} - \nu t) \tag{8}$$

is also a solution. This shows us that non-plane waves also are solutions of the wave equation (6).

Quantum opticians like to decompose electric fields into "positive" and "negative" frequency parts

$$\mathbf{E}(\mathbf{r},t) = \mathbf{E}^{+}(\mathbf{r},t) + \mathbf{E}^{-}(\mathbf{r},t)\ , \tag{9}$$

where $\mathbf{E}^{+}(\mathbf{r},t)$ has the form

$$\mathbf{E}^{+}(\mathbf{r},t) = \frac{1}{2}\sum_n \mathcal{E}_n(\mathbf{r})\, e^{-i\nu_n t}\ , \tag{10}$$

where $\mathcal{E}_n(\mathbf{r})$ is a complex function of $\mathbf{r}$, ν_n is the corresponding frequency, and in general

$$\mathbf{E}^{-}(\mathbf{r},t) = (\mathbf{E}^{+}(\mathbf{r},t))^{*}\ . \tag{11}$$

In itself this decomposition is just that of the analytic signal used in classical coherence theory (see Born and Wolf (1970)), but as we see in Chap. 12, it has deep foundations in the quantum theory of light detection. For now we consider this to be a convenient mathematical trick that allows us to work with exponentials rather than with sines and cosines. It is easy to see that since the wave equation (6) is real, if $\mathbf{E}^{+}(\mathbf{r},t)$ is a solution, then so is $\mathbf{E}^{-}(\mathbf{r},t)$, and the linearity of Eq. (6) guarantees that the sum (9) is also a solution.

In this book, we are concerned mostly with the interaction of monochromatic (or quasi-monochromatic) laser light with matter. In particular, consider a linearly-polarized plane wave propagating in the z-direction. Its electric field can be described by

$$\mathbf{E}^{+}(z,t) = \frac{1}{2}\,\hat{\mathbf{x}} E_0(z,t)\, e^{i(Kz-\nu t-\phi(z,t))}\ , \tag{12}$$

where $\hat{\mathbf{x}}$ is the direction of polarization, $E_0(z,t)$ is a real amplitude, ν is the central frequency of the field, and the wave number $K = \nu/c$. If $E(z,t)$ is truly monochromatic, E_0 and ϕ are constants in time and space. More generally, we suppose they vary sufficiently slowly in time and space that the following inequalities are valid

$$\left|\frac{\partial E_0}{\partial t}\right| << \nu E_0 \tag{13}$$

$$\left|\frac{\partial E_0}{\partial z}\right| << K E_0 \tag{14}$$

$$\left|\frac{\partial \phi}{\partial t}\right| << \nu \tag{15}$$

$$\left|\frac{\partial \phi}{\partial z}\right| << K\ . \tag{16}$$

These equations define the so-called *slowly-varying amplitude and phase approximation (SVAP)*, which plays a central role in laser physics and pulse propagation problems. Physically it means that we consider light waves whose amplitudes and phases vary little within an optical period and an optical wavelength. Sometimes this approximation is called the SVEA, for *slowly-varying envelope approximation.*

The SVAP leads to major mathematical simplification as can be seen by substituting the field Eq. (12) into the wave equation (6) and using Eqs. (13) through (16) to eliminate the small contributions $\ddot{E}_0$, $\ddot{\phi}$, E_0'', ϕ'', and $\dot{E}\dot{\phi}$. We find

$$\frac{\partial E_0}{\partial z} + \frac{1}{c}\frac{\partial E_0}{\partial t} = 0 \tag{17}$$

$$\frac{\partial \phi}{\partial z} + \frac{1}{c}\frac{\partial \phi}{\partial t} = 0\ , \tag{18}$$

where Eq. (17) results from equating the sum of the imaginary parts to zero and Eq. (18) from the real parts. Thus the SVAP allows us to transform the second-order wave equation (6) into first-order equations. Although this does not seem like much of an achievement right now, since we can solve Eq. (6) exactly anyway, it is a tremendous help when we consider Maxwell's equations in a medium. The SVAP is not always a good approximation. For example, plasma physicists who shine light on targets typically must use the second-order equations.

1-2. Maxwell's Equations in a Medium

Inside a macroscopic medium, Maxwell's equations (1) through (4) become

$$\nabla\cdot\mathbf{B} = 0 \tag{19}$$

$$\nabla\cdot\mathbf{D} = \rho_{\text{free}} \tag{20}$$

$$\nabla\times\mathbf{E} = -\frac{\partial\mathbf{B}}{\partial t} \tag{21}$$

$$\nabla\times\mathbf{H} = \mathbf{J} + \frac{\partial\mathbf{D}}{\partial t}\,. \tag{22}$$

These equations are often called the *macroscopic* Maxwell's equations, since they relate vectors that are averaged over volumes containing many atoms but which have linear dimensions small compared to significant variations in the applied electric field. General derivations of Eqs. (19) - (22) can be very complicated, but the discussion by Jackson (1962) is quite readable. In Eqs. (20) and (22), the displacement electric field **D** is given for our purposes by

$$\mathbf{D} = \epsilon\mathbf{E} + \mathbf{P}\,, \tag{23}$$

where the permittivity ϵ includes the contributions of the host lattice and **P** is the induced polarization of the resonant or nearly resonant medium we wish to treat explicitly. For example in ruby the Al_2O_3 lattice has an index of refraction of 1.76, which is included in ϵ. The ruby color is given by *Cr* ions which are responsible for laser action. We describe their interaction with light by the polarization **P**. Indeed much of this book deals with the calculation of **P** for various situations. The free charge density ρ_{free} in Eq. (20) consists of all charges other than the bound charges inside atoms and molecules, whose effects are provided for by **P**. We don't need ρ_{free} in this book. In Eq. (22), the magnetic field **H** is given by

$$\mathbf{H} = \frac{\mathbf{B}}{\mu} - \mathbf{M}\,, \tag{24}$$

where μ is the permeability of the host medium and $\mathbf{M}$ is the magnetization of the medium. For the media we consider, $\mathbf{M} = 0$ and $\mu = \mu_0$. The current density $\mathbf{J}$ is often related to the applied electric field $\mathbf{E}$ by the constitutive relation $\mathbf{J} = \sigma\mathbf{E}$, where σ is the conductivity of the medium.

The macroscopic wave equation corresponding to Eq. (6) is given by combining the curl of Eq. (21) with Eqs. (23) and (24). In the process we find $\nabla\times\nabla\times\mathbf{E} = \nabla(\nabla\cdot\mathbf{E}) - \nabla^2\mathbf{E} \simeq -\nabla^2\mathbf{E}$. In optics $\nabla\cdot\mathbf{E} \simeq 0$, since most light field vectors vary little along the directions in which they point. For example, a plane-wave field is constant along the direction it points, causing its $\nabla\cdot\mathbf{E}$ to vanish identically. We find

$$\boxed{-\nabla^2\mathbf{E} + \mu\frac{\partial\mathbf{J}}{\partial t} + \frac{1}{c^2}\frac{\partial^2\mathbf{E}}{\partial t^2} = -\mu\frac{\partial^2\mathbf{P}}{\partial t^2}}\,, \tag{25}$$

where $c = 1/\sqrt{\epsilon\mu}$ is now the speed of light in the host medium. In Chap. 6 we use the $\partial\mathbf{J}/\partial t$ term to simulate losses in a Fabry-Perot resonator. We drop this term in our present discussion.

For a quasi-monochromatic field, the polarization induced in the medium is also quasi-monochromatic, but generally has a different phase from the field. Thus as for the field (5) we decompose the polarization into positive and negative frequency parts

$$\mathbf{P}(z,t) = \mathbf{P}^+(z,t) + \mathbf{P}^-(z,t),$$

but we include the complex amplitude $\mathscr{P}(z,t) = N\mathscr{d}X(z,t)$, that is,

$$\begin{aligned}\mathbf{P}^+(z,t) &= \tfrac{1}{2}\hat{\mathbf{x}}\mathscr{P}(z,t)e^{i(Kz-\nu t-\phi(z,t))}\\ &= \tfrac{1}{2}\hat{\mathbf{x}}N(z)\mathscr{d}X(z,t)\,e^{i(Kz-\nu t-\phi(z,t))}\,.\end{aligned} \tag{26}$$

Here $N(z)$ is the number of systems per unit volume, $\mathscr{d}$ is the dipole moment constant of a single oscillator, and $X(z,t)$ is a complex dimensionless amplitude that varies little in an optical period or wavelength. In quantum mechanics, $\mathscr{d}$ is given by the electric dipole matrix element $\wp$. Since the polarization is real, we have

$$\mathbf{P}^-(z,t) = (\mathbf{P}^+(z,t))^*\,. \tag{27}$$

It is sometimes convenient to write $X(z,t)$ in terms of its real and imaginary parts in the form

$$X \equiv U - iV\,. \tag{28}$$

The classical real variables U and V have quantum mechanical counterparts that are components of the Bloch vector $U\hat{e}_1 + V\hat{e}_2 + W\hat{e}_3$, as discussed in Sec. 4-3. The slowly-varying amplitude and phase approximation for the polarization is given by

$$\left|\frac{\partial U}{\partial t}\right| << \nu|U| \tag{29}$$

$$\left|\frac{\partial V}{\partial t}\right| << \nu|V|\,. \tag{30}$$

or equivalently by

$$\left|\frac{\partial X}{\partial t}\right| << \nu|X|\,.$$

We generalize the slowly-varying Maxwell equations (17) and (18) to include the polarization by treating the left hand side of the wave equation (25) as before and substituting Eq. (26) into the right hand side of (25). Using Eqs. (29) and (30) to eliminate the time derivatives of U and V and equating real and imaginary parts separately, we find

$$\frac{\partial E_0}{\partial z} + \frac{1}{c}\frac{\partial E_0}{\partial t} = -\frac{K}{2\epsilon}\,\mathrm{Im}\{\mathscr{P}\} = \frac{K}{2\epsilon}\,N(z)\wp V \tag{31}$$

$$E_0\left[\frac{\partial \phi}{\partial z} + \frac{1}{c}\frac{\partial \phi}{\partial t}\right] = -\frac{K}{2\epsilon}\,\mathrm{Re}\{\mathscr{P}\} = -\frac{K}{2\epsilon}\,N(z)\wp\, U\,. \tag{32}$$

These two equations play a central role in quantum optics. They tell us how light propagates through a medium and specifically how the real and imaginary parts of the polarization act. Equation (31) shows that the field amplitude is driven by the *imaginary* part of the polarization. This *in-quadrature* component gives rise to absorption and emission.

Equation (32) allows us to compute the phase velocity with which the electromagnetic wave propagates in the medium. It is the real part of the polarization, i.e., the part *in-phase* with the field, that determines the phase velocity. The effects described by this equation are those associated with the *index of refraction* of the medium, such as dispersion and self focusing.

Equations (31) and (32) alone are not sufficient to describe physical problems completely, since they only tell us how a plane electromagnetic wave responds to a given polarization of the medium. That polarization must still be determined. Of course, we know that the polarization of a medium is influenced by the field to which it is subjected. In particular, for atoms or molecules without permanent polarization, it is the electromagnetic field itself that induces their polarization! Thus the polarization of the medium drives the field, while the field drives the polarization of the medium. In general this leads to a description of the interaction between the electromagnetic field and matter expressed in terms of coupled, nonlinear, partial differential equations that have to be solved *self-consistently*. The polarization of a medium consisting of classical simple harmonic oscillators is discussed in Sec. 1-3 and Chap. 2 discusses similar media with anharmonic (nonlinear) oscillators. Two-level atoms are discussed in Chaps. 3 through 6.

There is no known general solution to the problem, and the art of quantum optics is to make reasonable approximations in the description of the field and/or medium valid for cases of interest. Two general classes of problems reduce the partial differential equations to ordinary differential equations: 1) problems for which the amplitude and phase vary only in time, e.g., in a cavity, and 2) problems for which they vary only in space, i.e., a steady state exists. The second of these leads to Beer's law of absorption,* which we consider here briefly. We take the steady-state limit given by

$$\frac{\partial E_0}{\partial t} = 0$$

in Eq. (31). We further shine a continuous beam of light into a medium that responds linearly to the electric field as described by the slowly-varying complex polarization

$$\mathscr{P} = N(z)\wp(U - iV) \equiv N(z)\wp X \equiv \epsilon(\chi' + i\chi'')E_0(z), \tag{33}$$

where χ' and χ'' are the real and imaginary parts of the linear susceptibility χ. This susceptibility is another useful way of expressing the polarization. Substituting the in-quadrature part of $\mathscr{P}$ into Eq. (31), we obtain

---

* Beer's law is perhaps more accurately called Bouguier-Lambert-Beer's law. We call it Beer's law due to popular usage.

$$\frac{dE_0}{dz} = -\frac{K}{2}\chi'' E_0$$
$$= -\mathrm{Re}\{\alpha\}\, E_0\,, \tag{34}$$

where

$$\alpha = -\frac{iK}{2\epsilon}\frac{\mathcal{P}}{E_0} = -\frac{iK}{2\epsilon}\frac{N(z)\wp X}{E_0} = -K\frac{N(z)\wp}{2\epsilon E_0}(V+iU) \tag{35}$$

is called the complex amplitude absorption coefficient. We use an amplitude absorption coefficient instead of an intensity coefficient to be consistent with coupled-mode equations important for phase conjugation and other nonlinear mode interactions. If χ'' is independent of E_0, Eq. (34) can be readily integrated to give

$$E_0(z) = E_0(0)\, e^{-Re\{\alpha\}z}\,, \tag{36}$$

Taking the absolute square of Eq. (36) gives Beer's law for the intensity

$$I(z) = I(0)\, e^{-2Re\{\alpha\}z}\,. \tag{37}$$

We emphasize that this important result can only be obtained if α is independent of I, that is, that the polarization (33) of the medium responds linearly to the field amplitude E_0. Chapter 2 shows how to extend Eq. (33) to treat larger fields, leading to the usual discussion of nonlinear optics. Time dependent fields also lead to results such as Eq. (11.27) that differ from Beer's law. For these, Eq. (33) doesn't hold any more (even in the weak-field limit) if the medium cannot respond fast enough to the field changes. This can lead to effects such as laser lethargy, for which the field is absorbed or amplified according to the law

$$I(z) \propto \exp(-b\sqrt{z})\,, \tag{38}$$

where b is some constant.

The phase equation (32) allows us to relate the in-phase component of the susceptibility to the index of refraction n. As for the amplitude Eq. (34), we consider the continuous wave limit, for which $\partial\phi/\partial t = 0$. This gives

$$d\phi/dz = -K\,\chi'/2\,. \tag{39}$$

Expanding the slowly varying phase $\phi(z) \simeq \phi_0 + z d\phi/dz$, we find the total phase factor

$$\begin{aligned} Kz - \nu t - \phi &\simeq \nu[(K - d\phi/dz)z/\nu - t] - \phi_0 \\ &= \nu[(1+\chi'/2)z/c - t] - \phi_0 \\ &= \nu[z/v - t] - \phi_0. \end{aligned}$$

Noting that the velocity component* v is also given by c/n, we find the index of refraction (relative to the host medium)

$$n = 1 + \chi'/2\,. \tag{40}$$

In coupled-mode problems (see Secs. 2-2 and 9-2) and pulse propagation, instead of Eq. (12) it is more convenient to decompose the electric field in terms of a complex amplitude $\mathcal{E}(z,t) \equiv E_0(z,t)\exp(-i\phi)$, that is,

$$E(z,t) = \frac{1}{2}\,\mathcal{E}(z,t)\,e^{i(Kz-\nu t)} + \text{c.c.} \tag{41}$$

The polarization is then also defined without the explicit $\exp(i\phi)$ as

$$P(z,t) = \frac{1}{2}\,\mathcal{P}(z,t)\,e^{i(Kz-\nu t)} + \text{c.c.} \tag{42}$$

Substituting these forms into the wave equation (25) and neglecting small terms like $\partial^2\mathcal{E}/\partial t^2$, $\partial^2\mathcal{P}/\partial t^2$, and $\partial\mathcal{P}/\partial t$, and equating the coefficients of $e^{i(Kz-\nu t)}$ on both sides of the equation, we find the slowly-varying Maxwell's equation

$$\boxed{\frac{\partial\mathcal{E}}{\partial z} + \frac{1}{c}\frac{\partial\mathcal{E}}{\partial t} = i\,\frac{K}{2\epsilon}\,\mathcal{P}}\,. \tag{43}$$

Note that in equating the coefficients of $e^{i(Kz-\nu t)}$, we make use of our assumption that $\mathcal{P}(z,t)$ varies little in a wavelength. Should it vary appreciably in a wavelength due, for example, to a grating induced by an interference fringe, we would have to evaluate a projection integral as discussed for standing wave interactions in Sec. 5-3.

In a significant number of laser phenomena, the plane-wave approximation used in this chapter is inadequate. For these problems, a Gaussian beam may provide a reasonable description. A simple derivation of the Gaussian beam as a limiting case of a spherical wave $\exp(iKr)/r$ is given in Sec. 6-7.

---

* Note that the character v, which represents a speed, is different from the character ν, which represents a circular frequency (radians per second).

1-3. Linear Dipole Oscillator

As a simple and important example of the interaction between electromagnetic waves and matter, let us consider the case of a medium consisting of classical damped linear dipole oscillators. As discussed in Chap. 3, this model describes the absorption by quantum mechanical atoms remarkably well. Specifically we consider a charge (electron) cloud bound to a heavy positive nucleus and allowed to oscillate about its equilibrium position as

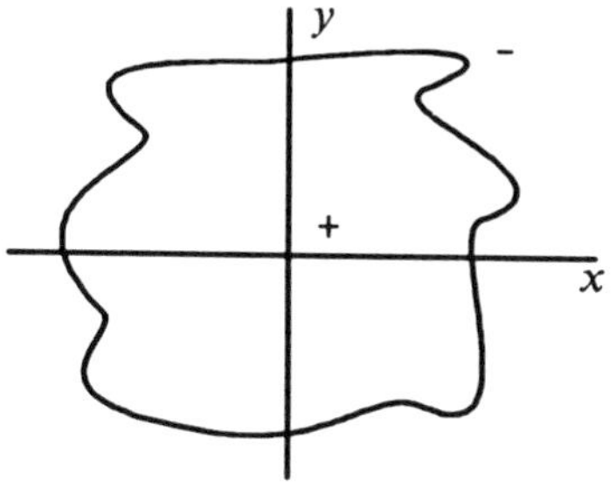

Fig. 1-1. Negative charge cloud bound to a heavy positive nucleus by Coulomb attraction. We suppose that some mysterious force prevents the charge cloud from collapsing into the nucleus.

shown in Fig. 1-1. We use the coordinate x to label the deviation from the equilibrium position with the center of charge at the nucleus. For small x it is a good approximation to describe the motion of the charged cloud as that of a damped simple harmonic oscillator subject to a sinusoidal electric field. Such a system obeys the Abraham-Lorentz equation of motion

$$\boxed{\ddot{x}(t) + 2\gamma\dot{x}(t) + \omega^2 x(t) = \frac{e}{m} E(t)}\ , \tag{44}$$

where ω is the natural oscillation frequency of the oscillator, and the dots stand for derivatives with respect to time. Note that since oscillating charges radiate, they lose energy. The end of this section shows how this process leads naturally to a damping constant γ. Quantum mechanically this decay is determined by spontaneous emission and collisions.

The solution of Eq. (44) is probably known to the reader. We give a derivation below that ties in carefully with the corresponding quantum mechanical treatments given in Chaps. 4 and 5. Chapter 2 generalizes Eq.

(44) by adding nonlinear forces proportional to x^2 and x^3 (see Eq. (2.1)). These forces lead to coupling between field modes producing important effects such as sum and difference frequency generation and phase conjugation. As such Eq. (44) and its nonlinear extensions allow us to see many "atom"-field interactions in a simple classical context before we consider them in their more realistic, but complex, quantum form.

We suppose the electric field has the form

$$E(t) = \frac{1}{2} E_0 \, e^{-i\nu t} + \text{c.c.}, \tag{45}$$

where E_0 is a constant real amplitude. In general the phase of $x(t)$ differs from that of $E(t)$. This can be described by a complex amplitude for x, that is,

$$x(t) = \frac{1}{2} x_0 X(t) \, e^{-i\nu t} + \text{c.c.}\,, \tag{46}$$

where $X(t)$ is the dimensionless complex amplitude of Eq. (26). In the following we suppose that it varies little in the damping time $1/\gamma$, which is a much more severe approximation than the SVAP. Our problem is to find the steady-state solution for $X(t)$.

Similarly as the discussion of Eqs. (33) and (34), we substitute Eqs. (45) and (46) into (44), neglect the small quantities $\ddot{X}$ and $\gamma\dot{X}$, and equate positive frequency components. This gives

$$\dot{X} = -[\gamma + i(\omega^2 - \nu^2)/2\nu]X + \frac{ieE_0}{2\nu m x_0}\,. \tag{47}$$

In steady state ($\dot{X} = 0$), this gives the amplitude

$$X = \frac{ieE_0/2\nu m x_0}{\gamma + i(\omega^2 - \nu^2)/2\nu}\,, \tag{48}$$

and hence the displacement

$$x(t) = \frac{i}{2} \frac{eE_0}{2m\nu} \frac{e^{-i\nu t}}{\gamma + i(\omega^2 - \nu^2)/2\nu} + \text{c.c.} \tag{49}$$

We often deal with the *near resonance* case, that is, the situation where $|\nu - \omega| \ll \nu + \omega$. For this case we can make the classical analog of the *rotating-wave approximation* defined in Sec. 3-2. Specifically we approximate $\omega^2 - \nu^2$ by

$$\omega^2 - \nu^2 \simeq 2\nu(\omega - \nu). \tag{50}$$

This reduces Eqs. (48) and (49) to

$$X = \frac{ieE_0/2\nu m x_0}{\gamma + i(\omega - \nu)}, \tag{51}$$

$$x(t) = \frac{i}{2}\frac{eE_0}{2m\nu}\frac{e^{-i\nu t}}{\gamma + i(\omega-\nu)} + \text{c.c.} \tag{52}$$

Equation (52) shows that in steady state the dipole oscillates with the same frequency as the driving field, but with a different phase. At resonance ($\nu = \omega$), Eq. (52) reduces to

$$x(t, \nu{=}\omega) = \frac{eE_0}{2m\nu\gamma}\sin\nu t, \tag{53}$$

that is, the dipole lags by $\pi/2$ behind the electric field (45), which oscillates as $\cos\nu t$.

The dipole moment $\mathscr{d}$ in Eq. (26) for the simple harmonic oscillator is ex_0. The corresponding polarization of the medium is $P = Nex(t)$, where N is the number of oscillators per unit volume. Substituting this along with Eq. (52) into Eq. (35), we find the complex amplitude Beer's law absorption coefficient

$$\alpha = K\frac{N}{2\epsilon\gamma}\frac{e^2}{2m\nu}\frac{\gamma}{\gamma + i(\omega-\nu)}$$

or

$$\boxed{\alpha = \frac{\alpha_0\gamma[\gamma - i(\omega-\nu)]}{\gamma^2 + (\omega-\nu)^2}}, \tag{54}$$

where the resonant absorption coefficient $\alpha_0 = KNe^2/4\epsilon\gamma m\nu$. The real part of this expression shows the Lorentzian dependence observed in actual absorption spectra (see Fig. 1-2). The corresponding quantum mechanical absorption coefficient of Eq. (5.29) differs from Eq. (54) in three ways:

1. $\gamma^2+(\omega-\nu)^2$ is replaced by $\gamma^2(1+I)+(\omega-\nu)^2$
2. N becomes negative for gain media
3. $e^2/2m\nu$ is replaced by $\wp^2/\hbar$

(55)

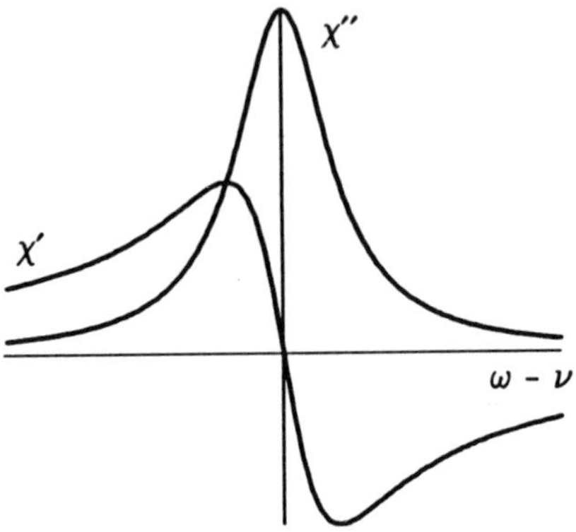

Fig. 1-2. Absorption (Lorentzian bell shape) and index parts of the complex absorption coefficient of Eq. (54).

For weak fields interacting with absorbing media, only the third of these differences needs to be considered and it just defines the strength of the dipole moment being used. Hence the classical model mirrors the quantum mechanical one well for linear absorption (for a physical interpretation of this result, see Sec. 3-2).

Identifying the real and imaginary parts of Eq. (47) and using Eq. (33), we have the equations of motion for the classical Bloch-vector components U and V

$$\dot{U} = -(\omega-\nu)V - \gamma U \tag{56}$$
$$\dot{V} = (\omega-\nu)U - \gamma V - eE_0/2m\nu x_0 \,. \tag{57}$$

Comparing Eq. (57) with Eq. (4.49) (in which $\gamma = 1/T_2$), we see that the E_0 term is multiplied by $-W$, which is the third component of the Bloch vector. This component equals the probability that a two-level atom is in the upper level minus the probability that it is in the lower level. Hence we see that the classical Eq. (57) is reasonable as long as $W \simeq -1$, i.e., so long as the atom is in the lower level.

From the steady-state value of X given by Eq. (51), we have the steady-state U and V values

$$U = \frac{eE_0}{2m\nu x_0} \frac{\omega-\nu}{\gamma^2 + (\omega-\nu)^2} \tag{58}$$

and

$$V = -\frac{eE_0}{2m\nu x_0}\frac{\gamma}{\gamma^2 + (\omega-\nu)^2}\,. \tag{59}$$

Since Eq. (44) is linear, once we know the solution for the single frequency field (45), we can immediately generalize to a multifrequency field simply by taking a corresponding superposition of single frequency solutions. The various frequency components in $x(t)$ oscillate independently of one another. In contrast the nonlinear media in Chap. 2 and later chapters couple the modes. Specifically, consider the multimode field

$$E(z,t) = \frac{1}{2}\sum_n \mathcal{E}_n(z)\, e^{i(K_n z - \nu_n t)} + \text{c.c.}, \tag{60}$$

where we allow the field amplitudes to be slowly varying functions of z and to be complex since they they do not in general have the same phases. The solution for the oscillator displacement $x(t)$ at the position z is a superposition of solutions like (46), namely

$$x(t) = \frac{1}{2}\sum_n x_{0n} X_n\, e^{i(K_n z - \nu_n t)} + \text{c.c.}, \tag{61}$$

where mode n's oscillator strength is proportional to x_{0n} and the coefficients

$$X_n = \frac{e\mathcal{E}_n/mx_{0n}}{\omega^2 - \nu_n{}^2 - 2i\nu_n\gamma}\,. \tag{62}$$

Here we don't make the resonance approximation of Eq. (50), since some of the modes may be off resonance. The steady-state polarization $P(z,t)$ of a medium consisting of such oscillators is then given by

$$P(z,t) = \frac{1}{2}\sum_n \mathcal{P}_n(z)\, e^{i(K_n z-\nu_n t)} + \text{c.c.}, \tag{63}$$

where $\mathcal{P}_n(z)$ is given by $N(z)ex_{0n}X_n$. In Sec. 2-1, we find that higher-order terms occur when nonlinearities are included in the equation of motion (44). These terms couple the modes and lead to anharmonic response. Finally, we note that the multimode field (60) and the polarization (63) have the same form in the unidirectional ring laser of Chap. 6, except that in a high-Q cavity the mode amplitudes $\mathcal{E}_n$ and polarization components $\mathcal{P}_n$ are functions of t, rather than z.

Radiative Damping

We now give a simple approximate justification for the inclusion of a damping coefficient γ in Eq. (44). As a charge oscillates it radiates electromagnetic energy and consequently emits a "self-field" E_s. We need to find the influence of this self field back on the charge's motion in a self-consistent fashion. We find that the main effect is the exponential damping of this motion as given by Eq. (44). Specifically, we consider the equation governing the charge's motion under the influence of the self-field E_s:

$$\ddot{x} + \omega^2 x = \frac{e}{m} E_s \ , \tag{64}$$

which is just Newton's law with the Lorentz force

$$\mathbf{F}_s = e(\mathbf{E}_s + \mathbf{v}\times\mathbf{B}_s) \tag{65}$$

in the limit of small charge velocities ($v << c$), where the magnetic part of the Lorentz force may be neglected.

While we don't know the explicit form of E_s, we can calculate its effects using the conservation of energy. We evaluate the force $\mathbf{F}_{rad}$ on the radiating charge by equating the work it expends on the charge (during a time interval long compared to the optical period $1/\omega$) to minus the energy radiated by the charge during that time

$$\int_t^{t+\Delta t} \mathbf{F}_{rad}\cdot\mathbf{v}\, dt' = - \int_t^{t+\Delta t} (\text{radiated power})\, dt' \ . \tag{66}$$

To calculate the radiated power, we note that the instantaneous electromagnetic energy flow is given by the Poynting vector

$$\mathbf{S} = \frac{1}{\mu_0} \mathbf{E}_s \times \mathbf{B}_s \ , \tag{67}$$

where for simplicity we suppose that the "host medium" is the vacuum. We note that the electric field radiated in the far field of the dipole is

$$\mathbf{E}_s(\mathbf{R},t) = \frac{e}{4\pi\epsilon_0 c^2}\left[\frac{\mathbf{n}\times(\mathbf{n}\times\dot{\mathbf{v}})}{R}\right]_{t-\mathrm{R/c}} \tag{68}$$

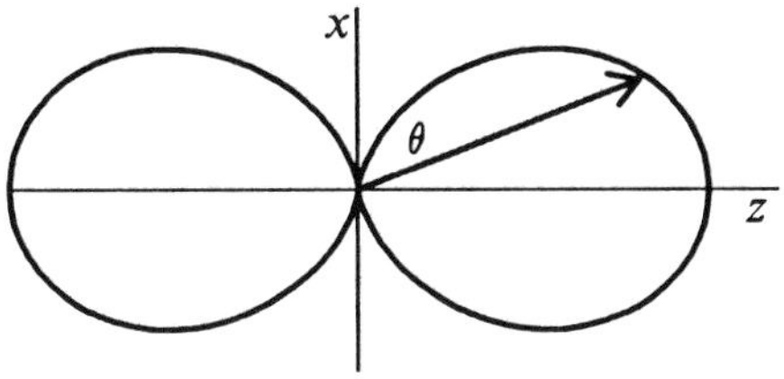

Fig. 1-3. Butterfly pattern given by Eq. (69) and emitted by an oscillating dipole. The vector gives the direction and relative magnitude of the Poynting vector **S** as a function of θ.

as shown in Fig. 1-3. The corresponding magnetic field is $\mathbf{B}_s(\mathbf{R},t) = c^{-1}\,\mathbf{n}\times\mathbf{E}_s(\mathbf{R},t)$. In both expressions the dipole acceleration $\dot{\mathbf{v}}$ is evaluated at the retarded time $t\text{-}R/c$ and $\mathbf{n}$ is the unit vector $\mathbf{R}/R$. Inserting these expressions into Eq. (66), we find the Poynting vector (Jackson 1962)

$$\begin{aligned}\mathbf{S} &= \frac{1}{\mu_0 c}(\mathbf{E}_s\cdot\mathbf{E}_s)\mathbf{n}\\ &= \frac{e^2}{16\pi^2\epsilon_0{}^2c^4}\,\frac{1}{\mu_0 c}\,\frac{1}{R^2}\,(\mathbf{n}\times\dot{\mathbf{v}})^2\mathbf{n}\\ &= \frac{e^2\dot{v}^2\sin^2\theta}{16\pi^2\epsilon_0 c^3R^2}\,\mathbf{n}\,.\end{aligned} \tag{69}$$

The total power radiated is given by integration of **S** over a sphere surrounding the charge. Noting that

$$\int_0^{2\pi} d\phi \int_0^{\pi} d\theta\, \sin^3\theta = -2\pi\int_1^{-1} \mathrm{d}(\cos\theta)\,[1 - \cos^2\theta] = 8\pi/3, \tag{70}$$

we find

$$\int \mathbf{S}\cdot d\mathbf{a} = \frac{2}{3}\frac{e^2}{4\pi\epsilon_0 c^3}\dot{v}^2, \tag{71}$$

which is the Larmor power formula for an accelerated charge. We now substitute Eq. (71) into Eq. (66) and integrate by parts. We encounter the integral

$$\int_t^{t+\Delta t} dt\ \dot{\mathbf{v}}\cdot\dot{\mathbf{v}} = \dot{\mathbf{v}}\cdot\mathbf{v}\Big|_t^{t+\Delta t} - \int_t^{t+\Delta t} dt\ \mathbf{v}\cdot\ddot{\mathbf{v}}$$

Since $\mathbf{v}$ and its derivatives are periodic, the constant of integration on the right hand side has a maximum magnitude, while the integrals continue to increase as Δt increases. Hence the constant can be dropped. Equating the integrands, we find the radiation force

$$\mathbf{F}_{rad} = \frac{2}{3}\frac{e^2}{4\pi\epsilon_0 c^3}\ddot{\mathbf{v}}, \tag{72}$$

A more detailed analysis of this problem is given in Sec. 17.3 of Jackson (1968), where the infinities associated with point-like charges are also discussed.

Assuming that the radiative damping is sufficiently small that the motion of the dipole remains essentially harmonic, Eq. (72) yields

$$\mathbf{F}_{rad} = m\ddot{\mathbf{x}} = -\frac{2}{3}\frac{e^2\omega^2}{4\pi\epsilon_0 c^3}\mathbf{v}, \tag{73}$$

which indicates that radiation reaction acts as a friction on the motion of the charge. This implies a damping rate constant

$$\gamma = \frac{1}{4\pi\epsilon_0}\frac{1}{3}\frac{e^2\omega^2}{c^3 m} = \frac{1}{3}\frac{\omega^2 r_0}{c}, \tag{74}$$

where the classical radius of the electron is

$$r_0 = \frac{e^2}{4\pi\epsilon_0 mc^2} \simeq 2.8\times10^{-15}\ \text{meters}. \tag{75}$$

For $1\mu m$ radiation, $\gamma = 2\pi\times1.8$ MHz, which is in the range of decay values found in atoms. In cgs units, remove the $4\pi\epsilon_0$ in Eqs. (74) and (75).

With the replacement of $e^2/2m\nu$ by $\wp^2/\hbar$ [see Eq. (55)], the classical decay rate Eq. (74) gives half the quantum mechanical decay rate (13.60). Here $\wp$ is the reduced position matrix element between the upper and lower levels of the two level transition. The other half of the decay rate is contributed by the effects of vacuum fluctuations missing in a classical description. Note that in both the classical and quantum mechanical cases, an ω^2 term appears. In the quantum case, this term results from the density of states of free space (13.46), while for the classical case it comes from the acceleration of the electron. In some sense the density of states for the field reflects the fact that the field itself is radiated by accelerating, oscillating charges. In free space the charge responsible for this field is the bound electron itself, radiating a field that acts back on the charge and causes it to emit radiation until no more downward transitions are possible. For further discussion, see P. W. Milonni (1986, 1984).

1-4. Coherence

Coherence plays a central role in modern physics. It is very hard to find a single domain of physics where this concept is not applied. In this book we use it a great deal, speaking of coherent light, coherent transients, coherent propagation, coherent states, coherent excitation, etc. Just what is coherence? The answer typically depends on whom you ask! In a very general sense, a process is coherent if it is characterized by the existence of some well-defined deterministic phase relationship, or in other words, if some phase is not subject to random noise. This is a very vague definition, but it is general enough to encompass all processes usually called "coherent." In this section and in section 12-5 we consider the coherence of classical light. Chapters 4 and 11 discuss coherence in atomic systems.

The classic book by Born and Wolf (1970) gives a discussion of coherent light in pre-laser terms. With the advent of the laser, a number of new effects have been discovered that have caused us to rethink our ideas about coherent light. In addition, the Hanbury Brown-Twiss experiment, which had nothing to do with lasers, plays an important role in this rethinking. Our discussion is based on the theory of optical coherence as developed by R. Glauber and summarized in his *Les Houches* lectures (1965).

We start with the famous Young double-slit experiment which shows how coherent light passing through two slits interferes giving a characteristic intensity pattern on a screen (see Fig. 1-3). Before going into the details of this experiment, we need to know how the light intensity is measured, either on a screen or with a photodetector. Both devices work by absorbing the light. The absorption sets up a chemical reaction in the case of film, and ionizes atoms or lifts electrons into a conduction band in the cases of two kinds of photodetectors. Chapter 12-5 shows by a quantum-

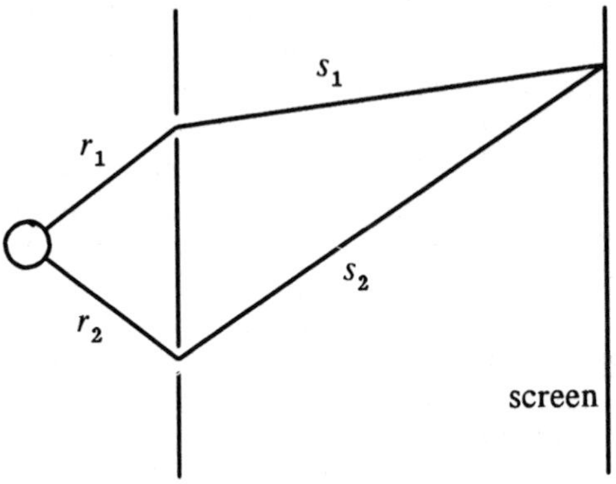

Fig. 1-4. Young double-slit experiment illustrating how coherent light can interfere with itself

mechanical analysis of the detection process that these methods measure $|E^+(\mathbf{r},t)|^2$, rather than $|E(\mathbf{r},t)|^2$. This is why we performed the decomposition in Eq. (10).

Returning to Young's double-slit experiment, we wish to determine $E^+(\mathbf{r},t)$, where $\mathbf{r}$ is the location of the detector. $E^+(\mathbf{r},t)$ is made up of two components, each coming from its respective slit:

$$E^+(\mathbf{r},t) = E^+(\mathbf{r}_1,t_1) + E^+(\mathbf{r}_2,t_2) , \tag{76}$$

where $\mathbf{r}_1$ and $\mathbf{r}_2$ are the locations of the slits and t_1 and t_2 are the retarded times

$$t_{1,2} = t - s_{1,2}/c , \tag{77}$$

s_1 and s_2 being the distances between the slits and the detector. From Eq. (76), the intensity at the detector is given by

$$|E^+(\mathbf{r},t)|^2 = |E^+(\mathbf{r}_1,t_1)|^2 + |E^+(\mathbf{r}_2,t_2)|^2 + 2\,\mathrm{Re}[E^-(\mathbf{r}_1,t_1)E^+(\mathbf{r}_2,t_2)] , \tag{78}$$

where we have made use of Eq. (9).

In general the light source contains noise. To describe light with noise we use a statistical approach, repeating the measurement many times and averaging the results. Mathematically this looks like

$$\langle|E^+(\mathbf{r},t)|^2\rangle = \langle|E^+(\mathbf{r}_1,t_1)|^2\rangle + \langle|E^+(\mathbf{r}_2,t_2)|^2\rangle + 2\,\mathrm{Re}\langle E^-(\mathbf{r}_1,t_1)E^+(\mathbf{r}_2,t_2)\rangle , \tag{79}$$

where the brackets $\langle\cdots\rangle$ stand for the ensemble average. Introducing the first-order correlation function

$$G^{(1)}(\mathbf{r}_1 t_1,\mathbf{r}_2 t_2) \equiv \langle E^-(\mathbf{r}_1,t_1)E^+(\mathbf{r}_2,t_2)\rangle, \tag{80}$$

we rewrite Eq. (79) as

$$\langle|E^+(\mathbf{r},t)|^2\rangle = G^{(1)}(\mathbf{r}_1 t_1,\mathbf{r}_1 t_1) + G^{(1)}(\mathbf{r}_2 t_2,\mathbf{r}_2 t_2) + 2\mathrm{Re}G^{(1)}(\mathbf{r}_1 t_1,\mathbf{r}_2 t_2)\,. \tag{81}$$

$G^{(1)}(\mathbf{r}_i t_i,\ \mathbf{r}_i t_i)$ is clearly a real, positive quantity, while $G^{(1)}(\mathbf{r}_i t_i,\ \mathbf{r}_j t_j)$ is in general complex.

With the cross-correlation function rewritten as

$$G^{(1)}(\mathbf{r}_1 t_1,\mathbf{r}_2 t_2) = |G^{(1)}(\mathbf{r}_1 t_1,\mathbf{r}_2 t_2)|\, e^{i\phi(\mathbf{r}_1 t_1,\mathbf{r}_2 t_2)}\,, \tag{82}$$

Eq. (81) becomes

$$\langle|E^+(\mathbf{r},t)|^2\rangle = G^{(1)}(\mathbf{r}_1 t_1,\mathbf{r}_1 t_1) + G^{(1)}(\mathbf{r}_2 t_2,\mathbf{r}_2 t_2) + 2|G^{(1)}(\mathbf{r}_1 t_1,\mathbf{r}_2 t_2)|\cos\phi\,. \tag{83}$$

The third term in Eq. (83) is responsible for the appearance of interferences.

We say that the highest degree of coherence corresponds to a light field that produces the maximum contrast on the screen, where contrast is defined as

$$V = \frac{I_{\max} - I_{\min}}{I_{\max} + I_{\min}}\,. \tag{84}$$

Substituting Eq. (83) with $\cos\phi = 1$, we readily obtain

$$V = \frac{2|G^{(1)}(\mathbf{r}_1 t_1,\mathbf{r}_2 t_2)|}{G^{(1)}(\mathbf{r}_1 t_1,\mathbf{r}_1 t_1) + G^{(1)}(\mathbf{r}_2 t_2,\mathbf{r}_2 t_2)}\,. \tag{85}$$

The denominator in Eq. (85) doesn't play an important role; $G^{(1)}(\mathbf{r}_i t_i,\ \mathbf{r}_i t_i)$ is just the intensity on the detector due to the ith slit and the denominator acts as a normalization constant. To maximize the contrast for a given source and geometry, we need to maximize the numerator $2|G^{(1)}(\mathbf{r}_1 t_1,\mathbf{r}_2 t_2)|$. To achieve this goal we note that according to the Schwarz inequality

$$G^{(1)}(\mathbf{r}_1 t_1,\mathbf{r}_1 t_1)\, G^{(1)}(\mathbf{r}_2 t_2,\mathbf{r}_2 t_2) \geq |G^{(1)}(\mathbf{r}_1 t_1,\mathbf{r}_2 t_2)|^2\,. \tag{86}$$

The coherence function is maximized when equality holds, that is when

$$|G^{(1)}(\mathbf{r}_1t_1,\mathbf{r}_2t_2)| = [G^{(1)}(\mathbf{r}_1t_1,\mathbf{r}_1t_1)G^{(1)}(\mathbf{r}_2t_2,\mathbf{r}_2t_2)]^{1/2}\ , \tag{87}$$

which is the coherence condition used by Born and Wolf. As pointed out by Glauber, it is convenient to replace this condition by the equivalent expression

$$G^{(1)}(\mathbf{r}_1t_1,\mathbf{r}_2t_2) = \mathscr{E}^*(\mathbf{r}_1t_1)\ \mathscr{E}(\mathbf{r}_2t_2), \tag{88}$$

where the complex function $\mathscr{E}(\mathbf{r}_1t_1)$ is some function not necessarily the electric field. If $G^{(1)}(\mathbf{r}_1t_1,\mathbf{r}_2t_2)$ may be expressed in the form (88), we say that $G^{(1)}$ *factorizes*. This factorization property defines first-order coherence: when Eq. (88) holds, the fringe contrast V is maximum.

This definition of first-order coherence can be readily generalized to higher orders. A field is said to have nth-order coherence if its mth-order correlation functions

$$\boxed{G^{(m)}(x_1\dots x_m, y_m\dots y_1) = \langle E^-(x_1)\cdots E^-(x_m)\ E^+(y_m)\cdots E^+(y_1)\rangle}\ . \tag{89}$$

factorize as

$$G^{(m)}(x_1\dots x_m, y_m\dots y_1) = \mathscr{E}^*(x_1)\cdots\mathscr{E}^*(x_m)\ \mathscr{E}(y_m)\cdots\mathscr{E}(y_1) \tag{90}$$

for all $m \le n$. Here we use the compact notation $x_j = (\mathbf{r}_j, t_j)$, $y_j = (\mathbf{r}_{m+j}, t_{m+j})$, and $G^{(m)}$ is a direct generalization of Eq. (80).

Before giving an example where second-order correlation functions play a crucial role, we point out that although a monochromatic field is coherent to all orders, a first-order coherent field is not necessarily monochromatic. One might be led to think otherwise because we often deal with stationary light, such as that from stars and *cw* light sources. By definition, the two-time properties of a stationary field depend only on the time difference. The corresponding first-order correlation function thus has the form

$$G^{(1)}(t_1,t_2) = G^{(1)}(t_1-t_2)\ . \tag{91}$$

If such a field is first-order coherent, then with Eq. (88), we find

$$G^{(1)}(t_1-t_2) = \mathscr{E}^*(t_1)\mathscr{E}(t_2)\ , \tag{92}$$

which is true only if

$$\mathscr{E}(t_1) \propto e^{-i\nu t_1}\ , \tag{93}$$

that is, *stationary* first-order coherent fields are monochromatic!

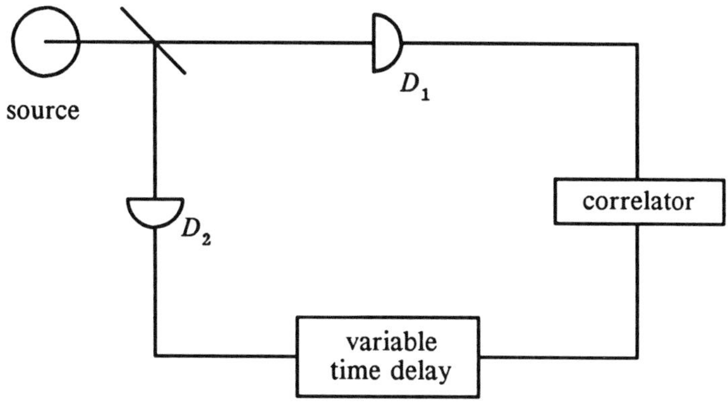

Fig. 1-5. Diagram of Hanbury Brown-Twiss experiment.

Let us now turn to the famous Hanbury Brown-Twiss experiment (Fig. 1-5), which probes higher-order coherence properties of a field. In this experiment, a beam of light (from a star in the original experiment) is split into two beams, which are detected by detectors D_1 and D_2. The signals are multiplied and averaged in a correlator. This procedure differs from the Young two-slit experiment in that light intensities, rather than amplitudes, are compared. Two absorption measurements are performed on the same field, one at time t and the other at $t+\tau$. It can be shown that this measures $|E^+(\mathbf{r},t+\tau)E^+(\mathbf{r},t)|^2$. Dropping the useless variable $\mathbf{r}$ and averaging, we see that this is precisely the second-order correlation function

$$G^{(2)}(t,t+\tau,t+\tau,t) = \langle E^-(t)E^-(t+\tau)E^+(t+\tau)E^+(t)\rangle \ , \tag{94}$$

or for a stationary process,

$$G^{(2)}(\tau) = \langle E^-(0)E^-(\tau)E^+(\tau)E^+(0)\rangle \ . \tag{95}$$

According to Eq. (89), the field is second-order coherent if Eq. (92) holds and

$$G^{(2)}(\tau) = \mathcal{E}^*(0)\mathcal{E}^*(\tau)\mathcal{E}(\tau)\mathcal{E}(0) \ . \tag{96}$$

It is convenient to normalize this second-order correlation function as

$$g^{(2)}(\tau) = \frac{G^{(2)}(\tau)}{|G^{(1)}(0)|^2} \ . \tag{97}$$

Since a stationary first-order coherent field is monochromatic and satisfies Eq. (93), second-order coherence implies that

$$g^{(2)}(\tau) = 1. \tag{98}$$

that is, $g^{(2)}(\tau)$ is independent of the delay τ.

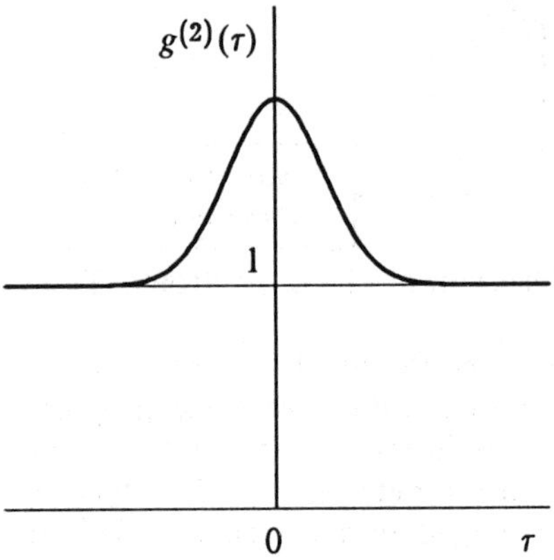

Fig. 1-6. Second-order correlation function (97) for starlight in original Hanbury Brown-Twiss experiment.

The original experiment of Hanbury Brown-Twiss was used to measure the apparent diameter of stars by passing their light through a pinhole. A second-order correlation function like that in Fig. 1-6 was measured. Although the light was first-order coherent, we see that it was not second-order coherent. The energy tended to arrive on the detector in *bunches*, with strong statistical correlations.

In contrast to the well-stabilized laser with a unity $g^{(2)}$ and the starlight with bunching, recent experiments in resonance fluorescence show *antibunching*, with the $g^{(2)}$ shown in Fig. 1-7. Chapter 15 discusses this phenomenon in detail; here we point out that such behavior cannot be explained with classical fields. To see this, note that

$$g^{(2)}(0) - 1 = \frac{G^{(2)}(0) - |G^{(1)}(0)|^2}{|G^{(1)}(0)|^2}. \tag{99}$$

In terms of intensities, this gives

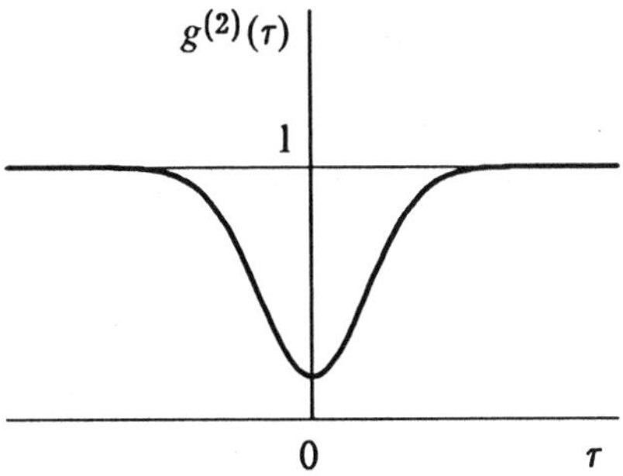

Fig. 1-7. Second-order correlation function showing antibunching found in resonance fluorescence.

$$g^{(2)}(0) - 1 = \frac{\langle I^2\rangle - \langle I\rangle^2}{\langle I\rangle^2} = \frac{\langle(I - \langle I\rangle)^2\rangle}{\langle I\rangle^2}\,, \tag{100}$$

where we do not label the times, since we consider a stationary system with $\tau = 0$. Introducing the probability distribution $P(I)$ to describe the average over fluctuations, we find for Eq. (100)

$$g^{(2)}(0) - 1 = \frac{1}{\langle I\rangle^2}\int dI\, P(I)\,(I - \langle I\rangle)^2\,. \tag{101}$$

Classically this must be positive, since $(I-\langle I\rangle)^2 \geq 0$ and the probability distribution $P(I)$ must be positive. Hence classically, $g^{(2)}$ cannot be less than unity in contradiction to the experimental result shown in Fig. 1-7. At the beginning of this chapter we say that the fields we use can usually be treated classically. Well we didn't say always! To use a formula like (101) for the antibunched case, we need to use the concept of *a quasi*probability function $P(I)$ that permits negative values. Quantum mechanics allows just that (see Sec. 12-6).

1-5. Free-Electron Lasers

At this point we already have all the ingredients necessary to discuss the basic features of free-electron lasers (FEL). They are extensions of devices such as klystrons, undulators, and ubitrons, which were well-known in the millimeter regime many years ago, long before lasers existed. In principle, at least, nothing should have prevented their invention 30 or 40 years ago.

As shown in Chap. 6, conventional lasers rely on the inversion of an atomic or molecular transition. Thus the wavelength at which they operate is determined by the active medium they use. The FEL eliminates the atomic "middle-man," and does not rely on specific transitions. Potentially, FEL's offer three characteristics that are often hard to get with conventional lasers, namely wide tunability, high power, and high efficiency. They do this by using a relativistic beam of free electrons that interact with a periodic structure, typically in the form of a static magnetic field. This structure exerts a Lorentz force on the moving electrons, forcing them to oscillate, similarly to the simple harmonic oscillators of Sec. 1-3. As discussed at the end of that section, oscillating electrons emit radiation with the field shown in Fig. 1-2. In the laboratory frame, this radiation pattern is modified according to Lorentz transformations. Note that in contrast to the case of radiative decay discussed in Sec. 1-3, the FEL electron velocity approaches that of light and the $\mathbf{v}\times\mathbf{B}$ factor in the Lorentz force of Eq. (65) cannot be neglected.

The emitted radiation is mostly in the forward direction, within a cone of solid angle $\theta = 1/\gamma$ (see Fig. 1-8). Here γ is the relativistic factor

$$\gamma = [1 - v^2/c^2]^{-1/2} , \tag{102}$$

where $\mathbf{v}$ is the electron velocity. For $\gamma = 200$, which corresponds to electrons with an energy on the order of 100 MeV, θ is about 5 milliradians, a very small angle.

In general for more than one electron, each dipole radiates with its own phase, and these phases are completely random with respect to one another. The total emitted field is $\mathbf{E}_T = \mathbf{E}_T^+ + \mathbf{E}_T^-$, where

$$\mathbf{E}_T^+ = \sum_{k=1}^{N} \mathbf{E}_k^+ \, e^{i\phi_k} , \tag{103}$$

and the sum is over all electrons in the system.

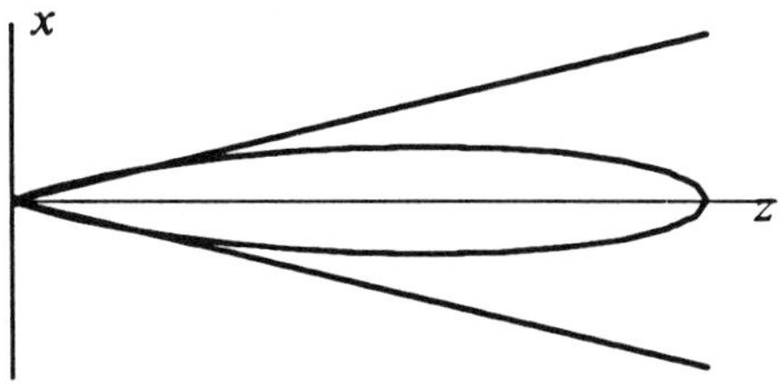

Fig. 1-8. Highly directional laboratory pattern of the radiation emitted by a relativistic electron in a circular orbit in the x-y plane while moving along the z axis at the speed $v = 0.9c$. The x axis is defined to be that of the instantaneous acceleration. Equation (14.44) of Jackson (1962) is used for an observation direction $\mathbf{n}$ in the x-z plane (the azimuthal angle $\phi = 0$). In the nonrelativistic limit ($v << c$), this formula gives the butterfly pattern of Fig. 1-3.

The total radiated intensity I_T is proportional to $|\mathbf{E}_T^+|^2$, which is

$$I_T = \left| \sum_{k=1}^{N} \mathbf{E}_k^+ \, e^{i\phi_k} \right|^2 . \tag{104}$$

Expanding the absolute value in Eq. (104), we obtain

$$I_T = \sum_{k=1}^{N} |\mathbf{E}_k^+|^2 + \sum_{k \neq j} \mathbf{E}_k^- \, \mathbf{E}_j^+ \, e^{-i(\phi_k - \phi_j)} . \tag{105}$$

Assuming that the amplitudes of the fields emitted by each electron are the same

$$|\mathbf{E}_k|^2 = I , \tag{106}$$

we obtain

$$I_T = NI + I \sum_{k \neq j} e^{-i(\phi_k - \phi_j)} . \tag{107}$$

For random phases, the second term in Eq. (107) averages to zero, leaving

$$I_T = NI , \tag{108}$$

that is, the total intensity is just the sum of the individual intensities. The contributions of the electrons add incoherently with random interferences, as is the case with synchrotron radiation.

However if we could somehow force all electrons to emit with roughly the same phase, $\phi_k \simeq \phi_j$ for all k and j, then Eq. (107) would become

$$I_T = NI + N(N-1)I = N^2 I. \tag{109}$$

Here the fields emitted by all electrons would add coherently, i.e., with constructive interference, giving an intensity N times larger than with random phases.

The basic principle of the FEL is to cause all electrons to have approximately the same phase, thereby producing constructive interferences (stimulated emission). A key feature of these lasers is that the wavelength of the emitted radiation is a function of the electron energy. To understand this, note that an observer moving along with the electrons would see a wiggler moving at a relativistic velocity with a period that is strongly Lorentz contracted. To this observer the field appears to be time-dependent, rather than static, since it flies by. In fact, the wiggler magnetic field appears almost as an electromagnetic field whose wavelength is the Lorentz-contracted period of the wiggler. It is well-known that an electron at rest can scatter electromagnetic radiation. This is called Thomson scattering. Because the electron energy is much higher than that of the photons at least in the visible range, we can neglect their recoil, and hence the wavelength of the scattered radiation equals that of the incident radiation

$$\lambda_s' = \lambda_w' . \tag{110}$$

Here we use primes to mean that we are in the electron rest frame. Going back to the laboratory frame, we examine the radiation emitted in the forward direction. As Prob. 1-16 shows, this is also Lorentz contracted with the wavelength

$$\lambda_s = \lambda_w / 2\gamma_z^2 , \tag{111}$$

where λ_w is the period of the wiggler and

$$\gamma_z = [1 - v_z^2/c^2]^{-1/2} . \tag{112}$$

Here we use γ_z rather than γ because the relevant velocity for the Lorentz transformation is the component along the wiggler (z) axis. Since **v** is directed primarily along this axis, λ_s is to good approximation given by $\lambda_w/2\gamma^2$.

An alternative way to obtain the scattered radiation wavelength λ_s of (111) is to note that for constructive interference of scattered radiation, $\lambda_s + \lambda_w$ must equal the distance ct the light travels in the electron transit time $t = \lambda_w/v_z$ it takes for the electrons to move one wiggler wavelength. This gives $\lambda_s + \lambda_w = c\lambda_w/v_z$, and Eq. (111) follows with the use of Eq. (112).

We see that two Lorentz transformations are needed to determine λ_s. Since $\gamma_z \simeq \gamma$ is essentially the energy of the electron divided by mc^2, we can change the wavelength λ_s of the FEL simply by changing the energy of the electrons. The FEL is therefore a widely tunable system. In principle the FEL should be tunable continuously from the infrared to the vacuum ultraviolet.

We now return to the problem of determining how the electrons are forced to emit with approximately the same phase, so as to produce constructive interference. We can do this with Hamilton's formalism in a straightforward way. For this we need the Hamiltonian for the relativistic electron interacting with electric and magnetic fields. We note that the energy of a relativistic electron is

$$E = \sqrt{m^2c^4 + p^2c^2} , \tag{113}$$

where p is the electron momentum. For an electron at rest, $p = 0$, giving Einstein's famous formula $E = mc^2$. For slow electrons ($p << mc$), we expand the square root in Eq. (113) finding $E \simeq mc^2 + p^2/2m$, which is just the rest energy of the electron plus the nonrelativisitic kinetic energy. For the relativistic electrons in the FEL, we need to use the exact formula (113).

To include the interaction with the magnetic and electric fields, we use the *principle of minimum coupling*, which replaces the kinetic momentum **p** by the canonical momentum

$$\mathbf{p} \rightarrow \mathbf{P} - e\mathbf{A} . \tag{114}$$

Here **A** is the vector potential of the field. Using the prescription (114) in Eq. (113), we find the required Hamiltonian

$$\mathcal{H} = c[(\mathbf{P} - e\mathbf{A})^2 + m^2c^2]^{1/2} \equiv \gamma mc^2 \,. \tag{115}$$

Hamilton's equations of motion are

$$\dot{P}_i = -\frac{\partial \mathcal{H}}{\partial q_i} \tag{116}$$

$$\dot{q}_i = \frac{\partial \mathcal{H}}{\partial P_i} \,, \tag{117}$$

where the three components of the canonical momentum, P_i, and the three electron coordinates, q_i, completely describe the electron motion. To obtain their explicit form, we need to know **A**. This consists of two contributions, that of the static periodic magnetic field, and that of the scattered laser field.

If the transverse dimensions of the electron beam are sufficiently small compared to the transverse variations of both fields, we can treat the fields simply as plane waves. **A** then has the form

$$\mathbf{A} = \frac{1}{2}\,\hat{\mathbf{e}}_-[A_w\, e^{-iK_w z} + A_s\, e^{-i(\omega_s t - K_s z)}] + \text{c.c.} \tag{118}$$

Here A_w and A_s are the amplitudes of the vector potential of the wiggler and the laser, respectively, and

$$\hat{\mathbf{e}}_- = \hat{\mathbf{e}}_+^{\,*} = (\hat{\mathbf{x}} - i\hat{\mathbf{y}})/\sqrt{2} \,, \tag{119}$$

where $\hat{\mathbf{x}}$ and $\hat{\mathbf{y}}$ are the unit vectors along the transverse axes x and y, respectively. This form of the vector potential is appropriate for the circularly polarized magnet used in the Stanford experiments. Also $K_w = 2\pi/\lambda_w$, where λ_w is the wiggler period, and ω_s and K_s are the frequency and wave number of the scattered light.

With this form of the vector potential, the Hamiltonian (115) doesn't depend explicitly on x and y. Hence from Eq. (116), we have

$$\dot{P}_x = -\frac{\partial \mathcal{H}}{\partial x} = 0$$

$$\dot{P}_y = -\frac{\partial \mathcal{H}}{\partial y} = 0 \,, \tag{120}$$

that is, the transverse *canonical* momentum is constant. Furthermore, this constant equals zero if the electrons have zero transverse canonical momentum upon entering the wiggler

$$\mathbf{P}_T = 0 . \tag{121}$$

This gives the kinetic transverse momentum

$$\mathbf{p}_T = -e\mathbf{A} , \tag{122}$$

which shows that the transverse motion of the electron is simply a circular orbit, as might be expected intuitively.

For the longitudinal motion, the Hamilton equations of motion reduce to

$$\dot{z} = p_z / m\gamma , \tag{123}$$

$$\dot{p}_z = - \frac{e^2}{2m\gamma} \frac{\partial}{\partial z}(A^2) . \tag{124}$$

Equation (123) just gives the usual formula $p_z = \gamma m v_z$ for relativistic particles. Equation (124) is more informative and states that the time rate of change of the longitudinal electron momentum is given by the spatial derivative of the square of the vector potential. Potentials proportional to A^2 are common in plasma physics where they are called *ponderomotive potentials*. Computing $\partial(A^2)/\partial z$ explicitly, we find

$$\frac{\partial(A^2)}{\partial z} = 2iKA_w{}^* A_s \, e^{i(Kz - \omega_s t)} + \text{c.c.}, \tag{125}$$

where

$$K = K_w + K_s . \tag{126}$$

Thus in the longitudinal direction, the electron is subject to a longitudinal force moving with the high speed

$$v_s = \frac{\omega_s}{K} . \tag{127}$$

Since according to Eq. (111) $K_s >> K_w$, v_s is almost the speed of light.

In the laboratory frame, both the electrons and the potential move at close to the speed of light. It is convenient to rewrite the equations of motion (123) and (124) in a frame moving at velocity v_s, that is, riding on the ponderomotive potential. For this we use

$$\xi = z - v_s t + \xi_0 - \pi, \tag{128}$$

which is the position of the electron relative to the potential and $K\xi$ is the phase of the electron in the potential. ξ_0 is determined by $A_w{}^* A_s =$

$|A_w A_s|\exp(iK\xi_0)$ and $K\xi$ is the phase of the electron relative to the ponderomotive potential at $z = t = 0$. This gives readily

$$\dot{\xi} = v_z - v_s , \qquad (129)$$

which is the electron velocity relative to the potential. To transform Eq. (124) we have to take into account that γ is *not* constant. First, taking A_w and A_s real, we readily find

$$\dot{p}_z = - \frac{2Ke^2}{m\gamma} |A_W A_s| \sin K\xi . \qquad (130)$$

This is a nonlinear oscillator equation that includes all odd powers of the displacement $K\xi$. Noting further that $\dot{p}_z = m\gamma\gamma_z{}^2\dot{v}_z$ (see Prob. 1-17) and that $\dot{v}_z = \ddot{\xi}$, we obtain

$$\ddot{\xi} = - \frac{2Ke^2}{M^2\gamma_s{}^4} |A_w A_s| \sin K\xi , \qquad (131)$$

where

$$M = m[1 + (eA/mc)^2]^{1/2} \qquad (132)$$

is the effective (or shifted) electron mass, and we have at the last stage of the derivation approximated γ_z by $\gamma_s = [1 - v_s{}^2/c^2]^{-1/2}$. Equation (131) is the famous *pendulum equation.* Thus in the frame moving at velocity v_s, the dynamics of the electrons is the same as the motion of a particle in a sinusoidal potential. Note that the shifted mass M is used rather than the electron mass m.

The pendulum equation describes the motion of particles on a corrugated rooftop. In the moving frame, the electrons are injected at some random position (or phase) ξ_0 with some relative velocity $\dot{\xi}(0)$. Intuitively, we might expect that if this velocity is positive, the electron will decelerate, transferring energy to the field, while if the velocity is negative, the electron will accelerate, absorbing energy from the field. However as we know from the standard pendulum problem, the relative phase ξ_0 with respect to the field also plays a crucial role. From Eq. (130), we see that $\dot{p}_z$ is negative if and only if $\sin K\xi$ is positive. Hence the electrons initially absorb energy for $0 \le \xi_0 \le \pi/K$ and give up energy for $\pi/K \le \xi_0 \le 2\pi/K$. This is illustrated in Fig. 1-9, which shows the phase space of the pendulum. The abscissa is the phase ξ of the electron relative to the potential, while the ordinate is the relative velocity $\dot{\xi}$. Initially all electrons have the same velocity (or energy) $\dot{\xi}$. Since there is no way to control their initial phases, they are distributed uniformly between $-\pi$ and π.

Electrons with phases between $-\pi$ and 0 accelerate, while others decelerate, so that after a small time, the phase-space distribution looks like that in Fig. 1-10. Three important things have occurred. First, the electrons now have different energies, more or less accelerated or decelerated, depending on their initial phases. Thus the initially monoenergetic electron beam now has an energy spread. Second, the average relative velocity $\langle\dot{\xi}\rangle$ of the electrons has decreased, giving an average energy loss by the electrons. Conservation of energy shows that the field energy has increased by the same amount. The *recoil* of the electrons leads to *gain.* Third, the electron distribution is no longer uniformly distributed between $\xi = -\pi/K$ and π/K. The electrons are now *bunched* in a smaller region. Instead of producing random interferences with an emitted intensity proportional to N, they are redistributed by the ponderomotive potential to produce constructive interference as discussed for Eq. (109). These three effects, recoil, bunching, and spread, are key to understanding FEL's. They always occur together, and a correct FEL description must treat them all.

What happens for even longer times is shown in Fig. 1-11. The bunching, recoil, and spread have all increased. Note that the spread increases much faster than the recoil. This is a basic feature of the FEL that makes it hard to operate efficiently. Since pendulum trajectories are periodic, still longer times cause electrons that first decelerated to accelerate and vice versa as shown in Fig. 1-12. For such times the average electron energy increases, that is, the laser *saturates.* To maximize the energy transfer from the electrons to the laser field, the length of the wiggler should be chosen just short enough to avoid this backward energy transfer. This kind of saturation is quite different from that for two-level media discussed in Chap. 5. For the latter, the gain is bleached toward zero, but does not absorb. Here the saturation results from the onset of destructive interference in a fashion analogous to the phase matching discussed in Sec. 2-2. To maintain the constructive interference required for Eq. (111) as the electrons slow down, some FEL's gradually decrease the wiggler wavelength along the propagation direction. This kind of wiggler is called *a tapered wiggler.*

In this qualitative discussion, we have assumed that the initial relative velocity of the electrons was positive, i.e., that the average electron moves faster than the ponderomotive potential. If the initial velocity is negative, the average electron initially absorbs energy. These trends are depicted in Fig. 1-13, which plots the small-signal gain of the FEL versus the relative electron velocity.

This elementary discussion of the FEL only considers the electrons, and uses conservation of energy to determine whether the field is amplified. A more complete FEL theory would be *self-consistent*, with the electrons and field treated on the same footing. Such a theory of the FEL is beyond the scope of this book and the reader is referred to the references

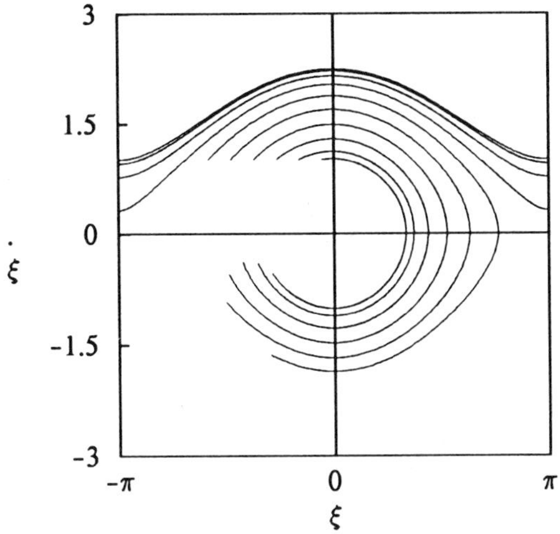

Fig. 1-11. As in Fig. 1-9, but at the instant of maximum energy extraction.

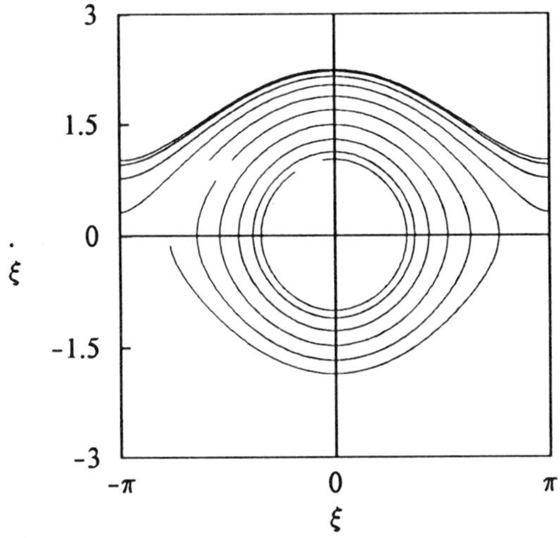

Fig. 1-12. As in Fig. 1-9, but for longer times such that the electron absorb energy from the laser field.

for further discussion. A self-consistent theory of conventional lasers is given in Chap. 6.

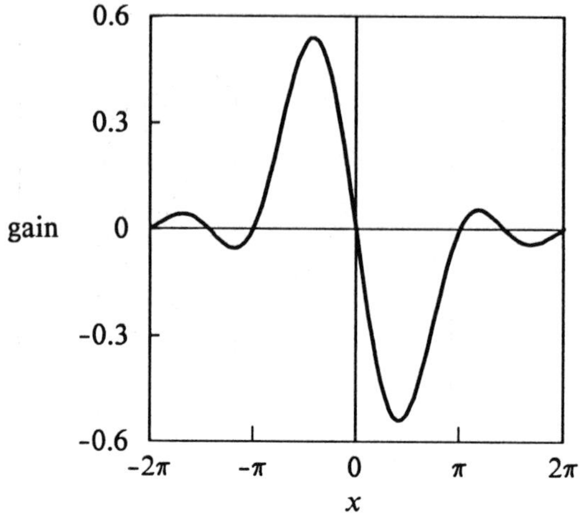

Fig. 1-13. FEL Gain function versus initial electron velocity relative to the ponderomotive potential.

References

Born, M., and E. Wolf (1970), *Principles of Optics*, Pergamon Press, Oxford, Great Britain. This is the standard reference on classical coherence theory.

Brau, C. (1990), *Free-Electron Lasers* Academic Press, San Diego, gives a detailed discussion of free-electron lasers both from the theoretical and the experimental points of view.

Glauber, R. J. (1965), in C. DeWitt, A. Blandin and C. Cohen-Tannoudji (Eds.) *Quantum Optics and Electronics*, Gordon and Breach, New York.

Hanbury Brown, H., and R. Q.Twiss (1954), Phil. Mag. **45**, 663; see also Hanbury Brown, H., and R. Q.Twiss (1956) Nature **178**, 1046.

Jackson, J. D. (1962), *Classical Electrodynamics*, John Wiley & Sons, Inc., New York. This is the classic book on classical electrodynamics.

Jacobs, S. F., M. O. Scully, M. Sargent III, and H. Pilloff (1978) - (1982), *Physics of Quantum Electronics*, Volumes 5 - 7, Addison Wesley Publishing Co., Reading, MA. These books give tutorial and advanced reviews of the theory and practice of free-electron lasers.

Milonni, P. W. (1976), Phys. Reports **25**, 1.

Milonni, P. W. (1984), Am. J. Phys. **52**, 340.

Portis, A. M. (1978), *Electromagnetic Fields: Sources and Media*, John Wiley & Sons, New York. This is another good reference on classical electromagnetic theory.

Problems

1-1. Derive the wave equation from microscopic Maxwell's equations that include a charge density and current. For this, Eqs. (2) and (4) become

$$\nabla\cdot\mathbf{E} = \frac{\rho}{\epsilon_0},$$

$$\nabla\times\mathbf{B} = \mu_0\mathbf{J} + \frac{1}{c^2}\frac{\partial\mathbf{E}}{\partial t},$$

respectively. Hint: First show that the conservation of charge equation

$$\frac{\partial\rho}{\partial t} + \nabla\cdot\mathbf{J} = 0$$

is solved by $\rho = -\nabla\cdot\mathbf{P} + \rho_{free}$, $\mathbf{J} = \partial\mathbf{P}/\partial t + \nabla\times\mathbf{M} + \mathbf{J}_{free}$. For our purposes, assume $\mathbf{M} = \mathbf{J}_{free} = \rho_{free} = 0$, and neglect a term proportional to $\nabla\rho$.

1-2. Show using the divergence theorem that

$$\int dV(\nabla\cdot\mathbf{P})\mathbf{r} = \oint(\mathbf{P}\cdot d\boldsymbol{\sigma})\mathbf{r} - \int\mathbf{P}dV .$$

Given the relation $\rho = -\nabla\cdot\mathbf{P}$ from Prob. 1-1, show that the polarization $\mathbf{P}$ can be interpreted as the dipole moment per unit volume $\mathbf{P} = \rho\mathbf{r}$.

1-3. Derive the slowly-varying amplitude and phase equations of motion (31) and (32) by substituting Eqs. (26) and (28) into the wave equation (25). Specify which terms you drop and why.

1-4. Derive the equations (56) and (57) of motion for the classical Bloch vector components U and V by substituting Eq. (28) into Eq. (44) and using the slowly-varying approximation. Calculate the evolution and magnitude of the classical Bloch vector in the absence of decay.

1-5. What are the units of $Ke^2/2m\nu\epsilon\gamma$ in Eq. (54), where γ is given by Eq. (74). What is the value of this quantity for the 632.8 nm line of the He-Ne laser (take $\omega = \nu$)? Calculate the absorption length ($1/\alpha$) for a 1.06 μm Nd:YAG laser beam propagating through a resonant linear medium with 10^{16} dipoles/m^3.

1-6. A field of the form $E(z,t) = E_0(z,t)\cos(Kz - \nu t)$ interacts with a medium. Using the "classical Bloch equations", derive an expression for the index of refraction of the medium. Assume the oscillator frequency ω is sufficiently close to the field frequency ν so that the rotating-wave approximation of Eq. (50) may be made.

1-7. In both laser physics and nonlinear optics, the polarization of a medium frequently results from the interaction of several separate fields. If $P(\mathbf{r},t)$ is given by

$$\mathcal{P}(\mathbf{r},t) = \sum_n \mathcal{P}_n(\mathbf{r},t)e^{i(\mathbf{K}_n\cdot\mathbf{r})},$$

solve for the polarization amplitude component $\mathcal{P}_n(\mathbf{r},t)$ in terms of $\mathcal{P}(\mathbf{r},t)$.

1-8. Find the magnetic field **B** corresponding to the electric field

$$\mathbf{E}(z,t) = \frac{1}{2}\,\hat{\mathbf{x}}\mathcal{E}U(z)\,e^{-i\nu t} + c.c.,$$

for running-wave [$U(z) = e^{iKz}$] and standing-wave [$U(z) = \sin Kz$] fields. Draw a "3-D" picture showing how the fields look in space at one instant of time.

1-9. The change of variables $z \to z'$ and $t \to t' = t - z/c$ transforms the slowly-varying Maxwell's equations from the laboratory frame to the so-called retarded frame. Write the slowly-varying Maxwell's equations in this frame. Discuss Beer's law in this frame.

1-10. In an optical cavity, the resonant wavelengths are determined by the constructive-interference condition that an integral number of wavelengths must occur in a round trip. The corresponding frequencies are determined by these wavelengths and the speed of light in the cavity. Given a cavity with a medium having anomalous dispersion, would it be possible to have more than one frequency resonant for a single wavelength? How?

1-11. Using Cartesian coordinates and using spherical coordinates show that the spherical wave $\exp(iKr-i\nu t)/r$ satisfies the wave equation for free space.

1-12. Calculate the magnitudes of the electric and magnetic fields for a 3-mW 632.8 nm laser focussed down to a spot with a $2\mu m$ radius. Assume constant intensity across the spot. How does this result scale with wavelength?

1-13. Derive the index of refraction (40) for the case that $\phi = \phi(t)$, i.e., not $\phi(z)$ as assumed in Eq. (39). The $\phi(t)$ case is generally more appropriate for lasers.

1-14. Solve Eq. (47) for $X(t)$, i.e., as a function of time.

1-15. Calculate the first and second-order coherence functions for the field

$$E^{+}(\mathbf{r},t) = \frac{E_0}{r}\, e^{-(\gamma + i\omega)(t-r/c)}\, \Theta(t-r/c) \,,$$

where Θ is the Heaviside (step) function. This would be the field emitted by atom located at $\mathbf{r} = 0$ and decaying spontaneously from time $t = 0$, if such a field could be described totally classically.

1-16. Derive the FEL Eq. (111) using the Lorentz transformation $z' = \gamma(z - \beta ct)$ and $t' = \gamma(t - \beta z/c)$, where γ is given by Eq. (102) and $\beta = v/c$.

1-17. Show that $\dot{p}_z = m\gamma\gamma_z{}^2\dot{v}_z$ and proceed to derive the pendulum equation (130). Use a personal computer to draw electron trajectories shown in Figs. 1-10 - 1-13 and discuss the trajectories.

1-18. The Kramers-Kronig relations allow one to calculate the real and imaginary parts of a linear susceptibility $\chi(\nu)$ as integrals over one another as follows:

$$\chi''(\nu) = -\frac{1}{\pi}\,\text{P.V.}\int_{-\infty}^{\infty} \frac{d\nu'\chi'(\nu')}{\nu' - \nu} \qquad (133)$$

$$\chi'(\nu) = \frac{1}{\pi}\, \text{P.V.} \int_{-\infty}^{\infty} \frac{d\nu' \chi''(\nu')}{\nu' - \nu}\,, \tag{134}$$

where P.V. means the principal value, i.e., the integral along the real axis excluding an arbitrarily small counterclockwise semicircle around the pole at $\nu' = \nu$. Equations (133) and (134) are based on the assumption that χ has no poles in the upper half plane; an equivalent set with a change in sign results for *a* χ that has no poles in the lower half plane. From Eqs. (33) and (51) we have the linear susceptibility

$$\chi(\nu) = \chi'(\nu) + i\chi''(\nu) = \frac{Nex_0 X(\nu)}{\epsilon E_0} = -\frac{Ne^2}{2\epsilon\nu m}\,\frac{1}{\nu - \omega + i\gamma}$$

$$= -\frac{Ne^2}{2\epsilon\nu m}\,\frac{\nu - \omega - i\gamma}{(\omega - \nu)^2 + \gamma^2}\,. \tag{135}$$

Show that this $\chi(\nu)$ satisfies Eqs. (133) and (134). Hint: for Eq. (133) use the residue theorem as follows: the desired principal part = the residue for the pole at $\nu' = \omega + i\gamma$ minus the half residue for the pole at $\nu' = \nu$. It is interesting to note that the power-broadened version of Eq. (135), namely Eq. (5.130), does not satisfy the Kramers-Kronig relations, since unlike Eq. (135), Eq. (5.130) does not reduce to a single pole in the lower half plane.

Chapter 2
CLASSICAL NONLINEAR OPTICS

Many problems of interest in optics and virtually all of those in this book involve nonlinear interactions that occur when the electromagnetic interaction becomes too large for the medium to continue to respond linearly. We have already seen how a nonlinearity plays an essential role in the free electron laser pendulum equation (1.131). In another example that we discuss in detail in Chap. 6, the output intensity of a laser oscillator builds up until saturation reduces the laser gain to the point where it equals the cavity losses. In situations such as second harmonic generation, one uses the fact that nonlinearities can couple electromagnetic field modes, transferring energy from one to another. Such processes can be used both to measure properties of the nonlinear medium and to produce useful applications such as tunable light sources.

In this chapter we extend Sec. 1-3's discussion of the simple harmonic oscillator to include quadratic and cubic nonlinearities, i.e. nonlinearities proportional to x^2 and x^3, respectively. Such nonlinearities allow us to understand phenomena such as sum and difference frequency generation, mode coupling, and even chaos in a simple classical context. Subsequent chapters treat these and related phenomena in a more realistic, but complex, quantized environment.

2-1. Nonlinear Dipole Oscillator

Section 1.3 discusses the response of a linear dipole oscillator to a monochromatic electric field. When strongly driven, most oscillators exhibit nonlinearities that can be described by equations of motion of the form [compare with Eq. (1.44)]

$$\boxed{\ddot{x}(z,t) + 2\gamma\dot{x}(z,t) + \omega^2 x(z,t) + ax^2(z,t) + bx^3(z,t) + \ldots = \frac{e}{m}E(z,t)}\ . \quad (1)$$

Here we include a z dependence, since the polarization modeled by $x(z,t)$ is a function of z. Specifically, x describes the position of an electron relative to the nucleus (internal degree of freedom) while z is the location of the dipole in the sample (external degree of freedom). Such oscillators are called "anharmonic". In many cases such as for isolated atoms, the coefficient a vanishes, leaving bx^3 as the lowest order nonlinear term.

We can determine the effect of the nonlinear terms on the response by a process of iteration that generates an increasingly accurate approximation to $x(t)$ in the form of a power series

$$x(t) \simeq x^{(1)}(t) + x^{(2)}(t) + \ldots + x^{(n)}(t) + \ldots \tag{2}$$

The leading term in this series is just the linear solution (1.49) itself or a linear superposition (1.61) of such solutions. To obtain the second-order contributions, we substitute the linear solution into Eq. (1), assume an appropriate form for the second-order contributions, and solve the resulting equations in a fashion similar to that for the linear solution. In the process, we find that new field frequencies are introduced. In general, the nth order term is given by solving the equation assuming that the nonlinear terms can be evaluated with the $(n-1)$th order terms. Many of the phenomena in this book require solutions that go beyond such a perturbative approach, since the corresponding series solution may fail to converge. Nevertheless, the subjects usually considered under the heading "nonlinear optics" are very useful and are typically described by second- and third-order nonlinearities.

We consider first the response of a medium with an ax^2 nonlinearity subjected to a monochromatic field of frequency ν_n, that is, Eq. (1.60) with a single amplitude $\mathcal{E}_n(z)$. Choosing $x^{(1)}$ to be given by the corresponding linear solution (1.49), we find for x^2 the approximate value

$$x^2 \simeq [x^{(1)}]^2 = \frac{1}{4}|x_n^{(1)}|^2 + \frac{1}{4}[x_n^{(1)}]^2 e^{2i(K_n z - \nu_n t)} + \text{c.c.}, \tag{3}$$

where the slowly varying amplitude $x_n^{(1)} = x_{0n} X_n$ and X_n is given by Eq. (1.48). We see that this nonlinear term contains both a dc contribution and one at twice the initial frequency. The dc term gives the intensity measured by a square-law detector and is the origin of the Kerr electro-optic effect in crystals, while the doubled frequency term leads to second harmonic generation. Observation of the latter in quartz subjected to ruby laser light kicked off the field known as nonlinear optics [see Franken, Ward, Peters, and Hill (1961)]. With an anharmonic forcing term proportional to Eq. (3), the second-order contribution $x^{(2)}(t)$ has the form

$$x^{(2)}(t) = \frac{1}{2}\,x_{dc}^{(2)} + \frac{1}{2}\,x_{2\nu_n}^{(2)}\,e^{i(2K_n z - 2\nu_n t)} + \text{c.c.} \tag{4}$$

According to our iteration method, we determine the second-order coefficients $x_{2\nu_n}^{(2)}$ and $x_{dc}^{(2)}$ by substituting Eq. (2) into Eq. (1) approximating x^2 by the first-order expression (3). By construction, the terms linear in $x^{(1)}$ cancel the driving force $e/m\,E$, and we are left with a simple harmonic oscillator equation for the $x^{(2)}$ coefficient, namely

$$\ddot{x}^{(2)}(t) + 2\gamma\dot{x}^{(2)}(t) + \omega^2 x^{(2)}(t) = -a[x^{(1)}(t)]^2\,. \tag{5}$$

Equating coefficients of terms with like time dependence, we find

$$\begin{aligned} x_{dc}^{(2)} &= -\frac{a}{2\omega^2}\,|x_n^{(1)}|^2 \\ &= -\frac{a}{2\omega^2}\left|\frac{e\mathcal{E}_n(z)/m}{\omega^2 - \nu_n{}^2 - 2i\nu_n\gamma}\right|^2 \end{aligned} \tag{6}$$

$$x_{2\nu_n}^{(2)} = -\frac{a}{2}\,[x_n^{(1)}]^2\,\frac{1}{\omega^2 - (2\nu_n)^2 - 2i(2\nu_n)\gamma}\,. \tag{7}$$

Note here that if the applied field frequency ν is approximately equal to the natural resonance frequency ω, both the dc and second-harmonic coefficients are divided by squares of optical frequencies. Hence these terms are usually very small, and second-order theory is a good approximation, for example in noncentrosymmetric crystals.

Now consider the response of this nonlinear oscillator to an electric field given by Eq. (1.60) with two frequency components at the frequencies ν_1 and ν_2. To lowest order (first order in the fields), we neglect the nonlinearities and find $x(t) \simeq x^{(1)}(t)$, which is given by the linear superposition (1.61) with two modes. The approximate second order nonlinearity $[x^{(1)}]^2$ has the explicit form

$$\begin{aligned} [x^{(1)}]^2 &= [x_1^{(1)}e^{i(K_1z-\nu_1t)} + x_2^{(1)}e^{i(K_2z-\nu_2t)}]^2 \\ &= \frac{1}{2}x_1^{(1)}[x_2^{(1)}]^*\,e^{i[(K_1-K_2)z-(\nu_1-\nu_2)t]} + \frac{1}{2}x_1^{(1)}x_2^{(1)}\,e^{i[(K_1+K_2)z-(\nu_1+\nu_2)t]} \\ &\quad + \frac{1}{4}\sum_{n=1}^{2}\left(|x_n^{(1)}|^2 + [x_n^{(1)}]^2 e^{2i(K_nz\,-\,\nu_nt)}\right) + \text{c.c.} \end{aligned} \tag{8}$$

As one expects from Eq. (4), each mode contributes a DC component and a component oscillating at its doubled frequency, $2\nu_n$. In addition, Eq. (8) contains components at the sum and difference frequencies $\nu_1 \pm \nu_2$. Hence the nonlinear dipoles can generate fields at these frequencies. More generally, when higher orders of perturbation are considered, the polarization of a medium consisting of such anharmonic oscillators can generate all frequency components of the form $m\nu_1 \pm n\nu_2$, with m and n being integers. When such combinations lead to frequencies other than harmonics of ν_1 or ν_2, they are called combination tones. Such tones are responsible for self mode locking in lasers (Sec. 10-3) and three- and four-wave mixing (rest of the present chapter and Chap. 9).

Generalizing the second-order contribution (4) to include the sum and difference frequencies, we have

$$x^{(2)}(t) = \frac{1}{2} x_s^{(2)} e^{i[(K_1+K_2)z - (\nu_1+\nu_2)t]} + \frac{1}{2} x_d^{(2)} e^{i[(K_1-K_2)z - (\nu_1-\nu_2)t]} + \frac{1}{2} \sum_{n=1}^{2} x_{2\nu_n}^{(2)} e^{i(2K_n z - 2\nu_n t)} + \frac{1}{2} x_{dc}^{(2)} + \text{c.c.} \tag{9}$$

Substituting Eqs. (8) and (9) into Eq. (5) and equating coefficients of like time dependence, we find that $x_{dc}^{(2)}$ is given by the sum of two terms like Eq. (6), the doubled frequency terms are given by Eq. (7), and the difference and sum frequency terms are given by

$$x_d^{(2)} = -a \frac{x_1^{(1)}[x_2^{(1)}]^*}{\omega^2 - (\nu_1-\nu_2)^2 - 2i(\nu_1-\nu_2)\gamma} \tag{10}$$

$$x_s^{(2)} = -a \frac{x_1^{(1)} x_2^{(1)}}{\omega^2 - (\nu_1+\nu_2)^2 - 2i(\nu_1+\nu_2)\gamma} . \tag{11}$$

Except for a factor of 2, for the degenerate frequency case $\nu_1 = \nu_2$, the difference frequency term (10) reduces to the dc term (6) and the sum frequency term (11) reduces to the second harmonic term (7).

Frequency combinations like those in Eq. (9) also appear in quantum mechanical descriptions of the medium, which typically involve more intricate nonlinearities. In particular, difference frequency generation induces pulsations of the populations in a medium consisting of two-level atoms irradiated by two beams of different frequencies. These pulsations play an important role in saturation spectroscopy, as discussed in Chap. 8.

2-2. Coupled-mode equations

So far we have obtained the steady-state response of the nonlinear dipole to second order in the field and have seen how combination tones at the frequencies $m\nu_1 \pm n\nu_2$ can be generated by such systems. The way in which the corresponding waves evolve is readily obtained from the wave equation (1.25) giving the propagation of an electromagnetic field $E(z,t)$ inside a medium of polarization $P(z,t)$. For a medium consisting of nonlinear oscillators, $P(z,t) = N(z)ex(z,t)$, where $N(z)$ is the oscillator density and z labels the position inside the medium. To analyze the growth of a wave at a frequency ν_3, we consider three modes in the field (1.60) and in the polarization (1.63). The slowly varying amplitude approximation allows us to derive coupled differential equations for the evolution of the field envelopes $\mathcal{E}_n$. These are called *coupled-mode equations*, and play an important role in multiwave phenomena such as phase conjugation (Sec. 2-4 and Chap. 9) and the generation of squeezed states (Chap. 18), in which case a fully quantum mechanical version of these equations is required. In general, coupled-wave equations form an infinite hierarchy of ordinary differential equations, and some kind of approximation scheme is needed to truncate them. For instance, if we are only interested in the small signal build-up of the wave at frequency ν_3, we can neglect its back-action on the nonlinear polarization $P(z,t)$ – first-order theory in $\mathcal{E}_3$. This is the procedure used in Sec. 2-1. Another common approximation assumes that the waves at frequencies ν_1 and ν_2 are so intense that their depletion via the nonlinear wave-mixing process can be neglected.

An important feature of coupled-mode equations is *phase matching*, which represents the degree to which the induced mode coupling terms in the polarization have the same phase as the field modes they affect. To the extent that the phases differ, the mode coupling is reduced. Phase matching involves differences in wave vectors and amounts to conservation of momentum. This is distinct from the frequency differences of the last section, which amount to conservation of energy.

Consider for instance the case of difference-frequency generation, where two incident waves at frequencies ν_1 and ν_2 combine in the nonlinear medium to generate a wave at the difference frequency ν_d. In the slowly-varying amplitude approximation and in steady state (we neglect the $\partial/\partial t$ terms), Eq. (1.43) becomes

$$\boxed{\frac{d\mathcal{E}_n}{dz} = \frac{iK_n}{2\epsilon_n}\mathcal{P}_n} \quad . \tag{12}$$

where ϵ_n is the permittivity at the frequency ν_n of the host medium in which our oscillators are found. From Eqs. (1.62) and (1.63), we use the linear solutions for modes 1 and 2

$$\frac{d\mathcal{E}_n}{dz} \simeq iK_n \frac{N(z)e^2}{2\epsilon_n m} \frac{\mathcal{E}_n}{\omega^2 - \nu_n{}^2 - 2i\nu_n\gamma} = -\alpha_\nu \mathcal{E}_n . \tag{13}$$

Note that in this example we assume that the host medium is purely dispersive; otherwise another absorption term would have to be included. Using (10) and (12), we find for the difference-frequency term at the frequency $\nu_d = \nu_1 - \nu_2$

$$\frac{d\mathcal{E}_d}{dz} = \frac{iK_d N(z)e}{2\epsilon_d} \frac{\frac{e}{m}\mathcal{E}_d - ax_1^{(1)}[x_2^{(1)}]^* e^{i(K_1-K_2-K_d)z}}{\omega^2 - \nu_d{}^2 - 2i\nu_d\gamma} . \tag{14}$$

The coupled-mode equations (13) and (14) take a rather simple form, since we have neglected the back action of $\mathcal{E}_d$ on $\mathcal{E}_1$ and $\mathcal{E}_2$. Equation (13) simply describes the linear absorption and dispersion of $\mathcal{E}_1$ and $\mathcal{E}_2$ due to the nonlinear oscillators, as does the first term on the right-hand side of Eq. (14) for $\mathcal{E}_d$. In many cases these are small effects compared to those of the host medium accounted for by ϵ_n, and we neglect them in the following. $\mathcal{E}_1$ and $\mathcal{E}_2$ then remain constant and Eq. (14) (without the leading term) can be readily integrated over the length L of the nonlinear medium to give

$$\mathcal{E}_d(L) = G\mathcal{E}_1\mathcal{E}_2{}^* \frac{e^{i\Delta KL} - 1}{i\Delta K} , \tag{15}$$

where

$$G = -\frac{iK_d aNe}{2\epsilon_d} \frac{x_1^{(1)}[x_2^{(1)}]^*/\mathcal{E}_1\mathcal{E}_2{}^*}{\omega^2 - \nu_d{}^2 - 2i\nu_d\gamma} ,$$

and the K-vector mismatch $\Delta K = K_1 - K_2 - K_d$. The resulting intensity $I_d = |\mathcal{E}_d|^2$ is

$$I_d(L) = |G|^2 I_1 I_2 \frac{\sin^2(\Delta KL/2)}{(\Delta K/2)^2} . \tag{16}$$

If $\Delta K = 0$, $I_d(L)$ reduces to $|G|^2 I_1 I_2 L^2$, but for $\Delta K \neq 0$, it oscillates periodically. To achieve efficient frequency conversion, it is thus crucial that $(K_1-K_2)L$ be close to $K_d L$. For $\Delta K \neq 0$ the maximum intensity I_d is reached for a medium of length $L = \pi/\Delta K$. For larger values of ΔKL, the induced polarization at the frequency ν_d and the wave propagating at that frequency start to interfere destructively, attenuating the wave. For still larger values of L, the interference once again becomes positive, and continues to oscillate in this fashion. Since nonlinear crystals are expensive, it is worth trying to achieve the best conversion with the smallest crystal, namely for $\Delta KL = \pi$. In the plane-wave, collinear propagation model described here, perfect phase matching requires that the wave speeds $v_n = \nu_n/K_n$ all be equal, as would be the case in a dispersionless medium. More generally we have the difference $\Delta K = K_1 - K_2 - K_d = (n_1\nu_1 - n_2\nu_2 - n_d\nu_d)/c \neq 0$ since the n's differ. For noncollinear operation the vectorial phase matching condition

$$\Delta KL = |\mathbf{K}_1 - \mathbf{K}_2 - \mathbf{K}_d|L \simeq \pi \tag{17}$$

must be fulfilled for maximum I_d. There are a number of ways to achieve this, including appropriate geometry, the use of birefringent media, and temperature index tuning.

2-3. Cubic Nonlinearity

We already mentioned that quadratic nonlinearities such as described in the preceding sections do not occur in isolated atoms, for which the lowest order nonlinear effects are cubic in the fields. These can be described in our classical model by keeping the bx^3 term instead of ax^2 in the nonlinear oscillator equation (1). In the presence of two strong pump fields at frequencies ν_1 and ν_2, the third-order polarization given by bx^3 includes contributions at the frequencies ν_1 and ν_2 *and* at the *sideband* frequencies $\nu_0 = \nu_1 - \Delta$ and $\nu_3 = \nu_2 + \Delta$ as well, where

$$\Delta = \nu_2 - \nu_1 \,. \tag{18}$$

The generation of these sidebands is an example of four-wave mixing. To describe the initial growth of the sidebands, we write the anharmonic term bx^3 to third-order in x_1 and x_2, and first-order in the small displacements x_0 and x_3, that is,

$$[x^{(1)}]^3 = \frac{1}{8}\Big[x_1^{(1)}e^{i(K_1z - \nu_1t)} + x_2^{(1)}e^{i(K_2z - \nu_2t)}$$
$$+ 3x_0^{(1)}e^{i(K_0z - \nu_0t)} + 3x_3^{(1)}e^{i(K_3z - \nu_3t)} + \text{c.c.}\Big]$$
$$\times\Big\{2x_1^{(1)}x_2^{(1)}\, e^{i[(K_1+K_2)z - (\nu_1+\nu_2)t]}$$
$$+ 2x_1^{(1)}[x_2^{(1)}]^*\, e^{i[(K_1-K_2)z - (\nu_1-\nu_2)t]}$$
$$+ \sum_{n=1}^{2} \{[x_n^{(1)}]^2e^{2i(K_nz - \nu_nt)} + |x_n^{(1)}|^2\} + \text{c.c.}\Big\}, \tag{19}$$

where the terms in {} are similar to those in Eq. (9). The factor of 3 results from the three ways of choosing the x_0 and x_3 from the triple product.

The curly braces in Eq. (19) contain two DC terms, a contribution oscillating at the difference frequency Δ, and three rapidly oscillating contributions oscillating at the frequencies $\nu_n + \nu_m$. These time-dependent terms are sometimes called (complex) index gratings, and the nonlinear polarization may be interpreted as the scattering of a light field $\mathcal{E}_n$ from the grating produced by two fields $\mathcal{E}_m$ and $\mathcal{E}_k$. In this picture, the DC terms are "degenerate" gratings produced by the fields $\mathcal{E}_m$ and $\mathcal{E}_m^*$. Equation (19) readily gives the third-order contributions to the components of the polarization $\mathcal{P}_n$ at the frequencies of interest.

One can interpret Eq. (19) as the scattering of components in the [] of the first line off the slowly varying terms in the {}. Specifically the $|x_n^{(1)}|^2$ terms in Eq. (19) contribute nonlinear changes at the respective frequencies of the components in the []. In contrast, the scattering off the "Raman-like" term $\exp[i(K_2-K_1)z - i\Delta t]$ and its complex conjugate contribute corrections at frequencies shifted by ±Δ. Taking $\nu_2 > \nu_1$, we see that the ν_2 term in the [] scatters producing components at both the lower frequency ν_1 (called *a Stokes* shift) and the higher frequency $\nu_3 = \nu_2 + \Delta$ (called an *anti-Stokes* shift). Similarly the ν_1 term in the [] leads to contributions at the frequencies $\nu_0 = \nu_1 - \Delta$ and at ν_2. The induced polarization components at the frequencies ν_0 and ν_3 are called *combination tones*. They are generated in the nonlinear medium from other frequencies. If the two pump beams at ν_1 and ν_2 are copropagating, the index grating represented by the K_2-K_1 term propagates at approximately the velocity of light in the host medium, but if the beams are counterpropagating, the grating propagates at the relatively slow speed $v = -\Delta/(K_1+K_2)$. In particular, it becomes stationary for the degenerate case $\nu_1 = \nu_2$. (Compare with the ponderomotive force acting on the electrons in the free electron laser, Eq. (1.124) ! -- Can you draw an analogy between the two situations?)

We are often only interested in induced polarizations near or at the fundamental frequencies ν_n, $n = 0, 1, 2, 3$. Keeping only these in Eq. (19) and neglecting combination tones involving x_0 and x_3 in the pump-mode polarizations (Prob. 2-7), we find

$$\begin{aligned}
[x^{(1)}]^3\Big|_{fund} &= \frac{3}{8}x_1^{(1)}(|x_1^{(1)}|^2 + 2|x_2^{(1)}|^2)e^{i(K_1z - \nu_1t)} \\
&+ \frac{3}{8}x_2^{(1)}(|x_2^{(1)}|^2 + 2|x_1^{(1)}|^2)e^{i(K_2z - \nu_2t)} \\
&+ \frac{6}{8}[|x_1^{(1)}|^2 + |x_2^{(1)}|^2][x_0^{(1)}e^{i(K_0z-\nu_0t)} + x_3^{(1)}e^{i(K_3z-\nu_3t)}] \\
&+ \frac{6}{8}x_1^{(1)}x_2^{(1)}[x_3^{(1)}]^*e^{i[(K_1+K_2-K_3)z - \nu_0t]} \\
&+ \frac{6}{8}x_1^{(1)}x_2^{(1)}[x_0^{(1)}]^*e^{i[(K_1+K_2-K_0)z - \nu_3t]} \\
&+ \frac{3}{8}[x_1^{(1)}]^2[x_2^{(1)}]^*e^{i[(2K_1-K_2)z - \nu_0t]} \\
&+ \frac{3}{8}[x_2^{(1)}]^2[x_1^{(1)}]^*e^{i[(2K_2-K_1)z - \nu_3t]} + \text{c.c.}
\end{aligned} \tag{20}$$

Combining the various terms, we find that the third-order polarization components are given by

$$\begin{aligned}
\mathcal{P}_0^{(3)} &= \frac{6}{8}Neb[|x_1^{(1)}|^2 + |x_2^{(1)}|^2]x_0^{(1)} \\
&+ \frac{6}{8}Nebx_1^{(1)}x_2^{(1)}[x_3^{(1)}]^*e^{i(K_1+K_2-K_3-K_0)z} \\
&+ \frac{3}{8}Neb[x_1^{(1)}]^2\,[x_2^{(1)}]^*e^{i(2K_1-K_2-K_0)z}
\end{aligned} \tag{21a}$$

$$\mathcal{P}_1^{(3)} = \frac{3}{8}Nebx_1^{(1)}[|x_1^{(1)}|^2 + 2|x_2^{(1)}|^2] \tag{21b}$$

$$\mathscr{P}_2^{(3)} = \frac{3}{8} Nebx_2^{(1)}[2|x_1^{(1)}|^2 + |x_2^{(1)}|^2] \tag{21c}$$

$$\begin{aligned}\mathscr{P}_3^{(3)} = &\frac{6}{8} Neb[|x_1^{(1)}|^2 + |x_2^{(1)}|^2]x_3^{(1)} \\ &+ \frac{6}{8} Nebx_1^{(1)} x_2^{(1)} [x_0^{(1)}]^* e^{i(K_1+K_2-K_0-K_3)z} \\ &+ \frac{3}{8} Neb[x_2^{(1)}]^2 [x_1^{(1)}]^* e^{i(2K_2-K_1-K_3)} \ .\end{aligned} \tag{21d}$$

The polarization components $\mathscr{P}_0^{(3)}$ and $\mathscr{P}_3^{(3)}$ are solely due to the existence of index gratings, which are also responsible for the factors of 2 in the cross coupling terms for $\mathscr{P}_1^{(3)}$ and $\mathscr{P}_2^{(3)}$. This asymmetry is sometimes called nonlinear nonreciprocity and was discovered in quantum optics by Chiao, Kelley, and Garmire (1966). It also appears in the work by van der Pol (1934) on coupled vacuum-tube tank circuits. In the absence of index gratings, the factors of 2 in Eqs. (21b) and (21c) are replaced by 1, and $|x_1^{(1)}|^2$ and $|x_2^{(1)}|^2$ play symmetrical roles in $\mathscr{P}_1^{(3)}$ and $\mathscr{P}_2^{(3)}$.

The polarizations $\mathscr{P}_n$ lead to coupled-mode equations for the field envelopes. The procedure follows exactly the method of Sec. 2-2 and we obtain (Prob. 2-2)

$$\begin{aligned}\frac{d\mathcal{E}_0}{dz} = &-\mathcal{E}_0[\alpha_0 - \theta_{01}|\mathcal{E}_1|^2 - \theta_{02}|\mathcal{E}_2|^2] + \vartheta_{0121}\mathcal{E}_1{}^2\mathcal{E}_2{}^* \, e^{i(2K_1-K_2-K_0)z} \\ &+ \vartheta_{0231}\mathcal{E}_2\mathcal{E}_3{}^*\mathcal{E}_1 e^{i(K_1+K_2-K_3-K_0)z} \ ,\end{aligned} \tag{22a}$$

$$\frac{d\mathcal{E}_1}{dz} = -\mathcal{E}_1[\alpha_1 - \beta_1|\mathcal{E}_1|^2 - \theta_{12}|\mathcal{E}_2|^2] \tag{22b}$$

$$\frac{d\mathcal{E}_2}{dz} = -\mathcal{E}_2[\alpha_2 - \beta_2|\mathcal{E}_2|^2 - \theta_{21}|\mathcal{E}_1|^2] \tag{22c}$$

$$\begin{aligned}\frac{d\mathcal{E}_3}{dz} = &-\mathcal{E}_3[\alpha_3 - \theta_{31}|\mathcal{E}_1|^2 - \theta_{32}|\mathcal{E}_2|^2] + \vartheta_{3212}\mathcal{E}_2{}^2\mathcal{E}_1{}^* \, e^{i(2K_2-K_1-K_3)z} \\ &+ \vartheta_{3102}\mathcal{E}_1\mathcal{E}_0{}^*\mathcal{E}_2 e^{i(K_1+K_2-K_0-K_3)z} \ .\end{aligned} \tag{22d}$$

Here $\mathcal{E}_n$ is the complex amplitude of the field at frequency ν_n, and the $-\alpha_n\mathcal{E}_n$ terms allow for linear dispersion and absorption.

Equations (22b) and (22c) for the pump modes amplitudes are coupled by the cross-coupling (or cross-saturation) coefficients θ_{nj}. To this order of perturbation, they are independent of the sidemode amplitudes $\mathcal{E}_0$ and

$\mathcal{E}_3$. Because $\mathcal{E}_1$ and $\mathcal{E}_2$ always conspire to create an index grating of the correct phase, the evolution of these modes is not subject to a phase matching condition. Equations of this type are rather common in nonlinear optics and laser theory. In Sec. 6-4, we obtain an evolution of precisely this type for the counterpropagating modes in a ring laser. We show then that the cross-coupling between modes can lead either to the suppression of one of the modes or to their coexistence, depending on the magnitude of the coupling parameter $C = \theta_{12}\theta_{21}/\beta_1\beta_2$ and relative sizes of the α_n.

In contrast, the sidemodes $\mathcal{E}_0$ and $\mathcal{E}_3$ are coupled to the strong pump fields $\mathcal{E}_1$ and $\mathcal{E}_2$ only, and not directly to each other. They have no back-action on the pump modes dynamics, and their growth is subject to a phase-matching condition.

2-4. Four-Wave Mixing with Degenerate Pump Frequencies

In many experimental situations, it is convenient to drive the nonlinear medium with two pump fields of the same frequency ν_2, but with opposite propagation directions given by the wave vectors $\mathbf{K}_{2\downarrow}$ and $\mathbf{K}_{2\uparrow}$. The pump waves cannot by themselves generate polarization components at sideband frequencies. However one can still take advantage of the index gratings produced by the pump beams with weak waves at frequencies symmetri-

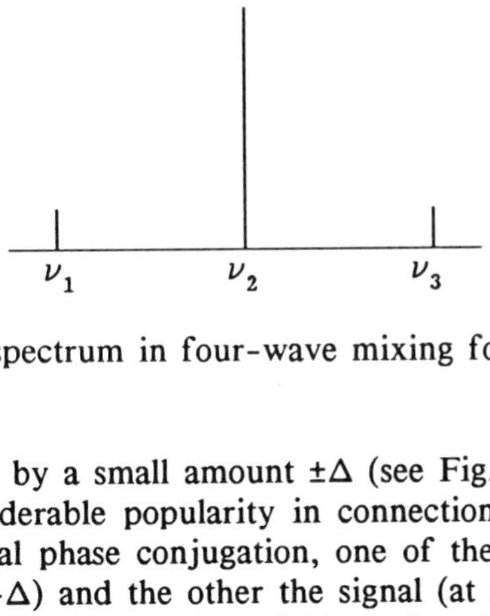

Fig. 2-1. Mode spectrum in four-wave mixing for optical phase conjugation

cally detuned from ν_2 by a small amount $\pm\Delta$ (see Fig. 2-1). This procedure has gained considerable popularity in connection with optical phase conjugation. In optical phase conjugation, one of the sidebands is called the probe (at $\nu_1 = \nu_2 - \Delta$) and the other the signal (at $\nu_3 = \nu_2 + \Delta$), and we adopt this notation here in anticipation of Chaps. 8 and 9.

We consider the wave configuration in Fig. 2-2 with two counterpropagating pump beams along one direction, and counterpropagating signal and "conjugate" waves along another direction, which we call the z axis. The electric field for these four waves has the form

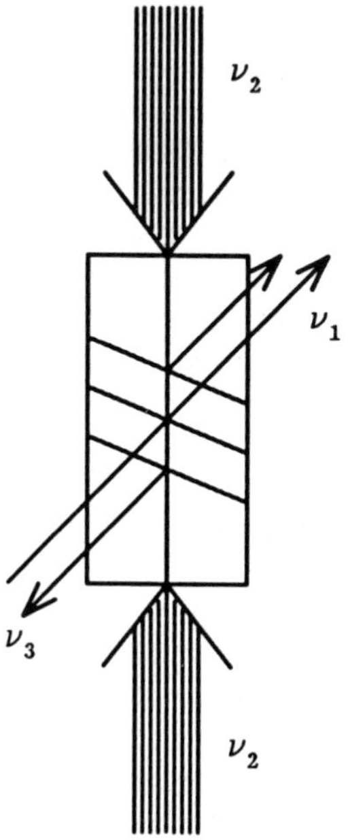

Fig. 2-2. Diagram of interaction between standing-wave pump beam (ν_2) with probe (ν_1) and conjugate (ν_3) beams used in phase conjugation

$$E(\mathbf{r},t) = \frac{1}{2}[\mathcal{E}_1 e^{i(K_1 z - \nu_1 t)} + \mathcal{E}_{2\downarrow} e^{i(\mathbf{K}_{2\downarrow}\cdot\mathbf{r} - \nu_2 t)} + \mathcal{E}_{2\uparrow} e^{i(\mathbf{K}_{2\uparrow}\cdot\mathbf{r} - \nu_2 t)}$$
$$+ \mathcal{E}_3 e^{i(-K_3 z - \nu_3 t)}] + \text{c.c.}\,, \qquad (23)$$

where we take $\mathbf{K}_{2\uparrow} = -\mathbf{K}_{2\downarrow}$. The field fringe patterns resulting from interference between the various waves can induce index gratings. The corresponding linear displacement $x^{(1)}(t)$ contains components proportional to each of the field amplitudes, and the third-order nonlinear displacement $x^{(3)}$ consists of the sum of all terms proportional to the products of three fields, each of which can be anyone of the four waves or their complex conjugates. This gives a grand total of $8\cdot 8\cdot 8 = 512$ terms. Fortunately, we're only interested in a relatively small subset of these terms, namely those with the positive frequency ν_1 linear in $\mathcal{E}_1$. This gives a third-order signal polarization $\mathcal{P}_1^{(3)}$ proportional to

$$\mathcal{E}_1\mathcal{E}_{2\downarrow}\mathcal{E}_{2\downarrow}^* + \mathcal{E}_1\mathcal{E}_{2\downarrow}^*\mathcal{E}_{2\downarrow} + \mathcal{E}_{2\downarrow}\mathcal{E}_1\mathcal{E}_{2\downarrow}^* + \mathcal{E}_{2\downarrow}^*\mathcal{E}_1\mathcal{E}_{2\downarrow}$$
$$+ \mathcal{E}_{2\downarrow}\mathcal{E}_{2\downarrow}^*\mathcal{E}_1 + \mathcal{E}_{2\downarrow}^*\mathcal{E}_{2\downarrow}\mathcal{E}_1 + \mathcal{E}_1\mathcal{E}_{2\uparrow}\mathcal{E}_{2\uparrow}^* + \mathcal{E}_1\mathcal{E}_{2\uparrow}^*\mathcal{E}_{2\uparrow}$$
$$+ \mathcal{E}_{2\uparrow}\mathcal{E}_1\mathcal{E}_{2\uparrow}^* + \mathcal{E}_{2\uparrow}^*\mathcal{E}_1\mathcal{E}_{2\uparrow} + \mathcal{E}_{2\uparrow}\mathcal{E}_{2\uparrow}^*\mathcal{E}_1 + \mathcal{E}_{2\uparrow}^*\mathcal{E}_{2\uparrow}\mathcal{E}_1$$
$$+ [\mathcal{E}_3^*\mathcal{E}_{2\uparrow}\mathcal{E}_{2\downarrow} + \mathcal{E}_3^*\mathcal{E}_{2\downarrow}\mathcal{E}_{2\uparrow} + \mathcal{E}_{2\uparrow}\mathcal{E}_3^*\mathcal{E}_{2\downarrow} + \mathcal{E}_{2\downarrow}\mathcal{E}_3^*\mathcal{E}_{2\uparrow}$$
$$+ \mathcal{E}_{2\uparrow}\mathcal{E}_{2\downarrow}\mathcal{E}_3^* + \mathcal{E}_{2\downarrow}\mathcal{E}_{2\uparrow}\mathcal{E}_3^*]\, e^{i(K_3-K_1)z} \tag{24a}$$

$$= 6\mathcal{E}_1(|\mathcal{E}_{2\downarrow}|^2 + |\mathcal{E}_{2\uparrow}|^2) + 6\mathcal{E}_3^*\mathcal{E}_{2\uparrow}\mathcal{E}_{2\downarrow}e^{i(K_3-K_1)z}\ . \tag{24b}$$

The various terms in (24a) have simple physical interpretations. For instance, the first term results from the product in $[x^{(1)}]^3$

$$\mathcal{E}_1 e^{i(K_1z-\nu_1t)}\ \mathcal{E}_{2\downarrow}e^{i(\mathbf{K}_{2\downarrow}\cdot\mathbf{r}-\nu_2t)}\ \mathcal{E}_{2\downarrow}^*\mathrm{e}^{-i(\mathbf{K}_{2\downarrow}\cdot\mathbf{r}-\nu_2t)}\ .$$

Note that the pump phase dependencies $\mathbf{K}_{2\downarrow}\cdot\mathbf{r}$ cancel identically, as they do for all terms in Eq. (24). The first and second terms represent contributions to the nonlinear refraction of the field $\mathcal{E}_1$ due the nonlinear index (Kerr effect) induced by the pump field intensity $I_{2\downarrow}$. The third term can be understood as originating from the scattering of the field $\mathcal{E}_{2\downarrow}$ off the grating produced by $\mathcal{E}_1$ and $\mathcal{E}_{2\downarrow}$, etc... Its effect on the polarization is precisely the same as that of the first two terms, but we have intentionally written it separately in anticipation of the quantum mechanical discussion of Chap. 9, where the order in which the fields are applied to the medium matters. Indeed in our classical model, the terms in the first three lines in (24a) are all proportional to the product of a pump field intensity and the probe field $\mathcal{E}_1$, and can be globally described as nonlinear absorption and refraction terms.

The last two lines in (24a) couple the sidemode $\mathcal{E}_3$ to $\mathcal{E}_1$ via the following scattering mechanism: the field $\mathcal{E}_3^*$ interferes with the pump fields $\mathcal{E}_{2\uparrow}$ and $\mathcal{E}_{2\downarrow}$ to induce two complex index gratings that scatter $\mathcal{E}_{2\downarrow}$ and $\mathcal{E}_{2\uparrow}$, respectively, into $\mathcal{E}_1$. This process, which is essentially the real-time realization of holographic writing and reading, is called phase conjugation and is discussed in detail in Chap. 9. The process retroreflects a wavefront, sending it back along the path through which it came (see Fig. 9-1). It can be used to compensate for poor optics. Note that although the pump phase dependencies cancel one another as they do for the terms in the first four lines of Eq. (24a), the induced polarization has the phase $\exp[i(K_3z-\nu_1t)]$, while Maxwell's equations require $\exp[i(K_1z-\nu_1t)]$. This gives the phase mismatch factor $\exp[i(K_3-K_1)z]$, which is important except in the degenerate frequency case $\nu_1 = \nu_2 = \nu_3$, for which $K_3 = K_1$.

Neglecting the depletion of the pump beams $\mathcal{E}_{2\uparrow}$ and $\mathcal{E}_{2\downarrow}$, we find the coupled-mode equations for $\mathcal{E}_1$ and $\mathcal{E}_3^*$

$$\frac{d\mathcal{E}_1}{dz} = -\alpha_1\mathcal{E}_1 + \chi_1\mathcal{E}_3^* e^{2i\Delta Kz} \tag{25a}$$

$$-\frac{d\mathcal{E}_3^*}{dz} = -\alpha_3^*\mathcal{E}_3^* + \chi_3^*\mathcal{E}_1 e^{-2i\Delta Kz} \quad , \tag{25b}$$

where

$$\Delta K = \frac{K_3 - K_1}{2} \; . \tag{26}$$

Here $-d\mathcal{E}_3^*/dz$ appears since $\mathcal{E}_3^*$ propagates along $-z$, and we use χ_n for the coupling coefficient to agree with later usage, although it is only a part of a susceptibility.

To solve these equations, we proceed by first transforming away the phase mismatch by the substitution

$$\mathcal{E}_1 = \mathcal{A}_1 e^{2i\Delta Kz} \tag{27}$$

into Eq. (25). In particular, Eq. (25a) becomes

$$\frac{d\mathcal{A}_1}{dz} = -(\alpha_1 + 2i\Delta K)\mathcal{A}_1 + \chi_1\mathcal{E}_3^* \; . \tag{28}$$

We seek solutions of (25b) and (28) of the form $e^{\mu z}$. Substituting $\mathcal{A}_1 = e^{\mu z}$ into Eq. (28), solving for $\mathcal{E}_3^*$, and substituting the result into Eq. (25b), we find the eigenvalues

$$\begin{aligned} \mu_\pm &= -\frac{1}{2}(\alpha_1 - \alpha_3^* + 2i\Delta K) \pm [(\alpha_1 + \alpha_3^* + 2i\Delta K)^2/4 - \chi_1\chi_3^*]^{1/2} \\ &= -a \pm [\alpha^2 - \chi_1\chi_3^*]^{1/2} = -a \pm w \; . \end{aligned} \tag{29}$$

Hence the general solutions are

$$\mathcal{A}_1(z) = e^{-az}\,[Ae^{wz} + Be^{-wz}] \tag{30a}$$

and

$$\mathcal{E}_3^*(z) = e^{-az}\,[Ce^{wz} + De^{-wz}]\,, \tag{30b}$$

where the coefficients A, B, C and D are determined by the boundary conditions of the problem.

We suppose here that a weak signal field $\mathcal{E}_1(0)$ is injected inside the nonlinear medium at $z = 0$, and we study the growth of the counterpropagating conjugate wave $\mathcal{E}_3^*$, which is taken to be zero at $z = L$. This means that $\mathcal{A}_1(0) = \mathcal{E}_1(0) =$ constant, and $\mathcal{E}_3^*(L) = 0$, in which case one has immediately $B = \mathcal{E}_1(0) - A$ and $D = -Ce^{wL}$. Matching the boundary conditions of Eqs. (25b) and (28) at $z = L$ yields

$$A = \frac{1}{2}\,\mathcal{A}_1(0)\,e^{-wL}\,(w-\alpha)/(w\cosh wL + \alpha\sinh wL)\,, \tag{31}$$

$$2wCe^{wL} = \chi_3^*(A\sinh wL + \mathcal{A}_1(0)\,e^{-wL})\,. \tag{32}$$

Further manipulation yields finally

$$\mathcal{E}_1(z) = \mathcal{E}_1(0)\,e^{-(a+w+2i\Delta K)z}\left[1 + \frac{(w-\alpha)e^{w(z-L)}\sinh wz}{w\cosh wL + \alpha\sinh wL}\right], \tag{33}$$

$$\mathcal{E}_3^*(z) = -\chi_3^*\mathcal{E}_1(0)\,\frac{e^{-az}\sinh w(z-L)}{w\cosh wL + \alpha\sinh wL}\,. \tag{34}$$

In particular the *amplitude reflection coefficient* $r = \mathcal{E}_3^*(0)/\mathcal{E}_1(0)$ is given by

$$r = \frac{\mathcal{E}_3^*(0)}{\mathcal{E}_1(0)} = -\chi_3^*\,\frac{\sinh wL}{w\cosh wL + \alpha\sinh wL}\,. \tag{35}$$

See Chap. 9 on phase conjugation for further discussion of these equations.

Coupled Modes and Squeezing

A popular current topic is the "squeezing", i.e., deamplifying, of noise in one quadrature of an electromagnetic wave at the expense of amplifying the noise in the orthogonal quadrature. One way to achieve such squeezing is through the use of mode coupling mechanisms such as described by Eqs. (25a) and (25b). To see under which conditions the χ_n coupling factors can lead to this quadrature-dependent amplification, let's drop the α_n terms in Eqs. (25) and put the time dependencies back in. We find for example the schematic equation

$$\{\chi^{(3)}\ \mathcal{E}_2{}^2\ e^{-2i\nu t}\}\ [\mathcal{E}_3 e^{i\nu t}]^* \rightarrow \mathcal{E}_1 e^{-i\nu t}\ , \tag{36}$$

where $\chi^{(3)}$ is a third-order susceptibility. Suppose that at a time t, $e^{i\nu t} = 1$ and that $\{\} = 1$. According to Eq. (36), this tends to amplify $\mathcal{E}_1$. Now wait until the orthogonal quadrature phasor $\exp(i\nu t - i\pi/2) = 1$. At this time, the second-harmonic (two-photon) phasor $\exp(-2i\nu t)$ has precessed through *two* times $\pi/2$, that is, $\{\} = -1$. Hence a two-photon coupling $\{\}$ flips the sign of the coupling between orthogonal quadratures. This is the signature of a coupling process that can lead to squeezing. It is equally possible that *a* $\chi^{(2)}$ process with a single pump photon having the value 2ν can cause squeezing. This $\chi^{(2)}$ process is known as parametric amplification. Chapter 16 discusses the squeezing of quantum noise by four-wave mixing.

2-5. Nonlinear Susceptibilities

So far we have used an anharmonic oscillator to introduce some aspects of nonlinear optics that are useful in the remainder of this book. Such a simple model is surprisingly powerful and permits us to understand numerous nonlinear optics effects intuitively. In general, however, first principle quantum mechanical calculations are needed to determine the response of a medium to a strong electromagnetic field. A substantial fraction of this book addresses this problem under resonant or near-resonant conditions, i.e., under conditions such that the frequency(ies) of the field(s) are near an atomic transition. Perturbative analyses such as sketched in this chapter are usually not sufficient to describe these situations.

In many cases, however, the incident radiation is far from resonance with any transition of interest, and/or the material relaxation rate is exceedingly fast. In such cases, perturbation theory based on the concept on nonlinear susceptibility may be of great advantage. This is the realm of conventional nonlinear optics, and the reader should consult the recent treatises by Shen (1984) and by Hopf and Stegeman (1986), as well as the classic book by Bloembergen (1965), for detailed descriptions of these fields. Here we limit ourselves to a brief introduction to the formalism of nonlinear susceptibility.

In linear problems, the polarization of the medium is (by definition) a linear function of the applied electric fields. The most general form that it can take is given by the space-time convolution of a linear susceptibility tensor $\boldsymbol{\chi}^{(1)}$ with the electric field:

$$\mathbf{P}(\mathbf{r},t) = \epsilon_0 \int d^3r \int_{-\infty}^{t} dt' \, \chi^{(1)}(\mathbf{r}-\mathbf{r}',t-t') : \mathbf{E}(r',t') \,. \tag{37}$$

Taking the four-dimensional Fourier transform of this expression for *a* monochromatic wave $\mathbf{E}(\mathbf{r},t) = \mathcal{E}(\mathbf{K},\nu)\, e^{i(\mathbf{K}\cdot\mathbf{r} - \nu t)}$, we find

$$\mathbf{P}(\mathbf{K},\nu) = \epsilon_0 \chi^{(1)}(\mathbf{K},\nu){:}\mathbf{E}(\mathbf{K},\nu) \,, \tag{38}$$

with

$$\chi^{(1)}(\mathbf{K},\nu) = \int d^3r \int_{-\infty}^{t} dt' \, \chi^{(1)}(\mathbf{r},\, t)\, e^{i(\mathbf{K}\cdot\mathbf{r}-\nu t)} \,. \tag{39}$$

The linear dielectric constant is related to $\chi^{(1)}(\mathbf{K},\nu)$ via

$$\epsilon(\mathbf{K},\nu) = \epsilon_0[1 + \chi^{(1)}(\mathbf{K},\nu)] \,. \tag{40}$$

In the nonlinear case, and for electric fields sufficiently weak that perturbation theory is valid, one gets instead

$$\begin{aligned}
\mathbf{P}(\mathbf{r},t) = {} & \epsilon_0 \int d^3r \int_{-\infty}^{t} dt' \, \chi^{(1)}(\mathbf{r}-\mathbf{r}',t-t') \cdot \mathbf{E}(\mathbf{r}',t') \\
& + \epsilon_0 \int d\mathbf{r}_1 dt_1 \, d\mathbf{r}_2 dt_2 \, \chi^{(2)}(\mathbf{r}-\mathbf{r}_1,t-t_1;\, \mathbf{r}-\mathbf{r}_2,t-t_2) : \mathbf{E}(r_1,t_1)\, E(r_2,t_2) \\
& + \epsilon_0 \int d\mathbf{r}_1 dt_1 \, d\mathbf{r}_2 dt_2 \, d\mathbf{r}_3 dt_3 \, \chi^{(3)}(\mathbf{r}-\mathbf{r}_1,t-t_1;\, \mathbf{r}-\mathbf{r}_2,t-t_2;\, \mathbf{r}-r_3,t-t_3) \\
& \vdots\, \mathbf{E}(\mathbf{r}_1,t_1)\, \mathbf{E}(\mathbf{r}_2,t_2)\, \mathbf{E}(\mathbf{r}_3,t_3) + \ldots \,,
\end{aligned} \tag{41}$$

where $\chi^{(n)}$ is the *n*th-order susceptibility. If $\mathbf{E}(\mathbf{r},t)$ can be expressed as a sum of plane waves,

$$\mathbf{E}(\mathbf{r},t) = \sum_n \mathbf{E}(\mathbf{K}_n,\, \nu_n)\, e^{i(\mathbf{K}_n\cdot\mathbf{r} - \nu_n t)} \,, \tag{42}$$

then as in the linear case, Fourier transform of (41) gives

$$\mathbf{P}(\mathbf{K},\nu) = \mathbf{P}^{(1)}(\mathbf{K},\nu) + \mathbf{P}^{(2)}(\mathbf{K},\nu) + \mathbf{P}^{(3)}(\mathbf{K},\nu) + \ldots \tag{42}$$

with $\mathbf{P}^{(1)}(\mathbf{K},\nu)$ given by Eq. (38) and

$$\mathbf{P}^{(2)}(\mathbf{K},\nu) = \boldsymbol{\chi}^{(2)}(\mathbf{K}=\mathbf{K}_n+\mathbf{K}_m,\, \nu=\nu_n+\nu_m) : \mathbf{E}(\mathbf{K}_n,\nu_n)\, \mathbf{E}(\mathbf{K}_m,\nu_m)$$

$$\mathbf{P}^{(3)}(\mathbf{K},\nu) = \boldsymbol{\chi}^{(3)}(\mathbf{K}=\mathbf{K}_n+\mathbf{K}_m+\mathbf{K}_\ell,\nu=\nu_n+\nu_m+\nu_\ell) \vdots\, \mathbf{E}(\mathbf{K}_n,\nu_n)\, \mathbf{E}(\mathbf{K}_m,\nu_m)\, \mathbf{E}(\mathbf{K}_\ell,\nu_\ell)$$

and

$$\boldsymbol{\chi}^{(n)}(\mathbf{K}=\mathbf{K}_1+\mathbf{K}_2+\ldots+\mathbf{K}_n,\, \nu=\nu_1+\nu_2+\ldots+\nu_n)$$

$$= \int d^3r_1 dt_1\, d^3r_2 dt_2 \ldots d^3r_n dt_n\, \boldsymbol{\chi}^{(n)}(\mathbf{r}-\mathbf{r}_1, t-t_1;\, \mathbf{r}-\mathbf{r}_2, t-t_2;\, \ldots\, \mathbf{r}-\mathbf{r}_n,\, t-t_n)$$

$$\times\, \exp[-i\mathbf{K}_1\cdot(\mathbf{r}-\mathbf{r}_1)-i\nu_1(t-t_1) + \ldots + i\mathbf{K}_n\cdot(\mathbf{r}-\mathbf{r}_n)-i\nu_n(t-t_n)]\,.$$

References

Chiao, R., P. Kelley, and E. Garmire (1966), Phys. Rev. Lett. 17, 1158.

Bloembergen, N. (1965), *Nonlinear Optics*, Benjamin, Inc., New York. The classic book in nonlinear optics.

Franken, P. A. , A. E. Hill, C. W. Peters, and G. Weinreich (1961), Phys. Rev. Letters 7, 118.

Hopf, F. A. and G. I. Stegeman (1986), *Applied Classical Electrodynamics. Vol. 2*, John Wiley & Sons, New York.

Shen, Y. R. (1984), *Principles of Nonlinear Optics*, John Wiley & Sons, New York. An excellent modern book on nonlinear optics.

van der Pol, B. (1934), Proc IRE 22, 1051.

Problems

2-1. Solve the coupled mode equations

$$\frac{d\mathcal{E}_1}{dz} = -\alpha_1\mathcal{E}_1 + \chi_1\mathcal{E}_3^*, \tag{43}$$

$$\frac{d\mathcal{E}_3^*}{dz} = -\alpha_3^*\mathcal{E}_3^* + \chi_3^*\mathcal{E}_1, \tag{44}$$

valid for phase-matched forward three-wave mixing. Ans:

$$\mathcal{E}_1(z) = e^{-az}\,[\mathcal{E}_1(0)\cosh wz + (-\alpha\mathcal{E}_1(0) + \chi_1\mathcal{E}_3^*(0))\sinh wz/w] \tag{45}$$

$$\mathcal{E}_3^*(z) = e^{-az}\,[\mathcal{E}_3^*(0)\cosh wz + (\alpha\mathcal{E}_3^*(0) + \chi_3^*\mathcal{E}_1(0))\sinh wz/w]\,, \tag{46}$$

where $a = (\alpha_1 + \alpha_3^*)/2$, $\alpha = (\alpha_1 - \alpha_3^*)/2$, and $w = \sqrt{\alpha^2 + \chi_1\chi_3^*}$.

2-2. Derive the coefficients in the coupled-mode equations (22).

2-3. Calculate all wavelengths generated in a $\chi^{(3)}$ nonlinear medium by a combination of 632.8 nm and 488 nm laser light.

2-4. Calculate the coupling coefficient χ_n for four-wave mixing based on an anharmonic oscillator.

2-5. Write the propagation equations for second-harmonic generation. Comment on phase matching.

2-6. Calculate the phase mismatch for a conjugate wave of frequency $\nu_3 = \nu_2 + (\nu_2 - \nu_1)$ generated by signal and pump waves with frequencies ν_1 and ν_2, respectively, and propagating in the same direction. Include the fact that the indices of refraction for the three waves are in general different, that is, $\eta(\nu_1) = \eta(\nu_2) + \delta\eta_1$ and $\eta(\nu_3) = \eta(\nu_2) + \delta\eta_3$.

2-7. Show that Eq. (20) contains all the fundamental contributions from the third-order expression (19).

2-8. Evaluate the reflection coefficient r of Eq. (35) in the limit of large L. Ans: $r = \mp\chi_3^*/(w \pm \alpha)$ for $\mathrm{Re}(w) \gtrless 0$.

Chapter 3
QUANTUM MECHANICAL BACKGROUND

Chapters 1 and 2 describe the interaction of radiation with matter in terms of a phenomenological classical polarization **P**. The question remains as to when this approach is justified and what to do when it isn't. Unexcited systems interacting with radiation far from the system resonances can often be treated purely classically. The response of systems near and at resonance often deviates substantially from the classical descriptions. Since the laser itself and many applications involve systems near atomic (or molecular) resonances, we need to study them with the aid of quantum mechanics.

In preparation for this study, this chapter reviews some of the highlights of quantum mechanics paying particular attention to topics relevant to the interaction of radiation with matter. Section 3-1 introduces the wave function for an abstract quantum system, discusses the wave function's probabilistic interpretation, its role in the calculation of expectation values, and its equation of motion (the Schrödinger equation). Expansions of the wave function in various bases, most notably in terms of energy eigenstates, are presented and used to convert the Schrödinger partial differential equation into a set of ordinary differential equations. Dirac notation is reviewed and used to discuss the state vector and how the state vector is related to the wave function. System time evolution is revisited with a short review of the Schrödinger, Heisenberg, and interaction pictures.

In Chaps. 4 through 11, we are concerned with the interaction of classical electromagnetic fields with simple atomic systems. Section 3-2 lays the foundations for these chapters by discussing wave functions for atomic systems and studying their evolution under the influence of applied perturbations. Time dependent perturbation theory and the rotating wave approximation are used to predict this evolution in limits for which transitions are unlikely. The Fermi Golden Rule is derived. Section 3-3 deals with a particularly simple atomic model, the two-level atom subject to a resonant or nearly resonant classical field. We first discuss the nature of the electric-dipole interaction and then use the Fermi Golden Rule to derive Einstein's A and B coefficients for spontaneous and stimulated emission. We then relax the assumption that the interaction is weak and derive the famous Rabi solution.

In Chaps. 12 through 18, we discuss interactions for which the electromagnetic field as well as the atoms must be quantized. In particular, Chap. 12 shows that electromagnetic field modes are described mathematically by simple harmonic oscillators. In addition, these oscillators can model the polarization of certain kinds of media, such as simple molecular systems. In preparation for such problems, Sec. 3-4 quantizes the simple harmonic oscillator. The section writes the appropriate Hamiltonian in terms of the annihilation and creation operators, and derives the corresponding energy eigenstates.

This chapter is concerned with the quantum mechanics of single systems in pure states. Discussions of mixtures of systems including the decay phenomena and excitation mechanisms that occur in lasers and their applications are postponed to Chap. 4 on the density matrix.

3-1. Review of Quantum Mechanics

According to the postulates of quantum mechanics, the best possible knowledge about a quantum mechanical system is given by its wave function $\psi(\mathbf{r},t)$. Although $\psi(\mathbf{r},t)$ itself has no direct physical meaning, it allows us to calculate the *expectation values* of all *observables* of interest. This is due to the fact that the quantity

$$\psi(\mathbf{r},t)^*\psi(\mathbf{r},t)d^3r$$

is the probability of finding the system in the volume element d^3r. Since the system described by $\psi(\mathbf{r},t)$ is assumed to exist, its probability of being somewhere has to equal 1. This gives the normalization condition

$$\int \psi(\mathbf{r},t)^*\psi(\mathbf{r},t)\, d^3r = 1\ , \tag{1}$$

where the integration is taken over all space.

An observable is represented by a *Hermitian* operator $\mathcal{O}$ and its expectation value is given in terms of $\psi(\mathbf{r},t)$ by

$$\langle \mathcal{O} \rangle = \int d^3r \, \psi(\mathbf{r},t)^* \mathcal{O} \psi(\mathbf{r},t) \, . \tag{2}$$

Experimentally this expectation value is given by the average value of the results of many measurements of the observable $\mathcal{O}$ acting on identically prepared systems. The accuracy of the experimental value for $\langle \mathcal{O} \rangle$ typically depends on the number of measurements performed. Hence enough measurements should be made so that the value obtained for $\langle \mathcal{O} \rangle$ doesn't change significantly when still more measurements are performed. It is crucial to note that the expectation value (2) predicts the average from many measurements; in general it is unable to predict the outcome of a single event with absolute certainty. This does not mean that quantum mechanics in other ways is unable to make some predictions about single events.

The reason observables, such as position, momentum, energy, and dipole moment, are represented by Hermitian operators is that the expectation values (2) must be real. Denoting by (ϕ, ψ) the inner or scalar product of two vectors ϕ and ψ, we say that a linear operator $\mathcal{O}$ is Hermitian if the equality

$$(\phi, \mathcal{O}\psi) = (\mathcal{O}\phi, \psi). \tag{3}$$

holds for all ϕ and ψ. In this notation, Eq. (2) reads $\langle \mathcal{O} \rangle = (\psi, \mathcal{O}\psi)$.

An important observable in the interaction of radiation with bound electrons is the electric dipole $e\mathbf{r}$. This operator provides the bridge between the quantum mechanical description of a system and the polarization of the medium $\mathbf{P}$ used as a source in Maxwell's equations for the electromagnetic field. According to Eq. (2), the expectation value of $e\mathbf{r}$ is

$$\langle e\mathbf{r} \rangle = \int d^3r \, \mathbf{r} \, e|\psi(\mathbf{r},t)|^2 \, , \tag{4}$$

where we can move $e\mathbf{r}$ to the left of $\psi(\mathbf{r},t)^*$ since the two commute (an operator like ∇ cannot be so moved). Here we see that the dipole-moment expectation value has the same form as the classical value if we identify $\rho = e|\psi(\mathbf{r},t)|^2$ as the charge density.

In nonrelativistic quantum mechanics, the evolution of $\psi(\mathbf{r},t)$ is governed by the Schrödinger equation

$$\boxed{i\hbar \frac{\partial}{\partial t}\psi(\mathbf{r},t) = \mathcal{H}\psi(\mathbf{r},t)}\ , \tag{5}$$

where $\mathcal{H}$ is the Hamiltonian for the system and $\hbar = 1.054\times10^{-34}$ joule-seconds is Planck's constant divided by 2π. The Hamiltonian of an unperturbed system, for instance an atom not interacting with light, is the sum of its kinetic and potential energies

$$\mathcal{H} = \frac{p^2}{2m} + V(\mathbf{r})\ , \tag{6}$$

where p is the system momentum, m is the system mass, and $V(\mathbf{r})$ the potential energy. The momentum $\mathbf{p}$ is in general given by the operator

$$\mathbf{p} = -i\hbar\nabla\ . \tag{7}$$

In view of Eq. (7), we see that the Schrödinger equation (5) is a partial differential equation. The time and space dependencies in Eq. (5) separate for functions having the form

$$\psi_n(\mathbf{r},t) = u_n(\mathbf{r})\, e^{-i\omega_n t} \tag{8}$$

for which the $u_n(\mathbf{r})$ satisfy the energy eigenvalue equation

$$\mathcal{H}u_n(\mathbf{r}) = \hbar\omega_n\, u_n(\mathbf{r})\ . \tag{9}$$

The eigenfunctions $u_n(\mathbf{r})$ can be shown to be orthonormal

$$\int u_n^*(\mathbf{r})u_m(\mathbf{r})d^3r = \delta_{n,m} = \begin{cases} 1 & n=m \\ 0 & n\neq m \end{cases} \tag{10}$$

and complete

$$\sum_n u_n^*(\mathbf{r})u_n(\mathbf{r}') = \delta(\mathbf{r}-\mathbf{r}')\ , \tag{11}$$

where $\delta_{n,m}$ and $\delta(\mathbf{r} - r')$ are the Kronecker and Dirac delta functions, respectively. The completeness relation Eq. (11) means that any function can be written as a superposition of the $u_n(\mathbf{r})$. Problem 3-1 shows that this definition is equivalent to saying that any wave function can be expanded in a complete set of states.

In particular the wave function $\psi(\mathbf{r},t)$ itself can be written as the superposition of the $\psi_n(\mathbf{r},t)$:

$$\psi(\mathbf{r},t) = \sum_n C_n(t)\, u_n(\mathbf{r})\, e^{-i\omega_n t} \,. \tag{12}$$

Here the expansion coefficients $C_n(t)$ are constants for problems described by a Hamiltonian satisfying the eigenvalue equation (9). We have nevertheless included the time dependence in anticipation of adding an interaction energy to the Hamiltonian. Such a modified Hamiltonian wouldn't quite satisfy Eq. (9), thereby causing the $C_n(t)$ to change in time.

Substituting Eq. (12) into the normalization condition (1) and using the orthonormality condition (10), we find

$$\sum_n |C_n|^2 = 1 \,. \tag{13}$$

The $|C_n|^2$ can be interpreted as the probability that the system is in the nth energy state. The C_n are complex probability amplitudes and completely determine the wave function. To find the expectation value (2) in terms of the C_n, we substitute Eq. (12) into Eq. (2). This gives

$$\langle \mathcal{O} \rangle = \sum_{n,m} C_n C_m^{\,*}\, \mathcal{O}_{mn}\, e^{-i\omega_{nm} t} \,, \tag{14}$$

where the operator matrix elements $\mathcal{O}_{mn}$ are given by

$$\mathcal{O}_{mn} = \int d^3r\, u_m^{\,*}(\mathbf{r})\, \mathcal{O}\, u_n(\mathbf{r}) \,, \tag{15}$$

and the frequency differences

$$\omega_{nm} = \omega_n - \omega_m \,. \tag{16}$$

Typically we consider the interaction of atoms with electromagnetic fields. To treat such interactions, we add the appropriate interaction energy to the Hamiltonian, that is

$$\mathcal{H} = \mathcal{H}_0 + \mathcal{V}. \tag{17}$$

If we expand the wave function in terms of the eigenfunctions of the "unperturbed Hamiltonian" $\mathcal{H}_0$, rather than those of the total Hamiltonian $\mathcal{H}$, the probability amplitudes $C_n(t)$ change in time. To find out just how, we substitute the wave function (12) and Hamiltonian (17) into Schröd-inger's equation (5) to find

$$\sum_n (\hbar\omega_n + \mathcal{V})\, C_n\, u_n(\mathbf{r})\, e^{-i\omega_n t} = \sum_n (\hbar\omega_n C_n + i\hbar\dot{C}_n)\, u_n(\mathbf{r})\, e^{-i\omega_n t} \ . \quad (18)$$

Canceling the $\hbar\omega_n$ terms, changing the summation index n to m, multiplying through by $u_n^*(\mathbf{r})\exp(i\omega_n t)$, and using the orthonormality property (10), we find the equation of motion for the probability amplitude $C_n(t)$

$$\boxed{\dot{C}_n(t) = -\frac{i}{\hbar} \sum_m \langle n|\mathcal{V}|m\rangle\, e^{i\omega_{nm} t}\, C_m(t)} \ , \quad (19)$$

where the matrix element

$$\langle n|\mathcal{V}|m\rangle = \int d^3r\, u_n^*(\mathbf{r})\, \mathcal{V}\, u_m(\mathbf{r}) \ . \quad (20)$$

Note that we can also write the superposition

$$\psi(\mathbf{r},t) = \sum_n c_n(t)\, u_n(\mathbf{r}) \ , \quad (21)$$

for which the $\hbar\omega_n$ time dependence in Eq. (18) doesn't cancel out. The $c_n(t)$ then obey the equation of motion

$$\dot{c}_n(t) = -i\omega_n c_n(t) - \frac{i}{\hbar} \sum_m \langle n|\mathcal{V}|m\rangle\, c_m(t) \ . \quad (22)$$

In terms of the c_n, the expectation value (2) becomes

$$\langle \mathcal{O} \rangle = \sum_{n,m} c_n c_m^* \mathcal{O}_{mn} \,, \tag{23}$$

Equation (19) and equivalently Eq. (22) show how the probability amplitudes for the wave function written as a superposition of energy eigenfunctions change in time. They are equivalent to the original Schrödinger equation (5), but are no longer concerned with the precise position dependence **r**. In particular if we're only concerned about how a system such as an atom absorbs energy from a light field, this development is completely described by the changes in the C_n or c_n.

The choice of using the relatively slowly varying C_n versus using the rapidly varying c_n is a matter of taste and convenience. The time dependence of the C_n is due to the interaction energy $\mathcal{V}$ alone, while that of the c_n is due to the total Hamiltonian $\mathcal{H}$. To distinguish between the two, we say that the C_n are in the *interaction picture*, while the c_n are in the *Schrödinger picture.* We discuss this more formally at the end of this section.

Armed with Eq. (19) or (22), you can skip directly to Sec. 3-2, which shows how systems evolve in time due to various interactions. Before going ahead, we review Dirac notation and some other aspects of the wave function and of its more abstract form, the state vector $|\psi\rangle$. This material is needed for our discussions involving quantized fields in Chaps. 12 through 18, and is useful in proving various properties of the density operator in Chap. 4.

Up to now we have used the position representation, that is, we have written the wave function as a function of **r**. Alternatively we can work in the momentum representation, writing the wave function as a function of **p**. This can be done simply by Fourier transforming $\psi(\mathbf{r},t)$:

$$\phi(\mathbf{p},t) = \frac{1}{(2\pi\hbar)^{3/2}} \int d^3r \; \psi(\mathbf{r},t)\, e^{-i\mathbf{p}\cdot\mathbf{r}/\hbar} \,. \tag{24}$$

Here $\phi(\mathbf{p},t)$ describes the same dynamical state as $\psi(\mathbf{r},t)$. It doesn't make any difference in principle which representation we choose to use, and as we see with the $\dot{C}_n$, we don't have to worry about the coordinate representation at all.

Dirac Notation

To formalize this flexibility, we use Dirac's notation. Roughly speaking Dirac's formulation is analogous to using vectors instead of coordinates. The notation has an additional advantage in that one can label the basis vectors much more conveniently than with ordinary vector notation. We start our discussion with a comparison between ordinary notation for a vector in a two-dimensional space and Dirac's version. As shown in Fig. 3-1, a vector **v** can be expanded as

$$\mathbf{v} = v_x \hat{\mathbf{x}} + v_y \hat{\mathbf{y}} \,, \tag{25}$$

where $\hat{\mathbf{x}}$ and $\hat{\mathbf{y}}$ are unit vectors along the x and y axes, respectively. In Dirac notation, this reads

$$|v\rangle = v_x |x\rangle + v_y |y\rangle \,. \tag{26}$$

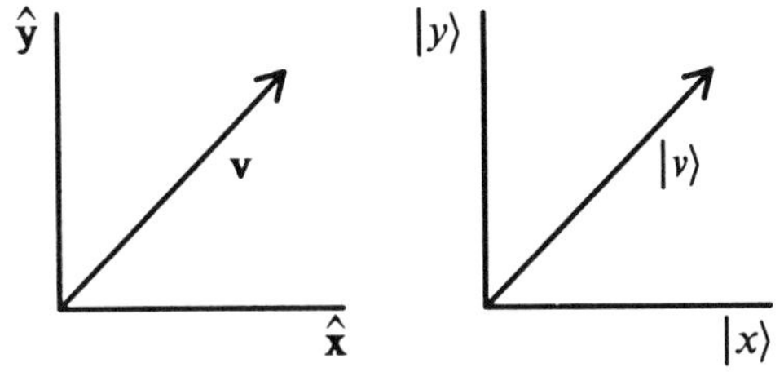

Fig. 3-1. A two dimensional vector written in ordinary vector notation and in Dirac notation.

The component v_x given in ordinary vector notation by the dot product $\hat{\mathbf{x}}\cdot\mathbf{v}$ is given in Dirac notation by

$$v_x = \langle x|v\rangle \,, \tag{27}$$

The Dirac vector $|v\rangle$ is called a "ket" and the vector $\langle v|$ a "bra", which come from calling the inner product (27) a "bra c ket". With the notation (27), Eq. (26) reads

$$|v\rangle = |x\rangle\langle x|v\rangle + |y\rangle\langle y|v\rangle \,. \tag{28}$$

This immediately gives the identity diadic (outer product of two vectors)

$$|x\rangle\langle x| + |y\rangle\langle y| = \mathcal{I} \,. \tag{29}$$

Equations (26) and (29) can be immediately generalized to many dimensions as in

$$|v\rangle = \sum_n |n\rangle\langle n|v\rangle \tag{30}$$

$$\mathcal{I} = \sum_n |n\rangle\langle n| \,, \tag{31}$$

where the $\{|n\rangle\}$ are a complete orthonormal set of vectors, i.e., a basis. The inner products $\langle n|v\rangle$ are the expansion coefficients of the vector $|v\rangle$ in this basis. The bra $\langle n|$ is the adjoint of the ket $|n\rangle$ and the expansion coefficients have the property

$$\langle k|v\rangle = \langle v|k\rangle^* \,. \tag{32}$$

Unlike the real spaces of usual geometry, quantum mechanics works in a complex space called a Hilbert space, where the expansion coefficients are in general complex.

The basis $\{|n\rangle\}$ is discrete. Alternatively we can expand vectors in terms of the coordinate basis $\{|\mathbf{r}\rangle\}$ which like the $\{|n\rangle\}$ forms a complete basis, albeit a continuous one. For such a situation we need to use continuous summations in Eqs. (30) and (31), that is, integrals. For example, the identity operator of Eq. (31) can be expanded as

$$\mathcal{I} = \int d^3r \; |\mathbf{r}\rangle\langle \mathbf{r}| \,. \tag{33}$$

One major advantage of the bra and ket notation is that you can label the vectors with as many letters as desired. For example, you could write $|r\theta\phi\rangle$ in place of $|\mathbf{r}\rangle$.

The vector of primary interest in quantum mechanics is the state vector $|\psi(t)\rangle$. The wave function is actually the expansion coefficient of $|\psi\rangle$ in the coordinate basis

$$|\psi\rangle = \int d^3r \; |\mathbf{r}\rangle\langle \mathbf{r}|\psi\rangle \,, \tag{34}$$

where the wave function

$$\psi(\mathbf{r},t) = \langle \mathbf{r}|\psi\rangle \ . \tag{35}$$

Hence the state vector $|\psi\rangle$ is equivalent to the wave function $\psi(\mathbf{r},t)$, but doesn't explicitly display the coordinate dependence.

Instead of using the position expansion of Eq. (34), we can expand the state vector in the discrete basis $\{|n\rangle\}$ as

$$|\psi\rangle = \sum_n c_n |n\rangle \ . \tag{36}$$

The most common basis to use consists of the eigenstates of the unperturbed Hamiltonian operator $\mathcal{H}_0$. For this basis, the expansion coefficient c_n are just those used in Eq. (21), and the energy eigenfunctions are related to the eigenvectors by

$$u_n(\mathbf{r}) = \langle \mathbf{r}|n\rangle \ . \tag{37}$$

A useful trick in transforming from one basis to another is to think of the vertical bar as an identity operator expanded either as in Eq. (31) or as in Eq. (33). Using the form of Eq. (33) in Eq. (36) on both sides of the equation along with Eq. (37), we recover Eq. (21).

The expectation value of the operator $\mathcal{O}$ is given in terms of the state vector by

$$\boxed{\langle \mathcal{O}\rangle = \langle \psi(t)|\mathcal{O}|\psi(t)\rangle} \ . \tag{38}$$

To see that this is the same as the position representation version of Eq. (2) is a little trickier. We substitute the identity expansion (33) in for the left bar and the same expansion with $\mathbf{r} \rightarrow \mathbf{r}'$ in for the right bar obtaining

$$\begin{aligned}\langle \mathcal{O}\rangle &= \int d^3r \int d^3r' \ \langle \psi|\mathbf{r}\rangle \ \langle \mathbf{r}|\mathcal{O}|\mathbf{r}'\rangle \ \langle \mathbf{r}'|\psi\rangle \\ &= \int d^3r \int d^3r' \ \psi^*(\mathbf{r},t) \ \mathcal{O}(\mathbf{r},\mathbf{r}') \ \psi(\mathbf{r}',t) \ . \end{aligned} \tag{39}$$

Hermitian operators in quantum mechanics usually turn out to be local, and hence

$$\mathcal{O}(\mathbf{r}, \mathbf{r}') = \mathcal{O}(\mathbf{r}, \mathbf{r})\, \delta(\mathbf{r}-\mathbf{r}') \,. \tag{40}$$

This fact reduces Eq. (39) to Eq. (2) as desired. More generally, we can interpret $\mathcal{O}\psi(\mathbf{r},t)$ on the right-hand side of Eq. (2) as

$$\mathcal{O}\psi(\mathbf{r},t) = \int d^3r'\, \mathcal{O}(\mathbf{r}, \mathbf{r}')\, \psi(\mathbf{r}',t) \,.$$

Similarly substituting the expansion (36) into Eq. (38), we find the expectation value (23), where the operator matrix elements written earlier as Eq. (15) can be written simply as

$$\mathcal{O}_{mn} = \langle m|\mathcal{O}|n\rangle \,. \tag{41}$$

Problem 3-2 shows that this equals Eq. (15). It is also useful to express the operator $\mathcal{O}$ directly in terms of the basis set. This then reads as

$$\mathcal{O} = \sum_{n,m} \mathcal{O}_{nm}\, |n\rangle\langle m| \,. \tag{42}$$

Finally we note that the state vector version of the Schrödinger equation (5) is

$$\boxed{i\hbar \frac{\partial}{\partial t} |\psi\rangle = \mathcal{H}|\psi\rangle} \,. \tag{43}$$

Using the fact that the Hamiltonian operator is local, you can easily show that this reduces to Eq. (5).

The basic problem we are faced with is how to find the expectation values (38) of observables at the time t. To do this we typically start with a system in a well defined state at an earlier time and follow the development up to the time t using the Schrödinger equation (43). It is possible to follow this evolution in three general ways and in many combinations thereof. The one we use primarily in the first part of this book is called the Schrödinger picture and puts all of the time dependence in the state vector. The interaction picture puts only the interaction-energy time dependence into the state vector, putting the unperturbed energy dependence into the operators. The Heisenberg picture puts all of the time dependence into the operators, leaving the state vector stationary in time. In the remainder of this section, we review the way in which these three pictures are tied together.

The Schrödinger equation (43) can be formally integrated to give

$$|\psi(t)\rangle = U(t)\,|\psi(0)\rangle\,, \tag{44}$$

where the evolution operator $U(t)$ for a time-independent Hamiltonian is given by

$$U(t) = \exp(-i\mathcal{H}t/\hbar)\,. \tag{45}$$

Substituting Eq. (44) into the expectation value (38) we obtain

$$\langle\mathcal{O}\rangle = \langle\psi(0)|U^{\dagger}(t)\;\mathcal{O}(0)\;U(t)|\psi(0)\rangle\,. \tag{46}$$

We can also find this same value if we can determine the time dependent operator

$$\mathcal{O}(t) = U^{\dagger}(t)\;\mathcal{O}(0)\;U(t)\,. \tag{47}$$

As Heisenberg first showed, it is possible to follow the time evolution of the quantum mechanical operators. In fact we can obtain their equations of motion by differentiating Eq. (47). This gives

$$\frac{d}{dt}\,\mathcal{O}(t) = \frac{dU^{\dagger}}{dt}\,\mathcal{O}\,U + U^{\dagger}\frac{\partial\mathcal{O}}{\partial t}U + U^{\dagger}\mathcal{O}\,\frac{dU}{dt}$$

i.e.,

$$\boxed{\frac{d}{dt}\,\mathcal{O}(t) = \frac{i}{\hbar}\,[\mathcal{H},\mathcal{O}] + U^{\dagger}\,\frac{\partial\mathcal{O}}{\partial t}\,U}\;, \tag{48}$$

where

$$[\mathcal{H},\mathcal{O}] \equiv \mathcal{H}\mathcal{O} - \mathcal{O}\mathcal{H} \tag{49}$$

is the *commutator* of $\mathcal{H}$ with $\mathcal{O}$. In deriving Eq. (48), we have used the fact that $\mathcal{H}$ commutes with U, which follows from the fact that U is a function of $\mathcal{H}$ only. The $\partial\mathcal{O}/\partial t$ accounts for any explicit time dependence of the Schrödinger operator $\mathcal{O}$.

In general when the system evolution is determined by integrating equations of motion for the observable operators, we say the Heisenberg picture is being used. When the evolution is determined by integrating the Schrödinger equation, we say that the Schrödinger picture is being used. In either case, Eq. (46) shows that we get the same answers. You ask, why use one picture instead of the other? The answer is simply, use the picture that makes your life easier. Typically the insights obtained with one differ somewhat from the other, but you get the same answers with either. Traditionally the Schrödinger picture is the first one taught to students and many people feel more comfortable with it. Most of this book is carried out in the Schrödinger picture.

On the other hand, the Heisenberg picture is a "natural" picture in the sense that the observables (electric fields, dipole moment, etc.) are time-dependent, exactly as in classical physics. As a result, their equations of motion usually have the same form as in the classical case, although they are operator equations, which modifies the way one can integrate and use them. Another aspect is that in the Schrödinger picture, one has to find $|\psi(t)\rangle$ (or its generalization the density operator ρ) before computing the desired expectation values. Since $|\psi(t)\rangle$ contains all possible knowledge about the system, you have to solve the complete problem, which may be more than you need. In many cases, you only want to know one or a few observables of the system. The Heisenberg picture allows you to concentrate on precisely those observables, and with some luck, you may not have to solve the whole problem to get the desired answers.

In discussing Eqs. (12) and (21), we hinted at another way of following the time dependence, namely we put only the time dependence due to the interaction energy into the $C_n(t)$, while the time dependence of the total Hamiltonian is contained in the $c_n(t)$. The state vector of Eq. (36) is the Schrödinger-picture state vector, while the state vector

$$|\psi_I(t)\rangle = \sum_n C_n(t)\,|n\rangle \tag{50}$$

is said to be the interaction-picture state vector. The thought behind using the interaction picture is to take advantage of the fact that we often face situations where we already know the solution of the problem in the absence of the interaction.

More formally, to eliminate the known part of the problem, we substitute the state vector

$$|\psi_S(t)\rangle = U_0(t)|\psi_I(t)\rangle\ , \tag{51}$$

where

$$U_0(t) = \exp(-i\mathcal{H}_0 t/\hbar) \tag{52}$$

into the Schrödinger equation (43). We include the subscript S in Eq. (51) to remind ourselves that $|\psi_S(t)\rangle$ is the Schrödinger picture state vector. We find

$$\boxed{\frac{d}{dt}|\psi_I(t)\rangle = -\frac{i}{\hbar}\mathcal{V}_I(t)\,|\psi_I(t)\rangle}\,, \tag{53}$$

where we have defined the interaction-picture interaction energy

$$\mathcal{V}_I(t) = U_0^\dagger(t)\,\mathcal{V}_S\,U_0(t) \tag{54}$$

and put a subscript S on the RHS to remind ourselves that $\mathcal{V}_S$ is in the Schrödinger picture. From Eqs. (51) and (38), we also immediately find that the expectation value of an operator $\mathcal{O}$ in the interaction picture is given by

$$\langle\mathcal{O}(t)\rangle = \langle\psi_I(t)|\,\mathcal{O}_I(t)\,|\psi_I(t)\rangle, \tag{55}$$

where

$$\mathcal{O}_I(t) = U_0^\dagger(t)\,\mathcal{O}_S\,U(t)_0\,. \tag{56}$$

Note that since we know the solution of the unperturbed problem, $\mathcal{O}_I(t)$ is already known. Comparing the equation of motion (53) with the original Schrödinger equation (43), we see that we have achieved our goal, namely that we have eliminated the part of the problem whose solution we already knew. We see in Chap. 4 that the interaction picture (or more precisely, an interaction picture) is particularly helpful in visualizing the response of a two-level atom to light.

3-2. Time-Dependent Perturbation Theory

To predict expectation values of operators, we need to know what the wave function is. Typically we know the initial value of the wave function, which then evolves in time according to the Schrödinger equation, or equivalently, the operators of interest evolve in time according to the Heisenberg equations. For some problems, these equations can be integrated exactly, giving us the values needed to compute the expectation values at the desired time t. More generally the equations can be integrated approximately using a method called time-dependent perturbation theory. The name comes from the introduction of a perturbation energy $\mathcal{V}$ as given in Eq. (17), which describes the interaction of the quantum system under consideration with some other system. An atom interacting with an electromagnetic field is the combination that we consider most often in this book. The perturbation energy forces the probability amplitudes in Eqs. (12) or (21) to be time dependent. The method of time dependent perturbation theory consists of formally integrating the Schrödinger equation, converting it into an integral equation, and then solving the integral equation iteratively. For example on the RHS of the equations of motion (19) for the C_n, we insert the initial values of the C_n and integrate, obtaining better values on the LHS. This first integration gives the "first-order" corrections to the C_n. That may be accurate enough for your purposes, and it is used in the famous Fermi Golden Rule. If it is not accurate enough, you substitute the improved values in on the RHS and integrate to obtain a second-order correction. One can iterate this procedure to obtain successively higher orders of perturbation. This section carries out this procedure to first order in the perturbation energy, i.e., one time integration. The answer is illustrated and then used to derive the Fermi Golden Rule. The section concludes with a general formulation of higher-order perturbation theory.

An important question is, given a quantum system initially in the state i, what are the probabilities that transitions occur to other states? This question asks, for example, what the probability is that an initially unexcited atom interacting with an electromagnetic field absorbs energy from the field. The wave function (12) has the initial value

$$\psi(\mathbf{r},0) = u_i(\mathbf{r}) \ ,$$

that is,

$$C_i(0) = 1, \ C_{n\neq i}(0) = 0 \ . \tag{57}$$

To find out the first-order correction to the $C_n(t)$, we use the initial values (57) on the RHS of the Schrödinger equations of motion (19) for the $C_n(t)$. This gives

$$\dot{C}_n(t) \simeq \dot{C}_n^{(1)}(t) = -i\hbar^{-1} \langle n|\mathcal{V}|i\rangle \, e^{i\omega_{ni}t} \ , \tag{58}$$

where $C_n^{(1)}$ is a special case of $C_n^{(k)}$, which means we have iterated Eq. (19) k times.

Equation (58) is easy to integrate for two important kinds of perturbation energies: one time independent, and one sinusoidal such that

$$\mathcal{V} = \mathcal{V}_0 \cos\nu t\ . \tag{59}$$

Integrating Eq. (58) from 0 to t for a time independent $\mathcal{V}$ (ν=0), we have

$$C_n(t) \simeq C_n^{(1)}(t) = -i\hbar^{-1}\mathcal{V}_{ni}\ \frac{e^{i\omega_{ni}t} - 1}{i\omega_{ni}}$$

$$= -i\hbar^{-1}\mathcal{V}_{ni}\ e^{i\omega_{ni}t/2}\ \frac{\sin(\omega_{ni}t/2)}{\omega_{ni}/2}\ ,$$

where we write $\mathcal{V}_{ni}$ for $\langle n|\mathcal{V}_0|i\rangle$. The probability that a transition occurs to level n is given by

$$|C_n^{(1)}|^2 = \frac{|\mathcal{V}_{ni}|^2}{\hbar^2}\ \frac{\sin^2(\omega_{ni}t/2)}{(\omega_{ni}/2)^2}\ . \tag{60}$$

It's interesting to note that we have already seen this kind of result in the phase matching discussion of Sec. 2-2, for which electromagnetic field amplitudes are used instead of the probability amplitudes used here. Problem 3-3 discusses the analogy between the two problems. The value of Eq. (60) is accurate so long as $C_i(t)$ doesn't change appreciably from the initial value $C_i(0) = 1$. In view of the normalization condition (13), this means that the total transition probability

$$P_T = 1 - |C_i^{(1)}|^2 = \sum_{n\neq i} |C_n^{(1)}|^2 \tag{61}$$

must be much less than unity.

Figure 3-2 plots the probability in Eq. (60) at the time t as a function of the frequency difference ω_{ni}. For short enough times we can expand the sine in Eq. (60) to find

$$|C_n^{(1)}|^2 \simeq \frac{|\mathcal{V}_{ni}|^2}{\hbar^2}\ t^2\ , \tag{62}$$

which shows that the center of the curve increases proportionally to t^2. We further see that for increasing frequency differences $|\omega_{ni}|$, the probability that the interaction induces a transition to level n becomes smaller rapidly. Thus transitions are much more likely if the energy is conserved between initial and final states.

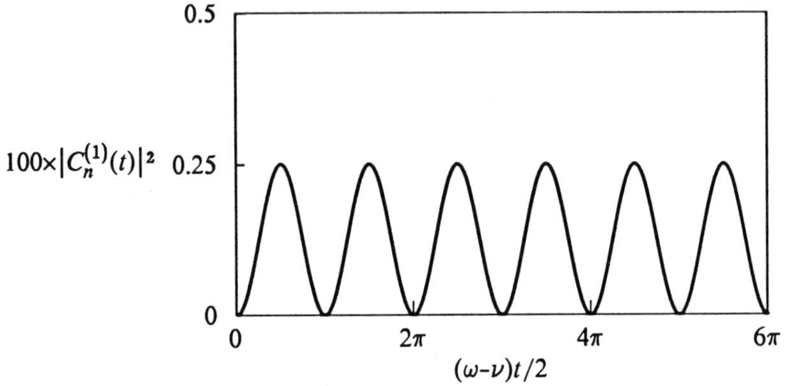

Fig. 3-2. Probability $|C_n^{(1)}(t)|^2$ of Eq. (60) versus $(\omega-\nu)t/2$.

Consider now the sinusoidal interaction energy (59), which can be used to model an atom interacting with a monochromatic electromagnetic field. For such a field, Eq. (59) is proportional to the electric field amplitude, as we see in Sec. 3-3. Integrating Eq. (58) accordingly, we have

$$C_n(t) \simeq C_n^{(1)}(t) = -i\frac{\mathcal{V}_{ni}}{2\hbar}\left[\frac{e^{i(\omega_{ni}+\nu)t}-1}{i(\omega_{ni}+\nu)}+\frac{e^{i(\omega_{ni}-\nu)t}-1}{i(\omega_{ni}-\nu)}\right]. \tag{63}$$

For the sake of definiteness, consider the case $\omega_{ni} > 0$. Then the denominator $\omega_{ni}+\nu$ is always positive and larger than ω_{ni}. This is not true for the denominator $\omega_{ni}-\nu$, which vanishes if the resonance condition

$$\nu \simeq \omega_{ni} \tag{64}$$

is satisfied. For interactions near resonance, the term with the relatively small denominator $\omega_{ni}-\nu$ is much larger than that with the $\omega_{ni}+\nu$, allowing us to neglect the latter. For the same reason, we can probably neglect transitions to levels with energies very different from $\hbar\nu$. This observation is used to justify the two-level atom approximation discussed in Sec. 3-3. Neglecting the term with the relatively large denominator $\omega_{ni}+\nu$ is called the *rotating-wave approximation*. It is used in much of the book.

Making the rotating-wave approximation in Eq. (63), we find the transition probability

$$|C_n^{(1)}|^2 = \frac{|\mathcal{V}_{ni}|^2}{4\hbar^2} \frac{\sin^2[(\omega_{ni} - \nu)t/2]}{(\omega_{ni} - \nu)^2/4} . \tag{65}$$

This result is formally the same as the dc case of Eq. (60), provided we substitute $\omega_{ni} - \nu$ for ω_{ni}. Thus Fig. 3-2 and the corresponding discussion apply to this case as well. In particular, we see that in the course of time, transitions are unlikely to occur unless the resonance condition (64) is satisfied, that is, unless the applied field frequency matches the transition frequency.

So far we have considered infinitely sharp energy levels. This is not realistic, since levels can be broadened by effects like spontaneous emission and collisions. Furthermore there may be a continuum of levels such as in the energy bands in solid-state media. For these situations, the summation in the total transition probability (61) can be replaced by an integral with a density of states factor $\mathcal{D}(\omega)$ to weight the distribution correctly. For example, there are typically more states per frequency interval for higher frequencies than for lower frequencies. The total transition probability P_T then has the value

$$P_T \simeq \int \mathcal{D}(\omega)\, |C_n^{(1)}|^2 \, d\omega ,$$

where the discrete frequency ω_{ni} is replaced by the continuous frequency ω. Substituting Eq. (65), we find

$$P_T = \int d\omega\, \mathcal{D}(\omega)\, \frac{|\mathcal{V}(\omega)|^2}{4\hbar^2}\, t^2\, \frac{\sin^2[(\omega-\nu)t/2]}{[(\omega-\nu)t/2]^2} . \tag{66}$$

It is interesting to evaluate the total transition probability integral (66) for two reasons. First, so long as it is small enough, we know that the first-order perturbation theory answer is valid. Secondly, dP_T/dt gives the rate at which transitions occur. The equation for this rate is called the Fermi Golden Rule, and can be used to find a variety of rates, such as those occurring in the photoelectric effect, spontaneous emission, and the Planck radiation law.

Equation (66) is a special case of the general integral

$$J = \int d\omega\, F(\omega)\, G(\omega) . \tag{67}$$

There are problems for which both the density of states factor $\mathcal{D}(\omega)$ and the matrix elements $\mathcal{V}_{ni}$ are known. But even in such situations the resulting integral for Eq. (66) is typically hard to solve. However, we can approximate J if either $F(\omega)$ or $G(\omega)$ varies little over the frequency range for which the other has an appreciable value. The extreme example is when one of the functions, say $G(\omega)$ is the delta function $\delta(\omega-\omega_0)$. Then $J = F(\omega_0)$. More generally, suppose $G(\omega)$ is sharply peaked about ω_0 and that $F(\omega)$ varies little in this interval. Then $J \simeq F(\omega_0) \int G(\omega)$. For the purposes of this problem, $G(\omega)$ *is* a delta function, and it is in this way that delta functions approximate natural behavior. This is an example of what one sometimes calls an "adiabatic elimination". This kind of elimination is equally important in the solution of coupled differential equations, for which one function varies slowly compared to another. Typically we consider problems in which atoms coupled to an electromagnetic field vary rapidly compared to the field envelope. In such cases, the technique of adiabatic elimination allows us to solve the atomic equations of motion assuming that the field envelope is constant, and then to substitute the resulting steady-state polarization of the medium into the correspondingly simplified slowly-varying field equations of motion (Chap. 5).

The integral (66) has the form of (67) with $F(\omega) = \mathcal{D}(\omega)\ |\mathcal{V}(\omega)|^2$ and $G(\omega) = \sin^2[(\omega-\nu)t/2]/[(\omega-\nu)t/2]^2$. Hence we can solve Eq. (66) when either F or G varies rapidly compared to the other. In particular, for times sufficiently small that all relevant values of $|\omega-\nu|t$ are much less than unity, the $G{=}\sin x/x$ function in Eq. (66) can be approximated by unity. By relevant values of $|\omega-\nu|t$, we mean those for which the density of states factor $\mathcal{D}(\omega)$ and the matrix element $\mathcal{V}(\omega)$ have appreciable values. This then gives

$$P_T \simeq t^2 \int \frac{\mathcal{D}(\omega)|\mathcal{V}(\omega)|^2}{4\hbar^2}\, d\omega\ . \tag{68}$$

Hence for such small times, the transition rate dP_T/dt is proportional to time, starting up from zero. Unless the density of states and the interaction energy matrix element have infinitely wide frequency response, i.e., infinite bandwidth, this limit implies a build-up time in the response of the system to the applied perturbation $\mathcal{V}(\omega)$. (This means, for example, that detectors have a nonzero response time.)

After this initial small time region, we suppose that the factor $\mathcal{D}(\omega)|\mathcal{V}_{ni}|^2$ varies little in the frequency interval for which the $\sin^2 x/x^2$ function in Eq. (66) has appreciable values (see Fig. 3-3). In this limit, we can evaluate the $\mathcal{D}(\omega)|\mathcal{V}_{ni}|^2$ at the peak $\omega = \nu$ of the $\sin^2 x/x^2$ function, finding

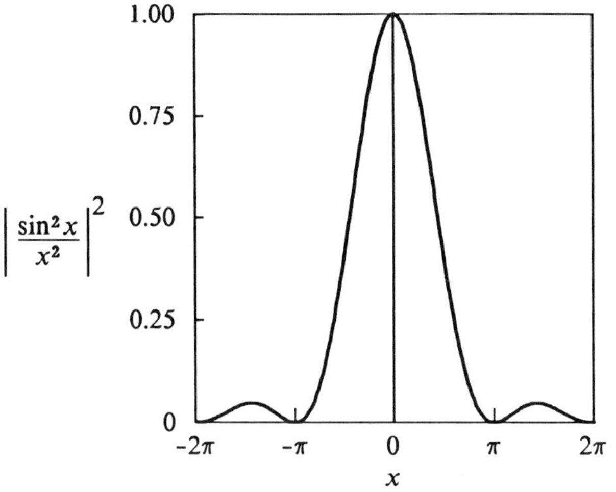

Fig. 3-3. $\sin^2 x/x^2$ term in the total transition probability integral (69) versus $x = (\omega-\nu)t/2$. As time increases, this function peaks up like the δ-function $\delta(\omega-\nu)$.

$$P_T = \mathscr{D}(\nu)\,\frac{|\mathcal{V}(\nu)|^2}{2\hbar^2}\,t\int_{-\infty}^{\infty} dx\,\frac{\sin^2 x}{x^2}$$

$$= \frac{\pi}{2\hbar^2}\mathscr{D}(\nu)\,|\mathcal{V}(\nu)|^2\,t\,. \tag{69}$$

Here we have extended the limits of the integral to $\pm\infty$ since in the present approximation this adds little to the integral and yields an analytic answer. Equation (69) gives then the Fermi Golden Rule rate

$$\boxed{\Gamma = \frac{dP_T}{dt} = -\frac{d}{dt}\,|C_i^{(1)}|^2 = \frac{\pi}{2\hbar^2}\,\mathscr{D}(\nu)\,|\mathcal{V}(\nu)|^2}\,, \tag{70}$$

which is a constant in time.

This constant rate proportional to the intensity of the incident radiation is what people typically observe in the photoelectric effect. Note that the rate vanishes if no transitions exist for the frequency ν. The photoelectric effect occurs in media that have an energy gap above the ground state. To be absorbed, the applied photon energy $\hbar\nu$ must be larger than this gap.

Summarizing the rate at which transitions occur, we see that for times short compared to the reciprocal of the width of the function $|\mathcal{V}_0|^2\mathcal{D}$, the rate increases linearly in time. For longer times the rate becomes constant. For still longer times, when P_T does not remain small, we cannot assume that the probability of the initial state is unity. Section 13-3 shows that to a good approximation, we can fix up the rate by multiplying it by the initial state probability $|C_i(t)|^2$. This generalizes the Fermi Golden Rule to

$$\frac{d}{dt}|C_i|^2 = -\Gamma|C_i|^2 , \tag{71}$$

which states that the probability for being in the initial level decays exponentially in time. Such a formula doesn't make a first-order perturbation theory approximation. This kind of time response is typical of certain important processes in quantum optics, such as spontaneous emission, which is described in Chap. 13.

Note that we get the same kind of integral (66) for two sharp levels interacting with a nonmonochromatic field with a spectral intensity distribution proportional to $\mathcal{V}(\nu)$. This yields a transition rate Γ given by Eq. (70) with ν replaced by ω, since it is the atomic frequency ω, rather than the field frequency ν, that determines the center of the $\sin^2 x/x^2$ distribution. This observation is important in the next section, where we derive the Planck black-body radiation formula.

Higher-Order Perturbation Theory

The iterative approach outlined at the start of this section can be written in an analytic form by using the formal solution of Eq. (44) in the interaction picture as

$$|\psi_I(t)\rangle = U_I(t)|\psi_I(0)\rangle . \tag{72}$$

Here we use the subscript $_I$ to remind ourselves that we are working in the interaction picture. Taking the time rate of change of Eq. (72) and using the Schrödinger equation (53), we obtain

$$i\hbar\,\frac{dU_I(t)}{dt} = \mathcal{V}_I(t)U_I(t) . \tag{73}$$

Remembering that $U_I(0) = 1$, we integrate this equation formally to get

$$U_I(t) = 1 - \frac{i}{\hbar}\int_0^t dt' \, \mathcal{V}_I(t')U_I(t') \, . \tag{74}$$

We can solve this equation by successive iterations, obtaining

$$U_I(t) = 1 - \frac{i}{\hbar}\int_0^t dt_1 \, \mathcal{V}_I(t_1) + \left[\frac{-i}{\hbar}\right]^2 \int_0^t dt_1 \, \mathcal{V}_I(t_1) \int_0^{t_1} dt_2 \, \mathcal{V}_I(t_2) + \dots . \tag{75}$$

Truncating this expression after the lowest order term in $\mathcal{V}_I$ gives first-order perturbation theory. Keeping higher-order terms gives second-order, third-order, etc., perturbation theory. Note that this iteration process implies a time ordering such that $t_2 \le t_1 \le t$.

By way of illustration, we calculate the first-order answer this way as

$$|\psi_I(t)\rangle \simeq \left[1 - \frac{i}{\hbar}\int_0^t dt_1 \, \mathcal{V}_I(t_1)\right] |\psi_I(0)\rangle \, . \tag{76}$$

Substituting this into the equation for the transition probability to level m

$$|C_m(t)|^2 = |\langle m|\psi_I(t)\rangle|^2 = |\langle m|\psi_S(t)\rangle|^2 \, , \tag{77}$$

we have

$$|C_m(t)|^2 \simeq |\langle m|i\rangle|^2 + \hbar^{-2}|\langle m|\textstyle\int dt_1 \mathcal{V}_I(t_1)|i\rangle|^2 .$$

Converting $\mathcal{V}_I$ back to the Schrödinger value using Eq. (54), we have

$$|C_m(t)|^2 \simeq |C_m^{(1)}(t)|^2 = \frac{1}{\hbar^2}\left|\int_0^t dt' \, e^{i\omega_{mi}t'} \langle m|\mathcal{V}_S|i\rangle\right|^2 . \tag{78}$$

This gives Eq. (65) as before. By including more terms in Eq. (76), we can calculate successively higher-order contributions.

3-3. Atom-Field Interaction for Two-Level Atoms

This section introduces the two-level atom, a concept we write a great deal about in this book. Such later consideration merits a careful introduction. Consider first the simplest of all atoms, hydrogen. This atom has an infinite number of bound levels, characterized by the energies

$$E_n = -\frac{e^2}{2a_0 n^2} = -\frac{R_\infty}{2n^2}, \tag{79}$$

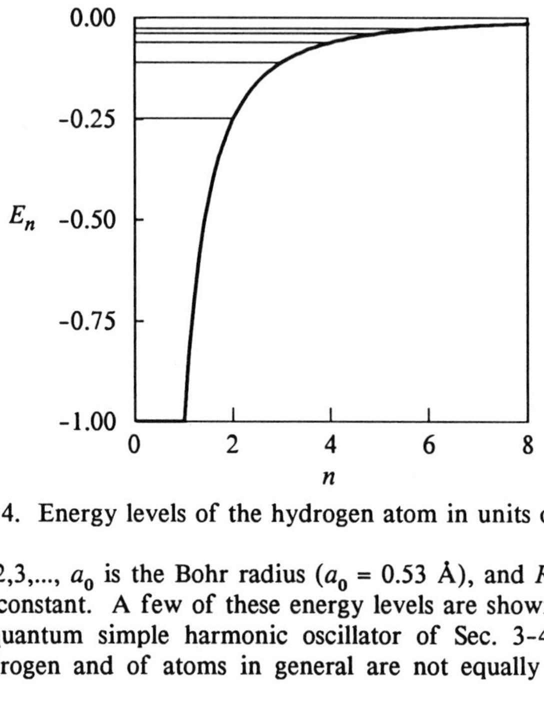

Fig. 3-4. Energy levels of the hydrogen atom in units of R_∞.

where $n = 1,2,3,\ldots$, a_0 is the Bohr radius ($a_0 = 0.53$ Å), and $R_\infty = 13.6$ eV is Rydberg's constant. A few of these energy levels are shown in Fig. 3-4. Unlike the quantum simple harmonic oscillator of Sec. 3-4, the energy levels of hydrogen and of atoms in general are not equally spaced. For example,

$$E_2 - E_1 = \frac{3}{4} R_\infty, \quad E_3 - E_2 = \frac{5}{36} R_\infty .$$

In quantum optics and in laser spectroscopy, we often shine monochromatic laser light on such an atom and study what happens. If the laser frequency almost matches a particular transition frequency, then the transition probability predicted by Eq. (63) for this transition is much larger than that for other transitions. The approximation is almost the same as that used in making the rotating-wave approximation: in both cases, one neglects terms with denominators large compared to the term with the resonant denominator. A particular frequency difference $\omega_{ni}-\nu$ is much smaller, say $2\pi\times10^8$ radians/sec, than the sum $\omega_{ni}+\nu$ ($\simeq 2\pi\times10^{14}$ radians/sec) for the antirotating wave or the difference $\omega_{mi}-\nu$, $m\neq n$, which might be $0.6\pi\times10^{14}$ radians/sec for some nonresonant transition. If this is the case, the problem reduces to two levels. Since the antirotating-wave contribution is actually smaller than many nonresonant contributions, it follows that the two-level atom approximation is usually only consistent if made simultaneously with the rotating-wave approximation. If one decides to keep the antirotating wave contribution, one must also keep all the nonresonant contributions as well. This is not as hard as it might seem, since nonresonant contributions can be usually treated using first-order perturbation theory. Also, we can account to some degree for transitions to levels other than the principal two by including various decay and pump rates.

A famous two-level system is the spin 1/2 magnetic dipole in nuclear magnetic resonance. This is a true two-level system with relatively simple decay mechanisms. It has a lot in common with its brother the two-level atom, but its response can differ significantly in cases where level decay rates play an important role.

In our treatment of atoms using the two-level approximation, we ignore the fact that levels usually have a number of sublevels that all can contribute to a resonant transition. This produces complications when experiments with real atoms are used to test theories based on the two-level approximation. In such cases, optical pumping techniques can sometimes be used to produce a true two-level atom.

We emphasize the two-level atom because we can often describe its interaction with the electromagnetic field in detail and obtain analytic solutions. It thus allows us to learn a great deal about the atom-field interaction, and hopefully this knowledge can be generalized to more realistic situations. Note that although the two-level atom includes the low-order $\chi^{(3)}$ type of nonlinearity of Sec. 2-3 as a special case, in general it provides for more complicated nonlinear responses, such as saturation.

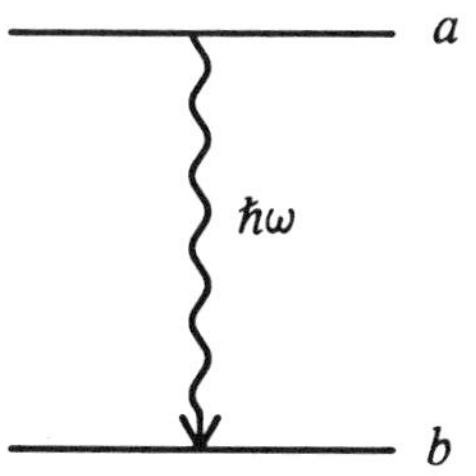

Fig. 3-5. Energy level diagram of two-level atom.

We label the upper level of our two-level atom by the letter a, and the lower by b as shown in Fig. 3-5. The corresponding wave function is

$$\psi(\mathbf{r},t) = C_a(t)e^{-i\omega_a t}\, u_a(\mathbf{r}) + C_b(t)e^{-i\omega_b t}\, u_b(\mathbf{r}) \,. \tag{80}$$

Before we see how this wave function evolves under the influence of an applied electromagnetic field, let us consider the kind of a charge distribution it represents. To be specific, suppose the lower level is the 1s ground state of hydrogen with the energy eigenfunction

$$u_b(\mathbf{r}) = u_{100}(r,\theta,\phi) = (\pi a_0{}^3)^{-1/2}\, e^{-r/a_0} \,, \tag{81}$$

and the upper level is the $2p$ state with the eigenfunction

$$u_a(\mathbf{r}) = u_{210}(r,\theta,\phi) = (32\pi a_0{}^3)^{-1/2}(r/a_0)\cos\theta\, e^{-r/2a_0} \,. \tag{82}$$

These eigenfunctions are plotted versus the z coordinate in Fig. 3-6, followed by the superposition $\psi(\mathbf{r},t)$ of Eq. (80) for two times separated by π/ω. For one of these times, the two probability amplitudes in Eq. (80) add. Half an period later, they subtract. Figure 3-6c shows the probability density $|\psi(z,t)|^2$ for these two points in time. We see that this probability density, and hence the "charge density" $e|\psi(z,t)|^2$, oscillates back and forth across the nucleus in a fashion analogous to the charge on the spring in Sec. 1-3. This similarity is the underlying reason why the classical model of Chapter 2 is so successful in describing the linear absorption of light by a collection of atoms. Chapter 5 derives this response quantum mechanically in detail, revealing where the classical model fails in laser physics.

We take our interaction energy to be that for an electric dipole interacting with an electromagnetic field

$$\boxed{\mathcal{V} = -e\mathbf{r}\cdot\mathbf{E}(\mathbf{R},t)} \,, \tag{83}$$

where $\mathbf{R}$ is the position of the center of mass of the atom. This gives a higher energy for the dipole $e\mathbf{r}$ aligned against the field $\mathbf{E}$ than along it, as it should. In Eq. (83), we have approximated $\mathbf{E}(\mathbf{r},t)$ by $\mathbf{E}(\mathbf{R},t)$, since we are interested in electromagnetic fields with wavelengths much larger than atomic dimensions. We can therefore approximate the electric field by a constant over the dimensions of the atom. This is called the dipole approximation. Typically we are also interested in plane waves, for which we write simply $E(z,t)$, where z is the axis of propagation. The dipole traditionally is written as the positive charge value times the distance

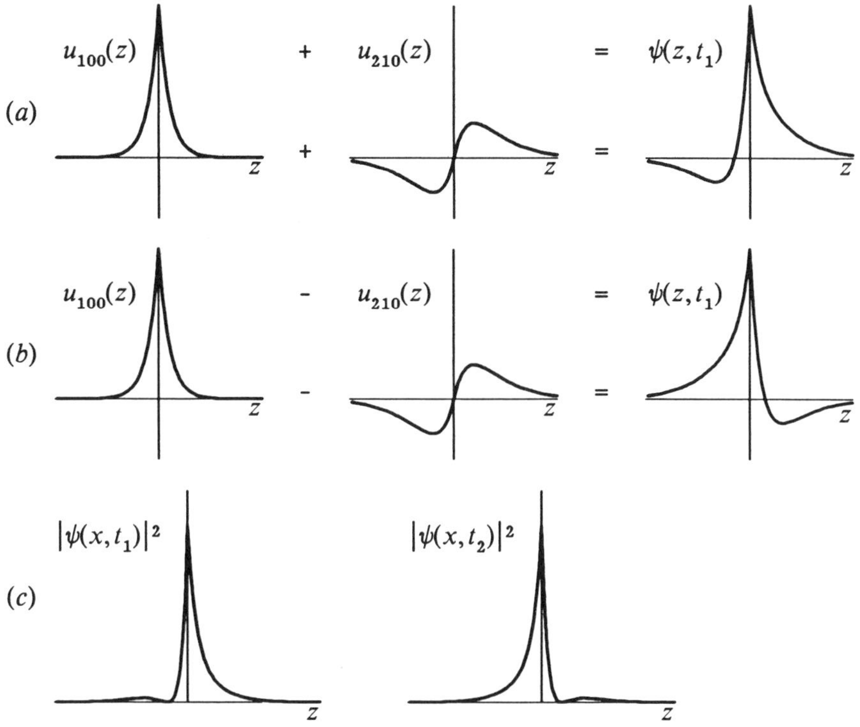

Fig. 3-6. (*a*) z dependence of $\psi(\mathbf{r},t_1) = u_{100}(\mathbf{r}) + u_{210}(\mathbf{r})$ for the time $t_1 = 2n\pi/\omega$. (*b*) z dependence of $\psi(\mathbf{r},t_2)\ u_{100}(\mathbf{r}) - u_{210}(\mathbf{r})$ for $t_2 = (2n+1)\pi/\omega = t_1 + \pi/\omega$. (*c*) Corresponding dependencies of the probability densities $|\psi(\mathbf{r},t_1)|^2$ and $|\psi(\mathbf{r},t_2)|^2$ (with a slightly different scale).

vector pointing from the negative to the positive charge. This gives the same answer as $e\mathbf{r}$, which is the negative charge value times the distance vector pointing from the positive charge to the negative charge. There has been substantial discussion over the years since Lamb (1952) first brought it up concerning the use of Eq. (83) versus a Hamiltonian involving $\mathbf{A}\cdot\mathbf{p}$, where $\mathbf{A}$ is the field vector potential and $\mathbf{p}$ is the electron momentum. For our purposes, Eq. (83) combines intuitive appeal with excellent accuracy.

The matrix element of the dipole operator between a level and itself [recall Eq. (15)]

$$e\mathbf{r}_{\alpha\alpha} = \int d^3r\ e\mathbf{r}\ |u_\alpha(\mathbf{r})|^2, \tag{84}$$

vanishes unless the system has a permanent dipole moment (like H_2O), since $|u_\alpha(\mathbf{r})|^2$ is inevitably a symmetrical function of $\mathbf{r}$ and $\mathbf{r}$ itself is antisymmetric. Matrix elements of $\mathbf{r}$ between different states can also vanish, but we are primarily interested in two levels a and b between which the matrix element does not vanish. We can then write the electric-dipole interaction energy matrix element

$$\mathcal{V}_{ab} = - \wp E(\mathbf{R},t), \tag{85}$$

where $\wp$ (pronounced "squiggle" and also used for the Weierstrass elliptic function) is the component of $e\mathbf{r}_{ab}$ along $\mathbf{E}$.

For the sake of simplicity, we ignore the spatial dependence altogether in the remainder of this section, and use

$$E(t) = E_0 \cos \nu t\ . \tag{86}$$

This gives the interaction energy matrix element

$$\mathcal{V}_{ab} = - \wp E_0 \cos \nu t\ . \tag{87}$$

Substituting this into Eq. (65) with a as the final state n and b as the initial state i, we have

$$|C_a{}^{(1)}|^2 = |\wp E_0/2\hbar|^2 \ \frac{\sin^2[(\omega-\nu)t/2]}{(\omega-\nu)^2/4}\ . \tag{88}$$

This is the probability that the two-level atom absorbs energy under the influence of a driving field, a phenomenon called (stimulated) *absorption*. Alternatively, by identifying the initial state as a and the final state as b, we describe a process called *stimulated emission*. It is easy to show that $|C_b{}^{(1)}|^2$ in this case is the same as $|C_a{}^{(1)}|^2$ in the case of absorption: the probabilities for stimulated emission and absorption are equal. This is an example of microscopic reversibility.

Blackbody Radiation

Now consider the probability of a transition due to a field that is not monochromatic, but rather has a continuous spectrum such as that for blackbody radiation. For this we replace the $E_0{}^2\mathcal{D}(\omega)$ that occurs in using Eq. (88) by $2\mathcal{U}(\nu)/\epsilon_0$, where $\mathcal{U}(\nu)$ is the energy density per radian/sec, and sum over all field frequencies ν. The 2 here comes from the fact that two polarizations are possible for each frequency. We then find the total transition probability

$$P_T = \int d\nu\, \mathcal{U}(\nu) \frac{t^2}{\hbar^2} \frac{\sin^2[(\omega-\nu)t/2]}{[(\omega-\nu)t/2]^2} \,. \tag{89}$$

This is the same kind of integral as that encountered for the Fermi Golden Rule in Eq. (66), except that for this two-level atom, we integrate over the field continuum frequency ν instead of the level continuum frequency ω. Here we factor the slowly-varying energy density $\mathcal{U}(\nu)$ outside the integral, evaluating it at the peak, $\nu=\omega$, of the $\sin^2 x/x^2$ curve. We find the transition rate

$$B(\omega)\mathcal{U}(\omega) = \frac{\pi}{3\hbar^2\epsilon_0} \wp^2\mathcal{U}(\omega) \,, \tag{90}$$

where the 3 comes from replacing $\wp^2$ by $\wp^2/3$ since the radiation can come from all directions: only 1/3 of the field components effectively couple to the dipole. Note that P_T of Eq. (89) has the same value if the atom is initially in the upper state, rather than in the lower state as taken for Eq. (88). Hence the stimulated emission rate equals the stimulated absorption rate of Eq. (90).

Another case in which a two-level atom interacts with a radiation continuum is spontaneous emission, which can be described as a combination of radiation reaction (see end of Sec. 1-3) and stimulated emission by vacuum fluctuations. This interpretation is clarified in Chap. 13, which derives the upper-level decay formula given by Eq. (71). For now just think of the radiation field as consisting of a continuum of modes, each of which acts like a quantized simple harmonic oscillator. Section 3-4 shows that such an oscillator has a zero-point energy, which is associated with fluctuations in the displacement variable. For the case of the electric field, this displacement becomes the field amplitude, which then has fluctuations. These fluctuations are called vacuum fluctuations, because they exist even in the vacuum, i.e., even when no classical field exists. The vacuum field has a continuous spectrum. When this is used with the quantized field version of Eq. (60) (see Sec. 13-3), we find the spontaneous emission rate constant called A by Einstein and Γ in Eq. (71). To find the value of A intuitively, we can use the rate given by Eq. (90) if we can guess what the energy density $\mathcal{U}(\omega)$ of the vacuum field is. We note that the number of field modes per unit volume between ω and $\omega+d\omega$ is ω^2/π^2c^3 (see Eq. (13.46)). Multiplying this number by the energy $\hbar\omega$ of one photon, we have $\mathcal{U}_{spon}(\omega) = \hbar\omega^3/\pi^2c^3$. Using this in Eq. (90) gives the spontaneous emission rate

$$A = B\,\frac{\hbar\omega^3}{\pi^2 c^3} = \frac{\wp^2\omega^3}{3\pi\epsilon_0\hbar c^3}\,. \tag{91}$$

The lifetime of the upper level is $1/A$. This is the same result (13.60) as derived in detail in Sec. 13-3. This section also shows that spontaneous absorption does not occur.

We can use these facts to derive informally the Planck blackbody spectrum

$$\boxed{\mathcal{U}(\omega) = \frac{\hbar\omega^3/\pi^2 c^3}{e^{\hbar\omega/k_B T} - 1}}\,, \tag{92}$$

where k_B is Boltzmann's constant and T is the absolute temperature. We describe the response of the atoms to the blackbody radiation in terms of the number of atoms n_a in the upper state and the number n_b in the lower state. Due to the three processes of spontaneous emission, stimulated emission, and stimulated absorption, these numbers change according to rate equations

$$\dot{n}_a = -An_a - B\mathcal{U}(\omega)(n_a - n_b) \tag{93}$$

$$\dot{n}_b = +An_a + B\mathcal{U}(\omega)(n_a - n_b)\,. \tag{94}$$

We solve these equations in steady state, defined by $\dot{n}_a = \dot{n}_b = 0$. Either equation (note that $\dot{n}_a = -\dot{n}_b$) gives

$$\mathcal{U}(\omega) = \frac{A/B}{n_b/n_a - 1}\,,$$

which with Eq. (91) becomes

$$\mathcal{U}(\omega) = \frac{\hbar\omega^3/\pi^2 c^3}{n_b/n_a - 1}\,, \tag{95}$$

Furthermore according to Boltzmann, in thermodynamic equilibrium the ratio of the number of atoms n_a in the upper state to that n_b in the lower state is given by

$$\frac{n_a}{n_b} = e^{-\hbar\omega/k_B T}\,. \tag{96}$$

Substituting this into Eq. (95), we find the Planck formula (92).

Rabi Flopping

Blackbody radiation is emitted by a collection of atoms in thermal equilibrium with the radiation field. On a microscopic basis the atoms constantly exchange energy with the field in such a way that macroscopically no change is noticed. As we see in Sec. 5-1, this limit is valid in the rate equation approximation, for which the field amplitude varies slowly (here not at all) compared to the atomic response.

Now let us consider the opposite extreme for which we ignore atomic damping altogether, and for simplicity we take the monochromatic field (86) with frequency ν approximately equal to the two-level transition frequency $\omega = \omega_a - \omega_b$. Examining the interaction energy (59) and Eq. (63), we see that in the rotating-wave approximation we keep the $e^{-i\nu t}$ term for $\omega_{ni} > 0$. In the present case, $\omega > 0$, and hence in the rotating-wave approximation we keep only

$$\boxed{\mathcal{V}_{ab} \simeq -\frac{1}{2}\wp E_0\, e^{-i\nu t}} \;. \tag{97}$$

For $\mathcal{V}_{ba}$ we use $e^{i\nu t}$. Because ν may differ somewhat from ω, it is convenient to write $\psi(\mathbf{r},t)$ slightly differently from Eq. (80), namely as

$$\psi(\mathbf{r},t) = C_a(t)\exp[i(\tfrac{1}{2}\delta-\omega_a)t]\,u_a(\mathbf{r}) + C_b(t)\exp[i(-\tfrac{1}{2}\delta-\omega_b)t]\,u_b(\mathbf{r}) \;, \tag{98}$$

where the frequency shift $\delta = \omega-\nu$. This choice places the wave function in the *rotating frame* used for the Bloch vector in Sec. 4-3. Substituting this expansion for ψ into the Schrödinger Eq. (5) and projecting onto the eigenfunctions u_a and u_b as in the derivation of Eq. (19), we find

$$\dot{C}_a = \tfrac{1}{2}i(-\delta C_a + \mathcal{R}_0 C_b) \;, \tag{99}$$

$$\dot{C}_b = \tfrac{1}{2}i(\delta C_b + \mathcal{R}_0^{*} C_a) \;, \tag{100}$$

where $|\mathcal{R}_0|$ is the *Rabi flopping frequency* defined by

$$\boxed{\mathcal{R}_0 \equiv \frac{\wp E_0}{\hbar}} \tag{101}$$

after Rabi (1936), who studied the similar system of a spin-$\frac{1}{2}$ magnetic dipole in nuclear magnetic resonance. Equations (99) and (100) provide a simple example of two coupled equations, a combination we see repeatedly in phase conjugation (Chaps. 2 and 9) and in linear stability analysis (Chap.

10). Equation (60) solves the n-level probability amplitude to first order in the interaction energy. Here we solve the two-level probability amplitudes to all orders in this energy.

Before solving Eqs. (99) and (100) generally, we can very quickly discover the basic physics by considering exact resonance, for which $\delta = 0$. We can then differentiate Eq. (100) with respect to time and substitute Eq. (99) to find

$$\ddot{C}_b = -\tfrac{1}{4}|\mathcal{R}_0|^2 C_b \ ,$$

i.e., the differential equation for sines and cosines. In particular if at time t=0 the atom is in the lower state ($C_b(0) = 1$, $C_a(0) = 0$), then

$$C_b(t) = \cos\tfrac{1}{2}|\mathcal{R}_0|t \tag{102}$$

which from Eq. (100) gives

$$C_a(t) = i\,\sin\tfrac{1}{2}|\mathcal{R}_0|t \ . \tag{103}$$

The probability that the system is in the lower level $|C_b(t)|^2 = \cos^2\tfrac{1}{2}|\mathcal{R}_0|t = (1 + \cos|\mathcal{R}_0|t)/2$, while $|C_a|^2 = \sin^2\tfrac{1}{2}|\mathcal{R}_0|t = (1 - \cos|\mathcal{R}_0|t)/2$. Hence the wave function oscillates between the lower and upper states sinusoidally at the frequency $|\mathcal{R}_0|$. In total contrast with blackbody radiation, instead of coming to an equilibrium with constant probabilities for being in the upper and lower levels, here the probabilities oscillate back and forth. In this case the atoms maintain a definite phase relationship with the inducing electric field, while for blackbody radiation any such relationship averages to zero. As we see in Chap. 4 on the density matrix, Rabi flopping preserves atomic coherence, while blackbody radiation destroys it. Further discussions on the irreversibility of coupling to a continuum are given in Sec. 13-3 on the theory of spontaneous emission, and more generally in Chap. 14.

To solve the coupled equations (99) and (100) including a nonzero detuning δ, we write them as the single matrix equation

$$\frac{d}{dt}\begin{bmatrix} C_a(t) \\ C_b(t) \end{bmatrix} = \frac{i}{2}\begin{bmatrix} -\delta & \mathcal{R}_0 \\ \mathcal{R}_0^* & \delta \end{bmatrix}\begin{bmatrix} C_a(t) \\ C_b(t) \end{bmatrix} . \tag{104}$$

This is a vector equation of the form $d\mathbf{C}/dt = \tfrac{1}{2}i\mathbf{MC}$, which has solutions of the form $\exp(\tfrac{1}{2}i\lambda t)$. Accordingly substituting $\mathbf{C}(t) = \mathbf{C}(0)\exp(\tfrac{1}{2}i\lambda t)$ into Eq. (104), we find that $\det(\mathbf{M} - \lambda\mathbf{I}) = 0$. This yields the eigenvalues $\lambda = \pm\mathcal{R}$, where $\mathcal{R}$ is the *generalized Rabi flopping frequency*

$$\boxed{\mathcal{R} \equiv \sqrt{\delta^2 + |\mathcal{R}_0|^2}}\,. \tag{105}$$

Equation (104) has simple sinusoidal solutions of the form

$$C_a(t) = C_a(0)\cos\tfrac{1}{2}\mathcal{R}t + A\sin\tfrac{1}{2}\mathcal{R}t$$
$$C_b(t) = C_b(0)\cos\tfrac{1}{2}\mathcal{R}t + B\sin\tfrac{1}{2}\mathcal{R}t\,.$$

Substituting these values into Eqs. (99) and (100) and setting $t = 0$, we immediately find the constants A and B. Collecting the results in matrix form, we have the general non-damped solution

$$\begin{bmatrix} C_a(t) \\ C_b(t) \end{bmatrix} = \begin{bmatrix} \cos\frac{1}{2}\mathcal{R}t - i\delta\mathcal{R}^{-1}\sin\mathcal{R}t & i\mathcal{R}_0\mathcal{R}^{-1}\sin\frac{1}{2}\mathcal{R}t \\ i\mathcal{R}_0^{*}\mathcal{R}^{-1}\sin\frac{1}{2}\mathcal{R}t & \cos\frac{1}{2}\mathcal{R}t + i\delta\mathcal{R}^{-1}\sin\frac{1}{2}\mathcal{R}t \end{bmatrix} \begin{bmatrix} C_a(0) \\ C_b(0) \end{bmatrix}. \tag{106}$$

The 2x2 matrix in this equation is precisely the Schrödinger evolution matrix $U(t)$ of Eq. (45) for the problem at hand. This U-matrix solution is valuable for the discussion of coherent transients in Chap. 11 and in general whenever damping can be neglected. It yields the first-order perturbation result Eq. (88) in the limit of a weak field ($\mathcal{R} \rightarrow \delta$). Section 4-1 shows how to account for possibly unequal level decay from both levels. More general decay schemes require the use of a density matrix as discussed in Sec. 4-2. Note that the matrix in Eq. (106) is a U matrix (see Eq. (44)). For further discussion, see Probs. 3-14 through 3-16 and Sec. 13-1.

Pauli Spin Matrices

In treating two-level atoms it is often handy to use a 2×2 matrix notation. The eigenfunctions u_a and u_b are then represented by the column vectors

$$u_a \longleftrightarrow \begin{bmatrix} 1 \\ 0 \end{bmatrix} \qquad u_b \longleftrightarrow \begin{bmatrix} 0 \\ 1 \end{bmatrix} \tag{107}$$

and the wave function by the column vector

$$\psi \longleftrightarrow \begin{bmatrix} C_a \\ C_b \end{bmatrix}. \tag{108}$$

The energy and electric dipole operators are conveniently written in terms of the Pauli spin matrices

$$\sigma_x = \begin{pmatrix} 0 & 1 \\ 1 & 0 \end{pmatrix} \quad \sigma_y = \begin{pmatrix} 0 & -i \\ i & 0 \end{pmatrix} \quad \sigma_z = \begin{pmatrix} 1 & 0 \\ 0 & -1 \end{pmatrix}. \tag{109}$$

While these matrices are Hermitian, the "spin-flip" operators

$$\sigma_- = \begin{pmatrix} 0 & 0 \\ 1 & 0 \end{pmatrix} \quad \sigma_+ = \begin{pmatrix} 0 & 1 \\ 0 & 0 \end{pmatrix} \tag{110}$$

are not Hermitian. σ_- flips the system from upper level to lower level

$$\sigma_- \begin{pmatrix} 1 \\ 0 \end{pmatrix} = \begin{pmatrix} 0 \\ 1 \end{pmatrix},$$

while σ_+ flips from lower to upper. This property is handy for representing transitions caused by the interaction energy.

By choosing the energy zero to be half way between the upper and lower levels, we can write the Schrödinger picture Hamiltonian (17) with the interaction energy (87) in the rotating wave approximation as

$$\mathcal{H} = \frac{1}{2} \begin{pmatrix} \hbar\omega & -\wp E_0 e^{-i\nu t} \\ -\wp E_0 e^{i\nu t} & -\hbar\omega \end{pmatrix}$$

or

$$\boxed{\mathcal{H} = \frac{\hbar\omega}{2}\sigma_z - \frac{1}{2}[\wp E_0 \sigma_+ e^{-i\nu t} + \text{adjoint}]}. \tag{111}$$

The pair of interaction-picture coupled equations (99) and (100) can be written on resonance (δ=0) as

$$\frac{d}{dt}\begin{pmatrix} C_a \\ C_b \end{pmatrix} = \frac{i}{2}\begin{pmatrix} 0 & \mathcal{R}_0 \\ \mathcal{R}_0^* & 0 \end{pmatrix}\begin{pmatrix} C_a \\ C_b \end{pmatrix}. \tag{112}$$

Section 4-1 uses such a matrix representation to solve a generalization of these equations including field detuning and atomic damping.

3-4. Simple Harmonic Oscillator

The simple harmonic oscillator plays a central role in quantum optics. Chapter 12 discusses its connection with the quantum theory of radiation in detail. It is also used in simple models of vibrational states of molecules, comes into the description of large ensemble of two-level atoms, etc...

The classical energy of a harmonic oscillator of frequency Ω and unit mass is

$$\mathcal{H}_c = p^2/2 + \Omega^2 q^2/2 . \tag{113}$$

where p is the momentum and q the displacement. Using the correspondence

$$p = -i\hbar \frac{d}{dq} , \tag{114}$$

we readily obtain the corresponding quantum-mechanical Hamiltonian

$$\mathcal{H} = -\frac{\hbar^2}{2} \frac{d^2}{dq^2} + \Omega^2 q^2/2 . \tag{115}$$

The eigenfunctions of this system are well-known and may be expressed in terms of Hermite polynomials $[H_n(\alpha q)]$

$$u_n(q) = \sqrt{\frac{\alpha}{\sqrt{\pi}2^n n!}} H_n(\alpha q) \exp(-\alpha^2 q^2/2) , \tag{116}$$

where $\alpha = \sqrt{\Omega/\hbar}$, with corresponding eigenenergies

$$\hbar\omega_n = \hbar\Omega(n + 1/2) \quad , n = 0, 1, 2, \ldots \tag{117}$$

Although this summarizes in principle the theory of the harmonic oscillator, it is useful to look at it from another point of view which provides further physical insight into its properties. We introduce two new operators a and $a^\dagger$ defined as

$$a = 1/\sqrt{2\hbar\Omega}\,(\Omega q + ip) \tag{118}$$

$$a^\dagger = 1/\sqrt{2\hbar\Omega}\,(\Omega q - ip) . \tag{119}$$

Inverting these expressions, we find the position and momentum

$$q = \sqrt{\hbar/2\Omega}\,(a + a^\dagger) \tag{120}$$

$$p = -i\sqrt{\hbar\Omega/2}\,(a - a^\dagger)\,. \tag{121}$$

Right now this is a purely mathematical exercise, but we shall see that a and $a^\dagger$ have important and simple physical interpretations. With the commutation relation $[q, p] = i\hbar$, we readily find that a and $a^\dagger$ obey the boson commutation relation

$$[a, a^\dagger] = 1\,. \tag{122}$$

Substituting Eqs. (120) and (121) into (113), we find the Hamiltonian in terms of a and $a^\dagger$ as

$$\boxed{\mathcal{H} = \hbar\Omega(a^\dagger a + \tfrac{1}{2})}\,. \tag{123}$$

In the Heisenberg picture, the time evolution of a and $a^\dagger$ are given by (Prob. 3-13)

$$\frac{da}{dt} = \frac{i}{\hbar}\,[\mathcal{H}, a] = -i\Omega a\,. \tag{124}$$

This has the solution

$$a(t) = a(0)\, e^{-i\Omega t}\,. \tag{125}$$

Similarly we find that

$$a^\dagger(t) = a^\dagger(0)\, e^{-i\Omega t}\,. \tag{126}$$

Consider now an energy eigenstate $|\mathcal{H}\rangle$ of the harmonic oscillator with eigenvalue $\hbar\omega$

$$\mathcal{H}|\mathcal{H}\rangle = \hbar\omega|\mathcal{H}\rangle\,, \tag{127}$$

and evaluate the energy of the state $|\mathcal{H}'\rangle = a|\mathcal{H}\rangle$. From Eq. (124), we have $\mathcal{H}a = a\mathcal{H} - \hbar\Omega a$, so that

$$\mathcal{H}a|\mathcal{H}\rangle = a\mathcal{H}|\mathcal{H}\rangle - \hbar\Omega a|\mathcal{H}\rangle = \hbar(\omega - \Omega)a|\mathcal{H}\rangle\,. \tag{128}$$

That is, $a|\mathcal{H}\rangle$ is again an eigenstate of the Hamiltonian, but of eigenenergy $\hbar\Omega$ lower than $|\mathcal{H}\rangle$. Because a lowers the energy, it is called an *annihilation* operator. Repeating the operation m times, we find

$$\mathcal{H} a^m |\mathcal{H}\rangle = \hbar(\omega - m\Omega)a^m |\mathcal{H}\rangle \ . \tag{129}$$

We can see that the lowest of these eigenvalues is positive as follows. For an arbitrary vector $|\phi\rangle$, the expectation value of $\mathcal{H}$ is

$$\langle\phi|\hbar\Omega(a^\dagger a + \tfrac{1}{2})|\phi\rangle = \hbar\Omega\langle\phi'|\phi'\rangle + \hbar\Omega/2,$$

where $|\phi'\rangle = a|\phi\rangle$. Calling the lowest eigenvalue $\hbar\omega_0$ with eigenstate $|0\rangle$, we have

$$a|0\rangle = 0 \tag{130}$$

and from Eq. (123)

$$\mathcal{H}|0\rangle = \hbar\Omega(a^\dagger a + 1/2)|0\rangle = \hbar\omega_0|0\rangle, \tag{131}$$

that is, the lowest-energy eigenvalue $\hbar\omega_0 = \hbar\Omega/2$.

Using the commutation relation (122), we find

$$\mathcal{H} a^\dagger |0\rangle = [a^\dagger \mathcal{H} + \hbar\Omega a^\dagger]|0\rangle = \hbar\Omega(1 + 1/2)a^\dagger |0\rangle \ ,$$

i.e., the eigenstate $|1\rangle$ has eigenvalue $\hbar\Omega(1+1/2)$. Because $a^\dagger$ raises the energy, it is called *a creation operator.* Substituting successively higher eigenstates into this equation, we find

$$\mathcal{H}(a^\dagger)^n |0\rangle = \hbar\Omega(n + 1/2)(a^\dagger)^n |0\rangle \ . \tag{132}$$

and hence that the eigenstate $|n\rangle \propto (a^\dagger)^n |0\rangle$ has the eigenvalue (117). To find the constant of proportionality, we note that

$$a|n\rangle = s_n |n-1\rangle \ , \tag{133}$$

where s_n is some scalar. This implies

$$\langle n|a^\dagger a|n\rangle = |s_n|^2 \langle n-1|n-1\rangle = |s_n|^2 \ . \tag{134}$$

Since $a^\dagger a|n\rangle = n|n\rangle$, this gives $s_n = \sqrt{n}$. Thus

$$\boxed{\begin{aligned} a|n\rangle &= \sqrt{n}\,|n-1\rangle & (135)\\ a^\dagger |n\rangle &= \sqrt{n+1}\,|n+1\rangle & (136)\end{aligned}}$$

With Eq. (132), this yields the normalized eigenstates

$$|n\rangle = \frac{1}{\sqrt{n!}}(a^\dagger)^n\,|0\rangle\ . \tag{137}$$

Since $a^\dagger a|n\rangle = n|n\rangle$, $a^\dagger a$ is called the *number operator*. It gives the number of quanta of excitation of the harmonic oscillator.

We can obtain the coordinate representation $u_0(q)$ of the ground state $|0\rangle$ by substituting Eq. (118) for a into Eq. (130) to find

$$(\Omega q + ip)u_0(q) = 0\ . \tag{138}$$

Using Eq. (7) in one dimension ($p = -i\hbar d/dq$), we find

$$\frac{d}{dq}u_0(q) = -\frac{\Omega}{\hbar}qu_0(q)\ ,$$

which has the normalized solution

$$u_0(q) = (\Omega/\pi\hbar)^{1/4}\,e^{-(\Omega/2\hbar)q^2}\ , \tag{139}$$

in agreement with Eq. (116). Similarly substituting Eq. (119) into (137) and using Eq. (139), we have

$$u_n(q) = \frac{1}{\sqrt{n!}}(a^\dagger)^n\,u_0(q) = \frac{1}{\sqrt{n!(2\hbar\Omega)^2}}\left(\Omega q - \hbar\frac{d}{dq}\right)^n u_0(q)\ , \tag{140}$$

which yields Eq. (116).

References

Cohen-Tannoudji, C., B. Diu, F. Laloë (1977), *Quantum Mechanics, Vol. I and II*, Wiley Interscience, New York.

Lamb, W. E., Jr. (1952), Phys. Rev **85**, 259.

Rabi, I. I. (1936), Phys. Rev. **49**, 324; (1937), Phys. Rev. **51**, 652.

Sargent, M. III, M. O. Scully, and W. E. Lamb, Jr. (1974), *Laser Physics*, Addison-Wesley Publishing Co., Reading, MA. This book introduces quantum mechanics from the point of view of lasers and applications.

For a good discussion of the **A·p** versus **E·r** forms of the electric-dipole interaction energy, see C. Cohen-Tannoudji, J. Dupont-Roc and G. Grynberg (1989), *Photons and Atoms, Introduction to Quantum Electrodynamics*, John-Wiley & Sons, New York.

Problems

3-1. Show that with Eq. (10), the completeness relation (11) is equivalent to the alternative definition that any function $f(\mathbf{r})$ can be expanded as

$$f(\mathbf{r}) = \sum_n d_n u_n(\mathbf{r}) \; .$$

3-2. Show that the operator matrix element of Eq. (15) written in terms of eigenfunctions has the same value as that of Eq. (41) written in terms of eigenvectors.

3-3. Compare and contrast Eq. (60) for the transition probability $|C_n^{(1)}(t)|^2$ with Eq. (2.16) for the phase matching of the generation of a difference frequency in a nonlinear medium.

3-4. What is the expectation value of the electric-dipole operator $e\mathbf{r}$ for a system in an energy eigenstate? Why?

3-5. Calculate the expectation value of the dipole moment operator $e\mathbf{r}$ for an atom with the wave function

$$\psi(\mathbf{r},t) = C_{210}u_{210}(\mathbf{r})e^{-i\omega_{210}t} + C_{100}u_{100}(\mathbf{r})e^{-i\omega_{100}t} \; .$$

Hint: Use spherical coordinates and write the position vector in the form

$$\mathbf{r} = \frac{1}{2}r\sin\theta\,[(\hat{x}-i\hat{y})e^{i\phi} + (\hat{x}+i\hat{y})e^{-i\phi}] + r\cos\theta\,\hat{z} \; .$$

3-6. What is the expectation value of the energy (6) for the wave function (12)? Is this value ever actually measured? What is the expectation value of the electric-dipole operator for a system in an energy eigenstate? Why?

3-7. Derive the wave-function Schrödinger equation (5) from the state-vector version (43) by appropriate projections.

3-8. Starting with the initial conditions $C_a(0) = 1$ and $C_b(0) = 0$, solve the equations of motion (99) and (100) to third-order in the electric-dipole interaction energy.

3-9. Given Eq. (106), *a*) What is the free evolution matrix? *b*) What are the matrices for π and $\pi/2$ pulses? *c*) How do you describe photon echo (a pulse, a free evolution, a second pulse, and a second free evolution) in terms of these matrices? (Don't solve, just set up) Ans: see Sec. 11-3.

3-10. Draw the level diagram for and write the general solution to the three-level equations of motion (taking $\mathcal{R}_0$ real)

$$\dot{C}_1 = \tfrac{1}{2} i\mathcal{R}_0 C_2$$
$$\dot{C}_2 = \tfrac{1}{2} i\mathcal{R}_0 (C_1 + C_3)$$
$$\dot{C}_3 = \tfrac{1}{2} i\mathcal{R}_0 C_2$$

3-11. Verify that the Pauli-spin operator commutation relations

$$[\sigma_x, \sigma_y] \equiv \sigma_x \sigma_y - \sigma_y \sigma_x = 2i\sigma_z$$
$$[\sigma_y, \sigma_z] = 2i\sigma_x ; \ [\sigma_z, \sigma_x] = 2i\sigma_y$$

Note that this can be written in the pseudovector form $\boldsymbol{\sigma} \mathbf{x} \boldsymbol{\sigma} = 2i\boldsymbol{\sigma}$.

3-12. Calculate the simple harmonic oscillator eigenfunction $u_1(q)$ using Eq. (140). Show that it is orthogonal to the ground-state eigenfunction $u_0(q)$.

3-13. Show that $[x, f(p)] = i\hbar df/dp$. Hint: use the momentum representation. Also show that $[p, e^{-\hbar K \partial/\partial p}] = -\hbar K$.

3-14. A useful alternative basis for the two-level system consists of *dressed states*, which are the eigenvectors of the matrix $\mathbf{M}$ in the equation of motion (104). Specifically, show that the eigenvectors satisfying the eigenvalue equation (taking $\mathcal{R}_0$ real)

$$\mathbf{M}\begin{pmatrix} u \\ v \end{pmatrix} = \begin{pmatrix} -\delta & \mathcal{R}_0 \\ \mathcal{R}_0 & \delta \end{pmatrix} \begin{pmatrix} u \\ v \end{pmatrix} = \lambda \begin{pmatrix} u \\ v \end{pmatrix}, \tag{141}$$

are given by

$$\begin{pmatrix} u_2 \\ v_2 \end{pmatrix} = \frac{1}{\sqrt{(\mathcal{R} - \delta)^2 + \mathcal{R}_0^{\,2}}} \begin{pmatrix} \mathcal{R} - \delta \\ \mathcal{R}_0 \end{pmatrix} = \begin{pmatrix} \cos\theta \\ \sin\theta \end{pmatrix} \tag{142}$$

for the eigenvalue $\lambda = \mathcal{R}$ and

$$\begin{pmatrix} u_1 \\ v_1 \end{pmatrix} = \begin{pmatrix} -\sin\theta \\ \cos\theta \end{pmatrix} \tag{143}$$

for $\lambda = -\mathcal{R}$. Hint: to find Eq. (142), in the bottom component equation given by Eq. (141) with $\lambda = \mathcal{R}$, equate u_2 to the coefficient of v_2 and vice versa, and normalize the resulting vector. Show that $\cos 2\theta = -\delta/\mathcal{R}$ and $\sin 2\theta = \mathcal{R}_0/\mathcal{R}$.

3-15. The dressed states of Eqs. (142) and (143) define the transformation matrix

$$U = \begin{pmatrix} \cos\theta & \sin\theta \\ -\sin\theta & \cos\theta \end{pmatrix}, \tag{144}$$

which diagonalizes the matrix **M** by

$$U\mathbf{M}U^{-1} = \begin{pmatrix} \mathcal{R} & 0 \\ 0 & -\mathcal{R} \end{pmatrix}.$$

This transformation matrix relates the dressed-state probability amplitudes to the "bare-state" amplitudes by

$$\begin{pmatrix} C_2(t) \\ C_1(t) \end{pmatrix} = U \begin{pmatrix} C_a(t) \\ C_b(t) \end{pmatrix}, \tag{145}$$

where the probability amplitudes C_2 and C_1 obey the equations of motion

$$\dot{C}_2 = \tfrac{1}{2} i\mathcal{R}C_2 \tag{146}$$

$$\dot{C}_1 = -\tfrac{1}{2} i\mathcal{R}C_1 . \tag{147}$$

Derive Eq. (106) by solving Eqs. (146) and (147), writing the initial values $C_1(0)$ and $C_2(0)$ in terms of $C_a(0)$ and $C_b(0)$ using Eq. (145), and then using the inverse of Eq. (145) to find $C_a(t)$ and $C_b(t)$.

3-16. Using the transformation matrix of Eq. (144), calculate the Pauli spin flip operators of Eq. (110) in the dressed-atom basis. Answer for σ_+:

$$\sigma_+ = \begin{pmatrix} \cos\theta\sin\theta & \cos^2\theta \\ -\sin^2\theta & -\cos\theta\sin\theta \end{pmatrix} . \tag{148}$$

3-17. Show by mathematical induction that

$$[a, a^{\dagger m}] = m(a^\dagger)^{m-1} , \tag{149}$$

$$[a^\dagger, a^m] = -ma^{m-1} . \tag{150}$$

Hint: show validity for $m = 1$, then write out commutator, assume relation is true for $m - 1$ and use Eq. (122).

3-18. Prove the operator identity

$$e^B X e^{-B} = X + [B, X] + \frac{1}{2!}[B, [B, X]] + \ldots + \frac{1}{n!}[B, [B,\ldots [B, X]\ldots]] + \ldots . \tag{151}$$

In particular, prove the Baker-Hausdorff

$$e^{A + B} = e^A e^B e^{-\frac{1}{2}[A, B]} \tag{152}$$

provided $[A, [A, B]] = [B, [A, B]] = 0$. Alternative method: show that the derivative of the operator $f(\lambda) = e^{\lambda B} X e^{-\lambda B}$ is $f'(\lambda) = [B, f(\lambda)]$. Using this derivative and its derivatives in turn, expand $f(\lambda)$ in a Maclaurin series. Setting $\lambda = 1$ in this series yields Eq. (151).

Chapter 4
MIXTURES AND THE DENSITY OPERATOR

In this chapter we generalize our treatment of two-level systems to include various kinds of damping. Some of these can be incorporated directly into the equations of motion for the probability amplitudes. However two important kinds cannot: upper to lower level decay, and more rapid decay of the electric dipole than the average level decay rate. For these two damping mechanisms, we need a more general description than can be provided by the state vector. Specifically, we need to consider systems for which we do not possess the maximum knowledge allowed by quantum mechanics. In other words, we do not know the state vector of the system, but rather the classical probabilities for having various possible state vectors. Such situations are described by the *density operator* ρ, which is a sum of projectors $|\psi_i\rangle\langle\psi_i|$ onto the possible state vectors $|\psi_i\rangle$, each weighted by a classical probability P_i.

We refer to a problem described by a single normalized state vector as a *pure case*, while a system described by a density operator consisting of an incoherent sum of pure-case contributions is a *mixed case* or a *mixture*. Such mixtures occur in particular when we consider only part of a total system. For example, with the two-level atom coupled to the vacuum field, we are often not interested in what happens to the field; instead we are only interested in what happens to the atom. Hence we write equations for the atom alone, and ignore what happens to the field. This "truncation" of the total problem automatically reduces our knowledge and usually results in a mixture. Chapter 14 discusses such problems in general.

The projectors $|\psi_i\rangle\langle\psi_i|$ used in the density operator formalism involve bilinear combinations of probability amplitudes, such as $C_{ia}C_{ia}^*$ and $C_{ia}C_{ib}^*$ for the case of two-level atoms. Although this might appear to be an added difficulty, it often simplifies mathematical analysis. To appreciate this point, note that the results of our discussions in Chap. 3, such as the probability of a transition or the value of the induced dipole moment are invariably expressed in terms of bilinear combinations of the amplitudes. In fact, the expectation value of any observable involves bilinear combinations.

The sum over all possible projectors performed in the density operator is analogous to the incoherent or partially incoherent addition of light fields. If two electric field amplitudes are added coherently, they interfere and the interference term is uniquely specified by the amplitudes. However if the addition is only partially coherent, the interference term is smaller than that specified by the individual amplitudes. For the two-level atom, the coherence term (electric dipole term) for the state vector $|\psi_i\rangle$ is given by $C_{ia}C_{ib}^*$. The polarization of the medium resulting from many such systems is given by a weighted sum of these individual $C_{ia}C_{ib}^*$. This sum is the matrix element $\rho_{ab} = \langle a|\rho|b\rangle$ of the density operator. For a number of systems with random phases between the upper and lower state probability amplitudes, ρ_{ab} tends to average to zero, even though the corresponding sum of probabilities $C_{ia}C_{ia}^*$ is unaffected by the random phases.

Section 4-1 shows how simple level decay can be incorporated into the two-level probability-amplitude equations of motion, and solves these equations for arbitrary tuning. This level decay causes the two-level probability amplitudes to decrease exponentially in time, thereby destroying the wave function's normalization. Such an unnormalized wave function actually describes a simple mixed case. Section 4-2 introduces the density matrix for two-level and more general cases. It also gives a simple derivation of the dipole decay constant, which due to collisions is in general larger than the average level decay constant. This phenomenon is an important manifestation of the partial coherence of a mixed case. Section 4-3 shows how the density matrix can be visualized in three dimensions by transforming to the Bloch vector. The Bloch-vector equations of motion provide an alternative to the Schrödinger equation and are popular in the literature and are particularly useful in studying coherent transients (Chap. 11). The results of this chapter are needed in our treatments of lasers, optical bistability, nonlinear spectroscopy, phase conjugation, optical instabilities, and coherent transients.

4-1. Level Damping

We have seen how the populations of excited atomic levels decay in time because of spontaneous emission. They can also decay because of collisions and other phenomena. In Fig. 4-1, we indicate one kind of such decay from both the a and the b levels, a situation which occurs in typical laser media. The loss of excited level probability corresponds to an increase of probability for lower-lying levels that we do not consider explicitly. The finite level lifetimes can be described very well by adding phenomenological decay terms to the equations of motion (3.99) and (3.100). We write

$$\dot{C}_a = -\tfrac{1}{2}(\gamma_a + i\delta)C_a + \tfrac{1}{2}i\mathscr{R}_0 C_b \tag{1}$$

$$\dot{C}_b = -\tfrac{1}{2}(\gamma_b - i\delta)C_b + \tfrac{1}{2}i\mathscr{R}_0 C_a\ , \tag{2}$$

where $\delta = \omega - \nu$ and the Rabi flopping frequency $\mathscr{R}_0 = \wp E_0/\hbar$ is assumed to be real. The factors of $\frac{1}{2}$ are included in the decay terms so that, for example, the probability $|C_a|^2$ decays as $\exp(-\gamma_a t)$ in the absence of E_0. The lifetimes are defined as the times at which the probabilities have decayed to $1/e$ of their original values. Hence they are given by the reciprocals of the decay constants γ_a and γ_b.

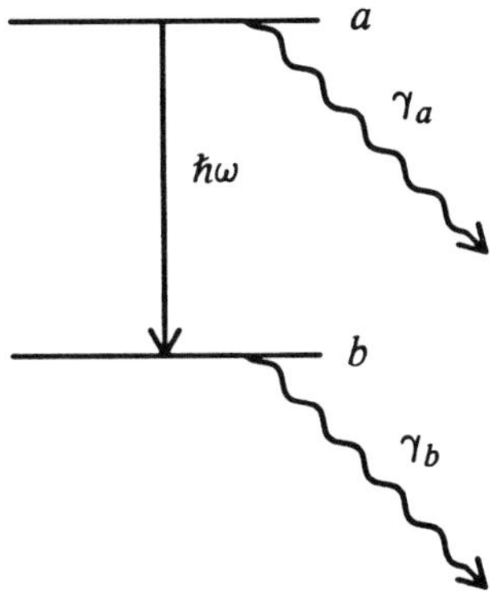

Fig. 4-1. Energy level diagram for two-level atom, showing decay rates γ_a and γ_b for the probabilities $|C_a|^2$ and $|C_b|^2$.

We can solve these equations by first-order perturbation theory as in Sec. 3-2, or by the more exact formulations of Sec. 3-3. We note here that for the simple case of equal decay constants, $\gamma_a = \gamma_b = \gamma$, the substitutions

$$\begin{aligned} C_b'(t) &= C_b(t)\, e^{\gamma t/2}\ , \\ C_a'(t) &= C_a(t)\, e^{\gamma t/2}\ , \end{aligned} \tag{3}$$

reduce Eqs. (1) and (2) to the undamped versions (3.99) and (3.100). The solutions for this damped case are, then, just those for the undamped case multiplied by the exponential decay factor $\exp(-\gamma t/2)$. In particular to lowest order in perturbation theory, the probability that stimulated absorption takes place changes from that given by Eq. (3.88) to the form

$$|C_a(t)|^2 \simeq |C_a{}^{(1)}(t)|^2 = \tfrac{1}{4}\mathcal{R}_0{}^2\, e^{-\gamma t} \left[\frac{\sin[(\omega-\nu)t/2]}{(\omega-\nu)/2}\right]^2 . \tag{4}$$

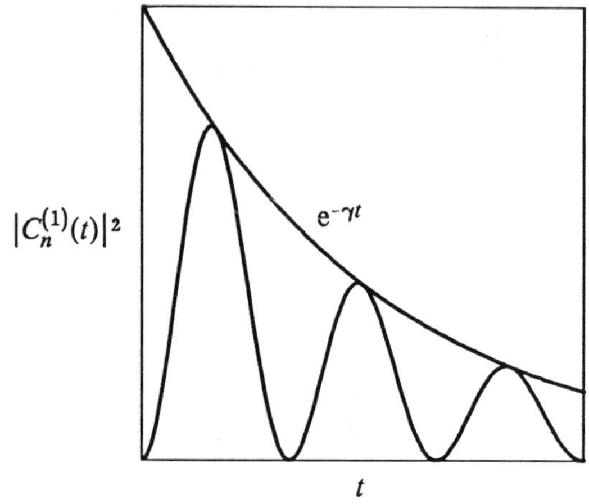

Fig. 4-2. Transition probability of Eq. (4) with $\gamma_a = \gamma_b = \gamma$.

This is illustrated in Fig. 4-2. Since the probability amplitudes decay away in time, the corresponding two-level wave function fails to remain normalized and as such describes a simple kind of mixed case.

We can find the spectral distribution for stimulated emission by starting with the atom in the upper level and calculating the total probability that it decays by spontaneous emission from the lower level to some other distant state(*s*). This follows because stimulated emission is needed to get from the *a* level to the *b* level in order that spontaneous emission from the *b* level can occur. The total probability of decay from the *b* level is given by

$$P_s = \gamma_b \int_0^\infty dt'\, |C_b(t')|^2 , \tag{5}$$

since $\gamma_b |C_b(t)|^2$ is the probability per unit time that the atom decays from the *b* level. For an initially excited atom, $|C_b(t)|^2$ is given for sufficiently small values of $|\mathcal{R}_0/\delta|^2$ by Eq. (4). Substituting this into Eq. (5), we find that the profile is Lorentzian with width γ, that is,

$$P_s = \frac{1}{2}\,\frac{\mathcal{R}_0^2}{(\omega-\nu)^2+\gamma^2}\,. \tag{6}$$

Of course, if the probability $|C_b(t)|^2$ fails to remain always much less than unity, Eq. (6) cannot be trusted. To generalize Eq. (6) accordingly and in anticipation of future need, we solve Eqs. (1) and (2) exactly.

As for Eq. (3.104), we seek solutions of the form $\mathbf{C} = \mathbf{C}(0)\,e^{i\lambda t}$ and find the eigenvalues

$$\lambda_{1,2} = \tfrac{1}{2}(i\gamma_{ab} \pm \mu)\,, \tag{7}$$

where the average decay rate constant

$$\gamma_{ab} = \tfrac{1}{2}(\gamma_a + \gamma_b)\,, \tag{8}$$

and where the complex Rabi flopping frequency

$$\mu = \sqrt{[\delta - \tfrac{1}{2}i(\gamma_a-\gamma_b)]^2 + \mathcal{R}_0^2} \tag{9}$$

Hence the solutions have the form

$$C_a(t) = \exp(-\tfrac{1}{2}\gamma_{ab}t)\,[C_a(0)\cos\tfrac{1}{2}\mu t + A\sin\tfrac{1}{2}\mu t]$$
$$C_b(t) = \exp(-\tfrac{1}{2}\gamma_{ab}t)\,[C_b(0)\cos\tfrac{1}{2}\mu t + B\sin\tfrac{1}{2}\mu t]\,.$$

Substituting these values into the equations of motion (1) and (2) at the time $t = 0$, we find

$$-\gamma_{ab}C_a(0) + \mu A = -(\gamma_a + i\delta)C_a(0) + i\mathcal{R}_0 C_b(0)$$
$$-\gamma_{ab}C_b(0) + \mu B = -(\gamma_b - i\delta)C_b(0) + i\mathcal{R}_0 C_a(0)\,.$$

This gives the integration constants $\mu A = -[\tfrac{1}{2}(\gamma_a-\gamma_b)+i\delta]C_a(0) + i\mathcal{R}_0C_b(0)$ and $\mu B = [\tfrac{1}{2}(\gamma_a-\gamma_b)+i\delta]C_b(0) + i\mathcal{R}_0C_a(0)$, that is,

$$\begin{pmatrix} C_a(t) \\ C_b(t) \end{pmatrix} = \begin{pmatrix} \cos\tfrac{1}{2}\mu t-\mu^{-1}[\tfrac{1}{2}(\gamma_a-\gamma_b)+i\delta]\sin\tfrac{1}{2}\mu t & i\mathcal{R}_0\mu^{-1}\sin\tfrac{1}{2}\mu t \\ i\mathcal{R}_0\mu^{-1}\sin\tfrac{1}{2}\mu t & \cos\tfrac{1}{2}\mu t+\mu^{-1}[\tfrac{1}{2}(\gamma_a-\gamma_b)+i\delta]\sin\tfrac{1}{2}\mu t \end{pmatrix}$$

$$\times\exp(-\tfrac{1}{2}\gamma_{ab}t)\begin{pmatrix} C_a(0) \\ C_b(0) \end{pmatrix}. \tag{10}$$

To find the corresponding profile for stimulated emission, we substitute $C_b(t)$ from Eq. (10) with the initial values $C_a(0) = 1$ and $C_b(0) = 0$ into Eq. (5) and find (with some algebra)

$$\boxed{P_s = \frac{1}{2}\frac{|\mathcal{R}_0|^2\gamma_{ab}/\gamma_a}{(\omega-\nu)^2+\gamma_b{}^2(1+I_0)}}\,, \tag{11}$$

where the dimensionless intensity I_0 is given by

$$I_0 = |\mathcal{R}_0|^2/\gamma_a\gamma_b\ . \tag{12}$$

Comparing Eqs. (6) and (11), we see that the frequency response of the atom to an applied field is broadened not only because of decay, but also because of saturation. This second effect is called *power broadening*. Note that on resonance ($\omega = \nu$), the stimulated emission profile of Eq. (11) approaches the constant value γ_{ab}/γ_b as I_0 becomes large.

4-2. The Density Matrix

The semiclassical situations discussed in the first part of this book (Chaps. 5 - 11) almost invariably use the Schrödinger picture (see Sec. 3-1). Hence for the two-level system we write the wave function with the Schrödinger picture amplitudes c_a and c_b defined by the general expression (3.21) for two levels, that is,

$$\psi(\mathbf{r},t) = c_a(t)u_a(\mathbf{r}) + c_b(t)u_b(\mathbf{r}) \tag{13a}$$

or equivalently by the state vector

$$|\psi(t)\rangle = c_a(t)|a\rangle + c_b(t)|b\rangle\ . \tag{13b}$$

The corresponding density operator is *defined* as the projector $\rho = |\psi\rangle\langle\psi|$ onto this state, and the density matrix elements $\rho_{ij} = \langle j|\rho|i\rangle$ are given by the bilinear products

$\rho_{aa} = c_a c_a{}^*$, probability of being in upper level

$\rho_{ab} = c_a c_b{}^*$, dimensionless complex dipole moment†

†Provided an electric-dipole transition is allowed between the a and b levels.

$$\rho_{ba} = c_b c_a^* = \rho_{ab}^* ,$$

$$\rho_{bb} = c_b c_b^* , \quad \text{probability of being in lower level} .$$

In matrix notation, the density operator ρ is therefore

$$\rho = \begin{bmatrix} c_a c_a^* & c_a c_b^* \\ c_b c_a^* & c_b c_b^* \end{bmatrix} = \begin{bmatrix} \rho_{aa} & \rho_{ab} \\ \rho_{ba} & \rho_{bb} \end{bmatrix} . \tag{14}$$

This density matrix is precisely the outer product

$$\rho = \begin{bmatrix} c_a \\ c_b \end{bmatrix} (c_a^* \quad c_b^*) .$$

In terms of the 2x2 density matrix of Eq. (14), the expectation value (3.23) of an operator $\mathcal{O}$ is given by

$$\langle \mathcal{O} \rangle = \rho_{aa} \mathcal{O}_{aa} + \rho_{ab} \mathcal{O}_{ba} + \rho_{ba} \mathcal{O}_{ab} + \rho_{bb} \mathcal{O}_{bb} . \tag{15}$$

In particular the dipole moment is given in the u_a, u_b basis by

$$\boxed{\langle er \rangle = \wp \rho_{ab} + \text{c.c.}} \tag{16}$$

Equation (15) and, more generally, Eq. (3.23) is just the trace of the matrix product $\rho\mathcal{O}$:

$$\langle \mathcal{O} \rangle = \sum_n \sum_m \rho_{nm} \mathcal{O}_{mn} = \sum_n (\rho\mathcal{O})_{nn} = \mathrm{tr}(\rho\mathcal{O}) . \tag{17}$$

We show at the end of this section that the expectation value of an operator is given by Eq. (17) even when the system is described by the most general density matrix.

We can derive the equations of motion for the elements of the density matrix from the Schrödinger equations of motion for the probability amplitudes $c_a(t)$ and $c_b(t)$ of Eq. (13). From Eq. (3.22) including phenomenological decays, we have

$$\dot{c}_a = -(i\omega_a + \gamma_a/2)c_a - i\hbar^{-1}\mathcal{V}_{ab} c_b \tag{18}$$

$$\dot{c}_b = -(i\omega_b+\gamma_b/2)c_b - i\hbar^{-1}\mathcal{V}_{ba}c_a\,, \tag{19}$$

where $\mathcal{V}_{ab} = \langle a|\mathcal{V}|b\rangle$. Proceeding one element at a time, we have

$$\begin{aligned}\dot{\rho}_{aa} &= \dot{c}_a c_a^{\,*} + c_a \dot{c}_a^{\,*}\\ &= (-i\omega_a c_a - \gamma_a c_a/2 - i\hbar^{-1}\mathcal{V}_{ab}c_b)c_a^{\,*}\\ &\qquad + c_a(i\omega_a c_a^{\,*} - \gamma_a c_a^{\,*}/2 + i\hbar^{-1}\mathcal{V}_{ba}c_b^{\,*})\\ &= -\gamma_a\rho_{aa} - [i\hbar^{-1}\mathcal{V}_{ab}\rho_{ba} + \text{c.c.}]\,.\end{aligned} \tag{20}$$

It is not surprising to find the complex conjugate in this equation, for probabilities are real. Similarly we find

$$\dot{\rho}_{bb} = -\gamma_b\rho_{bb} + [i\hbar^{-1}\mathcal{V}_{ab}\rho_{ba} + \text{c.c.}]\,. \tag{21}$$

Apart from the decay terms, this value is equal in magnitude and opposite in sign from that in Eq. (20). This expresses the fact that probability is transferred between the a and b levels by the interaction energy $\mathcal{V}_{ab}$. The off-diagonal element ρ_{ab} obeys the equation of motion

$$\begin{aligned}\dot{\rho}_{ab} &= \dot{c}_a c_b^{\,*} + c_a \dot{c}_b^{\,*}\\ &= (-i\omega_a c_a - \gamma_a c_a/2 - i\hbar^{-1}\mathcal{V}_{ab}c_b)c_b^{\,*}\\ &\qquad + c_a(i\omega_b c_b^{\,*} - \gamma_b c_b^{\,*}/2 + i\hbar^{-1}\mathcal{V}_{ba}c_a^{\,*})\\ &= -(i\omega+\gamma_{ab})\rho_{ab} + i\hbar^{-1}\mathcal{V}_{ab}(\rho_{aa} - \rho_{bb})\,.\end{aligned} \tag{22}$$

The example treated so far can be described equally well by an unnormalized state vector or by a density operator. However, the unnormalized state vector description becomes totally inadequate as soon as we consider more complex situations such as those encountered in the description of many-system phenomena. The phenomenological damping factors in Eqs. (1) and (2) actually result from the interaction of an atom with the many modes of the electromagnetic field. To treat more complicated cases, e.g., a decay of the dipole term ρ_{ab} independent of the decay of the level probabilities ρ_{ii}, the wave function becomes very cumbersome to use, or even incorrect. Such dipole decay results from an incoherent superposition of simple pure-case density matrices and can be cast in terms of system-reservoir coupling, a general approach followed in Chap. 14. Here we consider an important simple case.

Elastic collisions between atoms in a gas or between phonons and atoms in a solid can cause ρ_{ab} to decay separately from the diagonal elements. Specifically, if during an interaction the energy levels are merely shifted slightly without a change of state (e.g., distant van der Waals interactions), the decay rate for ρ_{ab} is increased without much change in γ_a and γ_b. This is due to the fact that the phase of the radiating atomic dipole is shifted in a somewhat random fashion, and the contributions of a collection of such dipoles tend to average to zero. We can gain a semiquantitative understanding of this process by considering the following discussion, couched in terms of phonon interactions in ruby.

The active atom in ruby is the Cr^{3+} ion, which is surrounded with O^{2-} atoms. At room temperature, all atoms vibrate, with the result that the energy levels in the Cr^{3+} ions experience random Stark shifts. For simplicity we assume that this phenomenon can be expressed mathematically by adding a random shift $\delta\omega(t)$ to the energy difference ω. Ignoring other perturbations for simplicity, we can write the equation of motion for the off-diagonal element ρ_{ab} as

$$\dot{\rho}_{ab} = -[i\omega + i\delta\omega(t) + \gamma_{ab}]\,\rho_{ab}\ . \tag{23}$$

Integrating this formally, we have

$$\rho_{ab}(t) = \rho_{ab}(0)\exp\left[-(i\omega+\gamma_{ab})t - i\int_0^t dt'\,\delta\omega(t')\right]. \tag{24}$$

We now perform *a classical* ensemble average of Eq. (24) over the random variations in $\delta\omega(t)$. This average affects only the $\delta\omega(t)$ factor. Expanding the second part of the exponential term by term, we have

$$\exp\left[-i\int_0^t dt'\,\delta\omega(t')\right] = 1 - i\int_0^t dt'\,\delta\omega(t') - \frac{1}{2}\int_0^t dt'\int_0^t dt''\,\delta\omega(t')\,\delta\omega(t'')$$
$$+ \frac{(-)^n}{(2n)!}\prod_{i=1}^{2n}\int_0^t dt_i\,\delta\omega(t_i) + \dots\ . \tag{25}$$

The function $\delta\omega(t)$ is as often positive as negative, as suggested in Fig. 4-3. Hence the ensemble average $\langle\delta\omega(t)\rangle$ is zero (a frequency shift as well as damping can occur, which would change ω). Furthermore, averages of products $\langle\delta\omega(t)\,\delta\omega(t')\rangle$ are zero as well, unless $t\simeq t'$, in which case the product is mostly positive since $(-1)^2=1$. Assuming that variations in $\delta\omega(t)$ are rapid compared to other changes (which occur in times like $1/\gamma_{ab}$), we take

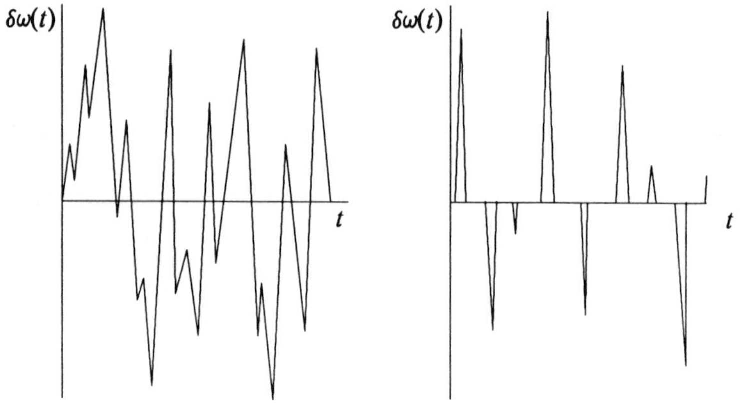

Fig. 4-3. (*a*) Possible time dependence of random frequency shift $\delta\omega(t)$ imposed upon frequency difference ω for Cr^{3+} in Al_2O_3 lattice. This shift could also occur for atoms in gaseous state due to "soft" (elastic) collisions. Because $\delta\omega(t)$ is as often positive as negative, the integral $\int_0^t dt'\langle\delta\omega(t')\rangle$ vanishes. (*b*) Possible dependence for collision fluctuations. Same random characteristics hold in our simple model.

$$\langle\delta\omega(t)\,\delta\omega(t')\rangle = 2\gamma_{ph}\,\delta(t-t')\ , \tag{26}$$

where the subscript *ph* stands for *ph*ase. This infinitely short memory approximation is called the *Markoff approximation.*

Similarly (for Gaussian statistics) the $2n$th correlation $\langle\delta\omega(t_1)...\delta\omega(t_{2n})\rangle$ is given by the sum of all distinguishable products of pairs like that in Eq. (26), for only when the random functions coincide in pairs is the entire product positive with nonvanishing ensemble average. The number of combinations of $2n$ terms in pairs is given by $\begin{pmatrix}2n\\2\end{pmatrix}$, that for the remaining $2n-2$ terms is $\begin{pmatrix}2n-2\\2\end{pmatrix}$, and so forth. Hence the total number of distinguishable ways of breaking $2n$ terms into products of n pairs is

$$\frac{1}{n!}\begin{pmatrix}2n\\2\end{pmatrix}\begin{pmatrix}2n-2\\2\end{pmatrix}\cdots\begin{pmatrix}2\\2\end{pmatrix} = \frac{(2n)!}{n!2^n}\ . \tag{27}$$

This gives for the 2*n*th term

$$\frac{(-1)^n}{(2n)!}(2\gamma_{ph})^n\frac{(2n)!}{2^n n!}\prod_{i=1}^{n}\int_0^t dt_{2i}\int_0^t dt_{2i-1}\,\delta(t_{2i}-t_{2i-1})=\frac{(-\gamma_{ph}t)^n}{n!}\,.$$

The (2*n*+1)th term vanishes since it is given by the sum of products of *n* pairs multiplied by a lone random function with zero average. Therefore

$$\exp\left[-i\int_0^t dt'\,\delta\omega(t')\right]=\sum_{n=1}^{\infty}\frac{(-\gamma_{ph}t)^n}{n!}=e^{-\gamma_{ph}t}\,,$$

which gives for the classical average over collisions of Eq. (24)

$$\overline{\rho}_{ab}(t)=\overline{\rho}_{ab}(0)\,e^{-(i\omega+\gamma_{ab}+\gamma_{ph})t}\,. \tag{28}$$

For typographical simplicity, from now on we drop the average bar, but we should always remember that ρ *typically includes such classical averages in addition to the quantum mechanical average.*

Defining the new decay rate

$$\gamma=\gamma_{ab}+\gamma_{ph}\,, \tag{29}$$

differentiating Eq. (28), and including $\mathcal{V}_{ab}$, we have the averaged equation of motion

$$\dot{\rho}_{ab}=-(i\omega+\gamma)\rho_{ab}+i\hbar^{-1}\mathcal{V}_{ab}(\rho_{aa}-\rho_{bb}) \tag{30}$$

$$\dot{\rho}_{aa}=-\gamma_a\rho_{aa}-[i\hbar^{-1}\mathcal{V}_{ab}\rho_{ba}+\text{c.c.}] \tag{20}$$

$$\dot{\rho}_{bb}=-\gamma_b\rho_{bb}+[i\hbar^{-1}\mathcal{V}_{ab}\rho_{ba}+\text{c.c.}] \tag{21}$$

where for convenience we repeat the equations of motion for ρ_{aa} and ρ_{bb}. We use Eq. (30) in place of Eq. (22) in our calculations. Equation (30) is an average equation with respect to collisions, whereas Eq. (23) includes fluctuations due to collisions. A fully quantal treatment reveals that the semiclassical Eqs. (20) - (22), too, are average equations, in which vacuum fluctuations have been averaged over, leading to the decay terms.

It is interesting to note that at low temperatures, the Cr^{3+} ions "freeze" irregularly into place in the crystal lattice and the decay rate γ becomes smaller. Different ions are subject to different Stark shifts and hence have different resonant frequencies. Hence the atomic medium as a whole responds to a range of frequencies considerably larger than that for a single

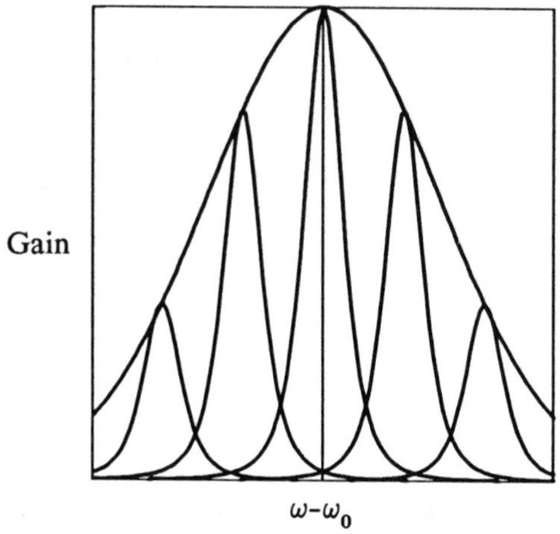

Fig. 4-4. Graph showing individual atomic response curves superimposed on inhomogeneously broadened line for possible laser medium. Here the homogeneous and inhomogeneous contributions to the linewidth are 150 MHz and 1000 MHz, respectively. Hence this medium is primarily inhomogeneously broadened.

atom, as suggested in Fig. 4-4. This kind of response is called *inhomogeneously broadened*, as contrasted to the kind found in ruby at higher temperatures (called *homogeneously broadened*, inasmuch as all active atoms experience statistically identical collisions and have essentially the same resonance frequency). The two kinds of broadening overlap to some degree in a real medium, but one kind is often responsible for most of the classical absorption (or emission) linewidth. An intermediate situation is met when the temperature of ruby is somewhere between that of liquid helium and room temperature, and the linewidth is due to approximately equal homogeneous and inhomogeneous contributions. We emphasize right away that the two sources of broadening are very different physically, the inhomogeneous one being a dynamical influence that can be effectively reversed (e.g., photon echo of Sec. 11-3), whereas the homogeneous is an irreversible influence. In Chap. 5 we see in greater detail just how differently the two sources are represented mathematically.

A transition similar to that occurring in ruby for temperature changes takes place in gaseous media for pressure changes. Although gas atoms such as neon generally have the same resonance frequencies in their rest frames, they see Doppler-shifted electric field frequencies (as discussed in Chap. 6) and hence respond as a group to a range of frequencies, that is, are inhomogeneously broadened. At low pressures, the linewidth of a single atom is almost completely due to spontaneous emission and is usually small compared to average Doppler shifts. As the pressure is increased, however, collisions broaden the atomic response homogeneously and ultimately mask out the Doppler effect altogether. We see in Chap. 6 on lasers, Chap. 8 on saturation spectroscopy, and Chap. 11 on coherent transients that the degree to which the atomic response is inhomogeneously broadened can be determined experimentally by measuring, for example, intensity *vs* tuning profiles in gases and photon echoes in general. Note that although the considerations discussed here apply qualitatively to many laser media, interesting counterexamples do exist, e.g., Dicke narrowing in gases and the Mössbauer effect.

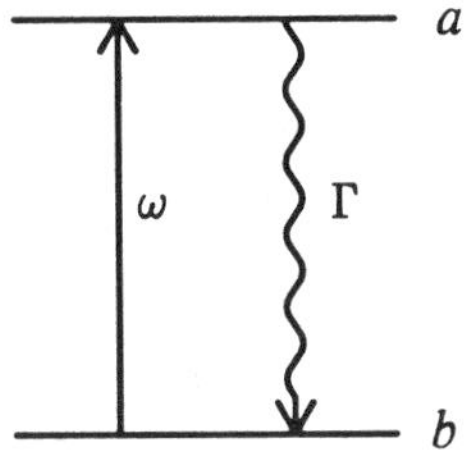

Fig. 4-5. Energy level diagram for two-level atom with upper-to-lower level decay with the ground state as the lower level. Unlike the level scheme in Fig. 4-1, only one level decay rate occurs.

The level decay scheme described in Sec. 4-1 is useful for many laser media, for which both levels are excited. For spectroscopy and for the ruby laser, the upper level decays to a ground-state lower level as depicted in Fig. 4-5. For that case, the ρ_{bb} equation of motion changes from (21) to

$$\dot{\rho}_{bb} = -\dot{\rho}_{aa} = \Gamma\rho_{aa} + [i\hbar^{-1}\mathcal{V}_{ab}\rho_{ba} + \text{c.c.}]\,, \tag{31}$$

where we use Γ for the upper-to-lower-level decay constant. Since the two levels are the only ones in the problem, the conservation of probability gives

$$\rho_{aa} + \rho_{bb} = 1 , \tag{32}$$

which is not the case for the two-level system in Sec. 4-1. Using Eq. (32), we can write Eq. (31) as

$$\dot{\rho}_{bb} = -\Gamma(\rho_{bb} - 1) + [i\hbar^{-1}\mathcal{V}_{ab}\rho_{ba} + \text{c.c.}] . \tag{33}$$

Note that due to the presence of the -1, this equation has no counterpart in wave function notation. Similar "excitation rate" constants occur for the level scheme of Sec. 4-1 when excitation processes are included as in Chap. 5. Equations (31) and (33) show that both ρ_{aa} and ρ_{bb} relax with the same rate constant Γ, but ρ_{bb} relaxes to the value 1, while ρ_{aa} relaxes to 0.

We now consider generalizations of our two-level density matrix to more levels and other statistical mixtures. Typically we do not know the state vector for a many-particle problem. Instead we know only certain statistical properties. These properties are conveniently incorporated into the density matrix formalism by definition of the general density operator

$$\boxed{\rho = \sum_{\psi} P_{\psi} \, |\psi\rangle\langle\psi|} \, . \tag{34}$$

Here the summation over ψ is the result of classical averages and can be discrete or continuous. P_{ψ} is that fraction of the systems represented by the state vector $|\psi\rangle$. The summation can take the form of several summations and integrals.

For state vectors of the form

$$|\psi\rangle = \sum_{n} c_n \, |n\rangle, \tag{35}$$

Eq. (34) reduces to

$$\rho = \sum_{\psi} P_{\psi} \sum_{n} \sum_{m} c_n c_m^{*} \, |n\rangle\langle m| = \sum_{n} \sum_{m} \rho_{nm} \, |n\rangle\langle m| \, . \tag{36}$$

The quantum mechanical expectation value of an operator $\mathcal{O}$ is still given by

$$\boxed{\langle \mathcal{O} \rangle = \mathrm{tr}(\rho\mathcal{O})}\,, \tag{37}$$

for in terms of P_ψ,

$$\langle \mathcal{O} \rangle = \sum_\psi P_\psi \langle\psi|\mathcal{O}|\psi\rangle = \sum_\psi P_\psi \sum_k \langle\psi|\mathcal{O}|k\rangle\,\langle k|\psi\rangle$$

$$= \sum_k \sum_\psi P_\psi \langle k|\psi\rangle\langle\psi|\mathcal{O}|k\rangle = \sum_k (\rho\mathcal{O})_{kk} = \mathrm{tr}(\rho\mathcal{O})\,.$$

The equation of motion of the density operator (and hence that of its matrix elements) is easily determined from the Schrödinger equation

$$\dot{\rho} = \sum_\psi P_\psi \left[|\dot{\psi}\rangle\langle\psi| + |\psi\rangle\langle\dot{\psi}|\right]$$

$$= -\frac{i}{\hbar} \sum_\psi P_\psi \left[\mathcal{H}|\psi\rangle\langle\psi| - |\psi\rangle\langle\psi|\mathcal{H}\right]$$

which in terms of the commutator $[\mathcal{H}, \rho] = \mathcal{H}\rho - \rho\mathcal{H}$ is given by

$$\boxed{\dot{\rho} = -\frac{i}{\hbar}\,[\mathcal{H}, \rho]}\,. \tag{38}$$

Equation (38) is valid only for collections of state vectors all having the same Hamiltonian $\mathcal{H}$. In much of our semiclassical discussions (Chaps. 5 - 11), we add contributions from atoms excited at different times, places, and states. Inasmuch as all these systems have the same equations of motion, we can use (38) for a density matrix (34) that includes summations over the times, places, and states of excitation (see Eq. (5.6)). By dealing with this sum of density matrices directly, we save considerable effort over both the single-system density matrix approach and the state vector method. Summations of statistical averages for systems with different equations of motion must be performed after the equations have been integrated. Note that the *nm*th matrix element of Eq. (38) is

$$\dot{\rho}_{nm} = -\frac{i}{\hbar}\langle n|\mathcal{H}\rho - \rho\mathcal{H}|m\rangle$$

$$= -\frac{i}{\hbar}\sum_k [\langle n|\mathcal{H}|k\rangle\langle k|\rho|m\rangle - \langle n|\rho|k\rangle\langle k|\mathcal{H}|m\rangle]$$

$$= -\frac{i}{\hbar}\sum_k [\mathcal{H}_{nk}\rho_{km} - \rho_{nk}\mathcal{H}_{km}]\,. \tag{39}$$

This formula is useful in the treatment of many-level problems.

A basic property of the more general density matrix ρ_{nm} of Eq. (36) is that the off-diagonal elements can be smaller than those given by a single state vector. Whereas the coherence between levels is preserved in the superposition given by the state vector, averages over the individual state vector coherences given by Eq. (36) can reduce their values substantially. Hence we see in Eq. (30) that the off-diagonal element ρ_{ab} decays with the value γ of Eq. (29), which leads to a smaller magnitude for ρ_{ab} than that given by $(\rho_{aa}\rho_{bb})^{1/2}$.

To see this important averaging process at work in a particularly simple context, consider a set of two-level state vectors $|\psi\rangle_j$ all having the same probabilities $|C_a|^2$ and $|C_b|^2$, but having a random relative phase ϕ_j between the probability amplitudes:

$$|\psi\rangle_j = c_a|a\rangle + e^{i\phi_j}c_b|b\rangle\,. \tag{40}$$

For this problem, the probability P_j of the jth state vector is $1/N$ given N systems, since all phase angles ϕ_j are equally likely. The off-diagonal element ρ_{ab} given by Eq. (36) for this problem is

$$\rho_{ab} = \sum_j P_j\, c_{aj}\, c_{bj}{}^* = c_a c_b{}^* \frac{1}{N}\sum_{j=1}^{N} e^{-i\phi_j} = c_a c_b{}^* \frac{1}{2\pi}\int_0^{2\pi} d\phi e^{-i\phi} = 0,$$

that is, the atomic coherence between the levels a and b is completely washed out by the averaging process. This destruction of the atomic coherence is analogous to the destruction of coherence in a light beam, for which a nonzero incoherent intensity is incapable of producing an interference fringe when passed through another light beam.

4-3. Vector Model of Density Matrix

It is possible to make the two-level density matrix equations of motion (20), (21), and (30) resemble those for a magnetic dipole undergoing precession in a magnetic field. This approach has value not only in solving the equations, but also in providing a physical picture of the density matrix in motion. The equations we derive here are equivalent to the Bloch equations for a spin 1/2 (two-level) system appearing in nuclear magnetic resonance. We note at the outset that atoms are only approximated by two levels and are typically characterized by three decay constants γ_a, γ_b and γ_{ab}, whereas the spin 1/2 Bloch equations have only two. In particular, laser media usually have $\gamma_b >> \gamma_a$, a limit in which these equations can be a poor approximation. On the other hand, for several useful situations (e.g., upper-to-ground-lower-level decay of Eq. (31)), a single level constant is a good approximation so that the Bloch model is accurate and may be easier to use. The solutions obtained are used in the study of coherent transients in Chap. 11.

We suppose that the perturbing energy $\mathcal{V}_{ab}$ is given in the rotating wave approximation by Eq. (3.97). We further go into an (not *the*) interaction picture by multiplying both sides of Eq. (30) by $e^{i\nu t}$, thereby obtaining

$$\frac{d}{dt}[\rho_{ab}\, e^{i\nu t}] = -[\gamma + i(\omega-\nu)]\rho_{ab}\, e^{i\nu t} - i(\wp E_0/2\hbar)\,(\rho_{aa} - \rho_{bb}). \tag{41}$$

This is the same transformation that we used for the wave function in Eq. (3.98). We introduce the real quantities

$$U = \rho_{ab} e^{i\nu t} + \text{c.c.} \tag{42}$$

$$V = i\rho_{ab} e^{i\nu t} + \text{c.c.} \tag{43}$$

$$W = \rho_{aa} - \rho_{bb}, \tag{44}$$

in terms of which

$$\rho_{ab} = \frac{1}{2}(U-iV)\, e^{-i\nu t}. \tag{45}$$

These quantities vary little in an optical frequency period and are the components of the vector **U** given by

$$\mathbf{U} = U\hat{\mathbf{e}}_1 + V\hat{\mathbf{e}}_2 + W\hat{\mathbf{e}}_3 = \mathrm{tr}(\rho'\boldsymbol{\sigma}), \tag{46}$$

where $\boldsymbol{\sigma}$ is the Pauli spin tensor

$$\boldsymbol{\sigma} = \hat{\mathbf{e}}_1\sigma_x + \hat{\mathbf{e}}_2\sigma_y + \hat{\mathbf{e}}_3\sigma_z,$$

and ρ' is an interaction-picture density matrix defined by

$$\rho' = \begin{bmatrix} \rho_{aa} & \rho_{ab}e^{i\nu t} \\ \rho_{ab}e^{-i\nu t} & \rho_{bb} \end{bmatrix}. \tag{47}$$

Taking derivatives of Eqs. (42), (43), and (44) and using Eqs. (41) and (31), we find the Bloch Equations (in this section, we take $\mathcal{R}_0$ to be real)

$$\dot{U} = -\delta V - U/T_2 \tag{48}$$

$$\dot{V} = \delta U - V/T_2 + \mathcal{R}_0 W \tag{49}$$

$$\dot{W} = -(W+1)/T_1 - \mathcal{R}_0 V \tag{50}$$

where $\delta = \omega - \nu$ and $\mathcal{R}_0$ is the (real) Rabi flopping frequency $\wp E_0/\hbar$. Here the $\dot{W}$ equation is written for the case of upper to lower level decay. This is a situation where a single level decay constant occurs (as in the spin 1/2 two-level system). In keeping with the literature on NMR we use T_2 for the induced dipole decay time in place of $1/\gamma$, and we use T_1 for the probability difference decay time. For the upper-to-lower-level decay of Eq. (31), $T_1 = 1/\Gamma$. As Problem 4-8 discusses, when two level decay constants γ_a and γ_b appear, then T_1 is the average level lifetime

$$T_1 = \frac{1}{2}\left[\frac{1}{\gamma_a} + \frac{1}{\gamma_b}\right], \tag{51}$$

but Eq. (50) is then only approximately correct, since it neglects the term $(\gamma_a - \gamma_b)(\rho_{aa} + \rho_{bb})/2$.

Comparing Eqs. (48) and (49) to their classical counterparts (1.49) and (1.50), we see that quantum mechanically a third variable W enters, which is the probability difference between being in the upper and lower levels. This difference variable has no classical meaning since classically no levels are involved. It makes the induced polarization of the two-level system nonlinear in contrast to the linear classical charge on a spring of Sec. 1-3.

If $T_1=T_2=\gamma^{-1}$, the three equations (48) - (50) have the simple, combined form

$$\dot{\mathbf{U}} = -\gamma(\mathbf{U} + \hat{\mathbf{e}}_3) + \mathbf{U}\times\boldsymbol{\mathscr{R}}, \tag{52}$$

where the effective Rabi precession "field" $\boldsymbol{\mathscr{R}}$ is given by

$$\boldsymbol{\mathscr{R}} = \mathscr{R}_0\hat{\mathbf{e}}_1 - \delta\hat{\mathbf{e}}_3\,. \tag{53}$$

The time dependence of such a vector is well known from classical mechanics. The **U** vector precesses clockwise about the effective field $\boldsymbol{\mathscr{R}}$ with diminishing magnitude. The precessions for resonance and slightly off resonance are depicted in Fig. 4-6. On resonance, **U** precesses about the $\hat{\mathbf{e}}_1$ axis in a major circle.

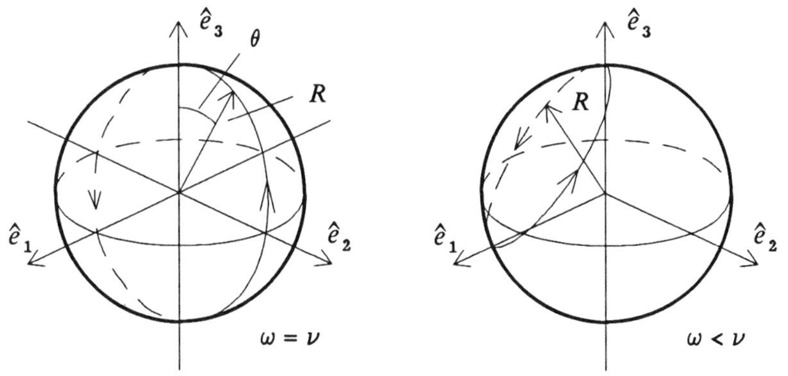

(*a*) (*b*)

Fig. 4-6. (*a*) For central tuning ($\omega=\nu$) and $\wp<0$ (for an electron), **U** precesses clockwise about $\hat{\mathbf{e}}_1$ at the angular frequency $\mathscr{R}_0$, as determined by Eq. (52). The electric-dipole interaction energy is directed along the $\hat{\mathbf{e}}_1$ axis in the rotating-wave approximation. For an initial $W(0)=-1$ (system in lower level), $W(\pi/\mathscr{R}_0) = 1$, that is, the atom makes a transition in a time $\pi/\mathscr{R}_0$ in agreement with the Rabi flopping formula (3.102). (*b*) For some detuning ($\omega\neq\nu$), **U** acquires a nonzero $\hat{\mathbf{e}}_1$ component and a complete transition (e.g., from upper to lower level) never occurs. The **U** precesses clockwise about the effective "field" $\boldsymbol{\mathscr{R}}$ of (53) with generalized Rabi flopping frequency $\mathscr{R}$, thus tracing out a cone in the rotating frame.

Physically **U** pointing along $\hat{\mathbf{e}}_3$ (W=1, U=V=0) represents a system initially in its upper level (ρ_{aa}=1, ρ_{bb}=0). Similarly **U** points along -$\hat{\mathbf{e}}_3$ for a system in its lower level. The on-resonance precession about $\hat{\mathbf{e}}_1$ is just a vector embodiment of the Rabi flopping of Eqs. (3.102) and (3.103).

To obtain an analytic feel for the time development of the Bloch vector, we solve Eqs. (48) through (50) first with no decay ($T_1 = T_2 = \infty$), second with free evolution (zero applied field), and third, in steady state. These solutions are used in the discussions of optical nutation (Sec. 11-1) and free induction decay (Sec. 11-2), photon echo (Sec. 11-3), and Ramsey fringes (Sec. 11-4). In the absence of decay, Eqs. (48) through (50) reduce to

$$\dot{U} = -\delta V \tag{54}$$

$$\dot{V} = \delta U + \mathcal{R}_0 W \tag{55}$$

$$\dot{W} = -\mathcal{R}_0 V , \tag{56}$$

In particular on resonance (δ=0), we find

$$\dot{U} = 0 \tag{57}$$

$$\dot{V} = \mathcal{R}_0 W . \tag{58}$$

Just as for the probability amplitudes C_a and C_b, Eqs. (56) and (58) are sinusoidal. For arbitrary initial conditions, we can write the solution as

$$\begin{pmatrix} U(t) \\ V(t) \\ W(t) \end{pmatrix} = \begin{pmatrix} 1 & 0 & 0 \\ 0 & \cos\mathcal{R}_0 t & -\sin\mathcal{R}_0 t \\ 0 & \sin\mathcal{R}_0 t & \cos\mathcal{R}_0 t \end{pmatrix} \begin{pmatrix} U(0) \\ V(0) \\ W(0) \end{pmatrix} , \tag{59}$$

which is a Rabi rotation of $\mathcal{R}_0 t$ radians about $\hat{\mathbf{e}}_1$. Knowing this we can solve the detuned case of Eqs. (54) - (56) by first rotating our coordinate system about $\hat{\mathbf{e}}_2$ so that $\hat{\mathbf{e}}_1'$ points along the effective field (53). This rotation is determined by the angle

$$\chi = \tan^{-1}(\delta/\mathcal{R}_0) . \tag{60}$$

In this new coordinate system we rotate an angle $\mathcal{R}t$ about $\hat{\mathbf{e}}_1'$, where the generalized Rabi flopping frequency is given by

$$\mathcal{R} = \sqrt{\delta^2 + \mathcal{R}_0^{\,2}} , \tag{61}$$

and rotate back to the original coordinate system. This set of three rotations is given by the product of the three matrices

$$\begin{bmatrix} \cos\chi & 0 & \sin\chi \\ 0 & 1 & 0 \\ -\sin\chi & 0 & \cos\chi \end{bmatrix} \begin{bmatrix} 1 & 0 & 0 \\ 0 & \cos\mathcal{R}t & -\sin\mathcal{R}t \\ 0 & \sin\mathcal{R}t & \cos\mathcal{R}t \end{bmatrix} \begin{bmatrix} \cos\chi & 0 & -\sin\chi \\ 0 & 1 & 0 \\ \sin\chi & 0 & \cos\chi \end{bmatrix} . \tag{62}$$

Multiplying these matrices out, we find the general solution without decay

$$\begin{bmatrix} U(t) \\ V(t) \\ W(t) \end{bmatrix} = \begin{bmatrix} \dfrac{\mathcal{R}_0{}^2+\delta^2\cos\mathcal{R}t}{\mathcal{R}^2} & -\dfrac{\delta}{\mathcal{R}}\sin\mathcal{R}t & -\dfrac{\delta\mathcal{R}_0}{\mathcal{R}^2}(1-\cos\mathcal{R}t) \\ \dfrac{\delta}{\mathcal{R}}\sin\mathcal{R}t & \cos\mathcal{R}t & \dfrac{\mathcal{R}_0}{\mathcal{R}}\sin\mathcal{R}t \\ -\dfrac{\delta\mathcal{R}_0}{\mathcal{R}^2}(1-\cos\mathcal{R}t) & -\dfrac{\mathcal{R}_0}{\mathcal{R}}\sin\mathcal{R}t & \dfrac{\mathcal{R}_0{}^2\cos\mathcal{R}t+\delta^2}{\mathcal{R}^2} \end{bmatrix} \begin{bmatrix} U(0) \\ V(0) \\ W(0) \end{bmatrix} \tag{63}$$

In particular, if $U(0) = V(0) = 0$, this gives

$$U(t) = (\mathcal{R}_0\delta/\mathcal{R}^2)\, W(0)\, [\cos\mathcal{R}t - 1] \tag{64}$$

$$V(t) = (\mathcal{R}_0/\mathcal{R})\, W(0)\, \sin\mathcal{R}t \tag{65}$$

$$W(t) = W(0)\, [1 + \mathcal{R}_0{}^2\mathcal{R}^{-2}\, (\cos\mathcal{R}t - 1)] . \tag{66}$$

These equations describe the motions depicted in Fig. 4-6 based on our knowledge of spin precession. In particular for exact resonance (δ=0), this solution reduces to Rabi flopping in the $\hat{\mathbf{e}}_2$ $\hat{\mathbf{e}}_3$ plane with $U(t) = 0$, $V(t) = W(0)\sin\mathcal{R}_0 t$, and $W(t) = W(0)\cos\mathcal{R}_0 t$, results which also follow immediately from Eq. (59). In optical nutation (Sec. 11-1), we integrate Eqs. (64) and (65) over δ weighted by an inhomogeneous broadening distribution (Sec. 5-2). This leads to a superposition of Rabi floppings that interfere destructively for times on the order of or greater than the reciprocal of the inhomogeneous broadening linewidth.

Another useful solution to the Bloch equations particularly for inhomogeneously broadened problems like free induction decay and photon echo is the free evolution with no applied field ($\mathcal{R}_0 = 0$). For simplicity we also take $T_1 = \infty$. In this limit, the $\dot{U}$ Eq. (48) remains the same, $W(t)$ is a constant of the motion, and Eq. (49) reduces to

$$\dot{V} = \delta U - V/T_2 \, . \tag{67}$$

The solution of these equations is also given by a rotation, namely

$$\begin{pmatrix} U(t) \\ V(t) \\ W(t) \end{pmatrix} = \begin{pmatrix} \cos\delta t\, e^{-\gamma t} & -\sin\delta t\, e^{-\gamma t} & 0 \\ \sin\delta t\, e^{-\gamma t} & \cos\delta t\, e^{-\gamma t} & 0 \\ 0 & 0 & 1 \end{pmatrix} \begin{pmatrix} U(0) \\ V(0) \\ W(0) \end{pmatrix}, \tag{68}$$

where we use $\gamma \equiv 1/T_2$.

Although the general time dependent solution of the Bloch equations (48) through (50) including T_1 and T_2 must be expressed in terms of the roots of a cubic equation, the steady state answer is both easy to obtain and useful for understanding the phenomenon of free induction decay (Sec. 11-2). For this, we simply set $\dot{U} = \dot{V} = \dot{W} = 0$ in Eqs. (48) through (50), and solve the resulting algebraic equations. From (48), we find

$$U = -\delta T_2 V\,. \tag{69}$$

Substituting this into Eq. (49), we obtain

$$T_2 \mathcal{R}_0 W = (\delta^2 T_2^{\,2} + 1)V\,. \tag{70}$$

Combining this with Eq. (50), we find the probability difference

$$W = -\frac{1}{1 + I\mathcal{L}(\delta)}\,, \tag{71}$$

where the dimensionless intensity

$$I = \mathcal{R}_0^{\,2} T_1 T_2 \tag{72}$$

and the dimensionless Lorentzian

$$\mathcal{L}(\delta) = \frac{1}{1 + \delta^2 T_2^{\,2}}\,. \tag{73}$$

Note that the dimensionless intensity (72) is the ratio of the square of Rabi flopping frequency $\mathcal{R}_0$ to the product of the dipole and probability decay rate constants. This combination inevitably enters calculations involving saturation of two-level systems. Combining Eqs. (70) and (69), we find

$$V = -\frac{\mathcal{R}_0 T_2 \mathcal{L}(\delta)}{1 + I\mathcal{L}(\delta)} = -\frac{\mathcal{R}_0 T_2}{1 + I + \delta^2 T_2^{\,2}}\,, \tag{74}$$

and inserting this into Eq. (69) gives

$$U = \frac{\delta \mathcal{R}_0 T_2{}^2 \mathcal{L}(\delta)}{1 + I\mathcal{L}(\delta)} = \frac{\delta \mathcal{R}_0 T_2{}^2}{1 + I + \delta^2 T_2{}^2} \,. \tag{75}$$

In Eqs. (74) and (75), we write the result in two forms, the first revealing how the saturation denominator has a Lorentzian dependence, the second showing how this effect leads to power broadening. Chapter 5 discusses these effects in substantially greater detail using an extension of the density matrix known as the population matrix. For now we note that in contrast to the Rabi flopping given by Eqs. (64) - (66), Eqs. (71), (74), and (75) give a stationary Bloch vector for which the coherent Rabi flopping process competes with the incoherent T_1 and T_2 decay processes to produce a constant partial rotation of the Bloch vector. Equations (74) and (75) correspond to the purely classical Eqs. (1.59) and (1.58), respectively. The classical versions use $eE_0/2m\nu x_0$ in place of the Rabi flopping frequency $\mathcal{R}_0$ and predict no power broadening ($I = 0$). The failure of the classical model to give power broadening is due to the absence of an equation for the inversion W, that is, to the lack of a probability difference, which is a quantum mechanical construct.

Our *UVW* discussion takes place in the interaction picture rotating at the optical frequency ν. For the vector model, this amounts to being on a merry-go-round rotating at frequency ν (!). We can return to the Schrödinger picture (get off the merry-go-round) by means of the transformation

$$\dot{\mathbf{u}} = \left[\frac{d\mathbf{U}}{dt}\right]_{space} = \left[\frac{d\mathbf{U}}{dt}\right]_{body} + \nu\hat{\mathbf{e}}_3 \times \mathbf{U} \,. \tag{76}$$

We then find the rapidly rotating vector $\mathbf{u}$, which obeys the equation of motion

$$\dot{\mathbf{u}} = -\frac{\mathbf{u}}{T_1} + \mathbf{u} \times \mathbf{b} \,, \tag{77}$$

where the effective field $\boldsymbol{\beta}$ is given by

$$\boldsymbol{\beta} = \mathcal{R}_0 \hat{\mathbf{e}}_1 - \omega \hat{\mathbf{e}}_3 \,. \tag{78}$$

The vector $\mathbf{u}$ rotates counterclockwise about $\hat{\mathbf{e}}_3$ at approximately the frequency ω.

In the case of the magnetic dipole (spin $\frac{1}{2}$) particle, $\wp E_0$ is replaced by μH in the interaction energy $\mathcal{V}_{ab}$. Otherwise the precession equations are the same. However unlike the electric-dipole case, the abstract axes $\hat{\mathbf{e}}_1$, $\hat{\mathbf{e}}_2$, and $\hat{\mathbf{e}}_3$ for the magnetic dipole actually coincide with the real-life axes **x**, **y**, and **z**, for an electron with spin up (along **z**) is in the upper state (along $\hat{\mathbf{e}}_3$). The reader should bear in mind that in the first instance the equations refer to the probabilistic density matrix, and only for an ensemble of magnetic dipoles can **U** itself be identified with a macroscopic classical dipole precessing in a magnetic field.

References

Allen, L. and J. H. Eberly (1975), *Optical Resonance and Two-Level Atoms*, John Wiley & Sons, New York, reprinted (1987) with corrections by Dover, New York. This book gives a detailed discussion of the optical Bloch equations.

Dicke, R. H. (1953), Phys. Rev. **89**, 472.

Feynman, R. P., F. L. Vernon, and R. W. Hellwarth (1957), J. Appl. Phys. **28**, 49. This classic paper showed that the Bloch equations of nuclear magnetic resonance apply to the two-level atom coupled to a single-mode electromagnetic field.

Sargent, M. III, M. O. Scully, and W. E. Lamb, Jr. (1974), *Laser Physics*, Addison-Wesley Publishing Co., Reading, MA.

Problems

4-1. Show that the expectation value $\langle d^\dagger d\rangle$ is real and ≥ 0, where d is an arbitrary operator. Assume a general mixture and $P_\psi > 0$.

4-2. Consider a collection of two-level atoms 30% of which are described by the wave function

$$\psi_1 = 2^{-1/2}\,[u_a e^{-i\omega_a t} + u_b e^{-i\omega_b t}],$$

50% are described by

$$\psi_2 = 10^{-1/2}\,[u_a e^{-i\omega_a t} - 3u_b e^{-i\omega_b t}]$$

and 20% are described by

$$\psi_3 = u_b e^{-i\omega_b t} .$$

Using the eigenfunctions u_a and u_b as a basis, determine the density matrix for this system. What is the probability that this system is in the state ψ_1? Show that $\rho^2 \neq \rho$.

4-3. Beginning at time $t = 0$, an optical field $E(t) = E_0 \cos\nu t$ interacts with an ensemble of two-level atoms. Using the optical Bloch equations in the rotating wave approximation, i.e., Eqs. (48) through (50), and considering times $<< 1/\gamma$, $1/\gamma_a$ so that decay processes may be neglected, show the field induces a polarization of the medium given by

$$P = -\frac{N\wp\mathcal{R}_0}{\mathcal{R}^2}[(\nu-\omega)(1 - \cos\mathcal{R}t)\cos\nu t - \mathcal{R}\sin\mathcal{R}t\sin\nu t] , \tag{79}$$

where N is the total number of atoms/volume, $\wp$ is the electric dipole matrix element, $\mathcal{R}$ is the generalized Rabi flopping frequency of Eq. (61), and ω is the atomic resonance frequency. Assume at t=0 that the Bloch vector $\mathbf{U} = (0, 0, -1)$.

4-4. An optical field $E(t) = E_0\cos\nu t$ interacting with a two-level medium induces the polarization of the medium given by Eq. (79). Calculate as a function of time the emission or absorption from an optically thin inhomogeneously broadened ensemble of these atoms characterized by the distribution

$$\mathcal{W}(\omega) = \frac{1}{Ku\sqrt{\pi}} e^{-(\omega-\nu)^2/(Ku)^2} ,$$

that is, calculate the expression $\int d\omega' \mathcal{W}(\omega')P(\omega')$. Assume the field is turned on instantaneously at t=0 and do your calculation for both the extremely short time regime just after t=0 (specify what this means) and for the time regime where the inhomogeneous distribution can be approximated by its value at the frequency ν. Use a table of integrals if necessary and ignore homogeneous damping processes. Plot the emission or absorption versus time and interpret your results. The time development is called optical nutation (see Sec. 11-1 for an alternative discussion).

4-5. Without making the rotating-wave approximation and including transitions between all of the levels, determine the equations of motion for all of the density matrix elements for the following three-level system interacting with the two optical fields:

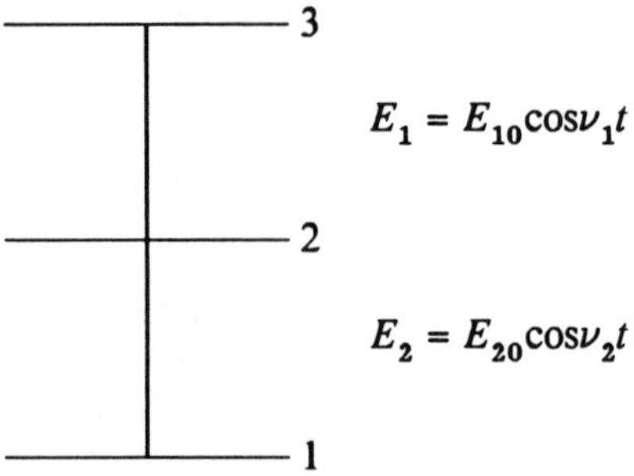

Now suppose $\nu_1 \simeq \omega_3 - \omega_2$ and $\nu_2 \simeq \omega_2 - \omega_1$ so that one may make the appropriate rotating wave approximations. How do the equations of motion simplify. Note that you are not required to solve the equations. This is an appropriate level scheme for three-level probe/saturation spectroscopy, as discussed in Sec. 8-3.

4-6. Some quickies: name two cases where the density matrix is required and why it is required. What three-dimensional object does a slightly detuned Bloch vector trace out in time? Can T_1 ever be smaller than T_2?

4-7. The Bloch vector corresponding to the wave function of Eq. (3.98) is defined by

$$\begin{aligned} U &= C_a C_b^* + \text{c.c.} \\ V &= iC_a C_b^* + \text{c.c.} \\ W &= |C_a|^2 - |C_b|^2 \end{aligned} \tag{80}$$

Using the equations of motion (1) and (2) for C_a and C_b, derive Eqs. (48) and (49) for $\dot{U}$ and $\dot{V}$. What value of T_2 do you find? Your derivation shows that the probability amplitudes in Eq. (3.98) are in the rotating frame, since by definition U and V of Eqs. (42) and (43) are in this frame.

4-8. Defining the probability sum $M = |C_a|^2 + |C_b|^2$ along with the usual Bloch-vector components in Eq. (80), find the equation of motion for W and M using Eqs. (1) and (2). In particular show that the population difference decay time T_1 is given by

$$T_1 \simeq \frac{1}{2}\left(\frac{1}{\gamma_a} + \frac{1}{\gamma_b}\right), \tag{81}$$

where $1/T_1$ is defined by the intensity-independent coefficient of $-W$ in the equation of motion for W. This problem shows that the three-component Bloch equations are in general inadequate to describe a two-level atom with differing level decay rates.

4-9. Derive P_s of Eq. (11) by combining Eqs. (5) and (10).

4-10. Using the coefficient $C_b(t)$ of Eq. (4.10) evaluated to first-order in $\wp E_0$, integrate Eq. (4.5) to the time t instead of ∞. Show that this leads to *a subnatural* linewidth.

4-11. Show that $\mathrm{tr}\{\rho^2\} \leq 1$. When does equality hold?

4-12. Quantum mechanically the entropy is defined as

$$S = -k_B\,\mathrm{tr}\{\rho\ln\rho\} \,. \tag{82}$$

Show that S vanishes for a pure state. What does this mean physically? Hint: $\ln\mathcal{O}$ is defined by a Taylor expansion.

4-13. The root-mean-square deviation for an operator $\mathcal{O}$ is defined by

$$\sigma = \sqrt{\langle\mathcal{O}^2\rangle - \langle\mathcal{O}\rangle^2} \,. \tag{83}$$

Evaluate this for the dipole moment operator

$$er = \begin{pmatrix} 0 & \wp \\ \wp & 0 \end{pmatrix}$$

using the density matrix of Eq. (14).

4-14. Calculate $|C_a(t)|^2$ for an atom interacting with a resonant but incoherent light source characterized by the intensity fluctuation function $P(I)$ given by

$$P(I) = I_0^{-1}\, e^{-I/I_0} \,.$$

Assume that $\gamma_a = \gamma_b = 0$ and that the field phase vanishes, i.e., $E = \sqrt{I}$. Discuss the results as $t \rightarrow \infty$.

4-15. Pure radiative decay is defined by an upper-to-ground-lower-level transition with the dipole decay constant $\gamma = \frac{1}{2}\Gamma$. By projecting onto the "bare-state" $|a\rangle\ |b\rangle$ basis, show that the density operator equation of motion

$$\dot{\rho} = -\frac{i}{\hbar}[\mathcal{H}, \rho] - \frac{\Gamma}{2}[\sigma_+\sigma_-\rho + \rho\sigma_+\sigma_-] + \Gamma\sigma_-\rho\sigma_+ \tag{84}$$

gives density matrix equations of motion like Eqs. (30), (31), and (33) for pure radiative decay. Here $\sigma_\pm$ are the Pauli spin-flip operators of Eq. (3.110). What term do you need to add to allow γ to vary independently of Γ?

Now project Eq. (84) onto the dressed-state basis defined by Eqs. (3.142) and (3.143). Use dressed-state Pauli spin-flip matrices like Eq. (3.148) and for simplicity, neglect decay-term coupling between diagonal and off-diagonal elements. Answer for ρ_{22}:

$$\dot{\rho}_{22} = -[i\hbar^{-1}\mathcal{H}_{21}\rho_{12} + \text{c.c.}] - \Gamma\cos^4\theta\rho_{22} + \Gamma\sin^4\theta\rho_{11} . \tag{85}$$

Note that inclusion of decay-term coupling between diagonal and off-diagonal elements makes the dressed-atom representation substantially more complicated than the bare-atom representation. Hence the dressed-atom representation is used almost exclusively in cases when this kind of decay can be neglected, such as when the Rabi frequency $\mathcal{R}_0$ greatly exceeds the decay constants γ and Γ.

4-16. Using the transformation matrix of Eq. (3.144), show that the dressed-atom density matrix is related to the bare-atom matrix by

$$\begin{pmatrix} \rho_{22} \\ \rho_{21} \\ \rho_{12} \\ \rho_{11} \end{pmatrix} = \begin{pmatrix} \cos^2\theta & \cos\theta\sin\theta & \cos\theta\sin\theta & \sin^2\theta \\ -\cos\theta\sin\theta & \cos^2\theta & -\sin^2\theta & \cos\theta\sin\theta \\ -\cos\theta\sin\theta & -\sin^2\theta & \cos^2\theta & \cos\theta\sin\theta \\ \sin^2\theta & -\cos\theta\sin\theta & -\cos\theta\sin\theta & \cos^2\theta \end{pmatrix} \begin{pmatrix} \rho_{aa} \\ \rho_{ab} \\ \rho_{ba} \\ \rho_{bb} \end{pmatrix}$$

How is the bare-atom matrix related to the dressed-atom matrix?

4-17. Show in the dressed-state basis that the electric-dipole expectation value is given by

$$\text{tr}\{\rho e r\} = \wp[\sin2\theta(\rho_{22} - \rho_{11}) + \cos2\theta(\rho_{21} + \rho_{12}) . \tag{86}$$

4-18. Show that the eigenvalues of a pure-case density matrix all vanish except for one, which equals 1. Hint: note that a pure-case density matrix can be written as the outer product of a column vector and a row vector [for example, see the equation following Eq. (4.14)]. Thinking about what this implies about the determinant and all subdeterminants, write the eigenvalue equation and simplify.

4-19. Write the Bloch equations corresponding to Eqs. (48) - (50) for a complex $\mathcal{R}_0$.

Chapter 5
CW FIELD INTERACTIONS

This chapter uses the density matrix methods of Chap. 4 to find the polarization induced by one or two continuous (*cw*) plane waves in two-level media. The density matrix is extended in a form known as the *population matrix*, which treats collections of atomic responses simply. Section 5-1 deals with homogeneously-broadened media, while Sec. 5-2 includes inhomogeneous broadening. The induced polarization is used as a source in the slowly-varying Maxwell equations to yield a nonlinear Beer's law for propagation. The population matrix equations of motion are solved in the important *rate equation approximation*, which assumes that the dipole lifetime T_2 is short compared to times for which the field envelope or population difference vary appreciably. The concepts of power-broadening and spectral hole burning are developed.

Section 5-3 treats the response of two-level media to two counterpropagating waves of possibly different amplitudes. The waves interfere with one another, producing interference patterns. These patterns burn spatial holes into the atomic population difference, which act in turn as a grating that scatters one wave back into the path of the other. The two waves are thus coupled both by saturating one another's gain/absorption and by scattering off the grating they induce together. Section 5-4 deals with the two-photon two-level model and Sec. 5-5 treats a semiconductor in quasi-equilibrium. Both models have similarities with the two-level model as well as significant differences. Section 5-6 discusses the effects of light forces on the center of mass motion of atoms, which leads to atomic trapping and cooling as well as atom optics.

The concepts of this chapter are important in laser theory (Chap. 6), optical bistability (Chap. 7), saturation spectroscopy (Chap. 8), phase conjugation (Chap. 9), and instability phenomena (Chap. 10).

5-1. Polarization of Two-Level Medium

In this section, we derive the polarization of a two-level medium subject to the plane running wave electric field

$$E(z,t) = \frac{1}{2}\,\mathcal{E}(z)\,e^{i(Kz - \nu t)} + \text{c.c.}, \tag{1}$$

where the complex field amplitude $\mathcal{E}(z)$ changes little in an optical wavelength, ν is the oscillation frequency and K is the wave number. This field induces in the medium a polarization of the similar form

$$P(z,t) = \frac{1}{2}\,\mathcal{P}(z)\,e^{i(Kz - \nu t)} + \text{c.c.}, \tag{2}$$

with a slowly-varying complex polarization $\mathcal{P}(z)$ that typically is out of phase from $\mathcal{E}(z)$. Eq. (1) is the steady-state case of the field (1.41), and hence the slowly-varying Maxwell equation of motion (1.43) reduces to

$$d\mathcal{E}(z)/dz = i(K/2\epsilon)\,\mathcal{P} = -\alpha\mathcal{E}\,, \tag{3}$$

where ϵ is the permittivity of the host medium, and where the complex absorption coefficient is

$$\alpha = -i(K/2\epsilon)\,\mathcal{P}/\mathcal{E}\,. \tag{4}$$

The dimensionless intensity $I = |\wp\mathcal{E}/\hbar|^2 T_1 T_2$ obeys the corresponding equation

$$\frac{dI}{dz} = -2Re(\alpha)I. \tag{5}$$

Here we include the factor $\wp^2 T_1 T_2/\hbar^2$ for reasons that become clear in connection with Eq. (21). Our goal in this chapter is to find the polarization $\mathcal{P}(z)$ induced by the field (1).

To do this we suppose that the medium consists of one of two kinds of two-level systems depicted in Figs. 4-1 and 4-5. An important new ingredient is that we now allow for external sources of excitation of the two levels. Such pump mechanisms become essential in the laser theory of Chap. 6. The first two-level system (Fig. 4-1) is characteristic of most laser media and has excitation to both upper and lower levels along with decay from these levels as depicted in Fig. 5-1.

The second kind (Fig. 4-5) has the ground state for its lower level, and decay of the upper level is inevitably to the lower level. The ruby-laser medium is approximated by this second model, as well as many cases of

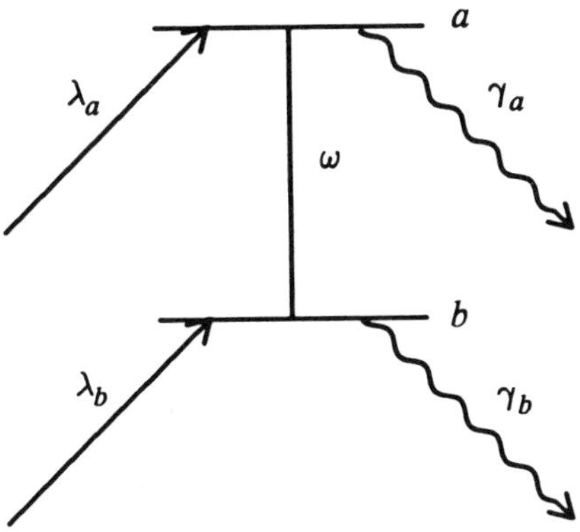

Fig. 5-1. Two-level system with pumping to and decay from both levels (typical of many laser media).

interest in saturation spectroscopy. The equations of motion for the corresponding two-level density matrices differ, although as we show, the formulas for the steady-state polarizations are very similar. Problems in Chaps. 6 and 8 treat a combination that includes decay from the a level to the b level as well as decays from these levels to lower lying levels.

The polarization of the medium $P(z,t)$ of Eq. (2) for the first kind of medium is contributed to by all atoms at the position z at the time t regardless of which state they were excited to and when they were excited. Hence we need to add up all these contributions. This is done conveniently in terms of *a population matrix* defined by

$$\rho(z,t) = \sum_{\alpha=a,b} \int_{-\infty}^{t} dt_0\, \lambda_\alpha(z,t_0)\, \rho(\alpha,t_0,z,t)\,. \tag{6}$$

Here $\lambda_a(z,t_0)$ is the pump rate per unit volume to the level α (α=a or b) for a homogeneously broadened medium, t_0 is the time of excitation and $\rho(\alpha,t_0,z,t)$ is the density matrix describing a system excited to the level α at the time t_0. $\rho(z,t)$ has diagonal elements giving the population densities (rather than probabilities) of the levels, and hence the name population matrix. The population matrix formalism leads directly to the popular rate equations (Eqs. (15) and (16)) for the population densities for sufficiently rapidly decaying dipole moments. It furthermore bypasses some of the algebra in summing contributions from all systems at the place z at the time t regardless of their initial times and levels of excitation.

In terms of $\rho(z,t)$ the polarization (2) is given by

$$P(z,t) = \wp\rho_{ab}(z,t) + \text{c.c.} \tag{7}$$

Equation (14) shows that $\rho_{ab} \propto e^{i(Kz - \nu t)}$, so that we can combine Eqs. (2) and (7) to find

$$\mathscr{P}(z) = 2\wp e^{-i(Kz - \nu t)}\rho_{ab}. \tag{8}$$

Hence we need to find $\rho_{ab}(z,t)$ as it evolves under the influence of the electric-dipole interaction energy.

To do this we calculate the equation of motion of $\rho(z,t)$ by differentiating (6) with respect to time. Two time dependencies exist: that of the upper limit of integration over the excitation time t_0 and that of the single-atom density matrix. We have

$$\frac{d\rho(z,t)}{dt} = \sum_\alpha \lambda_\alpha(z,t)\rho(\alpha,z,t,t) + \sum_\alpha \int_{-\infty}^{t} dt_0\, \lambda_\alpha(z,t_0)\, \dot\rho(\alpha,z,t_0,t).$$

Assuming no off-diagonal excitation, we have by construction $\rho_{ij}(\alpha,z,t,t) = \delta_{i\alpha}\delta_{j\alpha}$, and the first term can be replaced by the operator with the matrix representation

$$\begin{bmatrix} \lambda_a & 0 \\ 0 & \lambda_b \end{bmatrix}.$$

The second term has components identical to the right-hand sides of the equations of motion for the single-atom density matrix [see Eqs. (4.20), (4.21), and (4.30)]. This result follows because the Hamiltonian does not depend on a or b and the pump rates $\lambda_\alpha(z,t_0)$ are assumed to vary slowly enough to be evaluated at the time t. In component form, the equations of motion for the population matrix $\rho(z,t)$ become

$$\dot\rho_{ab} = -(i\omega + \gamma)\rho_{ab} + i\hbar^{-1}\mathcal{V}_{ab}(z,t)(\rho_{aa} - \rho_{bb}) \tag{9}$$

$$\dot\rho_{aa} = \lambda_a - \gamma_a\rho_{aa} - (i\hbar^{-1}\mathcal{V}_{ab}\rho_{ba} + \text{c.c.}) \tag{10}$$

$$\dot\rho_{bb} = \lambda_b - \gamma_b\rho_{bb} + (i\hbar^{-1}\mathcal{V}_{ab}\rho_{ba} + \text{c.c.}) \tag{11}$$

For the single-wave field (1) and in the rotating-wave-approximation, the perturbation energy $\mathcal{V}_{ab}$ has the value

$$\boxed{\mathcal{V}_{ab} = -\frac{1}{2}\wp\mathcal{E}(z)\, e^{i(Kz - \nu t)}}\ . \tag{12}$$

Rate Equation Approximation

We first integrate these equations using the *rate equation approximation*, so called because it leads to rate equations (15) and (16). The rate equation approximation consists of assuming that the dipole decay time $T_2 \equiv 1/\gamma$, is much smaller than times for which the population difference or the field envelope can change. For steady state, this approximation is exact since the population difference and field envelope are constant. We proceed by noting that Eq. (9) can be integrated formally to give

$$\rho_{ab}(z,t) = \frac{i}{\hbar}\int_{-\infty}^{t} dt'\, e^{-(i\omega+\gamma)(t-t')}\, \mathcal{V}_{ab}(z,t')\, [\rho_{aa}(z,t') - \rho_{bb}(z,t')]\ . \tag{13}$$

Using Eq. (12) for $\mathcal{V}_{ab}$, we can then factor both the population difference and the field envelope outside the t' integration, perform the integral over exponentials, and find

$$\rho_{ab}(z,t) = -i(\wp\mathcal{E}/2\hbar)e^{i(Kz - \nu t)}\,\frac{\rho_{aa} - \rho_{bb}}{\gamma + i(\omega-\nu)}\ . \tag{14}$$

Substituting this into the population equations of motion (10) and (11), we find the rate equations

$$\dot\rho_{aa} = \lambda_a - \gamma_a\rho_{aa} - R[\rho_{aa} - \rho_{bb}] \tag{15}$$

$$\dot\rho_{bb} = \lambda_b - \gamma_b\rho_{bb} + R[\rho_{aa} - \rho_{bb}] \tag{16}$$

where the rate constant R is given by

$$R = \frac{1}{2}\,|\wp\mathcal{E}/\hbar|^2\gamma^{-1}\,\mathcal{L}(\omega-\nu)\ , \tag{17}$$

and the dimensionless Lorentzian $\mathcal{L}(\omega-\nu)$ is given by

$$\boxed{\mathscr{L}(\omega - \nu) = \frac{\gamma^2}{\gamma^2 + (\omega-\nu)^2}} . \tag{18}$$

Solving Eqs. (15) and (16) in steady state ($\dot{\rho}_{aa} = \dot{\rho}_{bb} = 0$), we find the population difference

$$\rho_{aa} - \rho_{bb} = \frac{N(z)}{1 + I\,\mathscr{L}(\omega-\nu)} . \tag{19}$$

Here $N(z)$ is the *unsaturated population difference*, which we take to be the upper level population minus the lower to agree with the sign of the Bloch-vector component W [see Eq. (4.44)] and with laser theory. The unsaturated population difference for Eqs. (15) and (16) (set $R = 0$) is

$$N = \lambda_a \gamma_a{}^{-1} - \lambda_b \gamma_b{}^{-1} . \tag{20}$$

The dimensionless intensity I has the definition

$$I = |\wp\mathcal{E}/\hbar|^2 T_1 T_2 , \tag{21}$$

where for the level scheme of Fig. 5-1, the average level decay time T_1 is

$$T_1 = \frac{1}{2}\left[\frac{1}{\gamma_a} + \frac{1}{\gamma_b}\right] . \tag{22}$$

The dimensionless intensity of Eq. (21) is the average irradiance $c\epsilon_0|\mathcal{E}|^2$ given in units of the saturation intensity

$$I_s = c\epsilon_0|\hbar/\wp|^2/T_1T_2 . \tag{23}$$

The dimensionless intensity (21) is a generalization of Eq. (4.12), which takes $T_2 = 1/\gamma_{ab}$, i.e., Eq. (21) includes effects of dephasing collisions. For $I = 1$, the population difference is saturated down to the "half-saturation" point, that is, half its unsaturated value.

Combining the saturated population difference (19) with the dipole expression (14), we find

$$\rho_{ab} = -i(\wp\mathcal{E}/2\hbar)e^{i(Kz - \nu t)} \frac{N\,\mathscr{D}(\omega-\nu)}{1 + I\,\mathscr{L}(\omega-\nu)} , \tag{24}$$

where the complex Lorentzian denominator is defined as

$$\boxed{\mathscr{D}(\omega-\nu) = \frac{1}{\gamma + i(\omega-\nu)}} \; . \tag{25}$$

Using Eq. (8), we find the desired complex polarization

$$\mathscr{P}(z) = -i(\wp^2/\hbar)\, \frac{N\, \mathscr{D}(\omega-\nu)}{1 + I\, \mathscr{L}(\omega-\nu)}\, \mathscr{E} \, . \tag{26}$$

Substituting Eq. (26) into (4) gives the complex, nonlinear absorption coefficient

$$\boxed{\alpha = \alpha_0\, \frac{\gamma \mathscr{D}(\omega-\nu)}{1 + I\, \mathscr{L}(\omega-\nu)}} \; , \tag{27}$$

where the linear ($I = 0$), resonant ($\nu = \omega$) absorption coefficient

$$\alpha_0 = -K\, \frac{\wp^2 N}{2\epsilon\hbar\gamma} \tag{28}$$

is positive for an uninverted medium ($N < 0$). Equation (27) is a very important generalization of the linear case (28), and we use it in laser theory (Chap. 6), optical bistability (Chap. 7), and in saturation spectroscopy (Chap. 8). Figure 5-2 illustrates its form. The "half-height" value of α corresponds to an intensity equal to the saturation intensity I_s of Eq. (23). The complex absorption coefficient of Eq. (27) has the real and imaginary parts

$$\mathrm{Re}(\alpha) = \alpha_0\, \frac{\gamma^2}{\gamma^2(1+I) + (\omega-\nu)^2} \tag{29a}$$

$$\mathrm{Im}(\alpha) = -\alpha_0\, \frac{\gamma(\omega - \nu)}{\gamma^2(1+I) + (\omega-\nu)^2} \; , \tag{29b}$$

These parts are the same as those of Eq. (1.54), except that α_0 has a different definition and that here the natural width γ is *power broadened* by the factor $(1 + I)$. Equation (29a) is similar to the stimulated emission profile (4.11), and shows a power-broadened Lorentzian spectrum. Note that while the width of the Lorentzian in Eq. (29) increases as the intensity increases, the value for any given detuning decreases.

Writing the intensity equation (5) for the absorption coefficient (27), we have

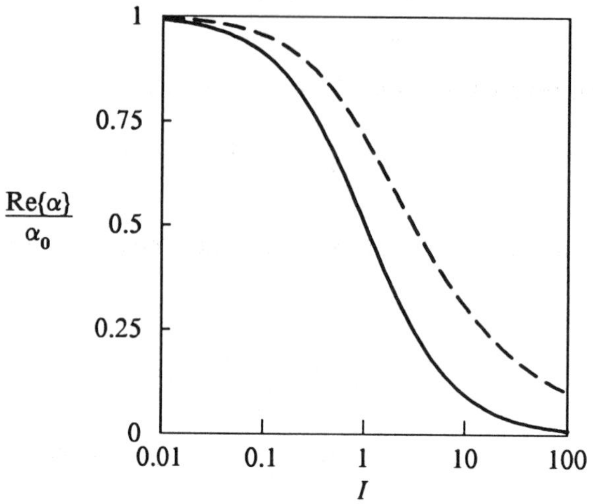

Fig. 5-2. (solid line) Resonant (ν=ω) absorption coefficient α of Eq. (27) versus intensity. The value I=1 corresponds to an intensity equal to the saturation intensity (23). (dashed line) Corresponding inhomogeneous-broadening case (40).

$$\frac{dI}{dz} = -\frac{2\alpha_0\mathcal{L}}{1 + I\mathcal{L}} I\,, \tag{30}$$

where $\mathcal{L} \equiv \mathcal{L}(\omega\text{-}\nu)$. For small $I\mathcal{L}$, I decays exponentially to 0. For large $I\mathcal{L}$, I decays linearly in z with the slope $-2\alpha_0$. In general the intensity absorption coefficient is given by twice the amplitude coefficient of Eq. (29a). Alternatively for a gain medium ($\alpha_0 < 0$), I grows exponentially at first and then approaches a linear growth rate.

For the upper-to-ground-lower-level-decay two-level system of Fig. 4-5, the pump rates in Eqs. (10) and (11) are given by $\lambda_a = \Lambda\rho_{bb}$ and $\lambda_b = \Gamma\rho_{aa}$, respectively, and the decay rate γ_b equals 0. The number of active systems per unit volume N' is a constant, so that the populations obey the conservation relation

$$N' = \rho_{aa} + \rho_{bb}\,. \tag{31}$$

This relation differs from Eq. (20), and in fact does not apply to the first kind of two-level medium. In view of this conservation of the number of systems, we have $\dot{\rho}_{aa} = -\dot{\rho}_{bb}$. With these observations, the population equations of motion (10) and (11) reduce to

$$\dot{\rho}_{aa} = -\dot{\rho}_{bb} = \Lambda\rho_{bb} - \Gamma\rho_{aa} - [i\hbar^{-1}\mathcal{V}_{ab}\rho_{ba} + \text{c.c.}]\,, \tag{32}$$

where atoms are pumped from the ground level b to the upper level a at the rate $\Lambda\rho_{bb}$. Forming the population difference

$$D = \rho_{aa} - \rho_{bb}\,, \tag{33}$$

we note that $2\rho_{aa} = N' + D$ and $2\rho_{bb} = N' - D$. Differentiating Eq. (33) with respect to t and using Eq. (32), we find

$$\frac{dD}{dt} = -(\Gamma + \Lambda)D - (\Gamma - \Lambda)N' - 2[i\hbar^{-1}\mathcal{V}_{ab}\rho_{ba} + \text{c.c.}]\,. \tag{34}$$

Here we see that the decay time for the population difference has the explicit value $T_1 = 1/(\Gamma+\Lambda)$. The rest of the calculation proceeds as for the first kind of two-level medium, and one obtains the saturated absorption coefficient (27) in which the dimensionless intensity is measured in units of I_s of (23) with the effective decay time $T_1=(\Gamma+\Lambda)^{-1}$ and with the further modification that α_0 of (28) contains now the unsaturated population difference $N = N'(\Lambda-\Gamma)/(\Lambda+\Gamma)$. It is interesting to note that the single time constant T_1 applies accurately to this second type of two-level medium, while as we see in Chaps. 6 and 8, it can give incorrect results for multi-mode fields interacting with the first type of two-level media or with a combination of the two.

5-2. Inhomogeneously Broadened Media

Many media are inhomogeneously broadened, that is, they are such that different atoms have different line centers as discussed in Sec. 4-2. This kind of broadening differs from the homogeneous variety considered in preceding sections in that it is a dynamical (not random) property and can be reversed. A famous way to achieve this reversal is the technique of photon echoes, which will be discussed in chapter 11 (and which, by the way, can be explained without reference to photons and has purely classical analogs). We consider here two basically different kinds of inhomogeneous broadening: static and Doppler. Static broadening occurs, for example in ruby at low temperatures, where each active atom (Cr^{3+}) experiences its own individual Stark shift based on local lattice characteristics. Doppler broadening occurs in gas media and results from the spread in Doppler shifts that atoms moving with different velocities experience. For unidirectional fields, the effects are the same: Doppler shifts yield essentially a spectrum of line centers. The standing-wave static case is also similar to the unidirectional Doppler case, with the addition of position complications discussed in Sec. 5-3. The standing-wave Doppler case is however more complicated, for each atom sees two frequencies, one Doppler up-shifted and one down-shifted. In this section, we consider the interaction of a monochromatic running wave with these various inhomogeneously broadened media, once again in the rate equation approximation. It is interesting to note at the outset that this approximation can be very accurate, a fact established by comparison with more exact theories (see Chap. 8).

Except for the Doppler standing-wave case, the spread of line centers in an inhomogeneously broadened medium is represented in the theory by a line-center dependence in the level populations, $\rho_{aa}(z,\omega',t)$ and $\rho_{bb}(z,\omega',t)$, and in the complex polarization $\rho_{ab}(z,\omega',t)$. The equations of motion are still given by (9), (10), and (11). The macroscopic polarization $\mathscr{P}(z)$, however, is contributed to by all systems at z at the time t regardless of ω' and hence includes an integral over the latter

$$\mathscr{P}(z) = \int d\omega' \, \mathscr{P}(z,\omega') \, . \tag{35}$$

Furthermore the linear population inversion includes a function specifying the nature of the inhomogeneity, namely

$$N(z,\omega',t) = \mathscr{W}(\omega') \, N(z,t) \, . \tag{36}$$

For many media, the Maxwellian distribution

$$\mathscr{W}(\omega') = \frac{1}{\Delta\omega\sqrt{\pi}} \, e^{-(\omega-\omega')^2/(\Delta\omega)^2} \tag{37}$$

is descriptive. Sometimes we consider a Lorentzian distribution, which typically leads to analytic results, or a distribution so broad that it can be evaluated at the peak of the homogeneous-broadening function (which acts like a δ function) and taken outside the ω' integral.

With these additions, we substitute the homogeneously-broadened result (26) into the frequency integral of (35) to find the complex polarization component

$$\mathscr{P}(z) = -\frac{i\wp^2 \mathscr{E} N}{\hbar} \int_{-\infty}^{\infty} d\omega' \, \frac{\mathscr{W}(\omega')\mathscr{D}(\omega'-\nu)}{1 + I\mathscr{L}(\omega'-\nu)} \, . \tag{38}$$

Suppose first that the width of the distribution $\mathcal{W}(\omega')$ is much greater than γ. This is called the inhomogeneously-broadened limit. As in the discussion of Eq. (3.67), we evaluate $\mathcal{W}(\omega')$ at ν and take it outside the integral. The imaginary part of $\mathcal{D}(\omega'-\nu)$ is antisymmetric about ν, thereby integrating to 0. Thus Eq. (38) reduces to

$$\mathcal{P}(z) = -i\frac{\wp^2\mathcal{E}N}{\hbar}\mathcal{W}(\nu)\int_{-\infty}^{\infty}\frac{dx}{1+x^2+I}$$

$$= -\frac{i\pi\wp^2 N\mathcal{W}(\nu)\mathcal{E}}{\hbar(1+I)^{1/2}}\,. \tag{39}$$

This gives the inhomogeneous-broadening coefficient

$$\boxed{\alpha = \alpha_0'/(1+I)^{1/2}}\,, \tag{40}$$

where the linear inhomogeneous-broadening coefficient

$$\alpha_0' = \alpha_0\gamma\pi\mathcal{W}(\nu). \tag{41}$$

Comparing this to the homogeneous-broadening case of Eq. (27), we see that the tuning dependence is gone, and that the saturation is weaker, i.e., Eq. (40) is proportional to $(1+I)^{-1/2}$, while Eq. (27) is proportional to $(1+I)^{-1}$. The reduced saturation of the inhomogeneously broadened case is due to the fact that contributions from detuned (and therefore less saturated) atoms are included in the average over ω.

For the Gaussian distribution (37), the frequency integral in Eq. (38) is conveniently expressed in terms of the plasma dispersion function defined by

$$Z(\gamma+i\omega-i\nu) = \frac{i}{\sqrt{\pi}}\int_{-\infty}^{\infty}d\omega'\, e^{-(\omega-\omega')^2/(\Delta\omega)^2}\,\mathcal{D}(\omega'-\nu)$$

$$= \frac{i}{\sqrt{\pi}}\int_{-\infty}^{\infty}d\omega'\, e^{-(\omega-\omega')^2/(\Delta\omega)^2}\,\frac{\gamma - i(\omega'-\nu)}{\gamma^2+(\omega'-\nu)^2}\,. \tag{42}$$

Specifically, multiplying the numerator and denominator of Eq. (38) by $\gamma - i(\omega'-\nu)$ and comparing the result with Eq. (42), we find

$$\mathcal{P}(z) = -\frac{\wp^2 N\mathcal{E}}{\hbar\Delta\omega}[Z_r(\gamma'+i\omega-i\nu) + i(\gamma/\gamma')Z_i(\gamma'+i\omega-i\nu)]\,, \tag{43}$$

where the power-broadened decay constant

$$\gamma' = \gamma(1+I)^{1/2}\,, \tag{44}$$

and Z_r and Z_i are the real and imaginary parts of Z. This gives the absorption coefficient

$$\alpha = \alpha_0\,\frac{\gamma}{\Delta\omega}\left[\frac{Z_i(\gamma'+i\omega-i\nu)}{(1+I)^{1/2}} - iZ_r(\gamma'+i\omega-i\nu)\right]. \tag{45}$$

In the limit $\Delta\omega >> \gamma$, this reduces to Eq. (40), since $Z_r \to 0$ and $Z_i \to \pi^{1/2}$. In general, Eq. (45) looks like the Lorentzian version of Fig. 5-2, but with a Gaussian influence near central tuning. The plasma dispersion function (42) is the convolution of a Gaussian with a complex Lorentzian.

A better approximation to the Gaussian than the constant value of the inhomogeneous broadening limit is the Lorentzian

$$\mathcal{W}(\omega') = \frac{1}{\pi}\,\frac{\Delta\omega}{(\omega'-\omega)^2 + (\Delta\omega)^2}\,. \tag{46}$$

This allows the polarization (38) to be integrated analytically, and hence allows us to study the transition from pure homogeneous broadening to pure inhomogeneous broadening. In terms of Eq. (46), Eq. (38) gives

$$\mathcal{P} = -\frac{\wp^2 N\mathcal{E}\gamma}{\hbar\gamma'}J(\Delta\omega,\,\gamma',\,\omega-\nu)\,, \tag{47}$$

where $J(\Delta\omega,\gamma',\omega-\nu)$ is a special case of the convolution between a real and a complex Lorentzian

$$J(\gamma'',\gamma',\Delta) = \frac{i}{\pi}\int_{-\infty}^{\infty} d\delta\,\frac{\gamma''}{\gamma''^2+\delta^2}\,\frac{\gamma'[1 - i(\delta+\Delta)/\gamma]}{\gamma'^2 + (\delta+\Delta)^2}$$

$$= \frac{i}{\pi}\int_{-\infty}^{\infty}\frac{d\delta\,\gamma''\gamma'[1 - i(\delta+\Delta)/\gamma]}{(\delta+i\gamma'')(\delta-i\gamma'')(\delta+\Delta+i\gamma')(\delta+\Delta-i\gamma')}\,. \tag{48}$$

This convolution is easily evaluated using the residue theorem. Closing the contour in the upper half plane around the poles $\delta = i\gamma''$, $-\Delta + i\gamma'$, we find

$$J(\gamma'',\gamma',\Delta) = \frac{i}{\Delta + i\gamma'' - i\gamma'}\left[\frac{\gamma'(1+\gamma''/\gamma - i\Delta/\gamma)}{\Delta + i\gamma'' + i\gamma'} + \frac{\gamma''(1+\gamma'/\gamma)}{\Delta - i\gamma'' - i\gamma'}\right]$$
$$= \frac{i(\gamma''+\gamma') + \Delta\gamma'/\gamma}{\Delta^2 + (\gamma''+\gamma')^2} \,. \tag{49}$$

Hence the convolution of two Lorentzians is itself a Lorentzian with width equal to the sum of the individual widths. Combining Eqs. (47), (49), and (4), we obtain

$$\alpha = \frac{\alpha_0 \gamma^2}{\gamma'} \frac{\Delta\omega + \gamma' - i(\omega - \nu)\gamma'/\gamma}{(\omega-\nu)^2 + (\Delta\omega + \gamma')^2} \,. \tag{50}$$

This reduces to Eqs. (29) and (40) in the homogeneous ($\Delta\omega = 0$) and inhomogeneous ($\Delta\omega >> \gamma$) broadening limits, respectively.

Having seen the results of inhomogeneous broadening, let us return to an earlier point in the calculation to look at a very useful picture. Figure 5-3 plots the population difference (19) with the inhomogeneous broad-

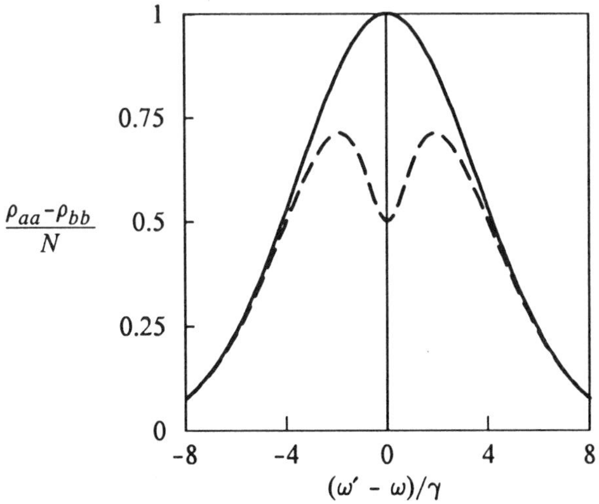

Fig. 5-3. Inhomogeneously broadened population difference given by Eq. (37) times Eq. (19) versus the normalized frequency difference $(\omega' - \omega)/\gamma$ with the inhomogeneous width $\Delta\omega = 5$, central tuning ($\nu = \omega$), and $I = 0$ (solid line) and $I = 1$ (dashed line). The $I = 1$ case shows spectral hole burning.

ened linear population inversion (36) *vs* detuning for the Gaussian distribution (37) and $\nu=\omega$. The I = 1 case reveals a "hole burned" into the inhomogeneous lineshape by the electromagnetic wave. The FWHM of this hole is about 2γ. This phenomenon was first discovered in the context of nuclear magnetic resonance by Bloembergen, Purcell, and Pound (1948) (who referred to it as "eating a hole"), and was developed for gas lasers by Bennett (1962). We use the hole burning concept in understanding features of gas laser operation (Sec. 6-3) and in saturation spectroscopy (Chap. 8).

An integral similar to the imaginary part of Eq. (48) arises when one tries to determine the hole width burned by a *saturator wave* of arbitrary intensity I_2 and frequency ν_2 by measuring the absorption of a weak *probe wave* of amplitude $\mathcal{E}_1$ and frequency ν_1. This is a form of saturation spectroscopy discussed in detail in Chap. 8 (see Fig. 8-1). Unlike in the treatments of Chap. 8, we ignore for now the response of the system to the field interference pattern created by the two waves. In this approach, the strong wave generates a saturated inversion according to Eq. (19) with N given by (36), which is then probed linearly by the weak field at the frequency ν_1. We then obtain the probe-wave polarization by replacing 1) the ν in the $\mathcal{D}(\omega' - \nu)$ factor of Eq. (38) by ν_1, and 2) the ν in the $I\mathcal{L}(\omega' - \nu)$ factor of Eq. (38) by ν_2. This gives

$$\mathcal{P}_1(z) = -i\wp^2\mathcal{E}_1\hbar^{-1} N \int_{-\infty}^{\infty} d\omega' \frac{\mathcal{W}(\omega')\,\mathcal{D}(\omega'-\nu_1)}{1 + I_2\mathcal{L}(\omega'-\nu_2)} \,. \tag{51}$$

In the inhomogeneous-broadening limit, this gives the absorption coefficient [using Eqs. (4) and (41) with $\mathcal{W}(\nu)\rightarrow\mathcal{W}(\nu_1)$]

$$\alpha_1 = \frac{\alpha_0'(\nu_1)}{\pi\gamma}\int_{-\infty}^{\infty}\frac{d\omega'\gamma\mathcal{D}(\omega'-\nu_1)}{1 + I_2\mathcal{L}(\omega'-\nu_2)} = \frac{\alpha_0'(\nu_1)}{\pi}\int_{-\infty}^{\infty}\frac{-id\delta'}{\delta'-i\gamma}\left[1 - \frac{\gamma^2 I_2}{(\delta'-\Delta)^2+\gamma'^2}\right]$$

$$= \alpha_0'(\nu_1)\left[1 - \frac{\gamma^2 I_2(\gamma+\gamma'-i\Delta)/\gamma'}{(\gamma+\gamma')^2 + \Delta^2}\right], \tag{52}$$

where $\delta' = \omega' - \nu_1$ and $\Delta = \nu_2 - \nu_1$. The integrals can be evaluated using contour integration (see Prob. 5-5). Note that the real part of the single-pole $i/(\delta' - i\gamma)$ converges much more readily (like $1/\delta'^2$) than the imaginary part (like $1/\delta'$). Hence for Eq. (52) to give an accurate calculation of $\mathrm{Im}\{\alpha_1\}$, $\mathcal{W}(\omega')$ must have a substantially larger width (or be symmetric about ν_1) than that needed for an accurate calculation of $\mathrm{Re}\{\alpha_1\}$.

We see that the probe samples the power-broadened Lorentzian spectral hole with its own Lorentzian. This gives a convolution of the probe Lorentzian with the power-broadened Lorentzian, and yields a Lorentzian with a width equal to the *sum* of the widths of the probe and power-broadened Lorentzians. Thus even though the waves are taken to be monochromatic, the probe absorption displays a spectral hole at least twice as wide as that burned into the population difference.

As discussed in Chap. 8, it is not in general valid to neglect the response of a nonlinear medium to the interference pattern between the two waves. The medium responds to the total field, which includes the probe field, and not just to the pump field alone. In fact as we see in the next section, pump scattering off a pump-probe interference pattern can cause the probe absorption coefficient [given by Eq. (70)] to be dramatically smaller than the single-mode absorption coefficient of Eq. (27).

5-3. Counterpropagating Wave Interactions

In laser theory, optical bistability, and phase conjugation, we often consider standing waves. Such a wave is the sum of two waves like Eq. (1) traveling in opposite directions

$$E(z,t) = \frac{1}{2}\,\mathcal{E}_+(z)e^{i(Kz-\nu t)} + \frac{1}{2}\,\mathcal{E}_-(z)e^{-i(Kz+\nu t)} + \text{c.c.}, \tag{53}$$

which induces a polarization of the form

$$P(z,t) = \frac{1}{2}\,\mathcal{P}_+(z)e^{i(Kz-\nu t)} + \frac{1}{2}\,\mathcal{P}_-(z)e^{-i(Kz+\nu t)} + \text{c.c.} \tag{54}$$

As in deriving Eq. (3), we substitute Eq. (53) and (54) without complex conjugates into the wave equation (1.25) and drop second derivatives of the slowly-varying quantities $\mathcal{E}_+$ and $\mathcal{E}_-$. Here, however, we wish to find *separate* equations like (3) for $\mathcal{E}_+$ and $\mathcal{E}_-$. To do this, we project both sides of our equation onto e^{iKz} for $\mathcal{E}_+$ and onto e^{-iKz} for $\mathcal{E}_-$. Specifically for $\mathcal{E}_+$, we multiply both sides by $[e^{iKz}]^*$ and integrate over a wavelength. This gives

$$\frac{K}{2\pi}\int_0^{2\pi/K} d\varsigma\, e^{-iK(z+\varsigma)}\left[\frac{d\mathcal{E}_+(z)}{dz}e^{iK(z+\varsigma)} - \frac{d\mathcal{E}_-(z)}{dz}e^{-iK(z+\varsigma)}\right] \simeq \frac{d\mathcal{E}_+}{dz}, \tag{55}$$

where we have used the fact that the $d\mathcal{E}_\pm/dz$ vary little in an optical wavelength. The corresponding projection for the polarization (54) together with Eq. (7) gives

$$\mathcal{P}_+(z) \simeq 2\wp \frac{K}{2\pi}\int_0^{2\pi/K} d\varsigma\, e^{-iK(z+\varsigma)}\, e^{i\nu t}\, \rho_{ab}(z+\varsigma,t)\,. \tag{56}$$

With this, we find

$$\pm\frac{d\mathcal{E}_\pm}{dz} = i(K/2\epsilon)\,\mathcal{P}_\pm = -\,\alpha_\pm\,\mathcal{E}_\pm. \tag{57}$$

Here the leading ± is present because $\mathcal{E}_-$ propagates along $-z$. As far as Eq. (55) is concerned, the two waves are "orthogonal" to one another, that is, they can be separated by projection. However this does not mean that they are uncoupled: the fringe pattern between the waves can induce a grating in ρ_{ab} that couples the waves.

To find ρ_{ab} for a homogeneously broadened medium, we write the interaction energy for the field of Eq. (53) in the rotating wave approximation as

$$\mathcal{V}_{ab} = -\frac{\wp}{2}\,[\mathcal{E}_+e^{iKz} + \mathcal{E}_-e^{-iKz}]\,e^{-i\nu t}\,. \tag{58}$$

Substituting this into Eq. (13) and making the rate equation approximation, we find

$$\rho_{ab}(z,t) = -i(\wp/2\hbar)[\mathcal{E}_+e^{iKz} + \mathcal{E}_-e^{-iKz}]\,e^{-i\nu t}\,\frac{\rho_{aa}-\rho_{bb}}{\gamma+i(\omega-\nu)}\,. \tag{59}$$

In turn substituting this into the population equations of motion (10) and (11), we find the rate equations (15) and (16), where the rate constant is now

$$R = \frac{1}{2}(\wp/\hbar)^2\gamma^{-1}\mathcal{L}(\omega-\nu)\,|\mathcal{E}_+e^{iKz} + \mathcal{E}_-e^{-iKz}|^2\,. \tag{60}$$

This gives the steady-state population difference

$$\rho_{aa} - \rho_{bb} = \frac{N(z)}{1 + [I_+ + I_- + 2(I_+I_-)^{1/2}\cos(2Kz-\Psi)]\mathcal{L}(\omega-\nu)}\,, \tag{61}$$

where $\mathcal{E}_+ = \mathcal{E}_-(I_+/I_-)^{1/2}e^{-i\Psi}$.

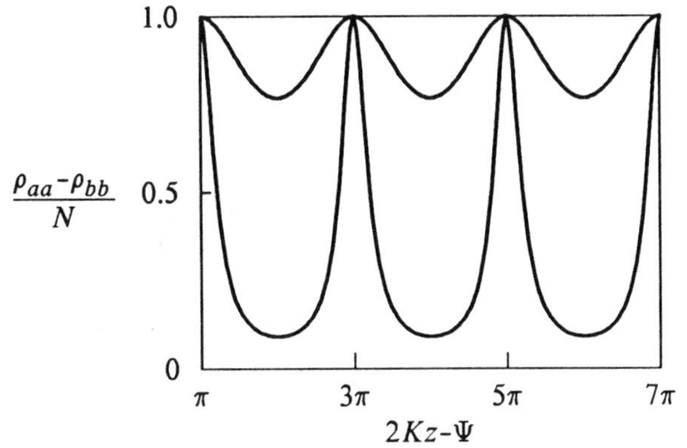

Fig. 5-4. Population difference (61) *vs* Kz for $I_+=I_-=.3$ and 10. Standing-wave fringe pattern burns spatial holes into the population difference. This creates a grating that scatters each wave into the other.

Before substituting Eq. (61) into Eq. (59) to find ρ_{ab}, consider the z variation of the population difference $\rho_{aa}-\rho_{bb}$. This is plotted in Fig. 5-4 for $I_+=I_-=.25$. Similarly to the spectral hole burning in Fig. 5-3, we see *spatial hole burning*, with a peak-to-peak separation of half a wavelength.

We are now in a position to see that this spatial pattern forms a grating that scatters one wave into the path of the other. Combining Eqs. (56), (59), and (61), we find

$$\mathscr{P}_+(z) = -i\,\frac{\wp^2 N(z)}{\hbar}\,\mathscr{D}(\omega-\nu)\,\frac{K}{2\pi}\int_0^{2\pi/K}\frac{[\mathscr{E}_+ + \mathscr{E}_- e^{-2iK\zeta-i\Psi}]d\zeta}{1+[I_++I_- + 2(I_+I_-)^{1/2}\cos(2K\zeta)]\mathscr{L}},$$

where for typographical simplicity we write $\mathscr{L}$ for $\mathscr{L}(\omega-\nu)$. We have chosen the z reference such that $\cos(2Kz-\Psi)=1$, a general choice since we assume that $N(z)$, $\mathscr{E}_+(z)$, and $\mathscr{E}_-(z)$ vary little in a wavelength [but see Prob. 6-18)]. In the absence of the $\cos 2K\zeta$ in the denominator, i.e., the grating contribution, the $\mathscr{E}_-$ term would vanish here as it does in Eq. (55). Further using that $\mathscr{E}_- e^{-i\Psi} = \mathscr{E}_+(I_-/I_+)^{1/2}$ and that $e^{-2iK\zeta}$ can be replaced by $\cos 2K\zeta$, since $\sin 2K\zeta$ integrates to 0 when multiplied by functions of $\cos 2K\zeta$ alone, we have

$$\mathscr{P}_+(z) = -i\frac{\wp^2 N}{\hbar}\mathscr{D}(\omega-\nu)\mathscr{E}_+\frac{K}{2\pi}\int_0^{2\pi/K}\frac{1+(I_-/I_+)^{1/2}\cos(2K\zeta)]d\zeta}{1+[I_++I_-+2(I_+I_-)^{1/2}\cos(2K\zeta)]\mathscr{L}}. \tag{62}$$

This integral can be evaluated using the formula

$$\frac{1}{2\pi}\int_0^{2\pi}\frac{d\theta}{a+b\cos\theta}=\frac{1}{(a^2-b^2)^{1/2}}, \tag{63}$$

a formula that appears in other problems in quantum optics as well (e.g., in mode locking of Sec. 6-5). Equation (62) becomes

$$\mathscr{P}_+(z) = -i(\wp^2/\hbar)\,\mathscr{D}(\omega-\nu)\,N\,\mathscr{E}_+\,\mathscr{S}(\theta)\,, \tag{64}$$

where the saturation factor

$$\mathscr{S}(\theta) = \frac{1}{2\pi}\int_0^{2\pi}d\theta\,\frac{1+c\cos\theta}{a+b\cos\theta}=\frac{1}{2\pi}\int_0^{2\pi}d\theta\,\frac{c}{b}\left[1-\frac{a-b/c}{a+b\cos\theta}\right]$$

$$=\frac{c}{b}\left[1-\frac{a-b/c}{(a^2-b^2)^{1/2}}\right]. \tag{65}$$

For the polarization (64), $a = 1+(I_++I_-)\mathscr{L}$, $b = 2(I_+I_-)^{1/2}\mathscr{L}$, and $c = (I_-/I_+)^{1/2}$. Using Eq. (57), we find the absorption coefficient

$$\alpha_+ = \alpha_0\gamma\mathscr{D}(\omega-\nu)\,\frac{1}{2I_+\mathscr{L}}\left[1-\frac{1+(I_--I_+)\mathscr{L}}{\sqrt{1+2(I_++I_-)\mathscr{L}+(I_+-I_-)^2\mathscr{L}^2}}\right]. \tag{66}$$

Note that for $I_-=0$, this reduces to the running-wave answer (27) as it should. For a standing wave, we set $I_-=I_+=I_{SW}$ to find

$$\alpha_{SW} = \alpha_0\gamma\mathscr{D}(\omega-\nu)\,\frac{1}{2I_{SW}\mathscr{L}}\left[1-\frac{1}{\sqrt{1+4I_{SW}\mathscr{L}}}\right]. \tag{67}$$

Expanding this to first-order in I_{SW}, we find

$$\alpha_{SW} \simeq \alpha_0\gamma\,\mathscr{D}(\omega-\nu)[1-3I_{SW}\mathscr{L}]. \tag{68}$$

The corresponding expansion for the unidirectional absorption coefficient (27) is

$$\alpha \simeq \alpha_0\gamma\, \mathscr{D}(\omega-\nu)[1 - I\mathscr{L}], \tag{69}$$

i.e., for small intensities, the standing-wave case (68) has three times the saturation for the same running wave intensity. The three is made up of the incoherent bleaching by the two running waves plus a third contribution due to the fact that the induced grating scatters *constructively*. To see that the scattering is constructive, note that the in-phase contributions to the polarization are more highly saturated than the out-of-phase contributions. The numerator in Eq. (62) weights the in-phase contributions more than those out-of-phase, thereby producing higher average saturation and thus reducing the absorption.

For large intensities, the square root term in Eq. (67) can be neglected, and the standing-wave saturation reduces to twice the running-wave saturation. These observations are important for lasers, since all else being equal, a running wave laser thereby has at least twice the output power of the standing-wave in a given direction.

For a weak nonsaturating probe wave $\mathscr{E}_+$ in the presence of an arbitrarily strong saturator wave $\mathscr{E}_-$, we expand Eq. (66) to find

$$\alpha_+ = \frac{\alpha_0\gamma\mathscr{D}(\omega-\nu)}{1 + I_-\mathscr{L}}\left[1 - \frac{I_-\mathscr{L}}{1 + I_-\mathscr{L}}\right]$$

or

$$\boxed{\alpha_+ = \frac{\alpha_0\gamma\mathscr{D}(\omega-\nu)}{(1 + I_-\mathscr{L})^2}}\ . \tag{70}$$

The denominator of this probe absorption coefficient is squared in contrast to that of the single wave case of Eq. (27). The increased saturation comes from the scattering of the saturator wave $\mathscr{E}_-$ off the weak grating induced by the saturator-probe fringe pattern. This effect is discussed for nondegenerate ($\nu_1 \neq \nu_2$) probe-saturator absorption in Sec. 8-1, where we see that the absorption coefficient as a function of the beat frequency $\nu_2 - \nu_1$ can have a very different shape from the pure Lorentzian of Eq. (70).

Expanding Eq. (66) to third-order in both field amplitudes, we find (see Prob. 6-3) $\alpha_+ = \alpha_0\gamma\mathscr{D}[1 - \mathscr{L}I_+ - 2\mathscr{L}I_-,]$, i.e., I_- saturates α_+ *twice* as much as I_+ does. This is the degenerate two-level version of the nonlinear nonreciprocity discussed in Sec. 2-3 [see Eq. (21b)]. It has important consequences for the operation of ring lasers as discussed in Sec. 6-4.

With a little more effort, one can write the propagation equations for the "distributed feedback" laser. This laser consists of a gain medium ($\alpha_0 < 0$) in which a gain and/or index grating is created by some external means. Like spatial holes, this extra grating also produces scattering, i.e., feedback, of one wave into the other. Chapter 6 treats a more conventional laser, in which feedback is accomplished by mirrors.

5-4. Two-Photon Two-Level Model

In this section we derive the "two-photon two-level model", a model that allows us to transfer much of the two-level understanding of previous sections immediately to the two-photon transition shown in Fig. 5-5. Here the transition between a and b is nearly resonant ($\omega_{ab} \simeq 2\nu$), but is not dipole allowed ($\wp_{ab}=0$). The transitions from a and b to the intermediate states j are dipole allowed, but are assumed to be sufficiently far from resonance that they can be treated using first-order perturbation theory. The model yields density matrix equations of motion that closely resemble those for the single-photon case. Two major differences occur between the two models. First, dynamic Stark shifts of the level frequencies that are ignored by construction in the one-photon situation can play an important role in the two-photon case. Second, the coherence induced between

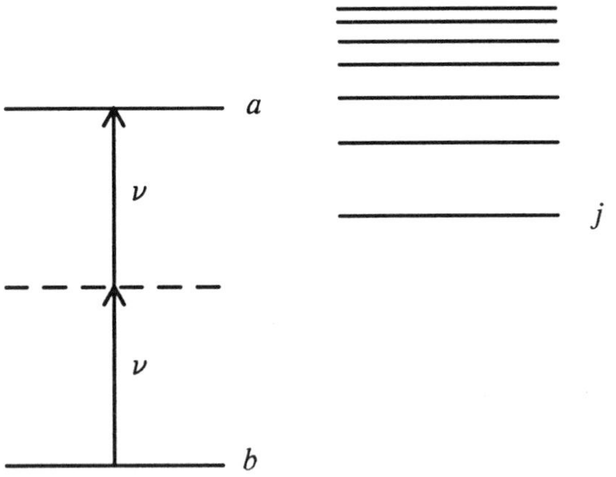

Fig. 5-5. Two-photon level scheme. The intermediate levels j are assumed to be sufficiently nonresonant that they acquire no appreciable population.

the two levels in the two-photon case does not contribute directly to the polarization; an additional atom-field interaction is required. These two differences cause the algebra for the two-photon problems to be about four times as involved as that for the one-photon problems, and the resulting physics has considerably more variety. We first derive the equations of motion for the two-photon density matrix and then find the complex absorption coefficient for a single running wave. The calculation is of value wherever the one-photon model can be used, e.g., in saturation spectroscopy, lasers, optical-bistability, and in phase conjugation.

In general the polarization of the medium with the level scheme in Fig. 5-5 is given by [Sargent *et al.* (1985)]

$$P(\mathbf{r},t) = N' \,\mathrm{tr}\{e\mathbf{r}\rho\} = N' \sum_j [\wp_{aj}\rho_{ja} + \wp_{bj}\rho_{jb}] + \text{c.c.}, \tag{71}$$

where N' is the number of interacting systems, $\wp_{aj}$ is the electric-dipole matrix element between the a and j states, and ρ_{ja} is the density-matrix element between j and a. Since $a \longleftrightarrow b$ is a two-photon transition, $\wp_{ab}$ vanishes. We consider cases in which the polarization (71) is induced by the electric field

$$E(\mathbf{r},t) = \frac{1}{2}\,\mathcal{E}(\mathbf{r})\, e^{-i\nu t} + \text{c.c.}, \tag{72}$$

where $\mathcal{E}(\mathbf{r})$ varies little in a time $1/\nu$, but may have rapid spatial variations like $\exp(i\mathbf{K}\cdot\mathbf{r})$. This field induces the polarization

$$P(\mathbf{r},t) = \frac{1}{2}\,\mathcal{P}(\mathbf{r})\, e^{-i\nu t} + \text{c.c.}, \tag{73}$$

where the complex polarization $\mathcal{P}(\mathbf{r})$ also varies little in the time $1/\nu$. Combining Eqs. (71) and (73), we find

$$\mathcal{P}(\mathbf{r}) = 2N' \sum_j [\wp_{aj}\rho_{ja} + \wp_{bj}\rho_{jb} + \text{c.c.}]\, e^{i\nu t}\ , \tag{74}$$

where in $\mathcal{P}(\mathbf{r})$ we keep only terms varying little in the optical period $1/\nu$.

The electric-dipole coherences ρ_{ja} are induced by the interaction energies

$$\mathcal{V}_{ja} = -\frac{1}{2}\wp_{ja}\mathcal{E}(\mathbf{r})\,e^{-i\nu t} + \text{c.c.} \tag{75}$$

with a similar formula for ρ_{jb}. Using the general Schrödinger equation of motion

$$\dot\rho_{ij} = -(\gamma_{ij} + i\omega_{ij})\rho_{ij} - i\hbar^{-1}[\mathcal{V}, \rho]_{ij}, \tag{76}$$

we have

$$\dot\rho_{ja} = -(\gamma_{ja} + i\omega_{ja})\rho_{ja} + i(2\hbar)^{-1}[\mathcal{E}e^{-i\nu t} + \mathcal{E}^* e^{i\nu t}][\wp_{ja}\rho_{aa} + \wp_{jb}\rho_{ba}] \tag{77}$$

$$\dot\rho_{jb} = -(\gamma_{jb} + i\omega_{jb})\rho_{jb} + i(2\hbar)^{-1}[\mathcal{E}e^{-i\nu t} + \mathcal{E}^* e^{i\nu t}][\wp_{jb}\rho_{bb} + \wp_{ja}\rho_{ab}], \tag{78}$$

where $\hbar\omega_{ij} = \hbar(\omega_i - \omega_j)$ is the energy difference between levels i and j, and γ_{ij} is the corresponding decay constant.

We integrate Eqs. (77) and (78) to first order in $\mathcal{E}$ *without making a rotating-wave approximation* (RWA), since ν differs substantially from all $\pm\omega_{ja}$ and $\pm\omega_{jb}$. Setting $\rho_{ba} = R_{ba}e^{2i\nu t}$, where R_{ba} varies little in an optical period, we find

$$\rho_{ja} = \frac{i}{2\hbar}\int_{-\infty}^{t} dt'[\mathcal{E}e^{-i\nu t'} + \mathcal{E}^* e^{i\nu t'}]\,e^{-(\gamma_{ja}+i\omega_{ja})(t-t')}[\wp_{ja}\rho_{aa} + \wp_{jb}R_{ba}e^{2i\nu t'}]$$

$$= \frac{1}{2\hbar}\left[\frac{\mathcal{E}e^{-i\nu t}}{\omega_{ja} - \nu} + \frac{\mathcal{E}^* e^{i\nu t}}{\omega_{ja} + \nu}\right]\wp_{ja}\rho_{aa} + \frac{1}{2\hbar}\left[\frac{\mathcal{E}e^{i\nu t}}{\omega_{ja} + \nu} + \frac{\mathcal{E}^* e^{3i\nu t}}{\omega_{ja} + 3\nu}\right]\wp_{jb}R_{ba}, \tag{79}$$

where we neglect the γ_{ja} in the nonresonant denominators. Since we assume $\omega_{ab} \equiv \omega \simeq 2\nu$, we have

$$\omega_{ja} + \nu \simeq \omega_{jb} - \nu\,, \tag{80}$$

which allows us to replace $\omega_{ja} + 3\nu$ in Eq. (79) by $\omega_{jb} + \nu$. Similarly integrating Eq. (78), we find

$$\rho_{jb} = \frac{1}{2\hbar}\left[\frac{\mathcal{E}e^{-i\nu t}}{\omega_{jb} - \nu} + \frac{\mathcal{E}^* e^{i\nu t}}{\omega_{jb} + \nu}\right]\wp_{jb}\rho_{bb} + \frac{1}{2\hbar}\left[\frac{\mathcal{E}e^{-3i\nu t}}{\omega_{ja} - \nu} + \frac{\mathcal{E}^* e^{-i\nu t}}{\omega_{jb} - \nu}\right]\wp_{ja}R_{ab}. \tag{81}$$

Substituting Eqs. (79) and (81) into the complex polarization (74) and keeping only terms that vary little in the time $1/\nu$, we have

$$\mathscr{P} = N'\mathscr{E}\,[k_{aa}\rho_{aa} + k_{bb}\rho_{bb}] + 2N'\,\mathscr{E}^* k_{ab}{}^* \rho_{ab}\, e^{2i\nu t}\,, \tag{82}$$

where the two-photon coefficients k_{ab}, k_{aa}, and k_{bb} are given by

$$k_{ab} = \frac{1}{\hbar}\sum_j \frac{\wp_{aj}\wp_{jb}}{\omega_{jb} - \nu} \simeq \frac{1}{\hbar}\sum_j \frac{\wp_{aj}\wp_{jb}}{\omega_{ja} + \nu} \tag{83}$$

$$k_{aa} = \frac{2}{\hbar}\sum_j \frac{|\wp_{ja}|^2\omega_{ja}}{\omega_{ja}{}^2 - \nu^2} \tag{84}$$

$$k_{bb} = \frac{2}{\hbar}\sum_j \frac{|\wp_{jb}|^2\omega_{jb}}{\omega_{jb}{}^2 - \nu^2}\,. \tag{85}$$

The k's are normalized such that $k_{ij}\mathscr{E}^2$ has units of energy in analogy with the one-photon $\wp\mathscr{E}$ energy. In these equations, we see that the $e^{\pm 3i\nu t}$ terms do not contribute to the polarization $\mathscr{P}$ directly. However they do contribute to ρ_{ab} via Eq. (87) and hence they contribute to $\mathscr{P}$ indirectly.

Equation (82) has a simple physical interpretation. $N'k_{bb}/\epsilon_0$ is the first order contribution to the induced susceptibility for a multilevel atom due to a probability ρ_{bb} for being in level b, as diagrammed in Fig. 5-6*a*. The term $2N'\mathscr{E}^* k_{ab}{}^*\rho_{ab}e^{2i\nu t}$ is the polarization component resulting from first-order electric-dipole interactions starting with the two-photon coherence ρ_{ab}. This contribution shows up in third and higher-order perturbation theory, since ρ_{ab} requires at least two interactions. It plays an important role in multiwave mixing in two-photon media. Such a perturbation process is depicted in Fig. 5-6*b*. It is convenient to write the polarization (82) in terms of the probability sum $\rho_{aa} + \rho_{bb} = 1$ and difference $D = \rho_{aa} - \rho_{bb}$ as

$$\mathscr{P} = \frac{1}{2}N'\mathscr{E}\,[k_{aa} + k_{bb} + (k_{aa} - k_{bb})D] + 2N'\mathscr{E}^* k_{ab}{}^*\rho_{ab}e^{2i\nu t}\,. \tag{86}$$

Using Eqs. (79) and (81), we also derive the "two-level" equations of motion for ρ_{aa}, ρ_{bb}, and ρ_{ab} using the two-photon rotating-wave approximation, i.e., we neglect terms like $1/[\gamma+i(\omega+2\nu)]$ compared to $1/[\gamma+i(\omega-2\nu)]$. According to Eq. (76), we have

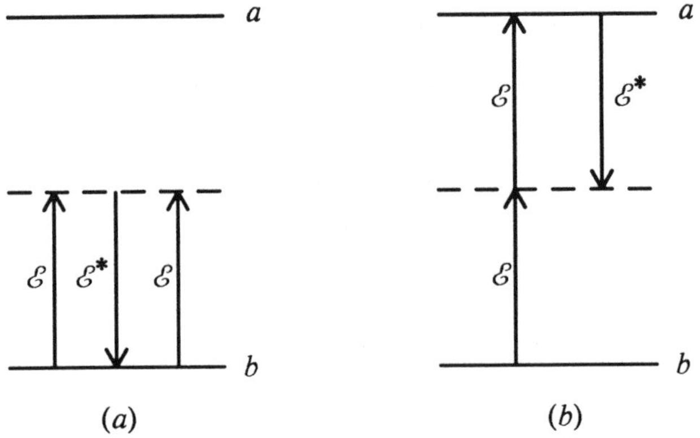

Fig. 5-6. (*a*) Perturbation process yielding a pure index contribution to the polarization of the medium (a similar process occurs for level *a*). (*b*) Perturbation process yielding a complex polarization due to scattering off the induced two-photon coherence ρ_{ab}.

$$\dot{\rho}_{ab} = -(\gamma + i\omega)\rho_{ab} - i\hbar^{-1} \sum_j [\mathcal{V}_{aj}\rho_{jb} - \rho_{aj}\mathcal{V}_{jb}] \tag{87}$$

$$\dot{\rho}_{aa} = -\gamma_a\rho_{aa} - \hbar^{-1} \sum_j [i\mathcal{V}_{aj}\rho_{ja} + \text{c.c.}]\,. \tag{88}$$

For simplicity we take upper to ground-lower level decay, for which $\dot{\rho}_{bb} = -\dot{\rho}_{aa}$, since we assume $\rho_{jj} = 0$. Problem 5-13 shows that cascade relaxation schemes, i.e., level *a* decays to level *b* via a cascade of intermediate states, are described by the same equations provided the population-difference decay time T_1 is appropriately defined. We find the population-difference equation of motion

$$\dot{D} = -2\gamma_a\rho_{aa} - \frac{2}{\hbar}\sum_j [i\mathcal{V}_{aj}\rho_{ja} + \text{c.c.}]$$

$$= -\frac{D+1}{T_1} - \frac{2}{\hbar}\sum_j [i\mathcal{V}_{aj}\rho_{ja} + \text{c.c.}]\,. \tag{89}$$

Substituting the dipole Eqs. (79) and (81) into (87), we have

$$\dot{\rho}_{ab} = -(\gamma + i\omega + i\omega_s I)\rho_{ab} - i[k_{ab}\,\mathcal{E}^2/4\hbar]e^{-2i\nu t}\,D\,, \tag{90}$$

where the two-photon dimensionless intensity

$$I = |k_{ab}\mathcal{E}^2|\sqrt{T_1T_2}/2\hbar \equiv |\mathcal{E}/\mathcal{E}_s|^2\,, \tag{91}$$

the two-photon coherence decay time $T_2 \equiv 1/\gamma$, and the Stark shift parameter

$$\omega_s = (k_{bb} - k_{aa})/2|k_{ab}|\sqrt{T_1T_2}\,. \tag{92}$$

Similarly substituting Eqs. (79) and (81) into (89), we find

$$\dot{D} = -(D + 1)/T_1 + \frac{1}{2\hbar}[ik_{ab}\mathcal{E}^2\,e^{-2i\nu t}\,\rho_{ba} + \text{c.c.}]\,. \tag{93}$$

Equation (90) is the same as that [Eq. (9) with (12)] for a one-photon two-level system with the substitutions

$$\omega \rightarrow \omega + \omega_s I;\quad \wp\mathcal{E}/\hbar \rightarrow k_{ab}\,\mathcal{E}^2/2\hbar;\quad \nu \rightarrow 2\nu\,. \tag{94}$$

Similarly, the corresponding Bloch equations (4.48) - (4.50) are given by these substitutions.

For single-frequency operation, we can solve Eqs. (90) and (93) in the Rate Equation Approximation of Sec. 5-1. Specifically, we assume that $\mathcal{E}$ and D vary little in the two-photon coherence decay time T_2, allowing Eq. (90) to be formally integrated with the value

$$\rho_{ab} = -i[k_{ab}\mathcal{E}^2/4\hbar]\,\mathcal{D}(\omega + \omega_s I - 2\nu)D\,e^{-i2\nu t}\,, \tag{95}$$

where the complex denominator $\mathcal{D}(\Delta)$ is given by (25). Substituting this into Eq. (93), we have

$$\dot{D} = -(D + 1)/T_1 - 2RD\,, \tag{96}$$

where the rate constant R is given by

$$R = \frac{1}{2\gamma}|k_{ab}\mathcal{E}^2/2\hbar|^2\,\mathcal{L}(\omega + \omega_s I - 2\nu) = \frac{I^2}{2T_1}\mathcal{L}(\omega + \omega_s I - 2\nu)\,, \tag{97}$$

and the Lorentzian $\mathcal{L}(\Delta)$ is given by Eq. (18).

Solving for D in steady-state ($\dot{D} = 0$), we have

$$D = -1/[1 + I^2\,\mathscr{L}(\omega + \omega_s I - 2\nu)]\ . \tag{98}$$

Substituting this into Eq. (95) gives

$$\rho_{ab} = i(k_{ab}\mathscr{E}^2/4\hbar)\,\mathscr{D}\,e^{-2i\nu t}/(1 + I^2\mathscr{L})\ , \tag{99}$$

where we have left off the frequency dependence on $\mathscr{D}$ and $\mathscr{L}$ for typographical simplicity. Finally substituting Eqs. (98) and (99) into the polarization (86) yields

$$\mathscr{P}(\mathbf{r}) = \frac{1}{2}\,N'\mathscr{E}\left[k_{aa} + k_{bb} + \frac{k_{bb} - k_{aa}}{1 + I^2\mathscr{L}} + i\,\frac{|k_{ab}\,\mathscr{E}|^2\mathscr{D}/\hbar}{1 + I^2\mathscr{L}}\right]. \tag{100}$$

In terms of the Stark shift parameter (92), this becomes

$$\mathscr{P}(\mathbf{r}) = N'\mathscr{E}(\mathbf{r})\left[\frac{1}{2}\,(k_{aa} + k_{bb}) + \frac{|k_{ab}|\sqrt{T_2/T_1}(\omega_s T_1 + iI\gamma\mathscr{D})}{1 + I^2\mathscr{L}}\right]. \tag{101}$$

For the two-photon polarization (101), the absorption coefficient (4) is

$$\alpha = -i\,\frac{KN'}{4\epsilon_0}\,(k_{aa} + k_{bb}) + \alpha_0\,\frac{I\gamma\mathscr{D} - i\omega_s T_1}{1 + I^2\mathscr{L}}\ . \tag{102}$$

The real part of α determines the absorption in the medium. This is given by

$$\mathrm{Re}(\alpha) = \alpha_0\,\frac{\gamma^2 I}{\gamma^2(1+I^2) + (\omega+\omega_s I-2\nu)^2}\ , \tag{103}$$

where the two-photon absorption parameter

$$\alpha_0 = KN'|k_{ab}|\sqrt{T_2/T_1}/2\epsilon_0\ . \tag{104}$$

The real part is totally due to the field scattering off the induced two-photon coherence ρ_{ab}, i.e., the D contribution in Eq. (86) produces only index of refraction changes. For small I, ρ_{ab} is linearly proportional to I. Hence $\mathrm{Re}(\alpha)$ is also linearly proportional to I, in contrast with the one-

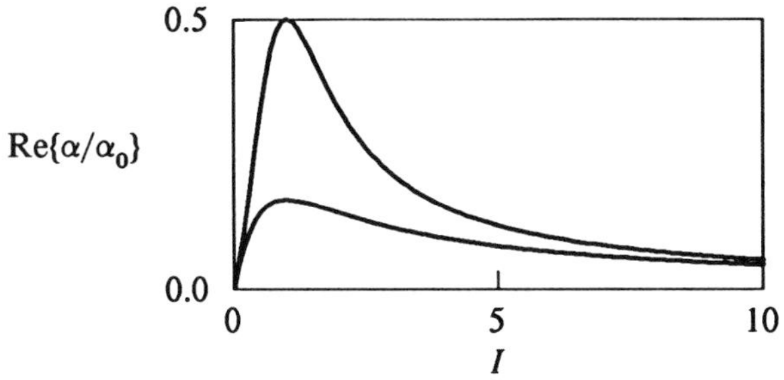

Fig. 5-7. $\mathrm{Re}\{\alpha\}/\alpha_0$ given by Eq. (103) for $\omega_s = \gamma$, and $\omega - 2\nu_2 = \pm\gamma$, where $-\gamma$ gives the larger values.

photon case given by (27), as shown in Fig. 5-7. The larger values occur for the negative detuning $-\gamma$, which is effectively tuned by the intensity I into resonance due to the dynamic Stark shift. In the limit of large I, $\mathrm{Re}(\alpha)$ reduces to $\alpha_0\, \mathscr{L}(\omega_s)/I$, which aside from the $\mathscr{L}(\omega_s)$ and the different definition of α_0 has the same value as in the one-photon case.

The imaginary part of α adds to the wave vector K and hence, changes the index of refraction. It has the value

$$\mathrm{Im}(\alpha) = -\frac{KN'}{4\epsilon_0}(k_{aa} + k_{bb}) - \alpha_0 \frac{I\gamma^{-1}(\omega+\omega_s I-2\nu)\mathscr{L} + \omega_s T_1}{1 + I^2\mathscr{L}} . \tag{105}$$

The term proportional to $\omega+\omega_s I-2\nu$ is similar to the one-photon index term and yields typical anomalous dispersion spectra. In addition, there is a Stark shift term (that proportional to $\omega_s T_1$) missing in the one-photon case. This term is purely positive and distorts the usual index spectra in a fashion similar to that noted by Fano (1961) for autoionization spectra.

In the limit of large $\omega_s I$, Eq. (105) approaches

$$\mathrm{Im}(\alpha) \to -\frac{KN'}{4\epsilon_0}(k_{aa} + k_{bb}) - \alpha_0 \frac{\gamma\omega_s + \omega_s{}^3 T_1}{\gamma^2 + \omega_s{}^2} . \tag{106}$$

Due to the dynamic Stark shift, this does not bleach to zero, unlike the single-photon case. This is important in phase conjugation, where we find that the reflection coefficient approaches a nonzero value for large $\omega_s I$ due to induced index gratings.

5-5. Polarization of Semiconductor Gain Media

This section derives the polarization of a semiconductor medium using a simple but effective two-band "quasiequilibrium" model. This model is fairly accurate for single-mode laser operation and reveals important ways in which semiconductor media are similar to and yet different from the inhomogeneously broadened two-level media of Sec. 5-2. Our discussion introduces the basic features of a two-band semiconductor, points out the most important assumptions of the model, and gives references to more precise treatments. Section 6-5 applies the model to semiconductor diode-laser operation. Section 10-1 ends with a discussion of the stability of this single-mode operation.

Our semiconductor model has an "upper-level" band called the *conduction* band, where electrons can flow, and a lower-level band, called the *valence* band, where *holes* can flow. Holes are the absence of electrons. When we say electron, we mean a conduction electron, although there are valence electrons, and when we say hole, we mean a valence hole. As such holes and electrons are charge carriers, and we often refer to them together simply as carriers. The two bands are diagrammed in Fig. 5-8. Relative to their respective bands, the electron and hole energies are given by

$$\varepsilon_e(\mathbf{k}) = \frac{\hbar^2 k^2}{2m_e}, \tag{107}$$

$$\varepsilon_h(\mathbf{k}) = \frac{\hbar^2 k^2}{2m_h}, \tag{108}$$

where m_e and m_h are the electron and hole effective masses, and $\hbar\mathbf{k}$ is the momentum. If an electron of momentum $\hbar\mathbf{k}$ in the valence band absorbs light, it is excited into the conduction band leaving behind a hole of momentum $-\mathbf{k}$ in the valence band. The energy of the photon inducing this transition is given by

$$\begin{aligned}\hbar\omega(\mathbf{k}) &= \varepsilon_e(\mathbf{k}) + \varepsilon_h(\mathbf{k}) + \varepsilon_g + \delta\varepsilon_g \\ &= \varepsilon(\mathbf{k}) + \varepsilon_g + \delta\varepsilon_g ,\end{aligned} \tag{109}$$

where ε_g is the zero-field band-gap energy, ε is the reduced-mass energy

$$\varepsilon = \frac{\hbar^2 k^2}{2m}, \tag{110}$$

m is the reduced mass defined by

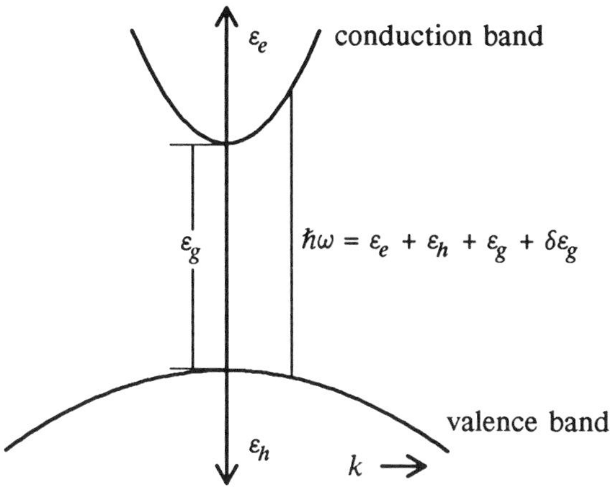

Fig. 5-8. Energy band diagram ε versus momentum k. The electron energy ε_e increases upward, and the hole energy ε_h increases downward. Optical transitions are vertical on this graph, since photons have negligible momentum compared to electrons. The semiconductor parameters illustrate GaAs, a common laser material. The bandgap energy ε_g = 1.462 meV, the dielectric constant ϵ = 12.35 giving an index of refraction of 3.5, the electron effective mass m_e = 1.127 m, where m is the reduced mass, and the hole effective mass m_h = 8.82 m. As seen from Eqs. (107) and (108), the curvatures of the bands are inversely proportional to their respective effective masses.

$$\frac{1}{m} = \frac{1}{m_e} + \frac{1}{m_h}, \tag{111}$$

and $\delta\varepsilon_g$ is a reduction in the bandgap energy due to the intraband (electron-electron and hole-hole) Coulomb repulsion and the fermion exchange correlation. Both of these effects are enhanced by increasing the carrier density. For our purposes, we express our results relative to the reduced (*renormalized*) band gap, so that we don't need to know the value of the reduction.

Interband (electron-hole) Coulomb attraction can also be important. For low (but nonzero) carrier densities, Coulomb attraction creates excitons, which are H-like "atoms" consisting of a bound electron-hole pair.

The exciton Bohr radius in GaAs is 124.3 Å, and the exciton Rydberg energy is 4.2 meV, which is tiny compared with 13.6 eV for the H atom and small compared to room-temperature $k_B T = 25$ meV. The nonlinear response of excitons can lead to optical bistability (see Chap. 7). As the carrier density increases (due to an injection current or optical absorption), the Coulomb potential becomes increasingly screened, and for densities above 10^{16} cm^{-3} the excitons are completely ionized. The Coulomb attraction still exists and reshapes the semiconductor absorption spectrum in a way called *Coulomb enhancement*. This is particularly important for media with one or two dimensions (quantum wires and wells), but it not so important for bulk gain media. Since we are basically interested in the latter, we neglect the Coulomb enhancement. For further discussion, see Sargent *et al.* (1988), Haug and Koch (1990), and the references therein.

The most important role of the Coulomb interaction is called *carrier-carrier* scattering. This has a counterpart in gas lasers known as velocity changing collisions, but is a much stronger effect in semiconductors and has consequences in addition to reequilibrating the carrier distributions. For densities high enough to get gain (2×10^{18} cm^{-3}), the excitons are ionized and two main effects remain from the carrier-carrier scattering: 1) The intraband carrier distributions can each be described by Fermi-Dirac distributions provided external forces like light fields vary little in the carrier-carrier scattering time of .1 picoseconds or less. 2) Spontaneous emission, called *radiation recombination*, is proportional to the product of electron and hole occupation probabilities, while for the pure radiative-decay two-level case it is proportional to the probability of upper-level occupation alone.

The rapid carrier equilibration into Fermi-Dirac distributions greatly simplifies the analysis, since instead of having to follow the carrier densities on an individual **k** basis, we only need to determine the total carrier density N. The individual k-dependent densities are then given by

$$\boxed{f_\alpha(k) = \frac{1}{e^{\beta[\varepsilon_\alpha(k)-\mu_\alpha]} + 1}}\,, \tag{112}$$

where $\alpha = e$ for electrons, $\alpha = h$ for holes, $\beta = 1/k_B T$, k_B is Boltzmann's constant, T is the absolute temperature, and μ_α is the carrier chemical potential, which is chosen to yield the total carrier density N. From Eq. (112), we see that μ_α equals the carrier energy ε_α for which f_α is precisely $\frac{1}{2}$. For intrinsic (undoped) semiconductors, the total (summed over **k**) electron number density equals total hole number density, that is

$$N = V^{-1}\Sigma_{\mathbf{k}} f_e(\mathbf{k}) = V^{-1}\Sigma_{\mathbf{k}} f_h(\mathbf{k})\ . \tag{113}$$

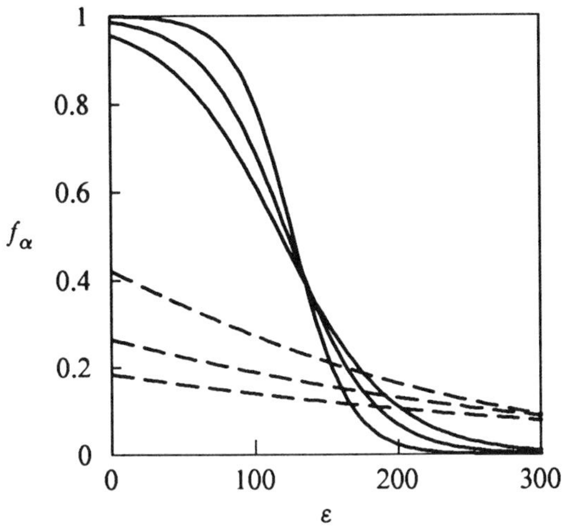

Fig. 5-9. Fermi-Dirac distributions given by Eq. (112) versus the "reduced" energy $\varepsilon = \hbar^2k^2/2m$ for electrons ($\alpha = e$, solid lines) and holes ($\alpha = h$, dashed lines) for the temperatures (high to low) T = 200, 300, and 400 K. The chemical potentials μ_α in Eq. (112) are chosen so that the total carrier density N of Eq. (113) is 3×10^{18} per cm^3 for all curves. This gives μ_e = 112.4, 109.4, and 104.8 meV and μ_h = -5.63, -26.5, and -51.5 meV for T = 200, 300, and 400 K, respectively.

Here for typographical simplicity, we include the two spin states as part of the $\Sigma_\mathbf{k}$. Three dimensions are summed for a bulk semiconductor, two for quantum wells, etc. The chemical potential μ_α that satisfies this equation can be determined numerically, but fairly accurate analytic formulas exist. The Fermi-Dirac distribution has neat properties such as

$$\frac{1}{e^x + 1} + \frac{1}{e^{-x} + 1} = 1 \ , \tag{114}$$

which we use at the end of this section to find the width of the laser gain region. Figure 5-9 plots electron and hole Fermi-Dirac distributions for several temperatures. Note that the hole distributions all have negative chemical potentials and look very much like the decaying tails of Maxwell-Boltzmann distributions.

We can describe the polarization of the medium in terms of a density matrix $\rho(\mathbf{k},z,t)$ similar in form to the $\rho(\omega,z,t)$ of Sec. 5-2. The polarization of the medium is

$$P(z,t) = V^{-1}\Sigma_{\mathbf{k}}\wp\rho_{ab}(\mathbf{k}) + \text{c.c.} \tag{115}$$

where $\wp$ is the electric-dipole matrix element which in general is **k**-dependent, V is the volume of the medium, ρ_{ab} is the off-diagonal element of the two-band density matrix whose elements are functions of the carrier wave vector **k**. The subscript a refers to the conduction band and b to the valence band. The volume V cancels out since it appears in the $\Sigma_{\mathbf{k}}$. Combining Eqs. (2) and (115), we find the slowly-varying complex polarization

$$\mathcal{P} = 2\wp e^{-i(Kz - \nu t)} V^{-1}\Sigma_{\mathbf{k}}\rho_{ab}(\mathbf{k}) \; . \tag{116}$$

The equation of motion for ρ_{ab} couples it to the probabilities that the electron is in the conduction or valence bands. It is convenient to use the probability of a hole instead of the probability for a valence electron. In this "electron-hole" picture, we define n_e to be the probability of having a conduction electron with momentum k, and n_h to be the probability of having a valence hole with momentum k. When a valence electron absorbs light, *both* a (conduction) electron and a hole are created. However due to carrier-carrier scattering, the probabilities that the electron and hole remain in the states of their creation decreases from unity to the values given by the electron and hole Fermi-Dirac distributions. Thus instead of two states for each **k**, we have the four states $|0_e\ 0_h\rangle_{\mathbf{k}}$, $|0_e\ 1_h\rangle_{\mathbf{k}}$, $|1_e\ 0_h\rangle_{\mathbf{k}}$, and $|1_e\ 1_h\rangle_{\mathbf{k}}$, where the first number is the occupation of the electron with momentum **k** and the second is that of the hole with momentum -**k**. Optically $|0_e\ 0_h\rangle_{\mathbf{k}}$ is connected to $|1_e\ 1_h\rangle_{\mathbf{k}}$, but carrier-carrier scattering connects them all. The states $|0_e\ 1_h\rangle_{\mathbf{k}}$ and $|1_e 0_h\rangle_{\mathbf{k}}$ do not occur in the two-level atom, since there's no way to take away the built-in hole of an excited two-level atom.

In general we need to define a 4×4 density matrix in this basis. However for semiclassical (classical field) problems, the problem reduces to three coupled equations. Specifically we identify $\rho_{aa} = \rho_{11,11} \equiv \langle 11|\rho|11\rangle$, $\rho_{bb} = \rho_{00,00}$, and $\rho_{ab} = \rho_{11,00}$. The commutator $[\mathcal{H}, \rho]$ tells us that ρ_{ab} is driven by $\rho_{aa} - \rho_{bb}$. Since $\mathrm{tr}\{\rho\} = 1$, i.e., $\rho_{11,11} + \rho_{10,10} + \rho_{01,01} + \rho_{00,00} = 1$, we eliminate ρ_{bb} to find

$$\rho_{aa} - \rho_{bb} = [\rho_{11,11} + \rho_{10,10}] + [\rho_{11,11} + \rho_{01,01}] - 1 = n_e + n_h - 1 \; , \tag{117}$$

where n_e is the probability of an electron with momentum **k** *independent* of whether there's a hole of momentum -**k**, and n_h is the corresponding probability of a hole. Hence ρ_{ab} is simply coupled to $n_e + n_h - 1$, and vice versa, and we don't have to treat a complicated four-level problem for a semiclassical theory. For spontaneous emission to occur, however, ρ_{aa} alone is involved, since the Pauli exclusion principle requires that both a hole of momentum -**k** and an electron of momentum **k** be occupied. We approximate ρ_{aa} for this purpose by $n_e n_h$, which is the value given by many-body theory.

From Eq. (117), we see that $\dot{n}_e$ is given by $\dot{\rho}_{aa} + \dot{\rho}_{10,10}$ and that $\dot{n}_h$ is given by $\dot{\rho}_{aa} + \dot{\rho}_{01,01}$. The optical contributions to $\dot{\rho}_{aa}$ are identical to those for the two-level system, since the commutator is the same. $\dot{\rho}_{10,10}$ and $\dot{\rho}_{01,01}$ result from carrier-carrier scattering alone. Including phenomenological terms to describe decay processes and the carrier-carrier scattering, we have the density matrix equations of motion

$$\dot{\rho}_{ab}(\mathbf{k}) = -(i\omega + \gamma)\rho_{ab}(\mathbf{k}) + i\hbar^{-1}\mathcal{V}_{ab}(z,t)[n_e(\mathbf{k}) + n_h(\mathbf{k}) - 1] \tag{118}$$

$$\dot{n}_e(\mathbf{k}) = \lambda_e - \gamma_{nr} n_e(\mathbf{k}) - \Gamma n_e(\mathbf{k}) n_h(\mathbf{k}) - \dot{n}_e\big|_{c\text{-}c} - (i\mathcal{V}_{ab}\rho_{ba}(\mathbf{k}) + c.c.)/\hbar \tag{119}$$

$$\dot{n}_h(\mathbf{k}) = \lambda_h - \gamma_{nr} n_h(\mathbf{k}) - \Gamma n_e(\mathbf{k}) n_h(\mathbf{k}) - \dot{n}_h\big|_{c\text{-}c} - (i\mathcal{V}_{ab}\rho_{ba}(\mathbf{k}) + c.c.)/\hbar \tag{120}$$

where λ_α, $\alpha = e$ or h, is the pump rate due to an injection current, γ_{nr} is the nonradiative decay constant for the electron and hole probabilities, Γ is the radiative recombination rate constant, and $\dot{n}_\alpha\big|_{c\text{-}c}$ is the carrier-carrier scattering contribution. This scattering drives the distribution n_α toward the Fermi-Dirac distributions of Eq. (112). In fact, the carrier-carrier scattering contribution found using many-body techniques vanishes when n_α is given by a Fermi-Dirac distribution. While rapidly suppressing deviations from the Fermi-Dirac distribution, the scattering does not change the total carrier density N of Eq. (113). Hence summing either Eq. (119) or (120), we find the equation of motion

$$\dot{N} = \lambda - \gamma_{nr} N - \frac{\Gamma}{V}\sum_{\mathbf{k}} n_e(\mathbf{k}) n_h(\mathbf{k}) - \left[\frac{i\hbar^{-1}\mathcal{V}_{ab}}{V}\sum_{\mathbf{k}} \rho_{ba}(\mathbf{k}) + \text{c.c.}\right], \tag{121}$$

where the injection current pump λ is given by

$$\lambda = \eta J/ed\ , \tag{122}$$

η is the efficiency that the injected carriers reach the active region, J is the current density, e is the charge of an electron, and d is the thickness of the active region.

The interaction energy matrix element $\mathcal{V}_{ab}$ for the field of Eq. (1) is given in the rotating-wave approximation by Eq. (12), where we assume that the electric-dipole matrix element $\wp$ varies little over the range of $\mathbf{k}$ values that interact. We suppose that the dipole decay rate γ is large compared to variations in $\mathcal{V}_{ab}$. For this we can solve Eq. (118) in the rate-equation approximation as

$$\rho_{ab}(\mathbf{k}) = -\frac{i\mathscr{D}\wp\mathcal{E}}{2\hbar} e^{i(Kz - \nu t)} d_0(\mathbf{k}) , \tag{123}$$

where the complex Lorentzian denominator $\mathscr{D}$ is defined by Eq. (25) and the probability difference $d_0(\mathbf{k})$ is given by

$$d_0(\mathbf{k}) = f_e(\mathbf{k}) + f_h(\mathbf{k}) - 1 . \tag{124}$$

We call this probability difference $d_0(\mathbf{k})$ in view of Chap. 8, which includes nonzero-frequency components ($d_{\pm 1}$) associated with the probability response to beat frequencies (population pulsations). Substituting Eqs. (12) and (123) into Eq. (121), we find the carrier-density equation of motion

$$\boxed{\dot{N} = \lambda - \gamma_{nr} N - \frac{\Gamma}{V} \sum_{\mathbf{k}} f_e(\mathbf{k}) f_h(\mathbf{k}) - \frac{|\wp\mathcal{E}/\hbar|^2}{2\gamma V} \sum_{\mathbf{k}} \mathscr{L}(\omega_{\mathbf{k}} - \nu) d_0(\mathbf{k})} \tag{125}$$

where $\mathscr{L}$ is the dimensionless saturator-wave Lorentzian of Eq. (18). Inserting Eq. (123) into Eq. (116), we have the complex polarization

$$\mathscr{P} = -\frac{i\wp^2\mathcal{E}}{\hbar V} \Sigma_{\mathbf{k}} \mathscr{D}(\omega_{\mathbf{k}} - \nu) d_0(\mathbf{k}) . \tag{126}$$

Substituting this into Eq. (4), we have the absorption coefficient

$$\boxed{\alpha = -\frac{K\wp^2}{2\epsilon\hbar\gamma V} \Sigma_{\mathbf{k}} \gamma\mathscr{D}(\omega_{\mathbf{k}} - \nu)[f_e(\mathbf{k}) + f_h(\mathbf{k}) - 1]} . \tag{127}$$

To evaluate α, we need to know the Fermi-Dirac distributions f_α of Eq. (112). These are implicit functions of N, since their chemical potentials μ_α must be chosen to satisfy the closure relation (113). We can determine N by solving Eq. (125) self-consistently using numerical techniques. Alternatively for any given N, we know the f_α and can evaluate Eq. (127) accordingly. To carry out the sums over $\mathbf{k}$, we convert them to integrals over $\mathbf{k}$ including a density of states factor specifying how the summation varies in $\mathbf{k}$. This factor depends greatly on the dimensionality of the system. For the three-dimensional case, the Im{α} of Eq. (127) diverges, due to an increasing density of states multiplied by a slowly decreasing index-like function. In Sec. 6-6 to obtain a satisfactory expression for the imaginary part, we subtract off the contribution in the absence of the carriers (the -1 in d_0), including it in the overall index factor ϵ. The α_i then contains only the carrier-induced contributions to the index. More discussion is given in Sec. 6-6, which calculates the semiconductor laser steady-state intensity and mode pulling.

Semiconductor Gain Media

We are about to turn to the first application of our theory of the interaction of light with matter, namely the laser. For this device, we need a gain medium, that is, the two-level Re{α} of Eq. (27) must be *negative.* This occurs for an inverted medium, i.e., one for which the N of Eq. (20) is positive. For a semiconductor laser, we need the Re{α} given by Eq. (127) to be negative. For simplicity suppose the linewidth function $\mathcal{L}$ = Re{$\gamma\mathcal{D}$} is a Dirac-δ function. Gain then occurs for $f_e + f_h > 1$, which according to Eq. (117) means that the electron-hole plasma has a population inversion. We can easily show that this inequality is valid provided the carrier reduced mass energy obeys the inequality

$$0 < \varepsilon < \mu_e + \mu_h \ . \tag{128}$$

To see that the *total chemical potential* $\mu_e + \mu_h$ is in fact the upper gain limit, note the identity of Eq. (114). Hence the condition $f_e + f_h = 1$ is satisfied for energies for which the sum of the electron and hole exponential arguments vanishes. Accordingly adding the exponential arguments implicit in Eq. (124) with the energy value $\varepsilon = \mu_e + \mu_h$ and using the reduced mass of Eq. (111), we find

$$[(\mu_e + \mu_h)m/m_e - \mu_e] + [(\mu_e + \mu_h)m/m_h - \mu_h] = 0 \ .$$

The total chemical potential $\mu_e + \mu_h$ is a crucial parameter in semiconductor laser theory since it defines the upper limit of the gain region. To the extent that the width $2\hbar\gamma$ of Lorentzian $\mathcal{L}$ is small compared to $\mu_e + \mu_h$, this sum in fact gives the width of the gain region. In general the gain width is somewhat smaller than $\mu_e + \mu_h$, since for sufficiently high energies $\mathcal{L}$ samples a medium with $f_e + f_h < 1$. As we see in the next chapter, if the gain exceeds the cavity losses, laser action can take place.

For carrier densities sufficiently high to give gain, the gain g = -Re{α} given by Eq. (127) is often replaced by $g = A_g(N - N_g)$, where A_g is a phenomenological gain coefficient and N_g is the carrier density at transparency (neither gain nor absorption). This simple "linear" gain formula leads to the same kind of gain saturation behavior (and laser output intensity) as that given by the *homo*geneously broadened two-level formula

(27), but without the tuning dependence. To show this, we substitute the real part of Eq. (127) into Eq. (125) and then approximate the gain by the phenomenological gain formula. Further including the Γ term approximately in the γ_{nr} term, and solving in steady state ($\dot{N} = 0$), we find

$$N - N_g = \frac{\lambda/\gamma_{nr} - N_g}{1 + I} ,$$

where the dimensionless intensity $I = \epsilon|\mathcal{E}|^2 A_g/K\hbar\gamma_{nr}$. This has the same form as Eq. (19) for D, but without any tuning dependence. This simple theory agrees remarkably well with observed laser operation. Since the semiconductor response involves a summation over a wide distribution of transition frequencies, one might think that an inhomogeneously broadened saturation behavior like that of Eq. (40) would be more appropriate. However the Fermi-Dirac quasiequilibria established by carrier-carrier scattering lead to a more homogeneously broadened saturation behavior.

5-6. Light Forces and Atomic Motion

So far our analysis has centered on the response of the internal degrees of freedom of the atoms to light. In this Section, we discuss how light fields also influence the center of mass motion. That this must be the case is evident on the grounds of momentum conservation: every time an atom interchanges energy with the electromagnetic field, the momentum of the absorbed or emitted light must be compensated for by the mechanical motion of the atom. As it stands now, our formalism cannot handle this problem, since it doesn't account for the center of mass of motion of the particles. This section generalizes our treatment of atom-light interactions to include this motion. Although we restrict our discussion to two-level atoms, our conclusions can readily be generalized to more complicated systems. These considerations lead to a number of important effects, such as the trapping of single atoms, atomic cooling, and atom optics [Kazantsev *et al.* (1990)].

We need to extend the Hamiltonian (3.111) in two major ways to discuss the mechanical effects of light on atoms. First, we must include the atomic kinetic energy, and second, we have to keep track of the spatial dependence of the fields explicitly. These generalizations give

$$\mathcal{H} = \frac{\hat{\mathbf{P}}^2}{2M} + \frac{\hbar\omega}{2}\sigma_z - [\wp E_0(\hat{\mathbf{R}},t)\sigma_+ + \text{adj.}] , \tag{129}$$

where M is the mass of the two-level atom and we use a ^ on the center of mass operators to avoid confusion. In this Section, we use capital letters to label the center of mass coordinates and momenta to distinguish them from the atomic internal degrees of freedom. In the dipole approximation, $\hat{\mathbf{R}}$ is the position of the atom, and is conjugate to the center of mass momentum operator $\hat{\mathbf{P}}$, with

$$[\hat{P}_i, \hat{R}_j] = -i\hbar\, \delta_{ij} \;, \tag{130}$$

The Hilbert space of the atom is the direct sum of the Hilbert spaces for the center of mass motion and for its internal degrees of freedom, and hence a general state vector has the form

$$|\Psi(t)\rangle = \int d^3R\; C_{\mathbf{R},i}(t)\; |R,i\rangle \;, \tag{131}$$

where $|R\rangle$ labels the position eigenstates, $\hat{\mathbf{R}}|R\rangle = \mathbf{R}|R\rangle$. Alternatively, we can use a momentum representation, such that

$$|\Psi(t)\rangle = \int d^3P\; C_{\mathbf{P},i}(t)\; |P,i\rangle \;, \tag{132}$$

with $\hat{\mathbf{P}}|P\rangle = \mathbf{P}|P\rangle$. From a mathematical point of view, the coordinate representation leads to a coupled set of second-order partial differential equations for probability amplitudes. This approach emphasizes the wave character of the problem. In contrast the momentum representation leads to a coupled set of ordinary differential equations that explicitly reveal the role of momentum transfer between the field and the atoms. This approach emphasizes the particle character of the problem.

Free Particle Motion

Consider first just the action of the kinetic energy part of the Hamiltonian (129) on the wave function

$$\Psi(X,t) = \psi(X,t)e^{iP_x X/\hbar} \;, \tag{133}$$

where $\psi(X,t)$ is a slowly varying function of X,

$$\left|\frac{\partial\psi}{\partial X}\right| << \left|\frac{P_x}{\hbar}\,\psi\right| \tag{134}$$

and P_x is the x-component of the center-of-mass momentum. The wave function (133) describes a plane wave propagating at velocity $V_x = P_x/M$ along the x-direction. Inserting this form of the wave function into Schrödinger's equation, we find

$$i\hbar \frac{\partial \psi}{\partial t} = -\frac{\hbar^2}{2M}\left[\frac{\partial^2 \psi}{\partial X^2} + \frac{2iP_x}{\hbar}\frac{\partial \psi}{\partial X} - \frac{P_x{}^2}{\hbar^2}\psi\right], \tag{135}$$

where we have used the correspondance $P_x \rightarrow -i\hbar\partial/\partial X$ and kept only the kinetic energy part of the Hamiltonian (129). The first term in parentheses can be ignored in the slowly varying approximation, while the last term only leads to a phase factor that can be straightforwardly transformed away. Under these conditions, Eq. (135) reduces to

$$\frac{\partial \psi}{\partial t} + V_x \frac{\partial \psi}{\partial X} = 0 , \tag{136}$$

that is, the partial time derivative in the Schrödinger equation is replaced by a convective derivative, which expresses explicitly the fact that the particle is moving at constant velocity V_x in the x-direction. This form of the Schrödinger equation plays an important role in the theory of Doppler-broadened lasers of Sec. 6-3. Equation (136) is similar to Eq. (1.17) for the electric field amplitude. Both are valid only if we neglect diffraction effects.

The derivation of Eq. (136) rests on two assumptions. The first one, expressed by the slowly varying amplitude condition (134), requires that the atomic velocity be large enough to ignore the first term in parentheses in Eq. (135). This indicates that the convective derivative approach fails for very low velocities. A more subtle difficulty is that by keeping only the kinetic energy part of the Hamiltonian, we have in fact assumed that the particle is a free particle, and hence that it moves at constant velocity. This is true provided that the interaction part of the Hamiltonian does not lead to significant changes in atomic velocities, which may or may not be a good approximation. To see that more clearly, we now discuss the effects of the dipole interaction on the atomic velocity.

Gradient Force and Scattering Force

It is convenient to work in the Heisenberg picture, where the equation for the center of mass momentum is

$$\frac{d\hat{\mathbf{P}}(t)}{dt} = \frac{i}{\hbar}[\mathcal{H}, \hat{\mathbf{P}}(t)] = \wp[\nabla E_0(\mathbf{R}(t))]\,[\sigma_+(t) + \text{adj.}] . \tag{137}$$

Taking the expectation value of this expression over the atomic internal degrees of freedom, we find

$$\langle d\hat{\mathbf{P}}(t)/dt\rangle = i\hbar^{-1}\langle[\mathcal{H}, \hat{\mathbf{P}}(t)]\rangle \simeq \wp[\nabla E_0(\mathbf{R}(t))]\,\langle\sigma_+(t) + \text{adj.}\rangle\ , \tag{138}$$

where $\langle\ \rangle$ indicates that we have also taken the expectation value of this expression over the internal degrees of freedom of the atom. The approximate equality in this expression is due to the fact that we have factorized $\nabla E_0(\mathbf{R}(t))$ outside the expectation value, i.e., we describe the center-of-mass motion of the atom semiclassically. For concreteness, we consider a running wave of the form

$$E_0(\mathbf{R}) = \frac{1}{2}E_0(X)e^{i(KZ - \nu t)} + \text{c.c.}\ , \tag{139}$$

where $E_0(X)$ might represent the transverse profile of a laser mode. Substituting this form into Eq. (138) we find in the rotating-wave approximation

$$\begin{aligned}\frac{d\langle\hat{\mathbf{P}}\rangle}{dt} &= \frac{\wp}{2}\,\hat{\mathbf{x}}\,\frac{dE_0(X)}{dX}\,\langle\sigma_+(t)e^{i(KZ-\nu t)} + \text{adj.}\rangle \\ &+ \frac{K\wp}{2}\,\hat{\mathbf{z}}E_0(X)\langle i\sigma_+(t)e^{i(KZ-\nu t)} + \text{adj.}\rangle\ .\end{aligned} \tag{140}$$

Noting that the expectation value $\langle\sigma_+(t)\rangle$ is just the density matrix element $\rho_{ba}(t)$, we use Eq. (8) to reexpress this result as

$$\begin{aligned}\frac{d\langle\hat{\mathbf{P}}\rangle}{dt} &= \frac{1}{2}\,\hat{\mathbf{x}}\,\frac{dE_0(X)}{dX}\,\text{Re}\{\mathcal{P}(Z)\} + \frac{K}{2}\hat{\mathbf{z}}E_0(X)\text{Im}\{\mathcal{P}(Z)\} \\ &\equiv \mathbf{F}_{grad} + \mathbf{F}_{scat}\ ,\end{aligned} \tag{141}$$

where $\hat{\mathbf{x}}$ and $\hat{\mathbf{z}}$ are unit vectors in the X and Z directions. We thus find two contributions to the mechanical force exerted by the electromagnetic field on the atoms. The first one, F_{grad}, is proportional to the real part of the polarization and depends on the field gradient. It is sometimes called the *gradient force.* The second force, F_{scat}, proportional to the imaginary part of the susceptibility, is sometimes called the *scattering force* or *light pressure force.*

Using the polarization of Eq. (26) written for a single atom ($N = 1$), we readily find

$$\mathbf{F}_{grad} = -\frac{\hbar}{4T_1}\,\frac{\gamma(\omega-\nu)}{\gamma^2(1+I) + (\omega-\nu)^2}\,\frac{dI_0(X)}{dX}\ , \tag{142}$$

where I is the dimensionless intensity (21) and $I_0(X) = |\wp E_0(X)/\hbar|^2 T_1 T_2$.

F_{grad} has a dispersive shape as a function of atom-field detuning, and changes sign at resonance. If the detuning is positive this force attracts the atom to the center of the light beam. This effect can be used to achieve stable optical traps.

Atomic Diffraction

The scattering force, in contrast, is responsible for Doppler shifts as well as for the atomic diffraction by light fields. To isolate this effect from those of the gradient force, we consider a constant field envelope $E_0(X) = E_0$, so that $dE_0/dX = 0$. It is convenient to work in the Schröd-inger picture and in the momentum representation. In this case, the Hamiltonian becomes

$$\mathscr{H} = \frac{\hat{\mathbf{P}}^2}{2M} + \frac{\hbar\omega}{2}\sigma_z - \frac{\wp E_0}{2}[\sigma_+ e^{-i\nu t}U(\hat{Z}) + \text{adj.}]\,, \tag{143}$$

where the (unnormalized) mode operator $U(\hat{Z})$ is

$$U(\hat{Z}) = \sin K\hat{Z} \tag{144}$$

for a standing wave and

$$U(\hat{Z}) = \exp(iK\hat{Z}) \tag{145}$$

for a running wave. These are the same as the mode functions that are used extensively in the laser theory of Chap. 6, except that here they have an operator character, since $\hat{Z}$ is the position operator for the atomic center of mass along the z-axis. In laser theory, in contrast, Z is simply a label for the position of the atom. With the correspondance $\hat{Z} \rightarrow i\hbar\partial/\partial\hat{P}_z$, we find readily that in the momentum representation, the mode operators become

$$U(\hat{P}_z) = \frac{1}{2i}[\exp(-\hbar K\partial/\partial\hat{P}_z) - \exp(\hbar K\partial/\partial\hat{P}_z)] \tag{146}$$

and

$$U(\hat{P}_z) = \exp(-\hbar K\partial/\partial\hat{P}_z) \tag{147}$$

for the standing-wave and running-wave cases, respectively.

Problem 5-15 shows that the action of the operator (147) on a momentum eigenstate $|P_z\rangle$ is

$$\exp(-\hbar K\partial/\partial\hat{P}_z)\;|P_z\rangle = |P_z - \hbar K\rangle\;, \tag{148}$$

that is, it decrements the z-component of the center-of-mass momentum by $\hbar K$. This provides a simple physical picture of the atom-field interaction. Each time the atom absorbs a quantum of energy from the wave propagating along the positive z direction, the center-of-mass momentum is increased by $\hbar K$. Conversely, each time the atom emits a quantum of energy $\hbar\nu$ in the positive z direction, its center of mass momentum is decreased by $\hbar K$. From the field dispersion relation $\nu = Kc$, we recognize that the momentum increment $\hbar K$ is precisely the momentum carried by a running wave of energy $\hbar\nu$. It is sometimes called the "photon momentum", a nomenclature which finds its origin in field quantization. (Note however that our description uses classical fields and never needs to invoke photons!) This transfer of momentum between the center-of-mass motion of the atom and the field is nothing but an expression of the conservation of momentum. This contrasts with the situation described by the Hamiltonian (3.111), which implicitly assumes a particle of infinite mass for which recoil effects vanish.

When the atom interacts with a standing wave, we see from the mode operator (146) that the excitation of the atom to its upper state can be accompanied by either a momentum increase or a decrease by $\hbar K$. This is because at the classical level, a standing wave can be understood as the sum of two counterpropagating running waves of equal amplitudes. To lowest order in perturbation theory, the atomic wave function after interaction with the field consists therefore of two components whose momenta differ by $2\hbar K$.

This discussion suggests that it is possible to reverse the roles of matter and light from the situation of conventional optics and for instance diffract an atom by a light beam. Several types of atomic mirrors, lenses, gratings and interferometers are readily conceivable. Indeed, all basic elements to perform atom optics, or "de Broglie optics," can be envisioned. Besides the interest in potential applications, these mechanical manifestations of light-matter interactions are also a fascinating area of fundamental research that combine the internal quantum structure of atomic particles with their translational degrees of freedom in an essential way.

To gain a more quantitative feeling for the scattering of a two-level atom off a standing wave, we consider an atom initially in its ground electronic state $|b\rangle$, with initial center of mass momentum $\mathbf{P}$. We neglect the gradient force, so that the component $\mathbf{P}_T$ of $\mathbf{P}$ perpendicular to the z-direction remains unchanged by the interaction with the field, and it is sufficient to keep track of P_z. In the momentum representation, the atomic wave function $|\Psi(t)\rangle$ reduces to

$$|\Psi(t)\rangle = \int d^3P \, C_{P_z,i}(t) \, |P_z \, i\rangle \, , \tag{149}$$

with

$$|\Psi(0)\rangle = |P_0 \, \mathrm{b}\rangle \, . \tag{150}$$

On resonance ($\nu = \omega$), and with Eq. (148), the equations of motion for the probability amplitudes $C_{P_{z,i}}$ are found by a straightforward generalization of Eqs. (3.99) and (3.100) as

$$\frac{d}{dt} C_{P_{z,a}} = -\frac{iP_z^2}{2M\hbar} C_{P_{z,a}} + \frac{\mathcal{R}_0}{2} [C_{P_z+\hbar K,b} - C_{P_z-\hbar K,b}] \tag{151}$$

$$\frac{d}{dt} C_{P_{z,b}} = -\frac{iP_z^2}{2M\hbar} C_{P_{z,b}} + \frac{\mathcal{R}_0}{2} [C_{P_z+\hbar K,a} - C_{P_z-\hbar K,a}] \, . \tag{152}$$

Since probability amplitudes are coupled via increments of $\hbar K$ in longitudinal momentum, it is convenient to express the momenta in units of $\hbar K$ as $P_0 = q\hbar K$ and $P_z = (q+n)\hbar K$ for n an integer. This allows to reexpress Eqs. (151) and (152) as

$$\frac{d}{dt} C_{n+q,a} = -\frac{i\hbar(n+q)^2K^2}{2M} C_{n+q,a} + \frac{\mathcal{R}_0}{2} [C_{n+q+1,b} - C_{n+q-1,b}] \tag{151}$$

$$\frac{d}{dt} C_{n+q,b} = -\frac{i\hbar(n+q)^2K^2}{2M} C_{n+q,b} + \frac{\mathcal{R}_0}{2} [C_{n+q+1,a} - C_{n+q-1,a}] \, . \tag{152}$$

The first term on the RHS of these equations contains the usual Doppler shift as can be seen from the expansion

$$\frac{\hbar(n+q)^2K^2}{2M} \simeq \frac{\hbar q^2K^2}{2M} + \frac{\hbar nqK^2}{M} + \frac{\hbar n^2K^2}{2M} \, . \tag{153}$$

The first term on the RHS is a global phase-shift that can be transformed away. Considering elementary absorption and emission processes, which are characterized by $n = 1$ and noting that $q\hbar K/M$ is the initial atomic velocity, the second term gives $\hbar qK^2/M = Kv = \nu v/c$, which is nothing but the first-order Doppler shift. This shift plays a considerable role in the physics of Doppler-broadened lasers of Chap. 6-3. The last term is the center-of-mass recoil energy.

Having reassured ourselves that Eqs. (151) and (152) contain the physical effects known to occur with moving particles, we now consider the case for atoms with no initial momentum along the z-axis, i.e., $q = 0$. In this case, Eqs. (151) and (152) reduce to

$$\frac{d}{dt} C_{n,a} = -\frac{i\hbar n^2 K^2}{2M} C_{n,a} + \frac{\mathcal{R}_0}{2} (C_{n+1,b} - C_{n-1,b}) \tag{154}$$

$$\frac{d}{dt} C_{n,b} = -\frac{i\hbar n^2 K^2}{2M} C_{n,b} + \frac{\mathcal{R}_0}{2} (C_{n+1,a} - C_{n-1,a}) \tag{155}$$

with the initial conditions

$$C_{n,a}(0) = \delta_{n,0} , \tag{156}$$

$$C_{n,b}(0) = 0 . \tag{157}$$

With these initial conditions, Eqs. (154) and (155) show that the only non-vanishing probability amplitudes are $C_{n,a}$ with n even and $C_{n,b}$ with n odd. It is therefore convenient to introduce the new variable

$$\begin{aligned} \zeta_n(t) &= C_{n,a}(t) \quad n \text{ even} \\ &= C_{n,b}(t) \quad n \text{ odd} , \end{aligned} \tag{158}$$

in terms of which Eqs. (154) and (155) reduce to the infinite set of difference-differential equations

$$\frac{d}{dt} \zeta_n = -\frac{i\hbar n^2 K^2}{2M} \zeta_n + \frac{\mathcal{R}_0}{2} (z_{n+1} - z_{n-1}) . \tag{159}$$

If the interaction time between the atom and the field is so short that only a few $\hbar K$ of transverse momentum can be transferred from the field to the atom, we can neglect the recoil energy part of this equation. This is the so-called Raman-Nath approximation, which holds provided that the recoil energy $\hbar^2 n^2 K^2/2M$ is negligible compared to the interaction energy $\hbar\mathcal{R}_0/2$ for all states $|n,i\rangle$ that become significantly populated. In this case, the solution of Eqs. (159) is given by (Prob. 5-16)

$$z_n(t) = i^n J_n(\mathcal{R}_0 t) , \tag{160}$$

where J_n is the Bessel function of integer order n. Problem 5-16 also shows that after an interaction time τ, $2n_{max}$ translational states are occupied, with $n_{max} \simeq \mathcal{R}_0\tau$. We can use Eq. (160) to compute the spread in transverse momentum after an interaction time τ. Simple properties of the Bessel functions show that

$$\langle p_z^2 \rangle^{1/2} = \hbar K \Big[\sum_n n^2 J_n{}^2(\mathcal{R}_0\tau) \Big]^{1/2} = \frac{\hbar K \mathcal{R}_0 \tau}{\sqrt{2}} \,. \tag{161}$$

The atomic spread in the transverse direction increases linearly with time and with the Rabi frequency. Increasing the interaction time τ, a situation is reached where translational states of the atom are occupied for which the recoil energy becomes comparable to the interaction energy $\hbar\mathcal{R}_0/2$. The conditions of the Raman-Nath approximation cease to be valid. The atomic spread is confined by the constraints of energy-momentum conservation and undergoes complicated oscillations. A numerical solution using a continued fraction of Eqs. (159) becomes necessary in this case [see Arimondo *et al.* (1981) and Bernhardt and Shore (1981)].

References

Arimondo, E., A. Bambini and S. Stenholm (1981), Phys. Rev. **A24**, 898.

Bernhardt A. F. and B. W. Shore (1981), Phys. Rev. **A23**, 1290.

Bennett, W. R., Jr. (1962), Appl. Opt. Suppl. **1**, 24.

Bloembergen, N., E. M. Purcell, and R. V. Pound (1948), Phys. Rev. **73**, 679.

Fano, U. (1961), Phys. Rev. **124**, 1866.

Gardiner, C. W. (1986), Phys. Rev. Lett. **56** (1917). This paper discusses the squeezed vacuum leading to Eqs. (129).

Kasantzev, A. P., G. I. Surdutovich, and V. P. Yakovlev (1990), *Mechanical Action of Light on Atoms*, World Scientific Publ., Singapore. This gives an up-to-date description of the mechanical action of light on near-resonant atoms.

Levenson, Marc D., and S. S. Kano, *Introduction to Nonlinear Laser Spectroscopy* (1988), Revised Edition, Academic Press, New York.

Sargent, M. III, S. Ovadia, and M. H. Lu, Phys. Rev. A**32**, 1596 (1985). This paper also calculates the two-photon sidemode absorption and coupling coefficients corresponding to Chaps. 8 and 9.

Stenholm, S. (1984), *Foundations of Laser Spectroscopy*, John-Wiley & Sons, New York. This is an excellent textbook on nonlinear laser spectroscopy.

The two-photon two-level model (Sec. 5-4) has been discussed in many papers starting with: M. Takatsuji (1971), Phys. Rev. A**4**, 808; B. R. Mollow (1971), Phys. Rev. A**4**, 1666.

Discussions of semiconductor media and lasers are given in

Agrawal, G. P., and N. K. Dutta (1986), *Long-Wavelength Semiconductor Lasers* Van Nostrand Reinhold Co., New York.

Chow, W. W., G. C. Dente, and D. Depatie (1987), IEEE J. Quant. Electron. E-23 1314.

Haug, H., and S. W. Koch (1990), *Quantum Theory of the Optical and Electronic Properties of Semiconductors*, World Scientific Publ., Singapore.

A derivation of "generalized Bloch equations" for semiconductor media is given by Lindberg, M. and S. W. Koch (1988), Phys. Rev. B**38**, 3342.

The specialization of these generalized Bloch equations to the simple model of Sec. 5-5 is given by Sargent, M. III, F. Zhou, S. W. Koch (1988), Phys. Rev. A**38**, 4673.

Yariv, A. (1989), *Quantum Electronics*, 3rd Ed., John-Wiley, New York.

Problems

5-1. An incident electric field $E = E_0 \cos\nu t$ interacts with a two-level medium whose lower level is the ground state.

a) If there are N atoms/volume, calculate the polarization of the medium by solving Eqs. (9) and (34) in the rate-equation approximation.

b) In general, the steady-state power absorbed by a medium is given by

$$P_{\rm abs} = \langle \dot{P}E \rangle_{\rm time\ average} ,$$

where P is the polarization of the medium and E is the electric field. Show that for a medium of two-level atoms the power absorbed is

$$P_{abs} = \frac{1}{2}\, N\Gamma\hbar\nu\, \frac{I\mathscr{L}}{1 + I\mathscr{L}} = \Gamma\hbar\nu f_a ,$$

where f_a is the saturated upper-level population/volume, $I = |\wp\mathcal{E}/\hbar|^2/\gamma\Gamma$, and the dimensionless Lorentzian $\mathscr{L}$ is given by (18).

5-2. Calculate the velocity of the fringe pattern for the field

$$E(\mathbf{r},t) = \frac{1}{2}A_1\exp(i\mathbf{K}_1\cdot\mathbf{r}-i\nu_1 t) + \frac{1}{2}A_1\exp(i\mathbf{K}_2\cdot\mathbf{r}-i\nu_2 t) + \text{c.c.}$$

5-3. List and describe eight approximations leading to the two-level rate equations.

5-4. Describe two phenomena that cannot be treated in the rate-equation approximation.

5-5. In the limit $\Delta\omega >> \gamma$, evaluate

$$J = \int_{-\infty}^{\infty} d\omega' \, \frac{\gamma e^{-(\omega-\omega')^2/(\Delta\omega)^2}}{\gamma^2 + (\omega'-\nu)^2} .$$

Using contour integration, evaluate the integral

$$J(\Delta) = i\int_{-\infty}^{\infty} d\delta \, \mathscr{D}(\delta)\mathscr{L}(\delta-\Delta) ,$$

where $\mathscr{D}$ and $\mathscr{L}$ are defined by Eqs. (25) and (18), respectively. Which half-plane yields the lesser algebra? Also evaluate this integral but with no $\mathscr{L}$, i.e., a single pole. Hint: close the contour in the lower-half plane, which has *no* poles! The answer is given by the *nonzero* contribution from the infinite circle. In this simple single-pole case, contour integration still works, while the usual residue theorem fails to apply.

5-6. Write a computer program in the language of your choice to evaluate the real part of the degenerate pump-probe absorption coefficient $\alpha_1 = \alpha_0\gamma\mathscr{D}_2/(1 + I_2\mathscr{L}_2)^2$. Have the program print out (or plot) a few values of Re$\{\alpha_1\}$ as the detuning is varied.

5-7. Write the complex absorption coefficient for a homogeneously broadened medium consisting of two isotopes with different line centers, ω_1 and ω_2. Weight the isotope contributions by their respective fractional abundances f_1 and f_2, where $f_1+f_2=1$.

5-8. In electrodynamics one learns that Rayleigh scattering results from light scattering by electric dipoles. By combining the results of Chaps. 1 and 5, calculate an expression for the intensity of Rayleigh scattering of a near-resonant electromagnetic wave of frequency ν interacting with an ensemble of two-level atoms with a resonance frequency ω. The lower level is the ground state and let Γ and γ denote the upper-level and dipole decay rates, respectively. What is the frequency of the scattered radiation? In the special case of central tuning ($\omega = \nu$), plot the scattered intensity versus the incident intensity and label any significant regions.

5-9. Consider the following three-level system. A monochromatic, near resonant field drives the transition from 1→2 ($\nu \simeq \omega_2 - \omega_1$). The excited level decays to the states $|1\rangle$ and $|3\rangle$ at the rates Γ and R, respectively, and the population returns from the metastable $|3\rangle$ state to the ground state at

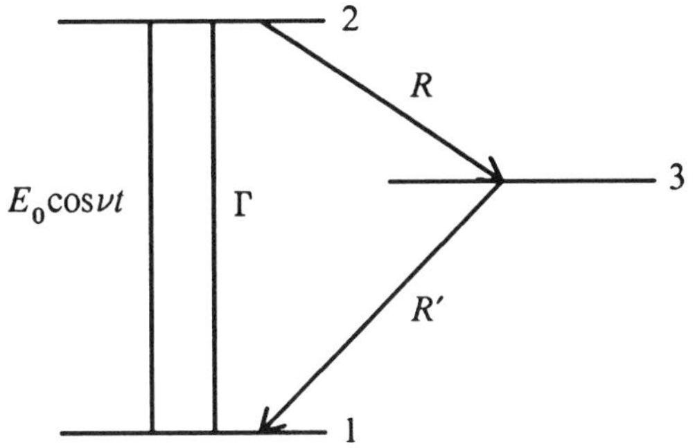

the rate R'. Determine the equations of motion for the relevant density matrix components for this system. Make the rotating wave approximation and let γ denote the linewidth of the transition (dipole decay) between 1 and 2. Note that there are only four nonzero components of ρ. Solve these equations for the upper-level population ρ_{22} in the rate-equation approximation. Show that in the limit of an infinitely large exciting resonant field that $\rho_{22} = R'/(2R' + R)$, which is smaller than the value of one-half predicted for a pure two-level system.

5-10. A "squeezed vacuum" is one for which the fluctuations in some optical quadrature are smaller than the average value permitted by the uncertainty principle. If the applied field is in phase with the squeezing field, $\mathcal{R}_0$ may be taken to be real, and the Bloch vector in such a squeezed vacuum may obey the equation of motion

$$\begin{aligned} \dot{U} &= -\gamma_u U - \delta V \\ \dot{V} &= -\gamma_v V + \delta U + \mathcal{R}_0 W \\ \dot{W} &= -(W+1)/T_1 - \mathcal{R}_0 V \,. \end{aligned} \tag{129}$$

In a normal vacuum, $\gamma_u = \gamma_v = T_2^{-1}$. *a*) Discuss what the constants δ, $\mathcal{R}_0$, T_1, and T_2 are. *b*) Solve these equations of motion in steady state. *c*) Find the real part of the absorption coefficient of a medium of such atoms with a density of N atoms per unit volume. *d*) What is the FWHM of the absorption spectrum?

5-11. Calculate the polarization

$$P(z,t) = \frac{1}{2}\sum_j \wp_{bj}\,\rho_{jb}\, e^{-i(Kz-\nu t)} + \text{c.c.}$$

between the ground-state level b and various levels j connected by electric-dipole transitions as induced by the field (1) which is assumed to be very nonresonant. Assume the system is initially in the ground state. Hint: use first-order perturbation theory.

5-12. The Kramers-Kronig relations allow one to calculate the real and imaginary parts of a linear susceptibility χ as integrals over one another as given by Eqs. (1.133) and (1.134). The power-broadened $\chi(\nu)$ corresponding to the complex polarization of Eq. (26) is

$$\chi(\nu) = \frac{\mathcal{P}}{\epsilon_0\mathcal{E}} = -\frac{iN\wp^2}{\hbar\epsilon_0}\,\frac{\mathcal{D}(\omega-\nu)}{1+I\mathcal{L}(\omega-\nu)} = \frac{N\wp^2}{\hbar\epsilon_0}\,\frac{\nu-\omega-i\gamma}{(\omega-\nu)^2+\gamma'^2}\,, \tag{130}$$

where $\gamma' = \gamma\sqrt{1+I}$ and N is the unsaturated population difference of Eq. (20) instead of the total number density used in Eq. (1.135). Show that unlike the linear susceptibility of Eq. (1.135), this power-broadened susceptibility does *not* satisfy Eqs. (1.133) and (1.134). What assumption do the Kramers-Kronig relations make that is violated by Eq. (130)? We see that these relations have to be applied with care in nonlinear situations.

5-13. Suppose level a decays to level b via an n-level cascade described by the equations of motion

$$\dot{\rho}_{aa} = -\gamma_a \rho_{aa} - [i\hbar^{-1}\mathcal{V}_{ab}\rho_{ba} + \text{c.c.}] .$$

$$\dot{\rho}_{jj} = \gamma_{j+1}\rho_{j+1,j+1} - \gamma_j \rho_{jj}$$

$$\dot{\rho}_{bb} = \gamma_1 \rho_{11} + [i\hbar^{-1}\mathcal{V}_{ab}\rho_{ba} + \text{c.c.}] .$$

Solve these equations in steady state, showing that $\rho_{aa} - \rho_{bb} = -N'/(1 + I\mathscr{L})$, where I is defined by Eq. (21) or (91) with

$$T_1 = \frac{1}{\gamma_a} + \frac{1}{2}\sum_{j=1}^{n} \frac{1}{\gamma_j} .$$

5-14. Above Eq. (79), it is asserted that R_{ba} varies slowly. Use Eq. (95) to verify this assertion.

5-15. Show that the action of the shift operator $\exp(-i\hbar K\partial/\partial\hat{P}_z)$ on a momentum eigenstate $|P_z\rangle$ is

$$\exp(-\hbar K\partial/\partial\hat{P}_z)|P_z\rangle = |P_z - \hbar K\rangle , \tag{148}$$

Hint: use the solution of Prob. 3-13 in a way analogous to the simple-harmonic-oscillator Eq. (3.128).

5-16. Show that the solution of Eq. (160) is $i^n\ J_n(\mathscr{R}_0 t)$.

Chapter 6
INTRODUCTION TO LASER THEORY

This chapter gives a simple theory of the laser using the classical electromagnetic theory of Chap. 1 in combination with Chap. 5's discussion of the interaction of radiation with two-level atoms. We consider arrangements of two or three highly reflecting mirrors that form cavities as shown in Fig. 6-1. Light in these cavities leaks out (decays to its $1/e$ value) in a time Q/ν, where ν is the frequency of the light and Q is the cavity quality factor (the higher the Q, the lower the losses). An active gain medium is inserted between the mirrors to compensate for the losses. In the simple cases we consider in this chapter, the electromagnetic field builds up until it saturates the gain down to the cavity losses. Chapter 10 considers some more complicated cases. Chapter 7 discusses a related cavity problem in which the medium in the cavity is not a gain medium, i.e., it has dispersion and/or absorption. This nonlinear cavity problem can lead to two or more stable output intensities for a given input intensity, and hence belongs to a class of problems called *optical bistability*. In the present chapter we also see a bistable configuration that involves active media, namely the homogeneously broadened ring laser.

Our theory is based on the principle of *self-consistency*, that is, we require that the electric field used to induce the polarization of the gain medium is identical to the one supported by that gain. To simplify the treatment, we assume that the electric field is plane polarized (e.g., by use of Brewster windows), and we ignore variations of the field transverse to the laser axis. Section 6-1 obtains multimode "self-consistency" equations, using the results of Chap. 1. These equations are interpreted in terms of energy conservation and mode pulling (as distinguished from anomalous dispersion). Section 6-2 substitutes the homogeneous and inhomogeneous broadening polarizations derived in Chap. 5 into the self-consistency equations to predict the laser steady-state amplitude and frequencies. Section 6-3 considers complications occurring from use of standing waves in Doppler-broadened media, and derives the very useful Lamb dip. Section 6-4 develops two-mode operation and the ring laser. Section 6-5 presents a simple theory of mode locking illustrated by frequency locking in the ring laser. Section 6-6 gives a simple single-mode theory of the semicon-

(a) (b)

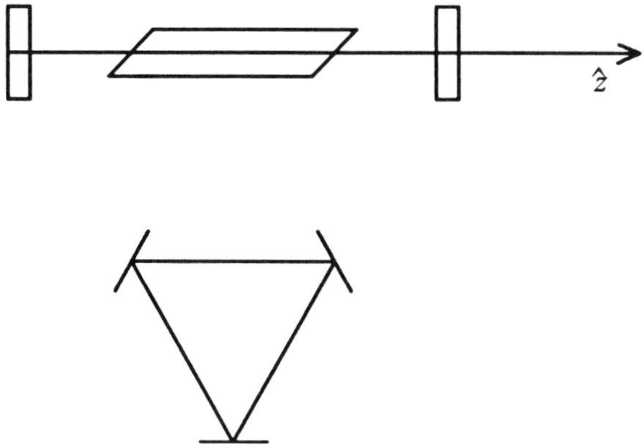

Fig. 6-1. (*a*) Diagram of laser showing reflectors in plane perpendicular to laser (z) axis and active medium between reflectors. Brewster windows are sketched on the ends of the active medium to help enforce conditions that only one polarization component of the electric field exists as is assumed in this chapter. (*b*) Corresponding ring laser configuration. Usually both running waves oscillate in a ring laser, although unidirectional operation is particularly easy to treat theoretically.

ductor diode laser based on the quasiequilibrium model of Sec. 5-5. Up to this point, the theory uses plane waves, although the fields in real lasers typically have Gaussian cross sections. Section 6-7 discusses the Gaussian beam and a simple way in which one can include some transverse variations in a laser theory.

6-1. The Laser Self-Consistency Equations

We suppose that the electromagnetic field in the laser cavity can be represented by the scalar electric field $E(z,t)$ written as the superposition of plane waves

$$E(z,t) = \frac{1}{2} \sum_n E_n(t) \exp[-i(\nu_n t+\phi_n)]\, U_n(z) + \text{c.c.} \tag{1}$$

Here the mode amplitudes $E_n(t)$ and phases $\phi_n(t)$ vary little in an optical period and the ν_n are the mode frequencies. The $U_n(z)$ specify the mode variations along the laser axis and consist of standing waves

$$U_n(z) = \sin(K_n z) \tag{2}$$

for the two-mirror laser and running waves

$$U_n(z) = \exp(iK_n z) \tag{3}$$

for the ring laser. The wave number $K_n = n\pi/L$ for the two-mirror laser and $2n\pi/L$ for the ring laser. The two cases are different since $2L$ is the round-trip length in the former, and L is the round-trip length in the ring. In Secs. 6-1 through 6-3, we consider standing-wave and unidirectional ring lasers, for which the K_n are positive. Section 6-4 considers the two-mode bidirectional ring laser, for which the two K_n are opposite in sign.

The superposition (1) differs from Eq. (1.60) used in Secs. 1-3 and Chap. 2 in that the amplitudes are functions of time and not of space. This is a basic characteristic of high-Q cavity problems, for which the *mode* amplitudes and phases are generally fairly uniform throughout the cavity (see Sec. 6-6 for discussion about transverse variations). This approximation doesn't preclude having rapid spatial variations in the *total* field envelope, such as a train of short pulses (see Sec. 10-4).

We take the polarization of the medium to be the corresponding superposition

$$P(z,t) = \frac{1}{2} \sum_n \mathcal{P}_n(t) \exp[-i(\nu_n t+\phi_n)]\, U_n(z) + \text{c.c.} \tag{4}$$

in which the complex polarization component $\mathcal{P}_n(t)$ also varies little in an optical period. They are complex inasmuch as in general the induced polarization has a different phase from the inducing field.

These quantities are then substituted into Maxwell's equations with careful attention paid to the slowly-varying properties of the E_n, ϕ_n, and $\mathcal{P}_n$, and with use of the orthogonality of the $U_n(z)$ [see the derivation of Eqs. (1.31) and (1.32)]. The result is the *self-consistency equations*

$$\dot{E}_n = -\frac{\nu}{2Q_n}E_n - \frac{\nu}{2\epsilon}\mathrm{Im}\{\mathcal{P}_n\} \tag{5}$$

$$\nu_n + \dot{\phi}_n = \Omega_n - \frac{\nu}{2\epsilon}\mathrm{Re}\{\mathcal{P}_n\}/E_n \tag{6}$$

so named because the field parameters ultimately appearing in the formulas for the $\mathcal{P}_n$ are taken to be the very same as the parameters in Eq. (1). Here the passive cavity frequencies $\Omega_n = cK_n$. In Eq. (5), we have included a phenomenological decay term $-\nu/2Q_n E_n$, where Q_n is the cavity quality factor for the nth mode. This is similar to the $\mu\partial\mathbf{J}/\partial t$ term in Eq. (1.25). Here Q_n is defined to be the ratio of the energy stored in the nth mode to the energy the nth mode loses per radian. The quantity ν/Q_n is thus the ratio of the energy stored in the nth mode to the energy the nth mode loses per second. As such $\nu/2Q_n$ is an amplitude cavity-loss coefficient. In the loss factor we have approximated the mode frequency ν_n by ν, independent of the mode number n. The Q_n can be adjusted to make up any difference as far as the cavity losses are concerned and the polarization contributions are only off by $|\nu_n - \nu|/\nu$, which is smaller than 10^{-6}. The advantage of approximating ν_n by ν is that the multimode polarization coefficients reduce to a simpler form.

It is worthwhile stopping at this point to gain a physical feel for these equations. Our discussion, although similar to that surrounding the classical case in Sec. 1-2 and 1-3, differs notably in the introduction of cavity losses and the use of a gain medium rather than an absorber. Chapter 2-5 shows how the complex polarization $\mathcal{P}_n$ can be related to the electric field component E_n by a complex susceptibility χ_n, that is,

$$\mathcal{P}_n = \epsilon\chi_n E_n = \epsilon(\chi'_n + i\chi''_n)E_n\ . \tag{7}$$

For our problem, this susceptibility is itself a decreasing function of the mode amplitude E_n inasmuch as the response of the laser medium saturates. With Eq. (7), Eqs. (5) and (6) simplify to

$$\dot{E}_n = -\frac{\nu}{2Q_n}E_n - \frac{\nu}{2}\chi''_n E_n \tag{8}$$

$$\nu_n + \dot{\phi}_n = \Omega_n - \frac{\nu}{2}\chi'_n . \tag{9}$$

Equation (8) expresses energy conservation (and could plausibly be postulated therefrom). To see this, note that the mode energy h_n is proportional to $E_n{}^2$. Hence multiplication of Eq. (8) by $2E_n$ yields the equation of motion for the nth component of energy h_n

$$\dot{h}_n = -\frac{\nu}{Q_n}h_n - \nu\chi''_n h_n$$

$$= -\frac{\text{cavity losses}}{\text{second}} + \frac{\text{medium gain } (\chi''_n < 0)}{\text{second}}$$

If the gain parameter $-\chi''_n$ saturates sufficiently in time to yield an energy gain/second equal in magnitude to the cavity losses/second, steady-state laser operation is achieved. This is the case in particular for single-mode operation.

Inasmuch as the susceptibility term χ'_n is small compared to unity, we can approximate the ν in Eq. (9) by Ω_n and interpret the resulting $1 - \frac{1}{2}\chi'_n$ factor as the first term in a Taylor expansion to get after "resumming"

$$\nu_n + \dot{\phi}_n = \frac{\Omega_n}{1 + \frac{1}{2}\chi'_n} = \frac{\Omega_n}{\eta},$$

where η is the index of refraction. This equation (or (9)) reveals an important difference between the gain problem and the classical absorption problem of Chap. 1, namely that the oscillation *frequency*, instead of the *wavelength*, is shifted by the medium. This results from the self-consistent nature of the laser field which requires an integral number of wavelengths in a round trip regardless of the medium characteristics.

Linear values of χ'_n and χ''_n are graphed in Fig. 6-2 for a homogeneously broadened medium having the line-center frequency ω ($\neq 2\pi\nu$!). Note that both curves are negative with respect to the classical index and absorption curves in Fig. 1-2. In addition to gain, this change of sign leads to mode pulling: ν_n is closer to ω than is the passive frequency Ω_n in contrast to the dispersive nature one expects of absorbers.

6-2. Steady-State Amplitude and Frequency

Chapter 5 derives steady-state polarizations of homogeneously- and inhomogeneously-broadened media subject to *cw* running waves, and finds complex Beer's law coefficients. The laser problem is very similar, but differs in two essential ways. First in lasers the unsaturated population difference N of Eq. (5.20) is typically positive, i.e., more systems are pumped to the upper state than to the lower state, thereby giving a gain medium. Second, rather than considering variations of the field envelope

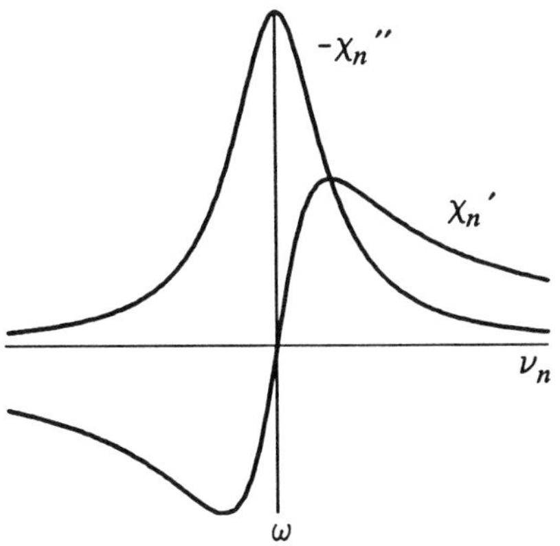

Fig. 6-2. Gain ($-\chi_n''$) and mode pulling (χ_n') parts of the complex susceptibility of Eq. (7) determined by Eq. (14) with no saturation [$\mathcal{S}(I_n) = 1$]. The medium is homogeneously broadened (Lorentzian gain).

in space, Eq. (5) determines its variation in time. This requires that we project the polarization onto the mode function $U_n(z)$ to obtain $\mathcal{P}_n(t)$. Specifically combining Eq. (4) with (5.7) and projecting both sides of the equation onto the mode $U_n(z)$, we find

$$\mathcal{P}_n(t) = 2\wp\, e^{i(\nu_n t+\phi_n)} \frac{1}{\mathcal{M}_n} \int_0^L dz\, U_n^*(z)\, \rho_{ab}(z,t)\ , \tag{10}$$

where $\mathcal{M}_n$ is the mode normalization factor

$$\mathcal{M}_n = \int_0^L dz\, |U_n(z)|^2\ . \tag{11}$$

In Eq. (5.24) for $\rho_{ab}(z,t)$, which was derived for a running wave (5.1), the e^{iKz} is just the mode function of the wave. Generalizing this result by replacing this factor by $U_n(z)$ and ν by ν_n, and inserting the outcome into Eq. (10), we find

$$\mathscr{P}_n(t) = -i\,\frac{\wp^2}{\hbar}\,E_n\,\mathscr{D}(\omega-\nu_n)\,\frac{1}{\mathscr{M}_n}\int_0^L dz\,\frac{N(z,t)|U_n(z)|^2}{1+I_n\,\mathscr{L}(\omega-\nu_n)|U_n(z)|^2}\,, \tag{12}$$

where the dimensionless intensity for the nth mode is

$$I_n \equiv |\wp E_n/\hbar|^2\,T_1T_2\,. \tag{13}$$

The polarization (12) can be integrated immediately for unidirectional operation since $|U_n(z)|^2 = 1$ in this case. This gives

$$\mathscr{P}_n(t) = -i\,\wp^2\hbar^{-1}\,E_n\,\mathscr{D}(\omega-\nu_n)\,\overline{N}\,\mathscr{S}(I_n)\,, \tag{14}$$

where the average population inversion density $\overline{N}$ is given by

$$\overline{N} = \frac{1}{L}\int_0^L dz\,N(z,t)\,, \tag{15}$$

and the unity-normalized saturation factor $\mathscr{S}(I_n)$ is

$$\mathscr{S}(I_n) = \frac{1}{1+I_n\,\mathscr{L}(\omega-\nu_n)}\quad\text{[running wave]}\,. \tag{16}$$

Equations (14) and (16) show that the gain $-\mathrm{Im}(\mathscr{P}_n)$ saturates as the intensity I_n increases. Substituting Eq. (5.18) for $\mathscr{L}$ and (5.25) for $\mathscr{D}$, we find

$$\mathscr{P}_n(t) = -\,i\wp^2\hbar^{-1}\,E_n\,\overline{N}\,\frac{\gamma - i(\omega-\nu_n)}{(\omega-\nu_n)^2+\gamma^2(1+I_n)}\,, \tag{17}$$

which reveals that the linear response width γ is power broadened to the value $\gamma(1+I_n)^{1/2}$. Equation (17) is very similar to Eq. (5.26), but refers to a time varying real field amplitude $E_n(t)$ in a cavity rather than to a spatially varying complex field amplitude $\mathscr{E}(z)$.

The polarization (12) can also be integrated for the standing-wave case as already discussed for Eq. (5.62). In fact, the bidirectional running-wave field (5.53) reduces to the single-mode standing wave field of Eq. (1) with $U_n(z) = \sin K_n z$ provided we set $\mathscr{E}_\pm = \pm E_n\exp(-i\phi_n)/2i$. The factors of $\pm 1/2i$ come from the $\sin K_n z$. Hence Sec. 5-3's standing-wave case $I_+ = I_-$ corresponds to $I_n/4$. We obtain the same saturation factor (5.65) except that now $a = 1+I_n\mathscr{L}/2$, $b = -I_n\mathscr{L}/2$, and $c = -1$, that is,

$$\mathcal{S}(I_n) = \frac{2}{I_n\ \mathcal{L}(\omega-\nu_n)}\left[1 - \frac{1}{[1 + I_n\ \mathcal{L}(\omega-\nu_n)]^{1/2}}\right] \quad \text{[standing wave]}\ . \tag{18}$$

This result is not simple to interpret, however, so we expand the square root in Eq. (18) up to order $I_n{}^2$. This expansion is valid for small laser intensities and is also directly derivable from third-order perturbation theory. We find

$$\mathcal{S}(I_n) = 1 - \tfrac{3}{4}\, I_n\ \mathcal{L}(\omega-\nu_n) \quad \left[\begin{matrix}\text{third-order}\\ \text{standing wave}\end{matrix}\right] . \tag{19}$$

This is the same as the third-order expansion of the unidirectional saturation factor (16) with I_n replaced by $3I_n/4$. Since the choice of mode function gives an intensity scale factor of four, we see (as for Eqs. (5.68) and (5.69)) that the low-intensity standing wave case has three times as much saturation as the unidirectional case. We get this factor of 3 because *two* waves saturate *and* the scattering off of the induced Bragg gratings (see Sec. 5-3) in the population inversion is destructive. Alternatively, we can interpret this result by noting that the standing-wave interacts with only part of the active medium and hence experiences less gain than a single running wave, yet this gain has to support two running waves. As discussed in connection with Eqs. (5.68) and (5.69), the standing-wave large-intensity saturation is only twice as large as the corresponding unidirectional saturation. We will see shortly that the steady-state laser operation is characterized by the fact that the saturated gain precisely equals the losses. Hence all else being equal, the running-wave laser gives from two to three times the output intensity of the standing wave laser.

Combining the amplitude self-consistency Eq. (5) with the polarization component (14) and multiplying through by the factor $2I_n/E_n$, we obtain the intensity equation of motion

$$\boxed{\dot{I}_n = 2I_n\left[g_n\ \mathcal{S}(I_n) - \frac{\nu}{2Q_n}\right]}\ , \tag{20}$$

in which the linear gain parameter

$$g_n(\nu_n) = (\nu\wp^2\overline{N}/2\epsilon\hbar\gamma)\ \mathcal{L}(\omega-\nu_n)\ , \tag{21}$$

and the unity-normalized saturation factor $\mathcal{S}(I_n)$ is given for various cases by Eqs. (16), (18), and (19).

Threshold operation is defined to be that for which the centrally tuned linear gain $g_n(\omega)$ equals the losses $\nu/2Q_n$. Below this value, Eq. (20) predicts exponential decay of I_n with the trivial steady-state intensity $I_n = 0$, that is, the laser doesn't oscillate. Threshold operation occurs for the threshold excitation $\overline{N}_T$, which allows the gain parameter g_n to be conveniently written in terms of the *relative excitation* $\mathcal{N}$ (excitation relative to threshold excitation) defined by

$$\mathcal{N} = \overline{N}/\overline{N}_T \ , \tag{22}$$

where $\overline{N}_T = \epsilon\hbar\gamma/\wp^2 Q_n$. In terms of $\mathcal{N}$, the gain is given by

$$g_n = \frac{\nu}{2Q_n} \mathcal{N} \mathscr{L}(\omega-\nu_n) \ . \tag{23}$$

In Eq. (20) we see quite simply the effect of saturation. For small intensity, there is exponential buildup with the factor $\exp[2(g_n - \nu/2Q_n)t]$. As I_n builds up, the gain term is reduced by the decreasing value of $\mathcal{S}(I_n)$, representing the fact that the atoms only have a finite amount of energy to offer. Eventually a steady-state ($\dot{I}_n = 0$) is reached when the *saturated gain* $g_n \mathcal{S}(I_n)$ equals the cavity losses, that is,

$$g_n \mathcal{S}(I_n) = \frac{\nu}{2Q_n} \ . \tag{24}$$

This is a particular case of the general single-mode, steady-state oscillation condition

$$\boxed{\text{Saturated Gain = Loss}} \tag{25}$$

valid for both classical and quantum oscillators. Steady-state intensities given by Eq. (24) for running and standing waves are summarized in Table 1. Figure 6-3 illustrates the detuning dependence for the standing-wave case. The intensity I_n of Eq. (13) is dimensionless, that is, it is given in units of the saturation intensity I_s of Eq. (5.23). Hence the laser output in watts/cm² is proportional to I_s. This is one role of I_s; another is power broadening, as seen, for example, in Eq. (17).

For extreme inhomogeneous broadening and a running-wave field, we use the polarization (5.39), which gives the intensity equation (20) with the saturation function

$$\mathcal{S}(I_n) = (1 + I_n)^{-1/2} \tag{26}$$

Steady-state Intensity I_n	Configuration
$\dfrac{\mathcal{N}\mathcal{L}-1}{\mathcal{L}}$	unidirectional ring [$\mathcal{S}$ of Eq. (16)]
$\dfrac{4}{3}\,\dfrac{\mathcal{N}\mathcal{L}-1}{\mathcal{N}\mathcal{L}^2}$	third-order, standing wave [from (19)]
$\dfrac{\mathcal{N}\mathcal{L}-\frac{1}{4}-\frac{1}{4}(10\mathcal{N}\mathcal{L}+1)^{1/2}}{\mathcal{L}}$	"exact", standing wave [from Eq. (18)]
$\mathcal{N}^2\exp\left[-\dfrac{2(\omega-\nu_n)^2}{(\Delta\omega)^2}\right]-1$	unidirectional ring with IHB $\Delta\omega>>\gamma$ [from Eq. (26)]

Table 6-1. Steady-state dimensionless intensities I_n for a number of laser configurations. Values are derived from Eq. (24) for various saturation functions. Here $\mathcal{L} = \mathcal{L}(\omega-\nu_n)$ for typographical simplicity.

and the linear gain

$$g_n = \frac{\nu\pi\wp^2\overline{N}\mathcal{W}(\nu)}{2\epsilon\hbar} = \frac{\nu}{2Q_n}\mathcal{N}\exp[-(\omega-\nu_n)^2/(\Delta\omega)^2] , \tag{27}$$

where in Eq. (22) we take $\overline{N}_T = \epsilon\hbar/\wp^2\mathcal{W}(\nu)Q_n$. Equation (5.43) describes the more general case of an arbitrary width Gaussian and gives the frequency dependence in terms of the plasma dispersion function. Hence the polarization cannot be written in the simple form of Eq. (14). Nevertheless the intensity equation is easily obtained (see Prob. 6-14).

Stability Analysis

Equation (20) is a simple example of a differential equation admitting more than one steady state solution. To be physically relevant, however, such a solution must be stable against small perturbations, otherwise, any noise will drive the system away from it. A powerful and general method to determine to which solution the laser will actually converge is to perform *a linear stability analysis*. For this we suppose that the intensity is given by the steady-state value in question plus a small deviation, that is,

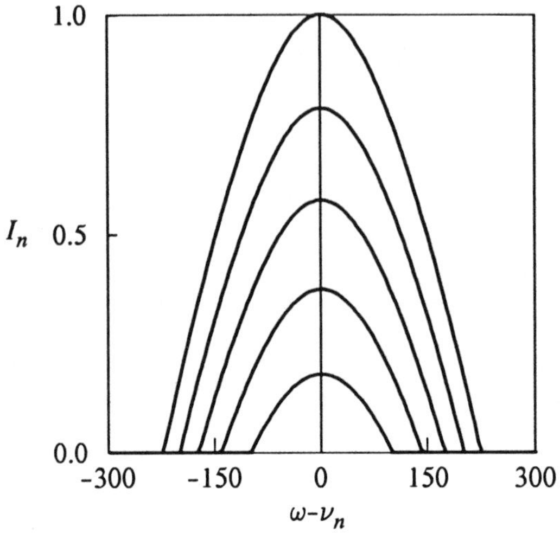

Fig. 6-3. Graphs of single-mode, strong-signal, standing-wave intensity (see Table 1) versus detuning ($\omega-\nu_n$) for homogeneously broadened ($\gamma=2\pi\times 100MHz$, $\gamma_{ab} = 2\pi\times 30MHz$) atoms. The relative excitations $\mathcal{N}$ of Eq. (22) are (in order of increasing maxima) 2, 3, 4, 5, and 6.

$I_n(t) = I_n^{(s)} + \epsilon_n(t)$. We substitute this into Eq. (20) and keep only terms that are linear in $\epsilon_n(t)$. This gives

$$\dot{\epsilon}_n = 2\epsilon_n \left[g_n \mathcal{S}(I_n^{(s)}) + I_n^{(s)} g_n \frac{d\mathcal{S}}{dI_n} - \frac{\nu}{2Q_n} \right] . \tag{28}$$

A steady-state solution is stable provided the small perturbation ϵ_n decays away in time. This requires the bracketed expression to be negative, since then both positive and negative values of ϵ_n will exponentially decay to zero. Hence the steady-state intensity $I_n^{(s)} = 0$ is stable provided $g_n - \nu/2Q_n < 0$, i.e., the laser is below threshold. Above threshold, $I_n = 0$ becomes unstable and the solution given by Eq. (26) is stable provided the slope $d\mathcal{S}/dI_n$ is negative, a condition satisfied by all the $\mathcal{S}$'s we have considered. Physically this means that a small intensity increase above the $I_n^{(s)}$ of Eq. (26) saturates the gain to a value less than the losses, which causes the intensity to decrease toward the steady-state value. Alternatively a small intensity decrease saturates the gain to a value greater than the losses, which causes the intensity to increase.

For purposes of comparison with the two-mode theory of Sec. 6-4, it is useful to write the third-order intensity equation of motion in the form

$$\dot{I}_n = 2I_n(g_n - \nu/2Q_n - \beta_n I_n) , \tag{29}$$

in which the self-saturation factor is

$$\beta_n = -g_n \left. \frac{d\mathcal{S}(I_n)}{dI_n} \right|_{I_n=0} . \tag{30}$$

The steady-state solutions to Eq. (29) are $I_n = 0$ below laser threshold and

$$I_n = \frac{g_n - \nu/2Q_n}{\beta_n} \tag{31}$$

above. Note that Eq. (31) is unphysical below threshold, since it predicts a negative intensity there.

Mode Pulling

So far we have studied Eq. (5), which determines the laser intensity. The other equation of interest is the frequency-determining equation. Combining the self-consistency Eq. (6) with the polarization (14), we have

$$\nu_n + \dot{\phi}_n = \Omega_n + \left[\frac{\omega-\nu_n}{\gamma}\right] g_n \mathcal{S}(I_n) . \tag{32}$$

This reveals an index variation as discussed in connection with Fig. 6-2. In contrast to the anomalous dispersion in Sec. 1-3, here we find an inverted index behavior for which the mode frequencies ν_n are pulled from their passive cavity values Ω_n toward the atomic line center ω. The reduction of the variation with increasing intensity is given by the saturation factor $\mathcal{S}(I_n)$. Using the steady-state solution (24) to eliminate $\mathcal{S}(I_n)$ and including ϕ_n in ν_n, we write Eq. (32) as

$$\boxed{\nu_n = \frac{\Omega_n \gamma + \omega(\nu/2Q_n)}{\gamma + \nu/2Q_n}} . \tag{33}$$

This equation is valid for all polarizations of the form of Eq. (14). It can be interpreted as a center of mass formula in which the oscillation frequency ν_n assumes a weighted average value of Ω_n and ω with weights γ and $\nu/2Q_n$, respectively. For high Q cavities, $\nu/2Q_n \ll \gamma$, and ν_n is pulled slightly from Ω_n toward the atomic line center.

6-3. Standing-Wave, Doppler-Broadened Lasers

A Doppler-broadened medium consists of a gaseous ensemble of atoms whose velocities Doppler shift incident light frequencies. Specifically, an atom moving with a velocity of z-component v interacting with a running wave of frequency ν effectively "sees" a wave with the Doppler shifted frequency $\nu' = \nu(1-v/c)$. The atom sees a resonant field when this shifted frequency coincides with ω, that is, when $\omega = \nu' = \nu(1-\nu/c)$. Alternatively we can say that the effective line center of such an atom is given by $\omega' = \omega/(1 - v/c) \simeq \omega(1 + v/c)$, since $|v/c| << 1$. Hence for the running wave, there is a one-to-one correspondence between axial velocity components and atomic line centers. Typically the distribution of axial velocity components is Maxwellian

$$\mathcal{W}(v) = \frac{1}{u\sqrt{\pi}} e^{-v^2/u^2} , \tag{34}$$

where u is the most probable speed. This corresponds to the inhomogeneous broadening distribution (5.37) with $\Delta\omega = \mathrm{K}u$. However with Doppler broadening an additional complication enters, namely that at a time t' of observation, the atom has moved from its initial position z_0 to the position z' given by $z' = z_0 + v(t' - t_0)$. Hence in adding up contributions to the polarization $\mathscr{P}(z,t)$, we have to ensure that the atoms arrive at the place z at the time t. This is accomplished by including the Dirac delta function $\delta(z - z_0 - vt + vt_0)$ in the population-matrix integrand for a gaseous medium. With this, the homogeneous-broadening population matrix of Eq. (5.6) is generalized to

$$\rho(z,v,t) = \sum_{\alpha=a,b} \int_{-\infty}^{t} dt_0 \int_0^L dz_0 \lambda_\alpha(z_0,t_0)\, \rho(\alpha,z_0,t_0,v,t) \times \delta(z - z_0 - vt + vt_0) , \tag{35}$$

The total complex polarization $\mathscr{P}_n$ results from integrating $\rho_{ab}(z,v,t)$ over the velocity distribution $\mathcal{W}(v)$. Problem 6-1 shows that this population matrix obeys the equations of motion (5.9) through (5.11) in which the time derivative is given by the *convective derivative*

$$\frac{d}{dt} = \frac{\partial}{\partial t} + v\frac{\partial}{\partial z} . \tag{36}$$

The population-matrix equations of motion can be integrated by noting that the equation

$$\left(\frac{\partial}{\partial t} + v\frac{\partial}{\partial z}\right) f(z,v,t) = g(z,v,t) \tag{37}$$

has the formal solution

$$f(z,v,t) = \int_{-\infty}^{t} dt' g(z',v,t'), \tag{38}$$

where z' is given by

$$z' = z - v(t - t') . \tag{39}$$

Accordingly integrating Eq. (5.9) using the convective derivative (36) and a single-mode field, we find

$$\rho_{ab}(z,v,t) = -\frac{i\wp}{2\hbar}\int_{-\infty}^{t} dt' e^{-(i\omega+\gamma)(t-t')} E_n(t') e^{-i[\nu_n t' + \phi(t')]} U_n(z')$$
$$\times [\rho_{aa}(z',v,t') - \rho_{bb}(z',v,t')] . \tag{40}$$

Comparing this with the homogeneous broadening case of Eq. (5.13), we see that the only major change is that the mode factor $U_n(z)$ is replaced by $U_n(z')$. For unidirectional operation, the mode factor of Eq. (3) becomes

$$U_n(z') = e^{iK_n z} e^{-iKv(t - t')} , \tag{41}$$

where for typographical simplicity we approximate $K_n v$ by a generic Kv, since any difference between K's leads to negligible differences in Doppler shifts. Substituting Eq. (41) into Eq. (40), we see that the frequency ω is replaced by the Doppler shifted value $\omega' = \omega + Kv$, which reduces to the ordinary inhomogeneous broadening of Sec. 5-2 and the formulas used there can be applied here.

The situation is more complicated for a standing-wave field. In this case $U_n(z) = \sin K_n z$ (Eq. (2)), which gives two oppositely directed running waves. An atom subject to this field and moving with an axial component v sees not one, but *two* Doppler-shifted frequencies, namely $\nu_n(1 \pm v/c)$, as indicated in Fig. 6-4. Instead of the single spectral hole burned as in Fig. 5-2, two holes are now burned in the population difference, as shown versus v in Fig. 6-5. As shown below, this leads to a famous effect known

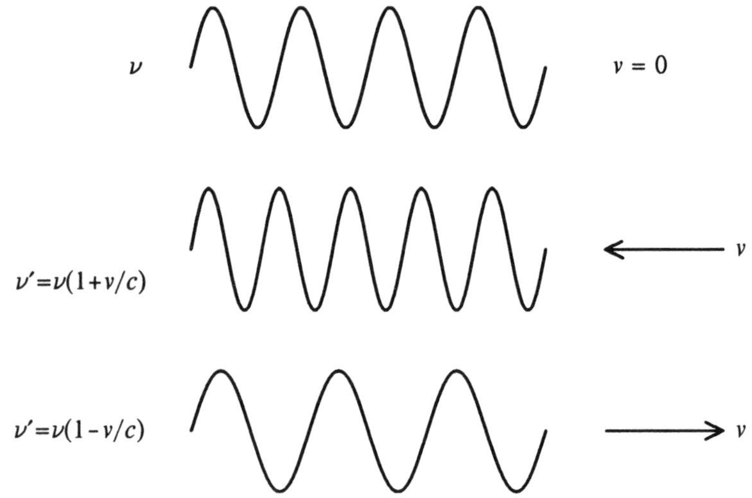

Fig. 6-4. Drawing showing how a traveling wave with oscillation frequency ν appears Doppler downshifted to an atom moving the same direction as the wave and upshifted to an atom moving the opposite direction.

as the Lamb dip. Furthermore the field fringe pattern experienced by a moving atom "walks", rather than stands, and each velocity experiences a different walking speed. The total polarization includes the contributions of all different walking speeds. The first paper able to treat all of these effects correctly was given by Stenholm and Lamb (1969) and involves integrals over continued fractions.

Here we limit ourselves to a simpler approach called the *Doppler rate equation approximation* (Doppler REA), which neglects the effects of the induced gratings. This is a good approximation provided the average Doppler shift is large compared to the homogeneous linewidth. Specifically we assume that the field amplitude and phase and the population difference vary little in the dipole lifetime $T_2 \equiv 1/\gamma$, and thereby can be evaluated at the time $t' = t$. Equation (40) becomes

$$\rho_{ab}(z,v,t) = -\frac{i\wp}{2\hbar}E_n(t)e^{-i[\nu t+\phi(t)]}[\rho_{aa}(z,v,t) - \rho_{bb}(z,v,t)]$$

$$\times\int_{-\infty}^{t} dt' e^{-(i\omega-i\nu_n+\gamma)(t-t')}U_n[z-v(t-t')]\,. \tag{42}$$

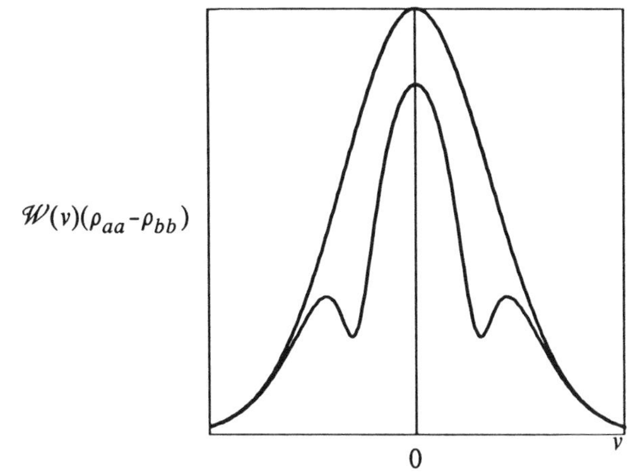

Fig. 6-5. Unsaturated (Gaussian) and saturated (with holes) population difference versus z component of velocity, with $R(v)$ given by Eq. (44). Holes are burned by the field intensity for $v = \pm c(1-\omega/\nu)$.

We substitute

$$U_n(z-v(t-t')) = \frac{1}{2i}[e^{i(K_n z - Kv(t-t'))} - e^{-i(K_n z - Kv(t-t'))}],$$

expand $\exp(iK_n z) = \cos(K_n z) + i\sin(K_n z)$, and then drop the cosines, since for standing-waves they vanish in the polarization integral of Eq. (12). This gives

$$\rho_{ab}(z,v,t) = -\frac{i\wp}{4\hbar}E_n(t)U_n(z)\exp[-i(\nu_n t+\phi_n)][\rho_{aa}(z,v,t) - \rho_{bb}(z,v,t)] \times [\mathscr{D}(\omega-\nu_n+Kv) + \mathscr{D}(\omega-\nu_n-Kv)] \,. \tag{43}$$

Substituting this, in turn, into the population equations of motion (5.10) and (5.11), we find the rate equations (5.15) and (5.16) with the rate constant

$$R(v) = \frac{1}{8}\,|\wp E_n/\hbar|^2\gamma^{-1}\,[\mathscr{L}(\omega-\nu_n+Kv) + \mathscr{L}(\omega-\nu_n-Kv)] \,. \tag{44}$$

Here we have replaced the $|U_n(z)|^2 = \sin^2 K_n z$ factor by the average value 1/2 that closely approximates the z dependence that rapidly moving atoms experience. This approximation amounts to neglecting the grating effects discussed in Sec. 5-3 for stationary atoms and hence is not as good near line center where $v \simeq 0$ atoms are involved, or, of course, in the homogeneously broadened limit. It is excellent for large Doppler broadening ($Ku >> \gamma$) due to the facts that rapidly moving atoms tend to see an average field, and that gratings corresponding to different velocity groups are shifted varying amounts along the axis, yielding a collection of scattered waves that tend to interfere destructively with one another.

The complex polarization component $\mathscr{P}_n(t)$ becomes

$$\mathscr{P}_n(t) = -i\wp^2\hbar^{-1}\overline{N}E_n\int_{-\infty}^{\infty}dv\,\frac{\mathscr{W}(v)\,\mathscr{D}(\omega-\nu_n+Kv)}{1+I_n[\mathscr{L}(\omega-\nu_n+Kv)+\mathscr{L}(\omega-\nu_n-Kv)]/4}\,. \quad (45)$$

For central tuning ($\nu_n=\omega$), the Lorentzians coincide yielding an integral just like that for the unidirectional case (14) for central tuning, but with a factor of 1/2 in the denominator. The integrals can be expressed easily in terms of the plasma dispersion function as in Eq. (5.43). Off line center, the algebra is considerably trickier and leads to a complicated combination of plasma dispersion functions (see Prob. 8-12 of Sargent, Scully, and Lamb (1977) for the result).

To obtain the Lamb-dip formula for the mode intensity, we expand the denominator in Eq. (45) to third-order in E_n. The first-order term is

$$\begin{aligned}\mathscr{P}_n^{(1)}(t) &= -\wp^2\overline{N}(\hbar Ku)^{-1}E_n\,Z[\gamma+i(\omega-\nu_n)]\\ &\simeq -i\sqrt{\pi}\,\wp^2\overline{N}(\hbar Ku)^{-1}E_n\,e^{-(\omega-\nu_n)^2/(Ku)^2}\,,\end{aligned} \quad (46)$$

where the superscript $^{(1)}$ means Eq. (46) is proportional to the first power of E_n. The third-order term is given by

$$\mathscr{P}_n^{(3)}(t) = \frac{1}{4}\wp^2\overline{N}\,(\hbar Ku\sqrt{\pi})^{-1}{}^*E_n\,I_n\,e^{-\delta^2/(Ku)^2}\,J(\delta)\,, \quad (47)$$

where $\delta = \omega-\nu_n$, we have evaluated the slowly-varying $\mathscr{W}(v)$ at the maximum of the $\mathscr{L}$ function, and (setting $x=Kv$)

$$J(\delta) = i\int_{-\infty}^{\infty}dx\,\mathscr{D}(\delta+x)\,[\mathscr{L}(\delta+x)+\mathscr{L}(\delta-x)]$$

$$= \gamma^2 \int_{-\infty}^{\infty} \frac{dx}{x+\delta-i\gamma} \left[\frac{1}{(x+\delta+i\gamma)(x+\delta-i\gamma)} + \frac{1}{(x-\delta+i\gamma)(x-\delta-i\gamma)} \right] . \tag{48}$$

This is easily integrated using the residue theorem. We close the contour in the lower half plane (which gives an overall minus sign) around the single poles $x=-\delta-i\gamma$ for the first product and $x=\delta-i\gamma$ for the second product. This gives

$$J(\delta) = \frac{\pi i}{2}[1 + \gamma\mathscr{D}(\delta)].$$

Inserting this into Eq. (47), we find the third-order contribution to the polarization

$$\mathscr{P}_n^{(3)}(t) = \frac{1}{8} i\sqrt{\pi}\,\wp^2\overline{N}(\hbar Ku)^{-1} E_n I_n\, e^{-(\omega-\nu_n)^2/(Ku)^2}\, [1 + \gamma\mathscr{D}(\omega-\nu_n)] . \tag{49}$$

Substituting $\mathscr{P}_n \simeq \mathscr{P}_n^{(1)} + \mathscr{P}_n^{(3)}$ into the amplitude self-consistency Eq. (5) and using the relative excitation

$$\mathscr{N} = \frac{2Q_n}{\nu}\,\frac{\sqrt{\pi}\,\wp^2\overline{N}\nu}{2\epsilon\hbar Ku} , \tag{50}$$

we find

$$\dot{E}_n = -\frac{\nu}{2Q_n}E_n + \frac{\nu}{2Q_n} E_n\, \mathscr{N} e^{-(\omega-\nu_n)^2/(Ku)^2} \left[1 - \frac{1}{8} I_n [1 + \mathscr{L}(\omega-\nu_n)]\right] . \tag{51}$$

In steady state, this yields

$$\boxed{I_n = 8\,\frac{\mathscr{N} - e^{(\omega-\nu_n)^2/(Ku)^2}}{\mathscr{N}[1 + \mathscr{L}(\omega-\nu_n)]}} . \tag{52}$$

For $\mathscr{N} > 1 + 2(\gamma/Ku)^2$ (see Prob. 6-6), this expression exhibits *a Lamb dip* versus detuning as illustrated in Fig. 6-6.

From a mathematical standpoint, we see that the Lamb dip results from the $\mathscr{L}(\omega-\nu_n)$ in Eq. (52). This term is missing in the intensity for ordinary inhomogeneous broadening. Tracing this $\mathscr{L}$ back, we see that it results from the $\mathscr{D}(\delta+Kv)\mathscr{L}(\delta-Kv)$ term in Eq. (48). This term is resonant when $\delta = \omega - \nu_n \simeq 0$ (central tuning), i.e., when *both* running waves saturate the response of the medium.

Pursuing this clue, we can understand the Lamb dip physically in terms of the saturation by the two running waves, each contributing a Lorentzian to Eq. (44). *Both* waves saturate the response of the $v \simeq 0$ atoms, since both Lorentzians are resonant, whereas only *one* running wave saturates the response for nonzero v atoms. Hence a given standing-wave intensity saturates less *off* line center than *on*. Since the Doppler-limit linear gain is the same in both cases, the *saturated-gain-equals-loss* condition (27) corresponds to a larger steady-state intensity off line center than on. This inten-

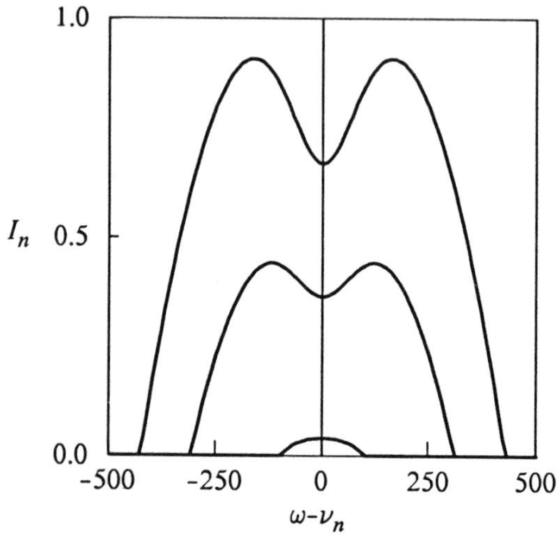

Fig. 6-6. Single-mode dimensionless intensity of Eq. (50). Doppler half-width (Ku) at $1/e$ point is $2\pi\times1010$ MHz, the decay constants $\gamma=2\pi\times80$ MHz, and the relative excitation $\mathcal{N}$ = 1.01, 1.05, 1.1, 1.15, and 1.2.

sity dip versus detuning (illustrated in Fig. 6-6) was predicted by Lamb in 1962 in letters to colleagues and was first explained intuitively by Bennett (1962) using a phenomenological spectral hole burning model. The Lamb dip, as it is called, has led to various laser stabilization schemes and plays an important role in the field of saturation spectroscopy (see Sec. 8-3).

One trouble with Eq. (52) is that for large $\mathcal{N}$, it saturates to the value $8/[1 + \mathscr{L}(\omega-\nu_n)]$, rather than increasing with $\mathcal{N}$. Following the example of the homogeneously broadened cases in Table 1, we might just drop the $\mathcal{N}$ in the denominator of (52), but this overestimates the depth of the dip and has no good justification. Respectable accuracy is obtained with the "Doppler REA" expression (45) even for central tuning. This might be surprising, since we assumed that grating effects could be ignored, that is, that the $\sin^2(K_n z)$ for the rate constant R could be replaced by the average value 1/2. A more careful analysis by Stenholm and Lamb (1969) reveals that the various atomic ensembles can have population inversion differences with considerable spatial dependence, although sufficiently rapidly moving atoms do, in fact, see an average field. An additional effect enters to help justify the Doppler REA, namely that due to the finite lifetimes, the gratings are shifted varying amounts in the direction of travel and tend to average to zero.

6-4. Two-Mode Operation and the Ring Laser

As discussed in further detail in Sec. 10-3, lasers often operate with two or more modes. In this section, we discuss a simple, very useful example of a two-mode laser, the bidirectional ring laser (Fig. 6-1*b*), often used as a laser gyro. This problem is interesting in its own right and serves to illustrate general two-mode formalism. We first show how the frequencies of the two oppositely-directed running waves can be split by rotation, a phenomenon called the Sagnac effect. We then derive general two-mode intensity equations using the specific homogeneously-broadened formulas of Sec. 5-3, and a ring-laser generalization of the Doppler standing-wave formulas in Sec. 6-3. The steady-state solutions to these two-mode equations are given along with a stability analysis revealing effects of mode competition. Mode locking, another kind of mode competition, is discussed in Sec. 6-5.

We describe the ring laser electric field by a special case of Eq. (1)

$$E(z,t) = \frac{1}{2} E_+(t)\, e^{-i(\nu_+ t + \phi_+ - K_+ z)} + \frac{1}{2} E_-(t)\, e^{-i(\nu_- t + \phi_- + K_- z)} + \text{c.c.} \tag{53}$$

This is a generalization of the single-mode standing wave in which the amplitudes, frequencies, and phases of the two oppositely-directed running waves can be independent of one another. It is the same as the bidirectional field (5.53) with $\mathcal{E}_\pm = E_\pm \exp(-i\phi_\pm)$. To see that rotation of the ring at rate Ω leads to unequal passive cavity frequencies, $\Omega_\pm = cK_\pm$, consider the distance a running wave travels in the time $\tau_\pm$ used to make a round trip (see Fig. 6-7). Since the starting point moves $\Omega r\tau_\pm$ in this time, we have $c\tau_\pm = L \pm \Omega r \tau_\pm$, i.e., $c\tau_\pm = L/(1 \mp \Omega r/c)$. For constructive interference, an integral number of wavelengths must fit into this distance, that is, $c\tau_\pm = n2\pi c/\Omega_\pm$. This gives

$$\Omega_\pm = \Omega_0(1 \mp \Omega r/c)\,. \tag{54}$$

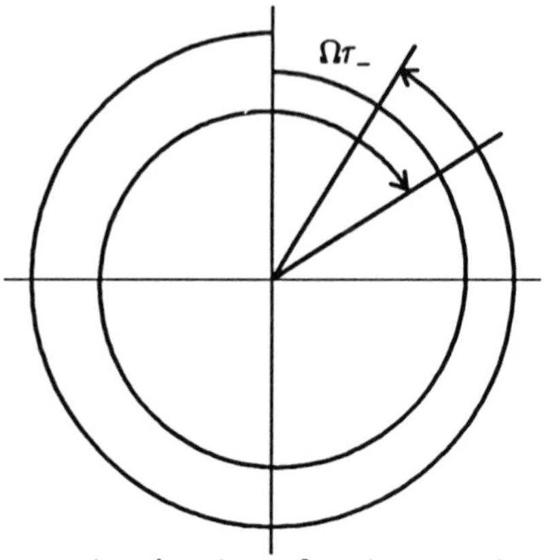

Fig. 6-7. Diagram showing how far the + and - running waves have to travel to make their respective round trips when the circular cavity rotates at rate Ω.

where $\Omega_0 = 2n\pi c/L$ is the Ω=0 degenerate frequency. The beat frequency $\Omega_+ - \Omega_-$ is

$$\Omega_+ - \Omega_- = -2\Omega_0 \Omega r/c = -(4A\Omega_0/Lc)\Omega = S\Omega \ , \tag{55}$$

where A is the enclosed area. The area form of the Sagnac formula (55) is clearly valid for a circle, since $A = \pi r^2$ and $L = 2\pi r$. It is actually valid for rings of arbitrary shape. Note that for a circle, the Sagnac constant S equals the number of nodes in the standing wave inside the ring cavity, since $S = 2\Omega_0 r/c = 2n$. By measurement of the beat frequency of the laser, we can thus determine the rotation rate. A complete theory must, of course, include effects of mode pulling introduced by the active medium.

To find the intensity equations of motion, we use the two-mode case of the multimode self-consistency Eq. (5) in combination with appropriate polarizations of the medium. One such polarization is that for the homogeneously-broadened case with equal frequencies $\nu_+ = \nu_- = \nu$ in Sec. 5-3. Note that approximating the frequencies of the two counterpropagating waves by the same value in the polarization calculations is an excellent approximation, since the Sagnac frequency shifts are typically extremely small compared to ν. We can change the linear propagation absorption coefficient α_0 of Eq. (5.28) into a cavity gain coefficient by replacing $-K$ by ν, i.e., here $\alpha_0 = \nu\wp^2 N/2\epsilon\hbar\gamma$. Using this α_0 in Eq. (5.66), we expand to third-order in the field amplitudes to find the gain coefficient

$$\alpha_+ = \alpha_0 \gamma \mathscr{D}(1 - I_+\mathscr{L} - 2I_-\mathscr{L}) \ . \tag{56}$$

Substituting this into Eq. (8) (set $\nu\chi''_n/2 = \mathrm{Re}\{\alpha_+\}$) and multiplying through by $2E_+(\wp/\hbar)^2T_1T_2$, we find the intensity equation

$$\dot{I}_+ = 2I_+[g_+ - \nu/2Q_+ - \beta_+I_+ - \theta_{+-}I_-]\ , \tag{57}$$

where for our current model the linear gain coefficient is $g_+=\alpha_0\mathscr{L}$, the self-saturation coefficient is $\beta_+ = \alpha_0\mathscr{L}^2$, and the cross-saturation coefficient is $\theta_{+-}=2\beta_+$. Similarly,

$$\dot{I}_- = 2I_-[g_- - \nu/2Q_- - \beta_-I_- - \theta_{-+}I_+]\ . \tag{58}$$

The classical amplitude equations (2.22b) and (2.22c) are closely related, but correspond to propagation.

Because these equations occur often in laser theory (and in other disciplines like biology and chemistry as well), it is worth spending some time analyzing them in some detail. Four steady-state solutions ($dI_\pm/dt = 0$) are possible in general, since either mode may or may not oscillate. (Note that the solutions are physically possible only if they are positive.) The case for which neither oscillates is trivial: both modes are below threshold. The two single-mode cases with solutions given by Eq. (31) are more interesting, since both modes might oscillate in the absence of the other, but might not be able to oscillate in the other's presence due to competition for the gain medium.

The solution for which both intensities are nonzero results from solving the pair of coupled equations given by setting the bracketed expressions in Eq. (57) and (58) separately equal to zero. We find

$$I_\pm = \frac{\alpha_{\pm eff}/\beta_\pm}{1-C}\ , \tag{59}$$

where the effective net gain coefficients

$$\alpha_{\pm eff} = g_\pm - \frac{\nu}{2Q_\pm} - \theta_{\pm\mp}\,\frac{g_\mp - \nu/2Q_\mp}{\beta_\mp}\ , \tag{60}$$

and the *coupling parameter*

$$C = \frac{\theta_{+-}\theta_{-+}}{\beta_+\beta_-}\ . \tag{61}$$

Here α_{+eff} is called an *effective net gain coefficient* since it is the linear net gain for I_+ *minus* the saturation induced by I_- oscillating alone.

To determine the stability of these solutions, we perform a linear stability analysis similar to that for the single-mode case of Eq. (29). Specifically we substitute $I_\pm(t) = I_\pm^{(s)} + \epsilon_\pm(t)$ into Eq. (57) to find

$$\begin{aligned}\dot{\epsilon}_+ = 2\epsilon_+[g_+ - \nu/2Q_+ - \beta_+ I_+^{(s)} - \theta_{+-} I_-^{(s)}] \\ - 2I_+^{(s)}(\beta_+\epsilon_+ + \theta_{+-}\epsilon_-) ,\end{aligned} \tag{62}$$

with a similar equation for ϵ_-. Consider first the stability of the $I_+^{(s)} = 0$ solution in the presence of I_- oscillating alone with the value $(g_- - \nu/2Q_-)/\beta_-$. In other words, we ask, can I_+ build up in the presence of I_-? For this case, Eq. (62) reduces to $\dot{\epsilon}_+ = 2\alpha_{+eff}\epsilon_+$. Hence if I_+'s effective net gain is positive, I_+ builds up. It may do so and then either suppress I_- or coexist with it.

To find out which of these scenarios actually takes place, we consider the stability of the two-mode solution of Eq. (59). In this case Eq. (62) reduces to

$$\dot{\epsilon}_+ = -\frac{2\alpha_{+eff}}{1-C}[\epsilon_+ + \epsilon_-\theta_{+-}/\beta_+] , \tag{63}$$

with a similar equation for $\dot{\epsilon}_-$. For two-mode stability, the coefficient matrix in these coupled equations has to have eigenvalues with negative real parts. Problem 6-7 shows that this is the case provided that both effective α's are positive and that the coupling parameter C is smaller than 1. In contrast, the solutions of Eq. (59) for which $C > 1$ and both effective α's are *negative* (to get $I_n > 0$) are unstable. For these cases, either single-mode solution is stable, that is, we have *bistability*. Which mode oscillates depends on the initial conditions.

Armed with these general solutions and their stability characteristics, we return to the homogeneously broadened ring laser, for which $\theta_{-+}/\beta_+ = \theta_{+-}/\beta_- = 2$. This gives $C = 4$ and $\alpha_{\pm eff} = g_\pm - \nu/2Q_\pm < 0$. This gives bistable operation and was the first kind of "optical bistability" to be predicted (see Chap. 7 for discussion of optical bistability in its usual contexts). This bistability is observed in ring dye lasers. The factor of 2 in the ratio of the cross-saturation to self-saturation coefficients results from the spatial-hole grating burned in the population difference by the field fringe created by the two waves. It is the same factor of two that occurs in the index grating discussions of Chap. 2, but here it occurs in a gain grating. This grating is out of phase with the fringe field that burns it and hence scatters out of phase destructively. Added to the ordinary saturation of the atoms, this gives twice as much cross saturation (one mode saturating the other) as self saturation.

Obviously a ring laser with a single running wave makes a bad gyro; one needs both counterpropagating waves to obtain a beat frequency. The gaseous medium of Sec. 6-3 averages out the grating thereby reducing the θ's relative to the β's. The appropriate density matrix element $\rho_{ab}(z,v,t)$ is given by Eq. (42) with the running-wave spatial factor $U_+(z)$ of Eq. (3). Evaluating the time integral and projecting the result onto U_+^*, we find the polarization

$$\mathscr{P}_+(t) = -i\wp^2\hbar^{-1}\overline{N}E_+\int_{-\infty}^{\infty}\frac{\mathscr{W}(v)\mathscr{D}(\omega-\nu_+ + Kv)\,dv}{1 + I_+\,\mathscr{L}(\omega-\nu_+ + Kv) + I_-\,\mathscr{L}(\omega-\nu_+ - Kv)}\,. \quad (64)$$

This yields the same first-order term (46) as the standing wave case with the subscript n replaced by +. The third-order term becomes (using contour integration as for Eq. (48))

$$\mathscr{P}_+^{(3)}(t) = \frac{1}{2}\sqrt{\pi}\,\wp^2\overline{N}(\hbar Ku)^{-1}E_+\,e^{-(\omega-\nu_+)^2/(Ku)^2}\,[I_+ + \gamma\mathscr{D}(\omega-\nu_0)I_-]\,, \quad (65)$$

where the average detuning $\nu_0 = (\nu_+ - \nu_-)/2$. This yields the intensity equations (57) and (58) with $g_+ = (\nu/2Q_+)\exp[-(\omega-\nu_+)^2(Ku)^2]$, $\beta_+ = g_+/2$, and $\theta_{+-} = \beta_+\,\mathscr{L}(\omega-\nu_0)$. For nonzero average detuning, the coupling ratio θ_{+-}/β_+ is reduced below unity and two-mode solution of Eq. (59) occurs if both effective net gains are positive.

In general in ring laser gyros, one avoids the strong mode competition at $\nu_0 \simeq \omega$ by using a mixture of two isotopes (Ne^{20} and Ne^{22}) whose line centers differ by about 800 MHz. By tuning in between these centers, the value of the Lorentzian $\mathscr{L}(\omega-\nu_0)$ is reduced to about 1/20, and the modes are essentially uncoupled from one another as far as saturation is concerned. In the next section, we see how backscattering from the mirrors can introduce a phase-dependent coupling that can lock the laser mode frequencies to the same value. This is the major problem with low rotation-rate operation of the laser gyro.

6-5. Mode Locking

The mode competition effects in the ring laser considered in Sec. 6-4 occur from the fact that the two modes share the same atoms. This competition is phaseless; the mode phases always cancel out in the nonlinear response (for the scalar ring laser). In this section, we consider competition effects in which the relative phase between the modes enters and can cause the mode frequencies to lock to the same value. In Sec. 10-3, we consider multimode operation with modes of different frequencies that lock into a regularly spaced comb leading to a periodic laser output. The frequency locking in the ring laser is the bête noire of the laser gyro, and people have worked very hard and successfully to minimize it.

The basic idea of mode locking is that modes quite capable of existing independently of one another are persuaded to follow the direction of some impressed force either internally or externally generated. In short, the modes are synchronized. As for mode competition, analogues of mode locking exist in everyday life. Examples of the externally locked variety include musicians following the direction of a conductor, a heart following the tempo of a pacemaker, a TV raster scan synced to the broadcasted signal, and synchronous transmission of computer data. An example of internally generated locking is musicians in a quartet or rock band who somehow stay together without the aid of a conductor.

A more scientific example was observed by Huygens and consisted of two pendulum clocks (Huygens first put the pendulum into the clock) that ticked at slightly different rates when apart, but ticked at the same rate when hung close together on the same wall. Later Lord Rayleigh studied forced tuning forks and similarly found that although a beat note existed between the forks when apart, the note vanished when both sat next to one another on a table. Van der Pol observed that a triode oscillator may lock to an injected signal.

A modern version of Huygen's clock experiment is to observe two computer clocks (MHz square-wave generators) placed near one another on the same printed circuit board. Due to imperfect decoupling of their common power supply, they are coupled in a phase-sensitive way and lock together when tuned sufficiently close to one another in frequency. This form of mode locking is easy to observe on an oscilloscope, and reveals the principal features of laser mode locking we talk about in this section.

Mode locking is caused by scattering or injecting some portion of one mode into the other mode. For example in the ring laser, imperfections in the mirrors can backscatter one mode into the other. For generality, we assume that the corresponding scattering constant $g+ih$ is complex: both index and absorption variations may scatter. To include this in two-mode operation, we replace the polarization $\mathscr{P}_n$ by

$$\mathscr{P}_n \rightarrow \mathscr{P}_n + 2\epsilon\nu^{-1}(g+ih)\, E_{3-n}\, e^{i\Psi_{n,3-n}} , \tag{67}$$

where $n = 1$ or 2 and the relative phase angle $\Psi_{n,3-m} = (\nu_n - \nu_{3-n})t + \phi_n - \phi_{3-n}$. The presence of Ψ_{nm} is crucial; it does not occur in coupling due to cross saturation.

Substituting Eq. (67) into the self-consistency equations (5) and (6), we see that the scattering affects both the amplitude and the frequency equations. We suppose the effect on the amplitudes is small enough to be neglected, which often is the case. The amplitude equations can then be solved and the resulting amplitudes substituted into the frequency equations. Using Eq. (6), we find that the relative phase angle

$$\Psi = (\nu_2-\nu_1)t + \phi_2-\phi_1 \tag{68}$$

has the equation of motion

$$\dot{\Psi} = a - g\left[\frac{E_1}{E_2} - \frac{E_2}{E_1}\right]\cos\Psi + h\left[\frac{E_1}{E_2} + \frac{E_2}{E_1}\right]\sin\Psi$$

or

$$\boxed{\dot{\Psi} = a + b\sin(\Psi-\Psi_0)}\ , \tag{69}$$

where

$$a = \Omega_2-\Omega_1 - \frac{\nu}{2\epsilon}\mathrm{Re}\{\mathscr{P}_2E_2^{-1} - \mathscr{P}_1E_1^{-1}\}$$

$$b^2 = g^2\left[\frac{E_1}{E_2} - \frac{E_2}{E_1}\right]^2 + h^2\left[\frac{E_1}{E_2} + \frac{E_2}{E_1}\right]^2$$

$$\tan\Psi_0 = g(E_1^2-E_2^2)/h(E_1^2+E_2^2)\ .$$

The solutions to the "mode locking" equation (69) are divided into two regimes by the ratio $|a/b|$. For $|a/b|>1$, no value of $b\sin(\Psi-\Psi_0)$ can cancel the a, and hence Ψ changes monotonically in time, decreasing if $a<0$ and increasing if $a>0$. On the other hand for $|a/b| \le 1$, one or two values of Ψ exist giving $\dot{\Psi} = 0$, i.e., mode locking. These values are

$$\Psi_s = \begin{cases} \Psi_0 - \sin^{-1}(a/b) \\ \Psi_0 + \pi + \sin^{-1}(a/b) \end{cases} . \tag{70}$$

To check their stability, we substitute $\Psi(t)=\Psi_s+\epsilon(t)$ into Eq. (70) and find

$$\dot{\epsilon} = b\cos(\Psi_s-\Psi_0)\,\epsilon, \tag{71}$$

i.e., Ψ_s is stable if $b\cos(\Psi_s-\Psi_0) < 0$.

If $|a/b|>>1$, the relative phase angle Ψ changes essentially linearly in time at the rate $\dot{\Psi}=a$. As $|a/b|$ decreases toward unity, one can observe that the b term in Eq. (69) starts to subtract from the a term in one half of the cycle and to add in the other half. This leads to a "slipping" behavior, as Ψ "slips" past the point where the b term tries to cancel the a term.

When $|a/b|$ reaches unity, Ψ gets to the slipping point and sticks. For $|a/b|>1$, it is useful to calculate the average frequency defined by the reciprocal of the $\Psi(t)$ period

$$\Delta\nu = \frac{2\pi}{t(\Psi_0+2\pi) - t(\Psi_0)} ,$$

where $t(\Psi_0)$ is the time when $\Psi = \Psi_0$. From Eq. (69), we have

$$\frac{1}{\Delta\nu} = \int_{t(\Psi_0)}^{t(\Psi_0+2\pi)} dt = \int_{\Psi_0}^{\Psi_0+2\pi} \frac{d\Psi}{a + b\sin\Psi} .$$

We have already seen this integral in connection with spatial hole burning in Sec. 5-3 (see Eq. (5.63)). Both cases involve interferences between two waves. For mode locking the interference is temporal and the average increases the beat frequency period. For spatial hole burning, the interference is spatial and the average increases the saturation. However, spatial hole burning has nothing that corresponds to locking: a/b is inevitably greater than unity. Using Eq. (5.63), we find

$$\Delta\nu = (a^2 - b^2)^{1/2} . \tag{72}$$

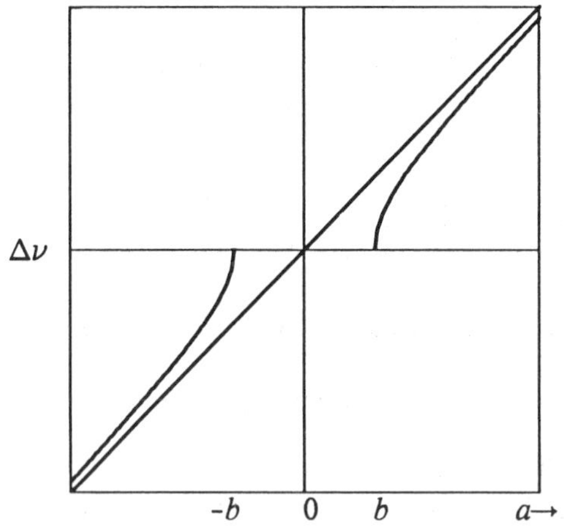

Fig. 6-8. Average frequency of Eq. (72) versus zero-locking beat frequency parameter a. Note that aside from some index corrections a of Eq. (69) is proportional to the rotation rate in laser gyroscopes.

This is plotted in Fig. 6-8 as a function of a, which shows the mode locking region as that for $\Delta\nu = 0$. In particular for the ring laser, one tries to minimize this region. The most popular technique is to use the best obtainable mirrors and to dither (shake about the rotation axis) the laser, thereby largely preventing it from locking.

6-6. Single-Mode Semiconductor Laser Theory

In its simplest form, the semiconductor laser consists of a diode, which is a p-type semiconductor joined to an n-type semiconductor. By itself the p-type material is neutral in charge, but contains doping atoms that need extra electrons to fill their outermost valence shell. By itself the n-type semiconductor is also neutral but contains doping atoms that have extra electrons outside their last complete shell. When the two media are brought into contact with one another, the extra electrons in the n-type medium move across the junction to the p-type medium, and corresponding holes from the p-type medium move to the n-type medium. The Coulomb attraction prevents the carriers from penetrating into the other medium very far. If a positive voltage is applied across the junction from the p-type to the n-type side, a current can flow and the conduction electrons recombine in the junction with the holes, emitting light. If this light can propagate in the plane of the junction and be partially reflected at the end facets, there may be enough gain to overcome the losses, and the device lases. To simplify our discussion we assume that the junction consists of a thin layer (typically $\leq .1\mu m$) of an intrinsic semiconductor, that is, an undoped semiconductor.

We wish to describe the diode gain and lasing conditions as functions of temperature, current, decay rates, and the carrier effective masses. We do this by using the semiconductor quasi-equilibrium model of Sec. 5-5. The appropriate complex polarization $\mathcal{P}_n$ is given by Eq. (5.126) with two changes: 1) the self-consistency equations (5) and (6) assume that the field phase ϕ_n is factored out, and 2) the constant background index contribution given by the -1 in the d_0 of Eq. (5.126) is included in the outside ϵ factor, since otherwise the long (and unphysical) wings of Lorentzian cause the sum over momentum to diverge. In a more complete model, the sum converges because the band structure is not parabolic for large k. This gives the slowly-varying complex polarization

$$\mathcal{P}_n = -\frac{i\wp^2 E_n}{\hbar\gamma V}\sum_k \mathcal{L}[d_0 - i(\omega-\nu_n)\gamma^{-1}(f_e + f_h)]\,.$$

Substituting this into Eqs. (5) and (6), we find

$$\dot{E}_n = (\alpha_{rn} - \nu/2Q_n)E_n , \tag{73}$$

$$\nu_n + \dot{\phi}_n = \Omega_n - \alpha_{in} , \tag{74}$$

where α_{rn} and α_{in} are the real and imaginary parts of the complex saturated gain coefficient

$$\alpha_n = \frac{\nu\wp^2}{2\epsilon\gamma\hbar V} \sum_k \mathcal{L}[d_0(k) - i(\omega-\nu_n)\gamma^{-1}(f_e(k) + f_h(k))] . \tag{75}$$

Steady-state operation occurs when $\dot{E}_n = 0$. Above the laser threshold, this implies the steady-state oscillation condition

$$\boxed{\alpha_{rn} = \frac{\nu\wp^2}{2\epsilon\gamma\hbar V} \sum_k \mathcal{L}d_0(k) = \frac{\nu}{2Q_n}} , \tag{76}$$

i.e., the saturated gain equals the cavity losses as in Eq. (27). According to Eq. (76), the difference $d_0(k)$ and hence the total carrier density N must remain at their respective laser-threshold values, a feature known as *gain clamping*. In particular, Eq. (76) determines the steady-state single-mode laser carrier density N_0 and the threshold pump value λ_{th} as functions of the cavity loss rate constant $\nu/2Q$ and temperature T. Since the intensity determines the total carrier density independently of the momentum k, we say that the semiconductor laser medium *saturates homogeneously*, in spite of the linear gain profile being inhomogeneously broadened.

In terms of the gain α_{rn}, the carrier-density equation of motion (5.125) becomes

$$\dot{N} = \lambda - \lambda_{th} - \alpha_{rn}\, n , \tag{77}$$

where the "photon-number" density $n(t)$ (note that this is a classical quantity in spite of the name we give it) is defined by

$$n(t) = \epsilon E_n{}^2/\hbar\nu , \tag{78}$$

and the threshold pump value λ_{th} (that for $n = 0$) is given by

$$\lambda_{th} = \gamma_{nr} N_0 + \Gamma V^{-1} \Sigma_k f_e(k) f_h(k) , \tag{79}$$

such that $V^{-1}\Sigma_k f_\alpha(k) = N_0$, for $\alpha = e$ or h. Equation (79) must be evaluated numerically. Multiplying the field amplitude Eq. (73) by $2\epsilon E_n/\hbar\nu$, we have the equation of motion

$$\dot{n} = 2n(\alpha_{rn} - \nu/2Q_n) \ . \tag{80}$$

The steady-state photon number density is determined by the condition $\dot{n} = \dot{N} = 0$, which gives

$$\boxed{n = (\mathcal{N} - 1)n_s} \ , \tag{81}$$

where the relative excitation $\mathcal{N}$ is defined by the ratio

$$\mathcal{N} = \frac{\lambda}{\lambda_{th}} \ , \tag{82}$$

and the "saturation photon number density" n_s is given by

$$n_s = \frac{\lambda_{th}}{\alpha_{rn}} \ . \tag{83}$$

Equation (81) shows that the laser intensity increases as a linear function of the pump rate with a slope proportional to n_s. According to Table 6-1, the unidirectional ring laser with a homogeneously broadened two-level active medium has the dimensionless intensity (intensity in units of the saturation intensity I_s) $I_n = (\mathcal{N}\mathcal{L} - 1)/\mathcal{L}$, which is similar to Eq. (81), although it displays its tuning dependence explicitly. For central tuning ($\mathcal{L} = 1$) and a relative excitation $\mathcal{N} = 2$, the two-level dimensionless intensity $I_n = 1$, i.e., the laser intensity equals the saturation intensity I_s. In this spirit, we call n_s the semiconductor-laser saturation photon number because it is the n value corresponding to the relative excitation $\mathcal{N} = 2$. We see that the semiconductor-laser steady-state oscillation intensity has similarities with the corresponding homogeneously broadened two-level laser intensity. However it differs in important ways, such as its dependencies on carrier density, temperature and effective masses, and on the cavity tuning. The semiconductor-laser n_s differs from the two-level I_s significantly in various respects, not the least of which is that n_s is negative, i.e., unphysical, for an absorbing medium. Section 10-1 gives further comparisons in discussing the transient response of Eqs. (77) and (80).

The shift in the laser oscillation frequency ν_n from the passive cavity value Ω_n is given by Eq. (74), where

$$\alpha_{in} = -\frac{\nu\wp^2}{2\epsilon\gamma^2\hbar V}\sum_k [f_e(k) + f_h(k)](\omega-\nu_n)\mathscr{L}\,. \tag{84}$$

Comparing Eq. (74) with Eq. (9), we see that Eq. (84) defines a negative change in the index of refraction $\delta\eta$ induced by the carriers of

$$\delta\eta \simeq \frac{\alpha_{in}}{\Omega_n} \simeq -\frac{\wp^2}{2\epsilon\gamma^2\hbar V}\sum_k (\omega-\nu_n)[f_e(k) + f_h(k)]\mathscr{L}\,. \tag{85}$$

Since the distributions f_α depend on intensity, Eq. (85) implies that the nonlinear medium has index antiguiding (defocussing) properties. These effects are ignored in the present plane-wave theory, but they would have consequences for a theory that includes transverse variations, such as the Gaussian beam theory of Sec. 6-7.

Numerical Evaluation of Laser Gain and Index Formulas

To evaluate the complex gain given by Eq. (75), we convert the sum $V^{-1}\Sigma_k$ into an integral. To do this we need to introduce the density of states, which determines how the k values are distributed as function of k. Specifically in a cavity of length L, the number of states up to a wavelength $2\pi/k$ is given by $L/(2\pi/k)$. In three dimensions, this becomes $Vk^3/(2\pi)^3$, where $V = L^3$. Writing this in spherical coordinates, we have the relation

$$\frac{1}{V}\sum_k \rightarrow \frac{2\cdot 4\pi}{(2\pi)^3}\int_0^\infty k^2 dk\,, \tag{86}$$

where the leading 2 comes from the two possible spin values, and the 4π comes from the integral over the solid angle. Converting to the reduced mass energy ε of Eq. (5.110), we have $d\varepsilon = (\hbar^2/m)kdk$ and $k = [2m\varepsilon/\hbar^2]^{1/2}$, which give

$$k^2 dk = \sqrt{2}[m/\hbar^2]^{3/2}\sqrt{\varepsilon}\,d\varepsilon\,. \tag{87}$$

Substituting this into Eq. (86), we find

$$\frac{1}{V}\sum_k \rightarrow \frac{(2m/\hbar^2)^{3/2}}{2\pi^2}\int_0^\infty \sqrt{\varepsilon}\,d\varepsilon = \frac{1}{2\pi^2 a_0{}^3 E_R{}^{3/2}}\int_0^\infty \sqrt{\varepsilon}\,d\varepsilon\,, \tag{88}$$

where the exciton Bohr radius and Rydberg energy are given by

$$a_0 = \frac{\hbar^2\epsilon}{me^2} \tag{89}$$

$$E_R = \frac{\hbar^2}{2ma_0{}^2}\,, \tag{90}$$

and ϵ is the permittivity of the medium. Quantities expressed in terms of a_0 and E_R are less dependent on the specific semiconductor medium. In these units, the carrier density of Eq. (5.113) is given by

$$N \simeq \frac{1}{2\pi^2 a_0{}^3 E_R{}^{3/2}}\int_0^\infty \frac{\sqrt{\varepsilon}\,d\varepsilon}{e^{\beta(\varepsilon/\tilde{m}_\alpha - \mu_\alpha)} + 1}\,, \tag{91}$$

where the effective mass ratio $\tilde{m}_\alpha \equiv m_\alpha/m$. In terms of ε, we have the energy detuning

$$\hbar(\omega - \nu) = (\hbar\omega - \varepsilon_g - \delta\varepsilon_g) - (\hbar\nu - \varepsilon_g - \delta\varepsilon_g) = \varepsilon - \hbar\delta\,, \tag{92}$$

where the laser detuning relative to the renormalized bandgap is given by $\hbar\delta = \hbar\nu - \varepsilon_g - \delta\varepsilon_g$.

To carry out integrations over carrier energy, we express all frequencies in meV. To express $\hbar\gamma$ in meV, we take advantage of the fact that γ is usually given in terms of its inverse, the carrier-carrier scattering time τ_s. Planck's constant $\hbar = 6.5817\times10^{-13}$ meV s, which gives

$$\hbar\gamma = \frac{\hbar}{\tau_s} = \frac{6.5817\times10^{-13}}{\tau_s}\,. \tag{93}$$

For example, a scattering time of $\tau_s = 10^{-13}$ s gives $\hbar\gamma = 6.58$ meV. Similarly we express the frequency difference $\hbar\delta$ in meV. For room temperature, $1/\beta \simeq 25$ meV.

The optical gain coefficient of Eq. (76) is given by

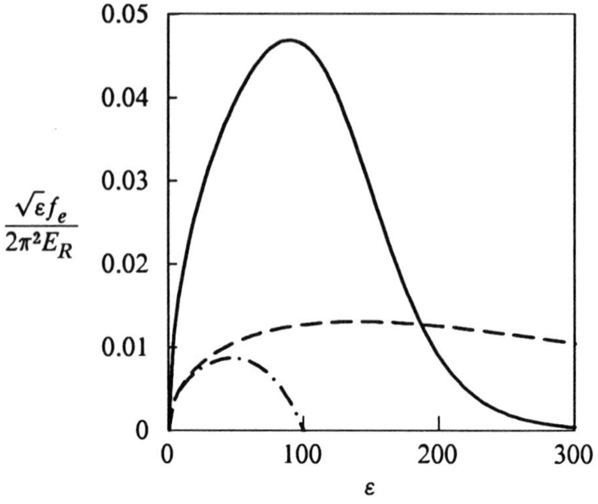

Fig. 6-9. $\sqrt{\varepsilon} f_e$ (solid line), $\sqrt{\varepsilon} f_h$ (dashed line), and $\sqrt{\varepsilon}(f_e + f_h - 1)$ (dot-dashed line), all in units of $2\pi^2 E_R$ versus energy ε in meV for $T = 300$K and a carrier density of 3.5×10^{18} carriers/cm^3, k_B = .086164 meV/K, and carrier parameters appropriate for GaAs, namely, $a_0 = 1.243\times10^{-6}$ cm, $E_R = 4.2$ meV, $m_e/m = 1.127$, $m_h/m = 8.82$. The gain formula used is Eq. (94) with the chemical potentials determined by the closure relation (5.113) to be $\mu_e = 4.73 k_B T$ and $\mu_h = -.86 k_B T$.

$$\alpha_r = \frac{\nu \wp^2}{2\epsilon\gamma\hbar(2\pi^2 a_0{}^3 E_R{}^{3/2})} \int_0^\infty \frac{\sqrt{\varepsilon}\, d\varepsilon}{1 + (\varepsilon - \hbar\delta)^2/(\hbar\gamma)^2}$$
$$\times \left[\frac{1}{e^{\beta(\varepsilon/\tilde{m}_e - \mu_e)} + 1} + \frac{1}{e^{\beta(\varepsilon/\tilde{m}_h - \mu_h)} + 1} - 1 \right]. \tag{94}$$

To get a feel for this gain function, we plot $\sqrt{\varepsilon} f_e$ (solid line) $\sqrt{\varepsilon} f_h$ (dashed line) and $\sqrt{\varepsilon}(f_e + f_h - 1)$ (dot-dashed line) in Fig. 6-9. The dot-dashed curve gives the gain spectrum for an optical probe for zero linewidth; linewidths ($2\hbar\gamma$) larger than zero sample a range of values, changing the gain spectrum somewhat. For the parameters chosen, the gain disappears below $N_0 \simeq 1.21\times10^{18}$ carriers/cm^3.

An increase in temperature reduces the gain by spreading the carriers out. This is shown in Fig. 6-10 for the three temperatures T = 200, 300,

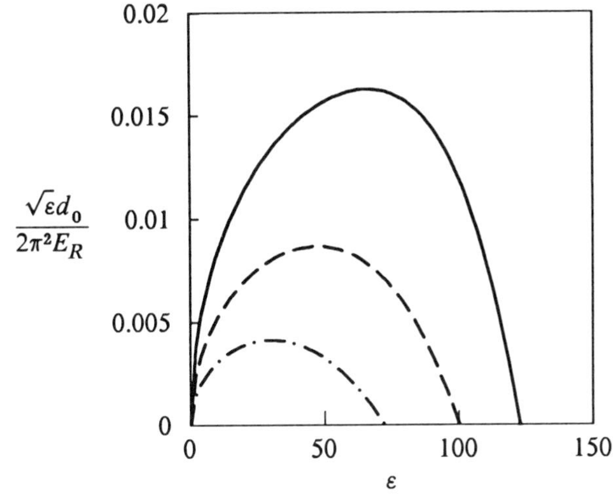

Fig. 6-10. $\sqrt{\varepsilon}d_0(k) = \sqrt{\varepsilon}(f_e(k) + f_h(k) - 1)$ versus reduced-mass energy ε for the temperatures T = 200K (solid line), 300K (dashed line), and 400K (dot-dashed line) and other parameters as in Fig. 6-9.

and 400K. The gain peak is defined by the condition $\partial\alpha_{rn}/\partial\delta = 0$. The laser threshold occurs for the minimum carrier density N_0 and the detuning δ_0 that allow this condition and Eq. (76) to be satisfied simultaneously. N_0 and δ_0 can be found by using a Newton-Raphson numerical procedure.

6-7. Transverse Variations and Gaussian Beams

Up to this point we have approximated the laser field by a plane wave. Observations and theoretical treatments reveal that the real modes in a laser are typically Hermite- or Laguerre-Gaussian functions, the simplest of which is the Gaussian beam. In this section, we show how a Gaussian beam is a solution of the Helmholz equation (spatial part of the wave equation), we give some of its properties, and we develop a simple single-mode laser theory that uses the Gaussian beam.

It is well-known that the spherical wave $\psi = e^{iKr}/r$ is a solution to the Helmholz equation $\nabla^2\psi + K^2\psi = 0$. The spherical wave is also a solution with arbitrary constant translations of the origin. The Gaussian beam has cylindrical symmetry about the z axis and decays as a Gaussian function of the radial distance from the z axis. We can obtain a wave with cylindrical symmetry about the z axis as well as a decay by choosing a spherical wave with an imaginary displacement along the z axis. Specifically we set $r = [x^2+y^2+(z-iz_0)^2]^{1/2}$ and suppose that $\rho^2 \equiv x^2+y^2 << z_0^2$. Then

$$\frac{e^{iKr}}{r} \simeq \frac{e^{iK(z-iz_0)}}{z-iz_0} e^{iK\rho^2/2(z-iz_0)}$$
$$= \frac{e^{Kz_0}}{z-iz_0} e^{iK[z+z\rho^2/2(z^2+z_0^2)]} e^{-Kz_0\rho^2/2(z^2+z_0^2)} . \tag{95}$$

Aside from the unimportant constant phase factor $\exp(Kz_0)$, this is the formula for a Gaussian beam. Its amplitude falls off to the $1/e$ point at $\rho^2 = w^2$, with the width w defined by

$$w^2 = w_0^2(1 + z^2/z_0^2); \quad w_0^2 = 2z_0/K . \tag{96}$$

This equation shows that the beam spreads to $\sqrt{2}$ times its minimum waist w_0 at $z=z_0$. The distance z_0 is called the *Rayleigh length.*

In addition, the Gaussian beam of Eq. (95) has a radius of curvature given by

$$R_c = z + z_0^2/z . \tag{97}$$

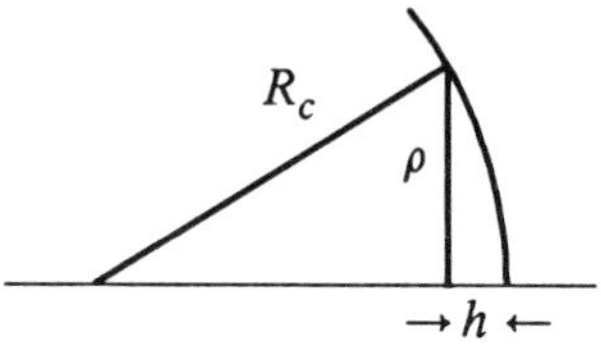

Fig. 6-11. Arc of a circle defining the "sag" h corresponding to off-axis distance ρ. The corresponding sag for the Gaussian beam defines a radius of curvature given by Eq. (96)

To see this, note in Fig. 6-11 that at a small off-axis distance ρ, a circle "sags" a distance h given by $R_c^2 = \rho^2+(R_c-h)^2$, which gives $h \simeq \rho^2/2R_c$. In Eq. (95) the sag $h = \rho^2 z/2(z^2+z_0^2)$, from which Eq. (96) follows. Figure 6-12 illustrates the Gaussian beam.

To make a laser resonator, we locate mirrors with appropriate curvature and separation to match the phase fronts given by Eq. (95). Provided some value of $z_0(w_0)$ subject to the condition $\rho^2<<z_0^2$ can yield appropriate phase fronts for the given mirrors and separation, the resonator is "stable," i.e., has low loss for highly reflecting mirrors.

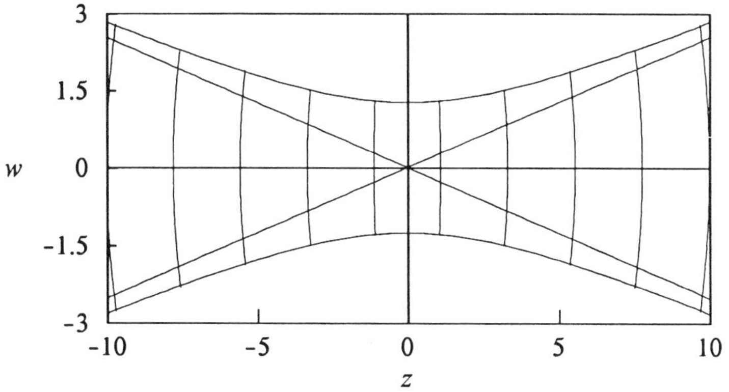

Fig. 6-12. Gaussian beam diagram showing minimum waist size w_0 and Rayleigh length z_0 (distance at which the beam spreads to $\sqrt{2}w_0$ in width and attains the minimum radius). The parameters used are $z_0 = 5$ and $\lambda = 1$.

One problem is that this simple recipe produces some focussing of the beam, and typically a substantial volume of potentially useful laser medium lies outside the interaction region. To get around this wasteful geometry, resonators that do not correspond to any Gaussian beam can be used, which lead to rays that walk off the mirrors instead of being focussed back onto them. The laser output is then taken from around the edges of the smaller mirror in the cavity as shown in Fig. 6-13. Such *unstable resonators* can give an efficient laser with high power output, although the beam has a hole in its middle. More elaborate schemes get rid of the hole as well.

To include a Gaussian beam in a single-mode laser theory, we replace $U_n(z)$ of Eqs. (2) and (3) by

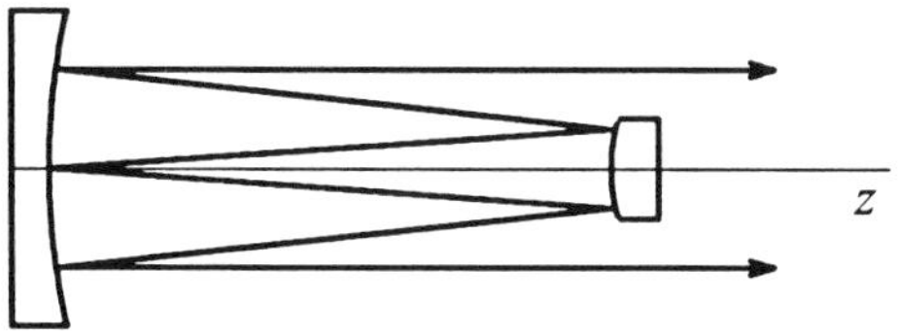

Fig. 6-13. Diagram of an unstable resonator.

$$U_n(r,z) = U_n(z)\, e^{-r^2/w_0^2} \tag{98}$$

and include corresponding integrals over r wherever projections on $U_n(z)$ occur. To simplify the analysis, we assume that the beam has a constant width w_0 throughout the interaction region. A more exact theory would allow for the fact that spreading does occur, leading to smaller peak intensities on the axis away from the beam waist. Still more elaborate theories would include higher-order Hermite-Gaussian functions that occur for other transverse modes found in lasers (see Fig. 6-14). We include an aperture of radius a, which serves both to prevent higher-order modes from oscillating, and can be used in theory to pass from the uniform-saturation, i.e., plane-wave limit (given by a=0, i.e., uniform across the aperture) to a full Gaussian limit (a=∞).

Under these approximations, the polarization (10) and normalization (11) integrals become, respectively

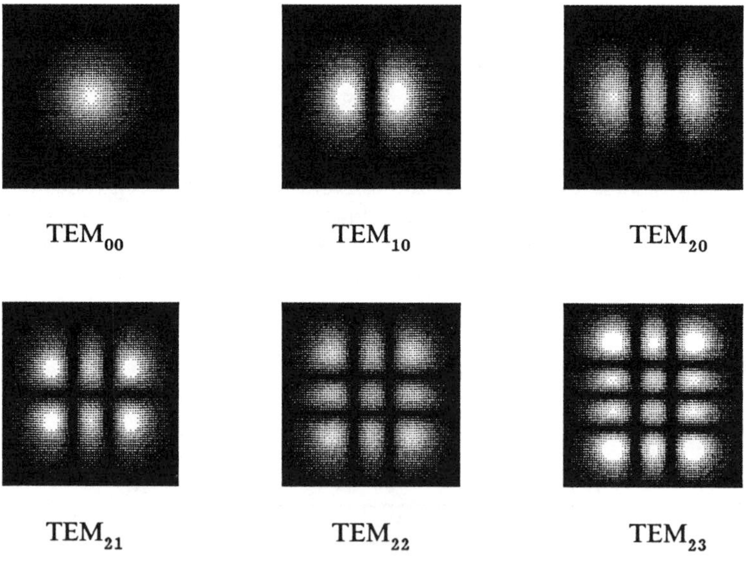

Fig. 6-14. Picture showing some of the transverse distributions for higher-order beams in lasers.

$$\mathscr{P}_n(t) = e^{i\nu_n t}\,\frac{2\wp}{\mathscr{M}_n}\int_0^a dr\int_0^{2\pi} r d\phi\, e^{-r^2/w_0^2}\int_0^L dz\, U_n^*(z)\rho_{ab}(r,z,t)\,, \tag{99}$$

where $\mathscr{M}_n$ is the mode normalization factor

$$\mathscr{M}_n = \int_0^a dr\int_0^{2\pi} r d\phi\, e^{-2r^2/w_0^2}\int_0^L dz\,|U_n(z)|^2\,. \tag{100}$$

As an example, consider the unidirectional ring cavity. The integrals can then be solved by the substitution

$$u = e^{-2r^2/w_0^2}\,, \quad du = -4rw_0^{-2}\,e^{-2r^2/w_0^2}\,dr\,. \tag{101}$$

The normalization factor (100) becomes

$$\mathscr{M}_n = 2(\pi L w_0^2/4)\int_{u_a}^1 du = \frac{\pi}{2}Lw_0^2(1-u_a)\,. \tag{102}$$

Substituting this and Eq. (5.24) into Eq. (99), we find

$$\mathscr{P}_n(t) = -i\wp^2 E_n \hbar^{-1}\overline{N}\,\frac{\mathscr{D}_n}{1-u_a}\int_{u_a}^1 \frac{du}{1+I_n\mathscr{L}_n u}$$

$$= -i\wp^2 E_n \hbar^{-1}\overline{N}\,\mathscr{D}_n\,\frac{1}{(1-u_a)I_n\mathscr{L}_n}\ln\left[\frac{1+I_n\mathscr{L}_n}{1+I_n\mathscr{L}_n u_a}\right]. \tag{103}$$

Note that for $a\to 0$, $u_a\to 1$, and the bracketed expression in Eq. (103) becomes $d\ln(1+I)/dI = 1/(1+I)$, where $I = I_n\mathscr{L}_n$. Hence for uniform saturation, i.e., a plane wave, we recover the plane-wave formula (14) with Eq. (16). In terms of Eq. (14), we obtain the saturation factor

$$\mathscr{S}(I_n) = \frac{1}{(1-u_a)I_n\mathscr{L}_n}\ln\left[\frac{1+I_n\mathscr{L}_n}{1+I_n\mathscr{L}_n u_a}\right]. \tag{104}$$

Substituting this into the intensity equation of motion (20) with g_n given by Eq. (23) (for which the linear loss terms have also had projections like Eq. (100)), we find the steady-state result

$$\frac{\mathcal{N}\mathcal{L}_n}{(1-u_a)I_n\mathcal{L}_n}\ln\left[\frac{1+I_n\mathcal{L}_n}{1+I_n\mathcal{L}_n u_a}\right]=1\ . \tag{105}$$

In particular for $a \gg w_0$, which corresponds to a full Gaussian beam, this gives

$$I_n = \mathcal{N}\ln(1+I_n\mathcal{L}_n)\ . \tag{106}$$

This gives a substantially softer saturation characteristic than the plane wave formula given in Table 1. Figure 6-15 compares the two (Eq. (106) is solved iteratively). Note that for a given relative excitation $\mathcal{N}$, the detuning values for the onset of oscillation are the same for Gaussian-beam and plane-wave theories, since the values are determined by linear theories.

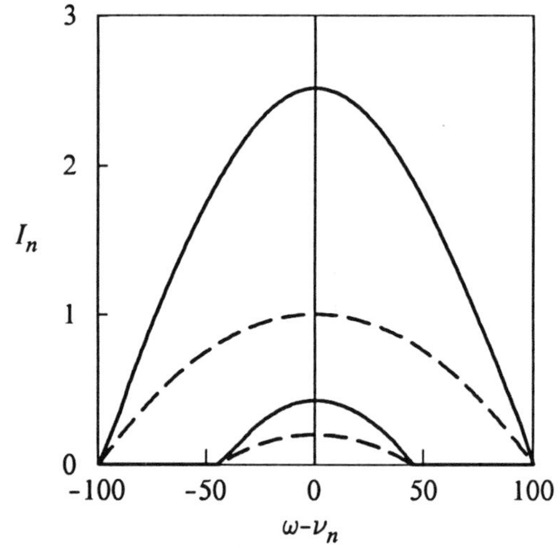

Fig. 6-15. Unidirectional ring intensity versus detuning for a plane wave (dashed lines - first entry in Table 6-1) and a Gaussian beam [solid lines -Eq. (106)]. The pair of higher intensity curves is for the relative excitation $\mathcal{N}$ = 2, and the lower pair is for $\mathcal{N}$ = 1.2. The dipole decay constant γ = 100.

References

Bennett, W. R., Jr. (1962), Appl. Opt. Suppl. **1**, 24.

Chow, W. W., J. B. Hambenne, T. J. Hutchings, V. E. Sanders, M. Sargent III, and M. O. Scully (1980), IEEE J. of Quant. Electronics **QE-16**, 918.

Milonni, P. W. and Eberly, J. H. (1988), *Lasers*, John Wiley & Sons, New York. This gives a broad coverage of lasers at a more introductory level than the present book.

Haken, H. (1970), *Laser Theory*, in *Encyclopedia of Physics*, **XXV/2c**, Ed. by S. Flügge, Springer-Verlag, Heidelberg. This gives a thorough compilation of the Haken school work up to 1970.

Lamb, W. E., Jr. (1964), Phys. Rev. **134**, A1429.

Sagnac, C. G. (1913), C. R. Acad. Sci. **157**, 708.

Sargent, M. III, M. O. Scully, and W. E. Lamb, Jr. (1977), *Laser Physics*, Addison-Wesley Publishing Co., Reading, MA. This book gives considerably more laser theory than the present chapter following the work of the Lamb school.

Sargent, M. III (1976), "Laser Theory" in *Applications of lasers to atomic and molecular physics, Proc. Les Houches Summer School*, eds. R. Balian, S. Haroche, and S. Liebermann, North-Holland, Amsterdam.

Siegman, A. E. (1986), *Lasers*, University Science Books, Mill Valley, CA. Excellent reference on lasers with thorough coverage of resonator theory.

Stenholm, S., and W. E. Lamb, Jr. (1969), Phys. Rev. **181**, 618.

Yariv, A. (1989), *Quantum Electronics*, 3rd Edition, John Wiley & Sons, New York. A standard reference in quantum electronics.

For a review of ring laser gyros, see F. Aronowitz (1978), Proc. SPIE **157**, 2, and W. W. Chow, J. B. Hambenne, T. J. Hutchings, V. E. Sanders, M. Sargent III, and M. O. Scully (1980), IEEE J. Quant. Electron. **QE-16**, 918.

See also references from Chap. 5 on semiconductor media.

Problems

6-1. Show that the population matrix $\rho(z, v, t)$ of Eq. (35) obeys the equations of motion (5.9) through (5.11) in which the time derivative is given by the convective derivative of Eq. (36). Show also that the formal solution (38) satisfies the equation of motion (37).

6-2. Consider a gas laser with $\gamma = 2\pi \times 100$ MHz and $Ku >> \gamma$. We insert a Doppler-broadened absorption cell with a natural linewidth of 10kHz and the same line center as the laser gain medium into the cavity. Suppose the steady-state electric field is sufficiently small that the induced polarization of the media can be treated to order E^3. Qualitatively, how do you expect the intensity to vary as a function of tuning? Write a formula for the combined polarizations of the media. Solve the resulting steady-state laser equation for the output intensity.

6-3. The gain for the I_+ wave in a bidirectional ring laser with both waves tuned to the atomic line center is

$$\alpha_+ = \frac{\alpha_0}{2I_+}\left[1 - \frac{1 + I_- - I_+}{[1 + 2I_+ + 2I_- + (I_+ - I_-)^2]^{1/2}}\right]$$

Show that to first-order in I_+ and I_-, the gain for I_+ is

$$\alpha_+ \simeq \alpha_0(1 - I_+ - 2I_-) \ .$$

Noting that α_- is given by the same expression with the subscripts + and -interchanged, what can you say about the steady-state solution(s)? What is the physical origin of the factor of 2?

6-4. Show that the fringe pattern in a bidirectional circular ring cavity remains stationary even when the cavity rotates. What two main problems occur in the ring laser gyro and what ways are used to overcome them?

6-5. What condition must be satisfied so that only a single cavity mode oscillates in a cavity?

6-6. By setting $\partial^2 I_n / \partial \nu_n^2 |_{\nu_n = \omega} = 0$, show that the Eq. (52) exhibits a Lamb dip provided the relative excitation $\mathcal{N}$ of Eq. (22) satisfies the inequality $\mathcal{N} > 1 + 2(\gamma/Ku)^2$.

6-7. Show that the coefficient matrix in the coupled equations of motion (62) and its ϵ_- counterpart has eigenvalues with negative real parts, provided that both effective α's are positive and that the coupling parameter C is smaller than 1. Alternatively apply the Hurwitz criterion of Prob. 10-4.

6-8. Show that the indefinite integral

$$J = \int d\theta \, \frac{1}{a + b\sin\theta} = \frac{2}{(a^2 - b^2)^{1/2}} \tan^{-1}\left\{\frac{a\tan(\theta/2) + b}{(a^2 - b^2)^{1/2}}\right\} + C$$

for $a^2 > b^2$. For $b^2 > a^2$, $\tan^{-1} \rightarrow \tanh^{-1}$ and $a^2-b^2 \rightarrow b^2-a^2$, and you get an overall minus sign. In what problem(s) does this integral occur in laser physics?

6-9. What is the radius of curvature R_c of a phase front of a Gaussian beam? What is the beam waist size w? *Ans*: $R_c = z + z_0^2/z$; $w = w_0\sqrt{1 + z^2/z_0^2}$, where the minimum waist size $w_0 = \sqrt{\lambda z_0/\pi}$.

6-10. The coupling between two longitudinal modes in a standing-wave laser described by the field

$$E(z,t) = \sum_n E_n(t) \cos(\nu_n t) \sin(K_n z)$$

is characterized in third-order perturbation theory by

$$\dot{E}_1 = E_1\left[g_1 - \frac{\nu}{2Q_1} - \beta_1 I_1 - \theta_{12} I_2\right]$$

with a similar equation for $\dot{E}_2$ with $_1$ and $_2$ interchanged. The self- and cross-saturation coefficients are given by

$$\beta_1 = F_3\, \mathscr{L}^2(\omega-\nu_1)$$

$$\theta_{12} = \frac{1}{3} F_3 (2 + N_2/\bar{N}) \,\mathrm{Re}\left\{\gamma\mathscr{D}_1\left[\mathscr{L}_2 + \mathscr{F}(\Delta)\frac{\gamma}{2}(\mathscr{D}_1 + \mathscr{D}_2^*)\right]\right\},$$

where the "cross excitation" function

$$N_2 = \frac{1}{L}\int_0^L dz\, N(z)\cos(2\pi z/L)\ .$$

Interpret these coefficients in terms of incoherent and coherent saturation effects, noting the role of population pulsations. Neglecting these pulsations, evaluate the coupling parameter $C = \theta_{12}\theta_{21}/\beta_1\beta_2$ when (*a*) $N(z)$ fills the cavity, (*b*) $N(z)$ is in the first quarter of the cavity, and (*c*) $N(z)$ is in the middle $1/10th$ of the cavity.

6-11. Consider the sidemode polarization $\mathscr{P}_1$ in a two-mirror laser given by the expression

$$\mathscr{P}_1 = -\frac{i\wp^2}{\hbar}\mathscr{E}_1\mathscr{D}_1\,\frac{1}{L}\int_0^L dz\,\frac{N(z)|U_1(z)|^2}{1 + I_2\mathscr{L}_2|U_2(z)|^2},$$

where $U_n(z) = \sin(K_n z)$. For what kind of a medium is this valid? Have population pulsations been neglected? Show that the answer is

$$\mathscr{P}_1 = -\frac{i\wp^2}{\hbar}\frac{\mathscr{E}_1\mathscr{D}_1}{\sqrt{1 + I_2\mathscr{L}_2}}\left[\overline{N} - N_2\frac{I_2\mathscr{L}_2/2}{1 + I_2\mathscr{L}_2 + \sqrt{1 + I_2\mathscr{L}_2}}\right],$$

where $\overline{N} = \frac{1}{L}\int_0^L dzN(z)$ and $N_2 = \frac{1}{L}\int_0^L dzN(z)\cos(2\Delta z/c)$. Hint:

$$\begin{aligned}\sin^2(K_1z) &= \frac{1}{2} - \frac{1}{2}\cos[2(K_2-\Delta/c)z]\\ &= \frac{1}{2} - \frac{1}{2}\cos(2K_2z)\cos(2\Delta z/c) - \frac{1}{2}\sin(2K_2z)\sin(2\Delta z/c)\end{aligned}$$

and the $\sin(2K_2z)$ term can be dropped since it is an odd function which is multiplied by an even function in $\mathscr{P}_1$ and integrated over an even interval. Also use the fact that $N(z)$ and $\cos(2\Delta z/c)$ vary little in a wavelength.

6-12. Using Eqs. (20), (24), and (25), calculate the steady-state intensity for a unidirectional ring laser with extreme inhomogeneous broadening. Ans: see Table 1.

6-13. Consider the three-level system shown below:

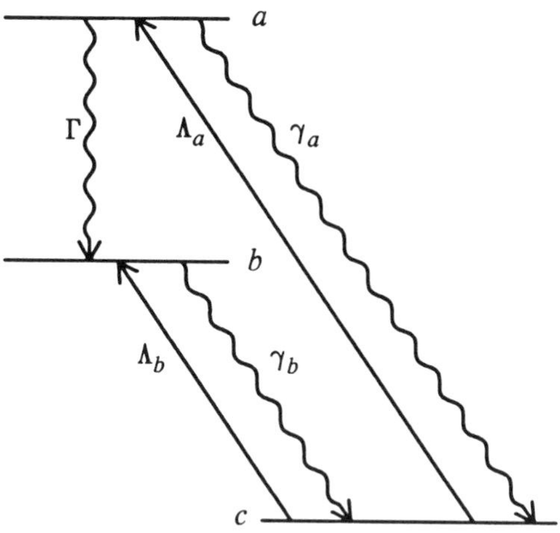

The applied electric field causes transitions between levels a and b while level c (the ground state) acts as a reservoir connected to a and b by level decays and pumps. γ_a and γ_b are the rates at which levels a and b decay to level c, and Λ_a and Λ_b are the pumping rates from level c to levels a and b. Γ describes the decay from level a to level b.

a. Assuming $E = E_0 \cos\nu t$ where $\nu \simeq \omega_a - \omega_b = \omega$ so that the rotating wave approximation may be made, determine the equations of motion for the population matrix elements. There are four equations, one for the dipole ρ_{ab} and one for each level population.

b. Solve these equations in steady state for ρ_{aa}, ρ_{bb}, and $\rho_{aa} - \rho_{bb}$ using the rate equation approximation. Keep in mind the trace condition $\rho_{aa} + \rho_{bb} + \rho_{cc} = N$, where N = total number of systems.

c. Defining $I = (\wp E_0/\hbar)^2 T_1 T_2$, $T_2 = 1/\gamma$, show that the population difference $\rho_{aa} - \rho_{bb} = N'/(1 + \mathrm{I}\mathscr{L})$, where N' = zero field population difference and $\mathscr{L} = \gamma^2/(\gamma^2 + (\omega - \nu)^2)$. Determine the expression for N' and show that T_1 is given approximately by Eq. (5.108).

d. The semiclassical laser theory of Chap. 6 assumes there is no decay between a and b and that the populations of levels a and b are small compared to the pumping reservoir. Set $\Gamma = 0$ and let $\gamma_a, \gamma_b >> \Lambda_a, \Lambda_b$. Show your expressions reduce to $N' = N(\Lambda_a/\gamma_a - \Lambda_b/\gamma_b)$ and $T_1 = 0.5(1/\gamma_a + 1/\gamma_b)$. In this limit we are assuming $\rho_{cc} \simeq N$ so that $N\Lambda_a \simeq \lambda_a$, $N\Lambda_b \simeq \lambda_b$.

e. This level scheme can also describe upper to ground state lower decay with no pumping. Show by setting $\gamma_a = \gamma_b = 0$ in your general expressions that $N' = -N$, $T_1 = 1/\Gamma$. Note that the pumps cancel out.

6-14. Show that the steady-state oscillation condition (26) with the inhomogeneous broadening saturation factor (24) yields the corresponding laser intensity given in Table 1.

6-15. Show that the width of the Lamb dip given by Eq. (52) is given approximately by $2\sqrt{2}\gamma$.

6-16. Show that in the limit $\Lambda_a, \Lambda_b << \gamma_a, \gamma_b$, the population-difference lifetime T_1 of Prob. 6-13 is given by

$$T_1 = \frac{1}{2}\,\frac{\gamma_a + \gamma_b}{\gamma_b(\gamma_a + \Gamma)} = \frac{1}{2}\,\frac{1}{\gamma_a + \Gamma}\left(1 + \frac{\gamma_a}{\gamma_b}\right).$$

In appropriate limits, this formula reduces to the excited two-level system of Fig. 4-1 and to the upper-to-ground-lower-level case of Fig. 4-5.

6-17. Calculate the polarization given by Eq. (12) for a two-mirror laser when the length of the medium is small compared to an optical wavelength, e.g., as producable by a molecular-beam epitaxy (MBE) machine. What is the corresponding steady-state laser intensity?

6-18. Using Eq. (5.61), calculate α_+ for a ring cavity with an active medium small compared to a wavelength. Note that for this problem, the medium determines the z reference, in contrast to the choice leading to the polarization of Eq. (5.62).

Chapter 7
OPTICAL BISTABILITY

Chapter 6 gives the theory of a laser, which is a self-sustained oscillator consisting of an active medium in a Fabry-Perot or ring cavity. The laser output frequencies, imposed by the self-consistent laser equations, are compromises between the atomic and cavity natural frequencies. In this chapter, we discuss another situation involving a nonlinear medium in Fabry-Perot and ring cavities, but with two major changes: 1) the cavity output depends on an *injected signal* for its energy and output frequency, and 2) the medium is passive, i.e., it absorbs and/or provides an index change - but for two-level media, the upper state is not pumped. The name optical bistability comes from the characteristic of such systems that for a single input intensity, two (or more) stable output intensities are often possible, one large and one small. The system is like an electronic flip-flop, except that it is all optical.

We have already encountered a situation leading to a bistable output in Sec. 6-4, which discusses optical bistability due to nonlinearities of the medium in the homogeneously-broadened ring laser. In this case, the bistability is directional, in that the laser likes to run in one direction or the other but not in both. Similarly a laser based on an atomic $J=2\longleftrightarrow 2$ transition likes to run with one circular polarization or the other, but not both. Both of these systems are lasers and hence self-sustained oscillators, and were understood in the mid 1960's.

Szöke *et al.* (1969) considered injecting a signal into a ring cavity with an unexcited two-level medium (see Fig. 7-1). They gave the theory presented in Sec. 7-1, which shows that the bleaching of the medium due to saturable absorption, combined with the feedback provided by the resonator, can lead to hysteresis: once large, the cavity field can be reduced below the value that led to the jump to the big value and the medium still remains sufficiently bleached to maintain the big field. Such "absorptive optical bistability" has been observed, but is relatively difficult to achieve.

In contrast, Gibbs *et al.* (1976) discovered and explained optical bistability for a cavity containing a medium with a nonlinear index and no absorption or gain, i.e., purely dispersive. For the simplest case of a Kerr nonlinearity, the cavity frequency is swept an amount proportional to the

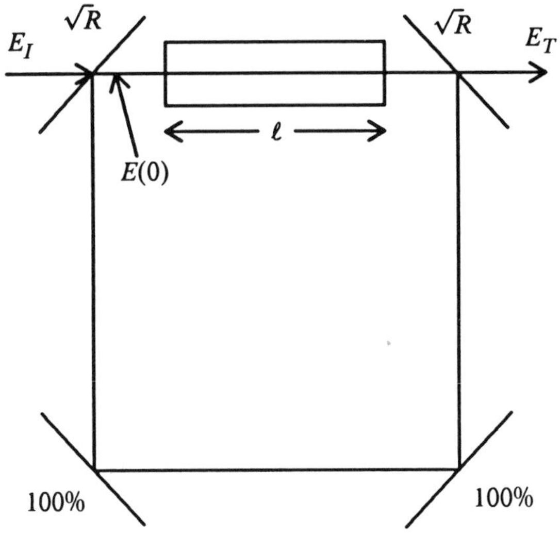

Fig. 7-1. Unidirectional ring cavity with input field E_I and output field E_T and a nonlinear medium. This configuration is often used in discussions of optical bistability.

light intensity in the cavity. By choosing a weak-field detuning with opposite sign from this sweep, larger fields can tune the cavity through resonance. Two stable output values are again possible, one for small cavity fields, one for large. In contrast to the absorptive case, the theory can be based on low-order perturbation theory, that is, on the first nonlinear index term n_2 or the susceptibility term $\chi^{(3)}$.

Dispersive optical bistability differs from the absorptive also in that increasing input intensities cause the cavity transmission to decrease after a certain point, since they keep tuning the cavity, while the absorptive case passes all intensity values past a certain point with nearly unit transmission. Furthermore, since the fields used in the dispersive case can tune substantial amounts, "multistability" is possible with large transmissions corresponding to input intensities appropriate for successive cavity resonances, again in contrast to the absorptive case.

Section 7-1 develops the simple theory of a cavity filled with a nonlinear medium and driven by an injected field, and considers the purely dispersive limit in some detail. Section 7-2 considers purely absorptive optical bistability and develops a simple stability analysis of the steady-state solutions. When bistability can occur, a third output value is a solution of the equations, but it is found to be unstable. Section 7-3 analyzes the stability of a more complicated situation in which a pair of sidemodes builds up in an optically bistable cavity containing a pure index $\chi^{(3)}$ medium [Ikeda (1989) instability]. This problem involves the three-wave mixing concepts developed in Chap. 2 and further in Chap. 9.

7-1. Simple Theory of Dispersive Optical Bistability

We consider the situation illustrated in Fig. 7-1 of a ring cavity with one leg filled by a nonlinear medium and driven by an injected field E_I. To determine the field inside the unidirectional ring cavity, we write the relation

$$\mathcal{E}_{n+1}(0) = \sqrt{T}\, E_I + f(\mathcal{E}_n(0)) \,, \tag{1}$$

which states that the field at z=0 just inside the input mirror on the (n+1)st "pass" around the cavity equals the transmitted portion of the input field E_I plus some function of the field present on the previous (nth) pass. Without loss of generality, we take the *cw* input field to be real and the cavity field $\mathcal{E}$ to be complex. In general the function $f(\mathcal{E})$ depends on the growth or decay of the field as it propagates through the nonlinear medium, but we take it to be the simple function

$$f(\mathcal{E}_n(0)) = Re^{-\alpha\ell} e^{iKL} \mathcal{E}_n(0) \,, \tag{2}$$

where L is the round-trip length of the ring cavity, ℓ is the length of the nonlinear medium, α is the complex absorption coefficient, and R = 1-T is the mirror intensity reflectivity coefficient. As such, we assume that the absorption coefficient depends only on the "uniform field" approximation (field envelope is position independent) and thereby neglect saturation or nonlinear index corrections due to field changes along the laser axis.

Possible steady-state solutions to Eq. (1) are given by setting $\mathcal{E}_{n+1}(0) = \mathcal{E}_n(0) = \mathcal{E}_0$. This gives

$$\mathcal{E}_0 = \sqrt{T} E_I + f(\mathcal{E}_0) \,, \tag{3}$$

which with Eq. (2) becomes

$$\mathcal{E}_0 = \sqrt{T} E_I + Re^{-\alpha\ell} e^{iKL} \mathcal{E}_0 . \tag{4}$$

Solving for $\mathcal{E}_0$, we have

$$\mathcal{E}_0 = \frac{\sqrt{T}\, E_I}{1 - R\, e^{-\alpha\ell + iKL}} . \tag{5}$$

Aside from a phase factor if the medium does not fill the first leg of the cavity, the output field is given by

$$E_T = \sqrt{T} E(\ell) = \sqrt{T} \mathcal{E}_0 e^{(-\alpha + iK)\ell} . \tag{6}$$

Combining this with Eq. (5), we have the amplitude transmission function

$$\frac{E_T}{E_I} = \frac{T e^{(-\alpha+iK)\ell}}{1 - R e^{-\alpha\ell + iKL}} = \frac{T e^{iK(\ell - L)}}{e^{\alpha\ell - iKL} - R} . \tag{7}$$

Similarly one can show for the two-mirror (Fabry-Perot) cavity that

$$\frac{E_T}{E_I} = \frac{T e^{\alpha\ell - iKL}}{e^{2\alpha\ell - 2iKL} - R} . \tag{8}$$

Up to this point, our equations apply to an arbitrary complex absorption coefficient, α, and hence can be used to study both purely dispersive and purely absorptive optical bistability. In the remainder of this section, we consider the purely dispersive case ($\mathrm{Re}(\alpha)=0$). This case is obviously an approximation, since a nonlinear dispersion implies the existence of a nonlinear absorption, but the latter decreases significantly faster as the laser frequency is detuned from the medium's resonances. Setting $\beta = \alpha_i \ell - KL$ - (nearest multiple of 2π)), we find that the amplitude transmission function (7) yields

$$\boxed{\frac{I_T}{I_I} = \frac{T^2}{|e^{i\beta} - R|^2} = \frac{1}{1 + 4R \sin^2(\beta/2)/T^2}} . \tag{9}$$

where $I_T = |E_T|^2$ and $I_I = |E_I|^2$ (here we suppose the E's are dimensionless fields corresponding to the usual dimensionless intensity definition). Near a cavity resonance, $|\beta| << 1$, which gives

$$\frac{I_T}{I_I} = \frac{1}{1 + R\beta^2/T^2} , \tag{10}$$

whereas for larger β, $I_T/I_I < T^2$. This gives the familiar Lorentzian transmission peaks (Fig. 7-2), which have FWHM $= 2T/\sqrt{R} = 2\pi/F$, where F is the cavity finesse. Ordinarily the transfer function is plotted versus the input frequency, while in Fig. 7-2 we have chosen the tuning axis to

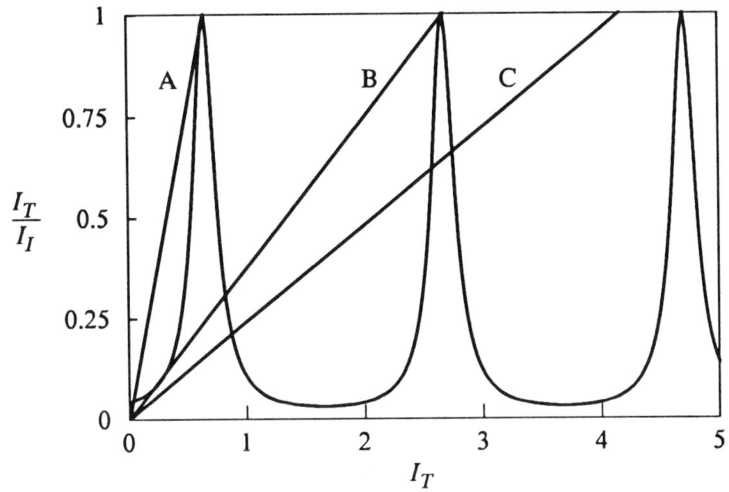

Fig. 7-2. Transmission peaks of a high Q cavity. Intersections of straight lines with the peaks correspond to various possible oscillation intensities for dispersive optical bistability. The reflectance $R = .7$, $\beta_0 = 2$, and $\beta_2 = -3.1$.

be the transmitted intensity I_T. This choice corresponds to dispersive optical bistability and is explained along with the straight lines below.

To understand dispersive bistability, we expand the phase shift β in Eq. (10) as

$$\beta = \beta_0 + \beta_2 I_T \ . \tag{11}$$

We take β_2 to have the opposite sign from β_0, so that the field can tune the cavity through resonance. Equation (10) then gives

$$I_I = I_T \ [1 + R(\beta_0 + \beta_2 I_T)^2/T^2] \ . \tag{12}$$

Figure 7-2 plots the cavity transmission I_T/I_I of Eq. (9) for cases with β given by Eq. (11). The figure also illustrates a graphical solution of Eq. (9). The straight lines are those of I_T/I_I *vs* I_T. The intersection points between these straight lines and the cavity resonance curve give the dispersive bistability solutions predicted by Eq. (9). This graphical approach gives one an immediate feel for the number of possible solutions corresponding to a given input intensity.

For bistability to occur, we must have a region for which $dI_T/dI_I < 0$, or equivalently $dI_I/dI_T < 0$. Setting $dI_I/dI_T = 0$ to find the boundary of this region. From Eq. (12) we obtain

$$\begin{aligned} 0 &= 1 + R(\beta_0+\beta_2 I_T)^2/T^2 + 2R\beta_2 I_T(\beta_0+\beta_2 I_T)/T^2 \\ &= (\beta_2 I_T)^2 + 4\beta_0\beta_2 I_T/3 + (T^2/R + \beta_0^2)/3. \end{aligned}$$

This equation has the roots

$$\beta_2 I_T = -\frac{2}{3}\beta_0 \pm \frac{1}{3}\sqrt{\beta_0^2 - 3T^2/R}\,. \tag{13}$$

For real solutions, $\beta_0^2 > 3T^2/R$. This bistability condition makes sense in terms of the cavity transmission diagram (Fig. 7-2), since for bistability, the cavity intensity must tune from a low transmission value on one side of a resonance to a high transmission value on the other side. Since the resonance has a FWHM $= 2T/\sqrt{R}$, $\sqrt{3}T/\sqrt{R}$ might be expected to do the trick. When plotting I_T as a function of I_I with I_I on the x axis (Fig. 7-3), we see how three output intensities can correspond to one input intensity.

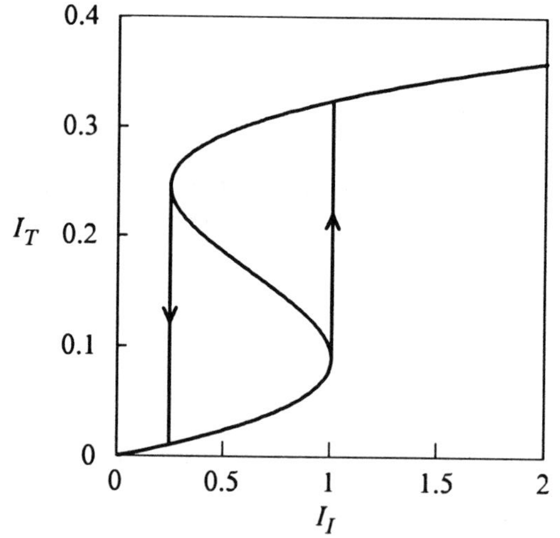

Fig. 7-3. Transmitted intensity I_T versus input intensity I_I given by Eq. (12) for dispersive optical bistability. The parameter values used are $R = .99$, $\beta_0 = .05$, and $\beta_2 = -.2$. The vertical lines show the sides of the hysteresis loop traced out by I_T as I_I is increased above its value at the lower turning point and then below its value at the upper turning point.

To predict the stability of these solutions intuitively, note that the upper branch corresponds to points on the right side of a resonance (see Fig. 7-2) and the negative slope region to points on the left side. A small increase of the cavity field for a point on the right side detunes the cavity, causing the self-consistent field to decrease. Similarly, a decrease tunes the cavity toward resonance, causing the field to increase. Hence such a point is stable. The opposite is true for points on the left side. From this argument, we see that for a given value of the input intensity, the line C in Fig. 7-2 has two such stable outputs, hence the name optical *bi*stability.

Figure 7-3 shows the *hysteresis loop* that the output follows as the input intensity I_I is successively increased and decreased within the appropriate range. Consider I_T initially on the lower branch as I_I increases past its value for the lower turning point. The output intensity I_T then "switches" up to the upper branch. So long as I_I remains larger than its value at this switch point, I_T is given by the upper branch solution. The remarkable point is that as I_I decreases below this value, I_T is still given by the upper-branch value until I_I decreases below its value for the upper turning point. At this point, I_T switches back down to the lower branch.

It is also possible to obtain *a* β_2 term purely classically by suspending one cavity mirror in a vacuum and letting the force from *radiation pressure* increase the cavity length. This force is proportional to the light intensity in the cavity and hence to I_T. In steady state, it is balanced by the force exerted by gravity on the mirror pendulum. Since the latter is approximately proportional to the mirror displacement ς from the vertical, we have $\varsigma \propto I_T$. Calling L_0 the cavity length for $\varsigma = 0$, we have $L = L_0 + \varsigma$. The two-mirror cavity modes are $\Omega_n = \pi nc/L$. Hence the round-trip phase shift is

$$\beta = \frac{2(\nu-\Omega_n)L}{c} = \frac{2(\nu-\Omega_{n\,0})L_0}{c} + \frac{2\nu\varsigma}{c} = \beta_0 + \beta_2 I_T. \tag{14}$$

Instead of changing the optical length of the cavity, this arrangement changes its physical length. Otherwise, everything remains the same. As for nonlinear dispersive optical bistability, this radiation pressure bistability can be multistable. Experimentally, noise plays an important role in this problem, ordinarily causing the mirror to move around in times on the order of .1 seconds. Some combinations of parameters can "cool" this motion to a value near a steady-state solution to Newton's law. For further information, see Dorsel *et al.* (1983).

7-2. Absorptive Optical Bistability

Let us now turn to the case of purely absorptive optical bistability for which the input field frequency coincides with both a cavity resonance and the atomic line center. Ignoring the unimportant phase factor in the numerator of Eq. (7), and supposing that $\alpha\ell << 1$ so that $e^{-\alpha\ell} \simeq 1-\alpha\ell$, we find that Eq. (7) reduces to

$$\frac{E_T}{E_I} \simeq \frac{1}{1 + \alpha\ell/T} \,. \tag{15}$$

If $\alpha\ell/T$ is large, i.e., for $T<<\alpha\ell<<1$, then E_T/E_I is small. However if the absorption can be bleached, E_T/E_I can approach unity transmission.

Specifically for a two-level atom, Sec. 5-1 shows that on resonance $\alpha=\alpha_0/(1+I)$, where I is given in units of the saturation intensity. For convenience, we take E_I and E_T also in the corresponding amplitude units, which gives $I=I_T/T$. Combining these formulas with Eq. (15) and solving for E_I, we find

$$\boxed{\frac{E_I}{\sqrt{T}} = \frac{E_T}{\sqrt{T}}\left[1 + \frac{\alpha_0\ell/T}{1 + I_T/T}\right]} \,. \tag{16}$$

Alternatively for a standing-wave cavity, $\alpha = \alpha_0\mathcal{S}(I_n)$, where the saturation factor $\mathcal{S}(I_n)$ is given by Eq. (6.16), here with $\nu_n=\omega$. This gives

$$\frac{E_I}{\sqrt{T}} = \frac{E_T}{\sqrt{T}}\left[1 + \frac{2\alpha_0\ell/T}{I_T/T}\left(1 - \frac{1}{(1 + I_T/T)^{1/2}}\right)\right] \,. \tag{17}$$

Figure 7-4 plots $E_I\sqrt{T}$ (on the x axis) versus $E_T\sqrt{T}$ for various values of $\alpha_0\ell/T$. We see that for sufficiently large $\alpha_0\ell/T$, three possible values of E_T occur for a single E_I value, just as for the dispersive case. As discussed below, both here as for the dispersive case the negative slope region $(E_T/E_I < 0)$ is unstable, so that only two solutions are stable.

To find the value of $\alpha_0\ell/T$ giving the onset of multiple solutions, we set dE_I/dE_T=0. This implies (we use the common notation $2C=\alpha_0\ell/T$ for typographical simplicity)

$$0 = 1 + \frac{2C}{1+I} - \frac{4CI}{(1+I)^2} \,.$$

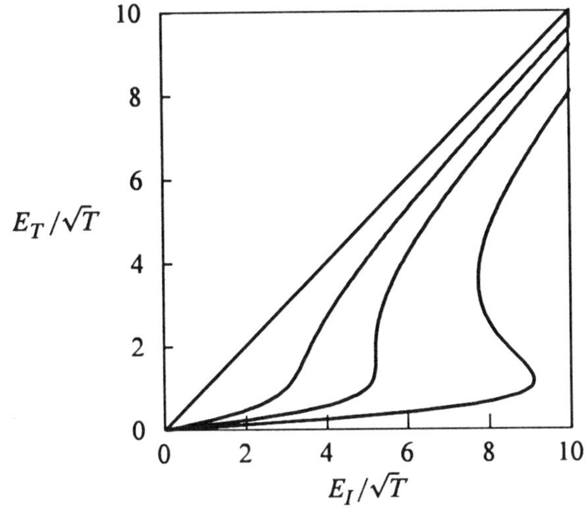

Fig. 7-4. Graphs of the output field versus input field given by Eq. (16) for centrally-tuned absorptive optical bistability for $\alpha_0\ell/T$ = 0, 4, 8, and 16, in order of decreasing initial slope.

This gives a quadratic equation for $I=I_T/T$ with the solutions

$$I = C - 1 \pm \sqrt{C}\,\sqrt{C-4}\,. \tag{18}$$

These solutions are real only if $C \geq 4$, i.e., only if $\alpha_0\ell/T \geq 8$.

Linear Stability Analysis

The question arises as to what solutions of Eq. (1) are stable. In Sec. 7-3, we present an analysis that leads to a simple interpretation in terms of sidemode build up. Here for simplicity, we consider purely absorptive bistability, for which the Eq. (2) is real and consider a real field E_n instead of the more general complex field $\mathcal{E}_n$. As in the treatments of Chap. 6, we perform a linear stability analysis by expanding E_n about E_0 as

$$E_n = E_0 + \epsilon_n \tag{19}$$

and ask that $\epsilon_n \to 0$ as $n \to \infty$. In principle ϵ_n could be complex and in extreme cases even lead to "phase switching", but we consider only real fluctuations. Substituting Eq. (17) into Eq. (1), we find

$$E_0 + \epsilon_{n+1} = \sqrt{T}\, E_I + f(E_0) + \epsilon_n \frac{df}{dE_0},$$

i.e.,

$$\epsilon_{n+1} = \frac{df(E_0)}{dE_0}\epsilon_n = \left[1 - \sqrt{T}\,\frac{dE_I}{dE_0}\right]\epsilon_n, \tag{20}$$

where Eq. (3) gives the second equality. Equation (20) converges provided

$$\left|1 - \sqrt{T}\,\frac{dE_I}{dE_0}\right| < 1. \tag{21}$$

Since by assumption the equations are real, Eq. (21) implies that dE_I/dE_T must be positive, but not too large. The negative sloped regions in Fig. 7-4 are therefore unstable. Intuitively one can understand this in terms of servo system concepts. Equation (3) is a self-consistent equation in which part of the output, $f(E_0)$, is fed back into the input. Equation (21) says that if an input increase leads to an output increase, the feedback is negative and the system is stable. However if a decrease in E_I produces an increase in E_0, the feedback is positive and therefore unstable. Note that if dE_I/dE_T is positive but too large, Eq. (21) also predicts instability.

More generally Eq. (20) predicts the onset of an instability when $|df(E_0)/dE_0| = 1$, that is when

$$\epsilon_{n+1} = \pm\,\epsilon_n\,,$$

We see immediately that the condition with the minus sign allows *period 2* or *period doubling*, since ϵ_n repeats after two round trips, ($\epsilon_{n+2} = \epsilon_n$). Section 7-3 considers a related but more complicated situation in terms of three-wave mixing.

It is sometimes useful to define a field time rate of change by

$$E_{n+1} \simeq E_n + \tau \frac{dE_n}{dt}, \tag{22}$$

where τ is the round-trip time. Substituting this into Eq. (1), we find

$$\tau \frac{dE_n}{dt} = \sqrt{T}\, E_I + f(E_n) - E_n\,. \tag{23}$$

Stability can also be studied in this case using Eq. (19), whereupon one finds

$$\frac{d\epsilon_n}{dt} = -\frac{1}{\tau}\left[1 - \frac{df(E_0)}{dE_0}\right]\epsilon_n = -\frac{T}{\tau}\frac{dE_I}{dE_T}\epsilon_n \ . \tag{24}$$

Note that this relation yields stability for all $dE_I/dE_T > 0$, i.e., is less stringent than Eq. (21). This is an indication that in stability problems, it can be dangerous to approximate a difference equation by a differential equation. It is easy to show that for the absorptive case (15), $df(E_0)/dE_0 > 0$, so that cases like $df/dE_0 < -1$ forbidden by Eq. (20) do not occur.

Absorptive optical bistability is easy to understand in terms of the bleaching of an absorber, but it is difficult to observe and to use. In contrast, purely dispersive nonlinear media lead to forms of bistability that are readily observed, and have greater promise for applications.

7-3. Ikeda Instability

The basic equations (1) and (2) can be written as

$$\mathcal{E}(t+\tau) = \sqrt{T}E_I + Re^{-\alpha\ell}e^{iKL}\mathcal{E}(t) \ , \tag{25}$$

where τ is the cavity round-trip time. This section studies the build-up of side-mode instabilities in this system for the simple case of a Kerr-type nonlinearity with instantaneous response time. For this case, the current method is a generalization of the simple analysis of Eqs. (19) - (21). With the dispersive bistability choice of Eq. (11), Eq. (25) becomes

$$\mathcal{E}(t+\tau) = \sqrt{T}\,E_I + Re^{i[\beta_0 + \beta_2 TI(t)]}\,\mathcal{E}(t) \ , \tag{26}$$

where the transmitted intensity $I_T = TI(t)$. In steady state, $\mathcal{E}(t+\tau) = \mathcal{E}(t)$ and as in Eq. (9), we have

$$\frac{I_T}{I_I} = \frac{1}{1 + 4(R/T^2)\sin^2[\frac{1}{2}(\beta_0 + \beta_2 I_T)]} \ . \tag{27}$$

We analyze the stability of this solution against small side-mode perturbations, i.e., we introduce a field with the envelope

$$\mathcal{E}(t) = \mathcal{E}_2 + \epsilon(t) = \mathcal{E}_2 + \mathcal{E}_1 e^{i\Delta t} + \mathcal{E}_3 e^{-i\Delta^* t} \tag{28}$$

into Eq. (26) and linearize the resulting equation about Eq. (27). Here $\mathcal{E}_2$ is the steady-state solution of Eq. (26) and Δ is *a complex frequency*

$\nu_2 - \nu_1 + ig$, where g is the net gain of the sidemodes. We need to determine Δ in a self-consistent fashion by subjecting the difference equation to a linear stability analysis which is related to the three-wave mixing in Chaps. 2 and 9. After one round trip, we find the field fluctuation

$$\mathcal{E}_1 e^{i\Delta(t+\tau)} + \mathcal{E}_3 e^{-i\Delta^*(t+\tau)} = R\exp[i(\beta_0+T\beta_2|\mathcal{E}_2|^2)]\,\{\mathcal{E}_1 e^{i\Delta t} + \mathcal{E}_3 e^{-i\Delta^* t}$$
$$+\, iT\beta_2\,[|\mathcal{E}_2|^2(\mathcal{E}_1 e^{i\Delta t} + \mathcal{E}_3 e^{-i\Delta^* t}) + \mathcal{E}_2^2(\mathcal{E}_1^* e^{-i\Delta^* t} + \mathcal{E}_3^* e^{i\Delta t})]\}\,. \tag{29}$$

Equating the terms in $e^{i\Delta t}$ in this equation and in its complex conjugate yields the coupled equations

$$\mathcal{E}_1 e^{i\Delta\tau} = B\mathcal{E}_1 + iBT\beta_2\,[|\mathcal{E}_2|^2\mathcal{E}_1 + \mathcal{E}_2^2\mathcal{E}_3^*]\,, \tag{30a}$$

$$\mathcal{E}_3^* e^{i\Delta\tau} = B^*\mathcal{E}_3^* - iB^*T\beta_2\,[|\mathcal{E}_2|^2\mathcal{E}_3^* + \mathcal{E}_2^{*2}\mathcal{E}_1]\,, \tag{30b}$$

where $B \equiv R\exp[i(\beta_0 + T\beta_2|\mathcal{E}_2|^2)]$. The first term in the square brackets is due to nonlinear refraction in the Kerr medium, while the second one is a four-wave mixing contribution: $\mathcal{E}_2$ is scattered off the index grating generated by $\mathcal{E}_2$ and $\mathcal{E}_3^*$ (or $\mathcal{E}_2^*$ and $\mathcal{E}_1$) to produce absorption or gain for the sideband amplitude $\mathcal{E}_1$ (or $\mathcal{E}_3^*$) one round-trip later. The instability resulting from this four-wave mixing process in dispersive bistability was discovered by Ikeda (1979) and its side-mode gain interpretation was given first by Firth *et al* (1984). Combining Eqs. (30a) and (30b), we find the eigenvalue equation

$$\begin{pmatrix} B(1+iT\beta_2|\mathcal{E}_2|^2) - e^{i\Delta\tau} & iBT\beta_2\mathcal{E}_2^2 \\ -iB^*T\beta_2\mathcal{E}_2^{*2} & B^*(1-iT\beta_2|\mathcal{E}_2|^2) - e^{i\Delta\tau} \end{pmatrix}\begin{pmatrix}\mathcal{E}_1 \\ \mathcal{E}_3^*\end{pmatrix} = 0. \tag{31}$$

The corresponding equation for the eigenvalues $\lambda = e^{i\Delta\tau}$ is

$$\lambda^2 - \lambda[B(1 + iT\beta_2|\mathcal{E}_2|^2) + \text{c.c.}] + R^2 = 0\,, \tag{32}$$

which gives

$$\lambda^2 - 2R\lambda[\cos(\beta_0+T\beta_2|\mathcal{E}_2|^2) - T\beta_2|\mathcal{E}_2|^2\sin(\beta_0+T\beta_2|\mathcal{E}_2|^2] + R^2 = 0. \tag{33}$$

The resonator is stable if the net gain $g < 0$ and unstable if $g > 0$. Thus the instability edges are given by the condition $|\lambda| = 1$. Consider first the case $\lambda = 1$. Substituting this value into Eq. (33) yields

$$\cos[\beta_0+T\beta_2|\mathcal{E}_2|^2] - T\beta_2|\mathcal{E}_2|^2 \sin[\beta_0+T\beta_2|\mathcal{E}_2|^2] = \frac{1+R^2}{2R} . \tag{34}$$

Using Eq. (27), we can show that this is precisely the same as the condition

$$\frac{dI_I}{dI_T} = 0 , \tag{35}$$

as can be seen readily from Eq. (27). This means that the instability edge corresponding to $\lambda = 1$ is just the usual "negative slope instability" of optical bistability and multistability. Regions of the transmission curve I_T/I_I with negative slope are always unstable.

The instability edge $\lambda = -1$ yields

$$\cos[\beta_0+T\beta_2|\mathcal{E}_2|^2] - T\beta_2|\mathcal{E}_2|^2 \sin[\beta_0+T\beta_2|\mathcal{E}_2|^2] = -\frac{1+R^2}{2R} . \tag{36}$$

which can be seen to correspond to instabilities occurring on the branches of positive slope.

The instabilities corresponding to $\lambda \equiv e^{i\Delta\tau} = \pm 1$ have quite distinct physical signatures. The first case gives $\Delta = 2q\pi$, q integer, i.e.,

$$\Delta_q = 2\frac{q\pi}{\tau} . \tag{37}$$

The negative branch instabilities are such that

$$\mathcal{E}(0,t+\tau) = \mathcal{E}(0,t) . \tag{38}$$

In this case the pump frequency ($\Delta = 0$) along with all other frequencies displaced from it by an integer multiple of the cavity mode spacing $2\pi/\tau$, is made cavity resonant. Therefore any small fluctuation in the input intensity I_I induces the device to go unstable.

In contrast, $\lambda = -1$ implies $\Delta = (2q+1)\pi$, q integer, or

$$\Delta_q = \frac{(2q+1)\pi}{\tau} . \tag{37}$$

Substituting Eq. (37) into Eq. (28) gives

$$\mathcal{E}(0,t+\tau) \neq \mathcal{E}(0,t) \tag{38a}$$

$$\mathcal{E}(0,t+2\tau) = \mathcal{E}(0,t) . \tag{38b}$$

This is an oscillatory solution with period 2τ, i.e., twice the round-trip time. After one round-trip time, $\mathcal{E}_1$ and $\mathcal{E}_3^*$ are interchanged, and after two times, they return to their original values. In this case the probe waves $\mathcal{E}_1$ and $\mathcal{E}_3$ are adjacent cavity modes separated by $2q\pi/\tau$ symmetrically placed about the pump wave $\mathcal{E}_2$. This gives rise to a 2τ modulation of the total field, the so-called Ikeda instability. This "period-2 bifurcation" is the first step in a route to deterministic chaos following the period doubling scenario [see Feigenbaum (1978)].

At the stability edges for the Ikeda instability, the parametric gain due to the four-wave mixing interaction exactly balances the transmission losses, and the pair of detuned probe fields give rise to a self-consistent set of sidebands. This situation is just like that of four-wave mixing and the instabilities discussed in Sec. 10-2, except here they occur far off any atomic line center and with difference, rather than differential, equations. Hence we see that the transient phenomena of Secs. 10-1, 10-2, and 10-3 can all be interpreted in terms of sidemode oscillations; in Sec. 10-1, they decay away, while in Secs. 10-2 and 10-3, they build up.

References

Bonifacio, R. and L. Lugiato (1976), Opt. Comm. **19**, 172.

Dorsel, A., J. D. McCullen, P. Meystre, E. Vignes, and H. Walther (1983), Phys. Rev. Lett. **51**, 1550.

Gibbs, H. M., S. M. McCall, and T. N. C. Venkatesan (1976), Phys. Rev. Lett. **36**, 1135.

Gibbs, H. M. (1985), *Optical Bistability: Controlling Light with Light*, Associated Press, Orlando, FA. This book reviews the theory and experiments of optical bistability and gives detailed references and the history of its discovery.

Feigenbaum, M. J. (1978), J. Stat. Phys. **19**, 25; (1979) **21**, 669.

Firth, W. J., E. M. Wright, and E. J. D. Cummins (1984), in *Optical Bistability II*, C. M. Bowden, H. M. Gibbs and S. L. McCall, eds (Plenum, New York).

Ikeda, K. (1979), Opt. Comm. **30**, 257.

Szöke, A., V. Daneu, J. Goldhar, and N. A. Kurnit (1969), Appl. Phys. Lett **15**, 376.

Problems

7-1. Show that

$$I_T = \frac{I_I T^2 \exp(-\alpha_B L/\cos\theta)}{(1-R')^2[1 + 4R'(1-R')^{-2} \sin^2(2\pi L\lambda^{-1}n_0\cos\theta)]}$$

for a Fabry-Perot with loss $A = 1 - R - T$ in each coating and unsaturable absorption coefficient (intensity) of α_B, $R' = R\exp(-\alpha_B L/\cos\theta)$. The Fabry-Perot makes an angle θ with respect to the normal. Note that the presence of an unsaturable absorber lowers the finesse and makes optical bistability more difficult to obtain.

7-2. From the equation relating the normalized incident and transmitted intensities, defined as $\mathcal{I}_I = I_I/T$ and $\mathcal{I}_T = I_T/T$, for absorptive optical bistability, calculate the maximum differential gain G. G is defined as $G = d\mathcal{I}_T/d\mathcal{I}_I$ evaluated at the inflection point $d^2\mathcal{I}_I/d\mathcal{I}_T{}^2 = 0$. What is the value of the normalized transmitted intensity $\mathcal{I}_T$ at that point?

7-3. Write the equation for absorptive optical bistability with an inhomogeneously broadened two-level medium. Prove that this equation can (or cannot) exhibit bistability.

7-4. Compare and contrast absorptive and dispersive optical bistability. How are the incident and transmitted field phases related?

7-5. Optical bistability yields two stable outputs for the same input. In purely dispersive optical bistability, for which there is no absorption, what happens to the light energy after the system switches from the upper to the lower branch?

7-6. Calculate the transmission equation $I_I = f(I_T)$ for a two-level medium allowing for the cavity detuning $\delta_c = KL$ - nearest multiple of 2π and atomic detuning $\delta = \omega - \nu$. Ans:

$$I_I = I_T[(1 + \alpha_r \ell/T)^2 + (1 + (\alpha_i \ell - \delta_c)/T)^2]\ .$$

Chapter 8
SATURATION SPECTROSCOPY

Nonlinear phenomena can be used in two general ways, one in applications such as second-harmonic generation, lasers, phase conjugation, optical bistability, etc. Alternatively one can use them to study the properties of the medium that generates them. The various kinds of nonlinear spectroscopy fall into the second category. Saturation spectroscopy deals typically with the *cw* absorption of waves passing through a medium to be studied. In the simplest case, *a probe wave* acts alone and one studies the probe absorption as a function of the intensity. The formulas for this are given earlier in Sec. 5-1, since this problem is of substantial use not only in spectroscopy, but also in laser theory and optical bistability.

A more complex setup is pictured in Fig. 8-1, where a weak (nonsaturating) probe wave passes through a medium saturated by an arbitrarily intense second wave, called the *saturator wave.* The theory predicts the absorption versus probe-saturator detuning, which reveals the dynamic Stark effect and various coherent dips. Section 5-2 already considers the measurement of a spectral hole this way, as well as *degenerate* probe absorption, that is, the situation where the probe and saturator waves have the same frequency. This chapter concentrates on the probe-saturation spectroscopy in two- and three-level homogeneously and inhomogeneously broadened media. The concept of population pulsations and its relationship to dynamic Stark splitting are presented. Chapter 9 on phase conjugation considers three and four waves with an emphasis on applications, but provides useful alternative probing configurations. Chapter 10 uses some of the results of Chaps. 8 and 9 to discuss the possible build up of laser modes in the presence of an oscillating mode, a further example of an optical instability. Chapter 11 studies fields with time varying, rather than *cw*, envelopes. Chapter 15 presents resonance fluorescence, which Sec. 15-4 shows to be closely related to the probe absorption problem. Absorption also plays a role in the generation of squeezed states discussed in Chap. 16.

Section 8-1 develops the two-mode theory of an arbitrarily intense saturator wave and a weak nonsaturating probe wave. This theory uses Fourier series to solve the Schrödinger equations of motion. Section 8-2 applies the resulting polarization of the medium to predict absorption coef-

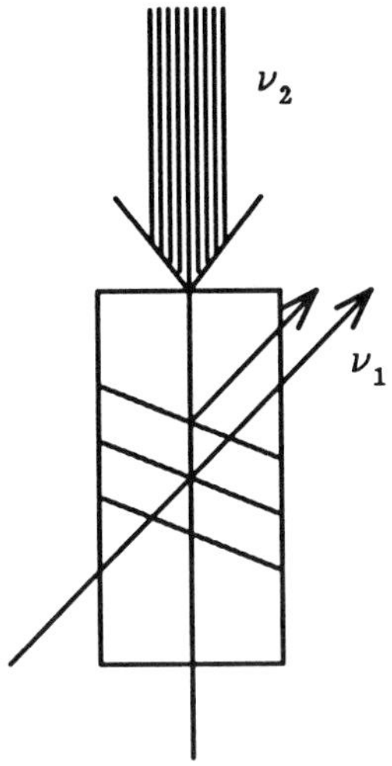

Fig. 8-1. Basic probe-saturator saturation spectroscopy configuration indicating the fringe pattern created by the interference between two waves. This fringe induces a grating that scatters some of wave 2 into the path of wave 1.

ficients in various homogeneously-broadened media and to illustrate coherent dips and the dynamic Stark effect. Here the strong wave fundamentally modifies the nature of the atomic interactions. Section 8-3 considers inhomogeneously-broadened and Doppler broadened media and shows how the Lamb dip can be used for spectroscopic purposes. Section 8-4 considers three-level phenomena.

8-1. Probe Wave Absorption Coefficient

We consider a medium subjected to a saturating wave and study the transmission of a weak (nonsaturating) probe wave as diagrammed in Fig. 8-1. We assume that the saturating wave intensity is constant throughout the interaction region and ignore transverse variations (see Sec. 6-6 for some discussion of these variations). We label the probe wave by the index 1 and the saturator by 2 as shown in Fig. 8-2. Our electric field has the form

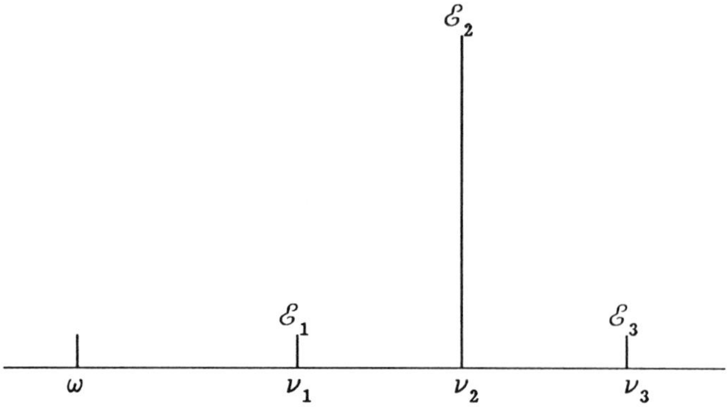

Fig. 8-2. Spectrum of two and three-wave fields used in Chaps. 8 and 9. Waves with frequencies ν_1 and ν_3 are usually taken to be weak (nonsaturating), while the ν_2 wave is allowed to be arbitrarily intense. The beat frequency $\Delta \equiv \nu_2 - \nu_1 = \nu_3 - \nu_2$.

$$E(\mathbf{r},t) = \frac{1}{2}\sum_n \mathcal{E}_n(\mathbf{r})\, e^{i(\mathbf{K}_n\cdot\mathbf{r} - \nu_n t)} + \text{c.c.}, \tag{1}$$

where the mode amplitudes $\mathcal{E}_n(\mathbf{r})$ are in general complex and $\mathbf{K}_n$ are the wave propagation vectors. This multimode field differs from that (Eq. (6.1)) used in laser theory in that the fields may not be collinear. In this chapter the mode index equals 1 or 2, while in Chap. 9, which considers two side-modes, it may equal 3 as well. The field (1) induces the complex polarization

$$P(\mathbf{r},t) = \frac{1}{2}\sum_n \mathcal{P}_n(\mathbf{r})\, e^{i(\mathbf{K}_n\cdot\mathbf{r} - \nu_n t)} + \text{c.c.}, \tag{2}$$

where $\mathcal{P}_n(\mathbf{r})$ is a complex polarization coefficient that yields index and absorption/gain characteristics for the probe and saturator waves. The polarization $P(\mathbf{r},t)$ in general has other components, but we are interested only in those given by Eq. (2). In particular, strong wave interactions induce components not only at the frequencies ν_1 and ν_2, but at $\nu_1 \pm k(\nu_2 - \nu_1)$ as well, where k is an integer. The classical anharmonic oscillator discussed in Chap. 2 has this property as well. To distill the components $\mathcal{P}_n(\mathbf{r})$ out of $P(\mathbf{r},t)$, we can use the mode factors $\exp(i\mathbf{K}_n\cdot\mathbf{r})$ as for Eq. (5.56), provided they differ sufficiently from one another in distances for which the amplitudes vary noticeably. For nearly parallel (or parallel) waves, the mode functions do not vary sufficiently rapidly, and one must separate components by their temporal differences, e.g., by heterodyne techniques.

The problem reduces to determining the probe polarization $\mathcal{P}_1(\mathbf{r})$, from which the absorption coefficient is determined from an equation like (5.4) with the subscript $_1$ on the $\mathcal{E}$, $\mathcal{P}$, and α. One might guess that the probe absorption coefficient is simply a probe Lorentzian multiplied by a population difference saturated by the saturator wave as done in Sec. 5-2 in measuring the width of a spectral hole. However an additional contribution enters due to *population pulsations*. Specifically, the nonlinear populations respond to the superposition of the modes to give pulsations at the beat frequency $\Delta = \nu_2 - \nu_1$. Since we suppose that the probe does not saturate, the pulsations occur only at $\pm\Delta$, a point proved below. These pulsations act as modulators (or like Raman "shifters"), putting sidebands onto the medium's response to the ν_2 mode. One of these sidebands falls precisely at ν_1, yielding a contribution to the probe absorption coefficient. The other sideband would influence the absorption at a frequency $\nu_2 + \nu_2 - \nu_1$, a point discussed in Sec. 9-3. As shown by Eq. (5.70) for degenerate counterpropagating probe and saturator frequencies ($\nu_+ = \nu_- = \nu$), the probe absorption coefficient can be substantially smaller than the simple saturated value, since the saturation denominator $(1+I_-\mathcal{L})$ is *squared*. The square is due to pump scattering off the population pulsations.

In this section we derive the complete nonsaturating probe absorption coefficient. The equations of motion for the population matrix are (5.9) - (5.11) for Type 1 two-level media (both levels excited - Fig. 4-1), and (5.9) and (5.32) for Type 2 media (ground-state lower level - Fig. 4-5). The interaction energy matrix element $\mathcal{V}_{ab}$ for Eq. (1) is given in the rotating-wave approximation by

$$\mathcal{V}_{ab} = -\frac{\wp}{2}\sum_n \mathcal{E}_n(\mathbf{r})\, e^{i(\mathbf{K}_n\cdot\mathbf{r} - \nu_n t)}\,, \tag{3}$$

where $n = 1$ or 2 for this chapter.

To determine the response of the medium to this multimode field, we Fourier analyze both the polarization component ρ_{ab} of the population matrix as well as the populations themselves. Specifically, we expand ρ_{ab} as

$$\rho_{ab} = Ne^{i(\mathbf{K}_1\cdot\mathbf{r} - \nu_1 t)} \sum_{m=-\infty}^{\infty} p_{m+1}\, e^{im[(\mathbf{K}_2-\mathbf{K}_1)\cdot\mathbf{r} - \Delta t]}, \tag{4}$$

where N is the unsaturated population difference. The population matrix elements $\rho_{\alpha\alpha}$ have the corresponding Fourier expansions

$$\rho_{\alpha\alpha} = N \sum_{k=-\infty}^{\infty} n_{\alpha k}\, e^{ik[(\mathbf{K}_2-\mathbf{K}_1)\cdot\mathbf{r} - \Delta t]}, \quad \alpha = a, b. \tag{5}$$

It is further convenient to define the population difference $D(\mathbf{r},t)$ with the expansion

$$D(\mathbf{r},t) \equiv \rho_{aa}(\mathbf{r},t) - \rho_{bb}(\mathbf{r},t) = N \sum_{k=-\infty}^{\infty} d_k\, e^{ik[(\mathbf{K}_2-\mathbf{K}_1)\cdot\mathbf{r} - \Delta t]}, \tag{6}$$

where $d_k \equiv n_{ak}-n_{bk}$. Note that $d_k^* = d_{-k}$. We now substitute these expansions into the population matrix equations of motion and identify coefficients of common exponential frequency factors. We suppose that $\mathcal{E}_1$ does not saturate, i.e., neglect terms where it appears more than once. We show that in this approximation that only p_1, p_2, and p_3 occur in the polarization expansion (4), and that only d_0 and $d_{\pm 1}$ appear in the population difference expansion (6). Physically this simplification occurs because a product of $\mathcal{E}_1$ and $\mathcal{E}_2$ creates the pulsations $d_{\pm 1}$, and from then on only $\mathcal{E}_2$ can interact. One obtains the polarization sidebands of ν_2 at frequencies ν_1 and ν_3, which subsequently combine with ν_2 only to give back $d_{\pm 1}$ components.

Consider first the component of the polarization driving the saturator wave. It is given by the coefficient of $\exp(i\mathbf{K}_2\cdot\mathbf{r}-i\nu_2 t)$, i.e. by p_2, or the $m = 1$ term in Eq. (4). We calculate this polarization by neglecting the nonsaturating probe field in $\mathcal{V}_{ab}$, substituting the expansions (4) and (5) and equating the terms in $\exp(i\mathbf{K}_2\cdot\mathbf{r}-i\nu_2 t)$ in Eq. (5.9). We find readily

$$-i\nu_2 p_2 = -(i\omega+\gamma)p_2 - i(\wp\mathcal{E}_2/2\hbar)d_0,$$

which gives

$$p_2 = -i(\wp/2\hbar)\, \mathcal{E}_2\mathcal{D}_2 d_0\,, \tag{7}$$

where for notational simplicity we have defined the complex Lorentzian

$$\mathcal{D}_n = 1/[\gamma + i(\omega-\nu_n)]\,. \tag{8}$$

Equation (7) is simply an alternate way of writing the single-mode density matrix element of Eq. (5.14) in which we include a subscript $_2$ to specify the saturator wave and have factored out the unsaturated population difference N and the rapidly-varying time/space factor $\exp(iKz-i\nu t)$.

In Eq. (7), we need the dc Fourier component of the population difference $d_0 = n_{a0} - n_{b0}$, saturated by the saturator wave $\mathcal{E}_2$ alone. Combining Eqs. (5) and (5.10), we find for the $k = 0$ term

$$0 = \lambda_a/N - \gamma_a n_{a0} + [i(\wp\mathcal{E}_2/2\hbar)p_2^* + \text{c.c.}].$$

Combined with (7), this yields

$$n_{a0} = \lambda_a/N\gamma_a - (2\gamma_a\gamma)^{-1}|\wp\mathcal{E}_2/\hbar|^2\,\mathcal{L}_2 d_0. \tag{9}$$

Here $\mathcal{L}_2$ is the saturator dimensionless Lorentzian (5.18) written in the form

$$\mathcal{L}_n = \gamma^2/[\gamma^2 + (\omega-\nu_n)^2]. \tag{10}$$

The $\mathcal{E}_1$ contributions are ignored here, since we assume that $\mathcal{E}_1$ doesn't saturate. The dc population component n_{b0} is given by Eq. (9) with $a \rightarrow b$ and a change of sign. As for $\rho_{aa} - \rho_{bb}$ of Eq. (5.19), this gives the population difference component

$$\begin{aligned} d_0 &= 1 - I_2\mathcal{L}_2 d_0\,, \\ &= 1/(1 + I_2\mathcal{L}_2)\,. \end{aligned} \tag{11}$$

Let us now turn to the probe wave, which is driven by the polarization component p_1 oscillating at $\exp(i\mathbf{K}_1\cdot\mathbf{r}-i\nu_1 t)$ ($m = 0$ term in Eq. (4)). Equating the terms with this phasor in Eq. (5.9) yields an equation for p_1 that includes an extra term proportional to $\mathcal{E}_2 d_{-1}$

$$-i\nu_1 p_1 = -(i\omega+\gamma)p_1 - i(\wp/2\hbar)[\mathcal{E}_1 d_0 + \mathcal{E}_2 d_{-1}]\,,$$

giving

$$p_1 = -i(\wp/2\hbar)\,\mathcal{D}_1\,[\mathcal{E}_1 d_0 + \mathcal{E}_2 d_{-1}]\,. \tag{12}$$

The $\mathcal{E}_2 d_{-1}$ term gives the scattering of $\mathcal{E}_2$ into the $\mathcal{E}_1$ mode by the population pulsation component d_{-1}. The polarization component p_0 remains zero when only d_0 and $d_{\pm 1}$ are nonzero, since it is proportional to $\mathcal{E}_1 d_{-1}$, which involves at least two $\mathcal{E}_1$'s.

Even though the saturator and one probe wave only are present, the polarization component p_3 has the nonzero value

$$p_3 = -i(\wp/2\hbar)\mathscr{D}_3\mathscr{E}_2 d_1 \,, \tag{13}$$

while $p_{j>3}$ vanishes since $d_{(k>1)}$ would be involved. The component p_3 effects the probe absorption coefficient, as will be seen shortly, and also plays an essential role in three and four-wave mixing, as discussed in Chap. 9.

Proceeding with the population pulsation terms $n_{a,-1}$, $n_{b,-1}$, and d_{-1}, we have

$$i\Delta n_{a,-1} = -\gamma_a n_{a,-1} + i(\wp/2\hbar)\,[\mathscr{E}_1 p_2^* + \mathscr{E}_2 p_3^* - \mathscr{E}_2^* p_1],$$

where we take $\wp$ to be real without loss of generality. Solving for $n_{a,-1}$ and subtracting a similar expression for $n_{b,-1}$, we obtain the difference

$$d_{-1} = -2iT_1\mathscr{F}(\Delta)(\wp/2\hbar)\,[\mathscr{E}_2^* p_1 - \mathscr{E}_1 p_2^* - \mathscr{E}_2 p_3^*], \tag{14}$$

where the dimensionless complex population-pulsation factor

$$\mathscr{F}(\Delta) = (2T_1)^{-1}[\mathscr{D}_a(\Delta) + \mathscr{D}_b(\Delta)] \,, \tag{15}$$

where $\mathscr{D}_a = [\gamma_a + i\Delta]^{-1}$ and T_1 is given by Eq. (5.22). The $\mathscr{F}$ factor approaches unity as $\Delta \to 0$. Note that in Eq. (14), both p_1 and p_3^* contain d_{-1} contributions, since $d_1^* = d_{-1}$. These contributions lead to saturation in the following expression for d_{-1}. Substituting Eqs. (7), (12) and (13) for the p_n, and solving for d_{-1}, we have

$$d_{-1} = -\frac{(\wp/\hbar)^2\mathscr{E}_1\mathscr{E}_2^* T_1 T_2 \mathscr{F}(\Delta)\,\frac{\gamma}{2}(\mathscr{D}_1 + \mathscr{D}_2^*)}{1 + I_2\mathscr{F}(\Delta)\,\frac{\gamma}{2}(\mathscr{D}_1 + \mathscr{D}_3^*)}\, d_0 \,, \tag{16}$$

Here the sum $\mathscr{D}_1 + \mathscr{D}_2^*$ comes from the $\mathscr{E}_2^* p_1 - \mathscr{E}_1 p_2^*$ factor in Eq. (14) and $\mathscr{D}_1 + \mathscr{D}_3^*$ comes from $\mathscr{E}_2^* p_1 - \mathscr{E}_2 p_3^*$. In general, $\mathscr{D}_n$ comes from the polarization component p_n.

Our calculation is self-consistent, since only d_0 and $d_{\pm 1}$ can obtain nonzero values from p_1, p_2, p_3, and vice versa. Combining the pulsation component (16) with the polarization component (12), setting $\mathscr{P}_1 = 2\wp N p_1$, and using Eq. (5.4), we find the complex absorption coefficient

$$\alpha_1 = \frac{\alpha_0 \gamma \mathscr{D}_1}{1 + I_2 \mathscr{L}_2} \left[1 - \frac{I_2 \mathscr{F}(\Delta) \frac{\gamma}{2}(\mathscr{D}_1 + \mathscr{D}_2^*)}{1 + I_2 \mathscr{F}(\Delta) \frac{\gamma}{2}(\mathscr{D}_1 + \mathscr{D}_3^*)} \right] \tag{17}$$

which can be written as

$$\alpha_1 = \alpha_{inc} + \alpha_{coh} \cdot$$

We refer to the term including only the 1 inside the [] as the *incoherent* contribution α_{inc} to the probe absorption coefficient α_1, since it does not involve the response of the medium to the coherent superposition of the probe and saturator fields. We refer to the term containing the $\mathscr{F}(\Delta)$ factor as the *coherent* contribution α_{coh} to α_1 since it results from the scattering of the saturator wave off the population pulsations induced by the probe/saturator fringe field. A formula equivalent to Eq. (17) was derived for the first time by Mollow (1972).

It is instructive to interpret the incoherent and coherent contributions in terms of transitions. By restricting the intensity of the probe to nonsaturating values, we have obtained an expression valid for arbitrarily large values of the saturator intensity I_2. The saturation factor $1/(1+I_2\mathscr{L}_2)$ appearing in Eq. (17) expands to $1 - I_2\mathscr{L}_2$ in the third-order approximation ($\mathscr{E}_1\mathscr{E}_2\mathscr{E}_2^*$ is involved).

For much of saturation spectroscopy this value is inadequate, for I_2 is typically as large as unity or larger, and the geometric series fails to converge! Hence we interpret Eq. (17) in a nonperturbative fashion as follows: The saturator interacts with the unsaturated population difference N an effective number of times giving the "summed series" saturation factor $1/(1+I_2\mathscr{L}_2)$. Given an effective dc saturated population difference $N/(1+I_2\mathscr{L}_2)$, the probe then interacts producing a polarization at the probe frequency. This yields the incoherent contribution and in addition gives the numerator of the $\mathscr{F}(\Delta)\mathscr{D}_1$ term in Eq. (17). For the latter, the saturator in turn interacts with the probe polarization to yield a population pulsation. Alternatively to this probe interaction, the saturator interacts with the effective dc saturated population difference to generate a polarization at the frequency ν_2 (giving the $\mathscr{D}_2^*$ term in Eq. (17) without its denominator), followed by a probe interaction, a sequence also yielding a population pulsation (the $\mathscr{E}_1 p_2^*$ contribution in Eq. (14)). The saturator then interacts an additional amount represented by the factor $1/(1+I_2\mathscr{F}(\Delta)...)$ in Eq. (17), and corresponding to successive generations of probe polarizations (at ν_3 and ν_1, i.e., at $\nu_2 \pm \Delta$) and population pulsations at Δ. These sequences give the scattering of the saturator into the probe wave, i.e., the coherent $\mathscr{F}(\Delta)$ term of Eq. (17). For a saturating probe, higher-order population pulsations (at $n\Delta$, $n>1$) occur, forcing one to use a continued fraction (see Sargent (1978)). This fraction truncates ultimately due to the finite bandwidth of the medium. For small Δ, a saturating probe can generate a substantial number of higher-order pulsations.

An interesting property of α_1 is that the integrated area under the $\alpha_1(\Delta)$ curve is independent of the coherent contribution α_{coh}. Whatever decrease the population pulsations cause for one Δ must be made up in increases for other values of Δ. The population pulsations merely redistribute the absorption as a function of Δ and do not modify the medium's broadband absorption. To see this, we note that Eq. (17) has no poles in the lower-half plane for the beat frequency Δ. Therefore the integral

$$\boxed{\int_{-\infty}^{\infty} d\Delta\ \alpha_{coh}(\Delta) = 0}\ . \tag{18}$$

This fact has analogs in three-level probe absorption spectra.

8-2. Coherent Dips and the Dynamic Stark Effect

In this section we illustrate the probe absorption coefficient of Eq. (17) for a variety of relaxation rates and saturator intensities that lead to simple formulas. We first recover the degenerate ($\nu_1 = \nu_2$) probe absorption coefficient of Eq. (5.70), which reveals that the degenerate probe absorption is substantially less than that for single-wave saturation. We then consider level lifetimes long compared to dipole lifetimes ($T_1 >> T_2$). This leads to *a coherent dip* in absorption versus probe detuning, caused by the inability of the population inversion to follow a probe-saturator beat frequency much larger than its decay rate. Hence the coherent contribution to the probe absorption coefficient falls off as the beat frequency is increased. Such dips allow one to measure the population decay times, a fact particularly valuable for situations in which that decay is nonradiative, e.g., picosecond decays in liquids or semiconductors. We also discuss detuned operation, which introduces a dipole phase shift that turns the coherent dip into a dispersive-like lineshape (like the χ_n' curve in Fig. 6-2).

We then allow the population and dipole decay times to approach one another, and find that the coherent interaction leads to a dynamic Stark splitting with dispersive-like lineshape resonances at the Rabi sidebands. Hence in our model using the unperturbed eigenstates of the Hamiltonian, it is population pulsations that are responsible for the dynamic Stark effect.

Alternatively, one can diagonalize the total Hamiltonian to find new energy levels of shifted value (see dressed atom picture of Sec. 13-1). In detuned operation, the extra dipole phase shift turns the dispersive-like lineshapes into Lorentzian shapes, one with gain (the "Raman" resonance), and the other absorptive. For ease of reading, we divide this section into a sequence of six subsections going from the limit $T_1 >> T_2$ to $T_1 \simeq T_2$.

Degenerate probe absorption coefficient

When the saturator and probe wave frequencies are all equal, the $\mathscr{D}_n$'s of Eq. (17) are all equal. Accordingly writing the detunings as $\omega - \nu_n \simeq \omega - \nu_2 = \delta$, we find that the sums of the $\mathscr{D}_n$ functions collapse to a single Lorentzian of width δ given by Eq. (5.25) and that the population response function $\mathscr{F}(\Delta)$ equals unity. The absorption coefficient (17) reduces to

$$\alpha_1 = \frac{\alpha_0\gamma\mathscr{D}_2(\delta)}{1 + I_2\mathscr{L}_2(\delta)}\left[1 - \frac{I_2\mathscr{L}_2(\delta)}{1 + I_2\mathscr{L}_2(\delta)}\right] = \frac{\alpha_0\gamma\mathscr{D}_2(\delta)}{[1 + I_2\mathscr{L}_2(\delta)]^2}, \tag{19}$$

where the dimensionless Lorentzian $\mathscr{L}_2(\delta)$ is given by Eq. (5.18). Equation (19) is the same as Eq. (5.70). Noting that the pump scattering off the fringe induced population pulsations gives the $-I_2\mathscr{L}_2/(1 + I_2\mathscr{L}_2)$ contribution, we see very clearly that this scattering is responsible for the square of the saturation denominator and hence for the increased saturation of the probe absorption coefficient. Stated in other words, the additional saturation is due to the increased transmission of the probe wave created by constructive scattering of the pump wave off the fringe-induced grating. This scattering is constructive because the projection of the total polarization onto the probe wave preferentially weights contributions for which the probe and pump are in phase, and these contributions have greater saturation than those for which the waves are out of phase.

Short T_2 limit

In media dominated by dephasing collisions such as ruby at room temperature, the population difference lifetime T_1 is much greater than the dipole lifetime T_2. For simplicity, we consider beat frequencies Δ small compared to γ ($\equiv 1/T_2$) and restrict ourselves to the upper-to-ground-lower-level decay scheme of Fig. 4-5, for which $T_1 = 1/\Gamma$. We remind the reader that for this case the lower level matrix element ρ_{bb} relaxes with the same rate constant (Γ) as the upper level element ρ_{aa}, but instead of decaying to 0 as ρ_{aa} does, it relaxes to N', where N' is the number of systems/volume. As for the degenerate case of Eq. (19), in this limit the $\mathscr{D}_n$'s are all equal. However as shown in Prob. 8-7, the population response function $\mathscr{F}$ is given by

$$\mathscr{F}(\Delta) = \frac{\Gamma}{\Gamma + i\Delta} . \tag{20}$$

This reduces α_1 of Eq. (17) to

$$\alpha_1 = \frac{\alpha_0\gamma(\gamma - i\delta)}{\gamma'^2 + \delta^2}\left[1 - \frac{I_2\mathscr{L}_2\Gamma(\Gamma' - i\Delta)}{\Gamma'^2 + \Delta^2}\right], \tag{21}$$

where the power-broadened decay constants γ' and Γ' are given by $\gamma' = \gamma\sqrt{1 + I_2}$ and $\Gamma' = \Gamma(1 + I_2\mathscr{L}_2)$, respectively.

In particular for *a resonant pump wave* ($\delta = 0$), we recover the coherent-dip (grating-dip) formula of Sargent (1976)

$$\mathrm{Re}\{\alpha_1\} = \alpha_0\left[\frac{1}{1 + I_2} - \frac{I_2\Gamma^2}{\Gamma^2(1+I_2)^2 + \Delta^2}\right] . \tag{22}$$

This formula predicts a set of power-broadened Lorentzians with FWHM widths given by $2\Gamma'$ as illustrated in Fig. 8-3. As such it provides a way to measure the population-difference lifetime $T_1 \equiv 1/\Gamma$ in long T_1 media. Equation (22) cannot give gain for any value of the pump intensity I_2.

Physically the coherent dip of Eq. (22) is due to pump scattering off the population pulsations induced by the probe/saturator field fringe component. If the waves propagate in opposite directions, this fringe has a spacing equal to one half the light wavelength and moves with a "walking speed" of approximately $\Delta/(K_1+K_2)$. If they propagate in the same direction, the fringe spacing goes to infinity, but the fringe component still oscillates at the beat frequency Δ. For $\Delta = 0$, the $\mathrm{Re}\{\alpha_1\}$ given by Eq. (22) reduces to the degenerate value $\alpha_0/(1+I_2)^2$ implied by Eq. (19). The atomic populations act like nonlinear anharmonic oscillators with a vanishing resonance frequency. Accordingly when driven by a nonzero beat frequency component they respond with reduced amplitude and a phase shift given by the complex Lorentzian response function $\mathscr{F}$ appropriately power broadened. When $\Delta >> \Gamma$, the populations lag behind by $\pi/2$, thereby scattering no pump energy into the probes, and the absorption increases to the single-wave value $\alpha_0/(1 + I_2)$ implied by Eq. (5.27). A related reduction in scattering occurs in the reflection coefficient produced by nondegenerate four-wave mixing in two-level media (see Sec. 9-4). This case yields a

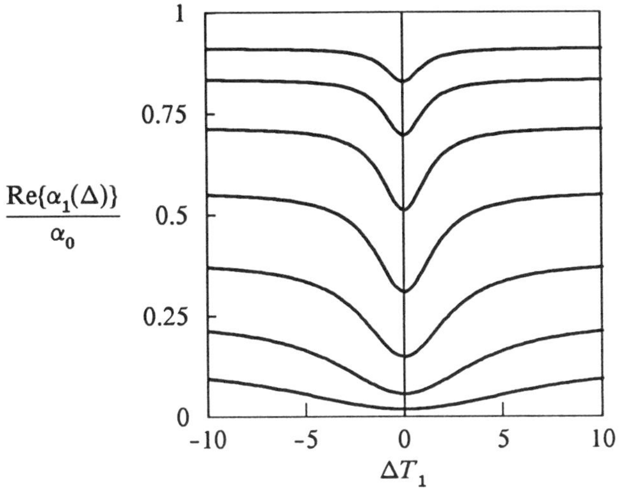

Fig. 8-3. Graphs of the real part of the normalized probe absorption coefficient of Eq. (22) showing power-broadened Lorentzian coherent dips as the saturator intensity is varied. The medium is homogeneously broadened. From top to bottom, the curves are for the saturator intensities I_2 = .1, .2, .4, .8, 1.6, 3.2, and 6.4.

narrow-band retroreflection with a spectral width determined by $1/T_1$. The situation is analogous to a car driving over a "washboard" road. At slow speeds, the car bounces up and down, following the road's vertical variations. At higher speeds these variations arrive too fast for the the car chassis to follow them and the driver gets a smooth ride.

Suppose that the system is driven off the dipole (atomic) line center. Then the dipole response contributes an additional phase shift determined by the $\gamma - i\delta$ factor in Eq. (21). In particular if the dipole is driven far off its line center ($|\delta| >> \gamma$) in the "pure-index" regime, the dipole contributes a $\pi/2$ phase shift yielding a dispersive-like absorption profile that can give gain. In this limit Eq. (21) reduces to

$$\alpha_1 = \frac{\alpha_0 I_2 \gamma^3}{\delta^3} \frac{\Gamma(\Delta + i\Gamma)}{\Delta^2 + \Gamma^2} . \tag{23}$$

This formula is very interesting because in a simple way it shows that two waves can exchange energy in what appears to be a pure index medium. Ordinarily such media only allow two waves to affect one

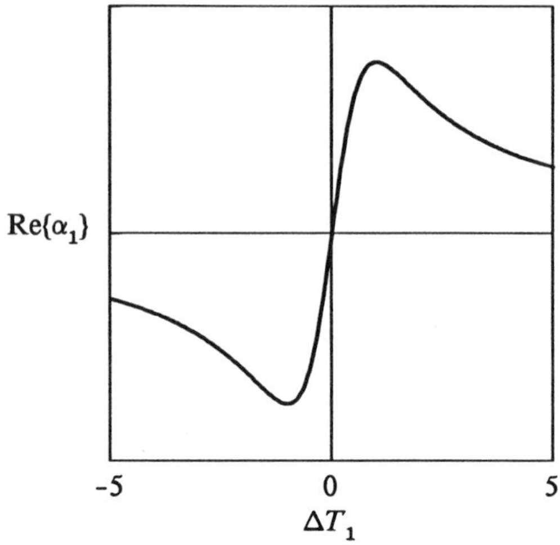

Fig. 8-4. Graph of the detuned probe absorption coefficient given by the real part of Eq. (23). For negative probe/saturator detunings, the probe experiences gain.

another's index of refraction. The real absorption given by Eq. (23) is *negative* if $\Delta < 0$, that is, gain occurs for $\Delta < 0$ as shown in Fig. 8-4. The width of the interaction region is characterized by $T_1 \equiv 1/\Gamma$.

Physically the reason that two waves can exchange energy in this highly detuned case is exactly the same as the reason for the existence of the coherent dip of Eq. (22). Specifically the grating or population pulsations induced by the probe/saturator fringe lag behind the forcing fringe, thereby giving a phase shift. However in the highly detuned case of Eq. (23), this phase shift gives the pure-index coupling a gain/absorptive coupling character, which is just the opposite from its effect on the coherent dip.

Transition from $T_2 \ll T_1$ to Comparable T_1 and T_2

To get a feel for the transition between Eq. (21) and Eq. (25) valid for pure radiative decay, we plot in Fig. 8-5 two sets of curves of the absorption coefficient α_1 of Eq. (17), one for $T_1 = 100T_2$, and one for $T_1 = T_2$. By lengthening the response time T_2 of the dipole relative to that T_1 of the population difference, we can prevent both the dipole and population difference from following the field fringe component oscillating at sufficiently large values of the beat frequency Δ. To understand the changes

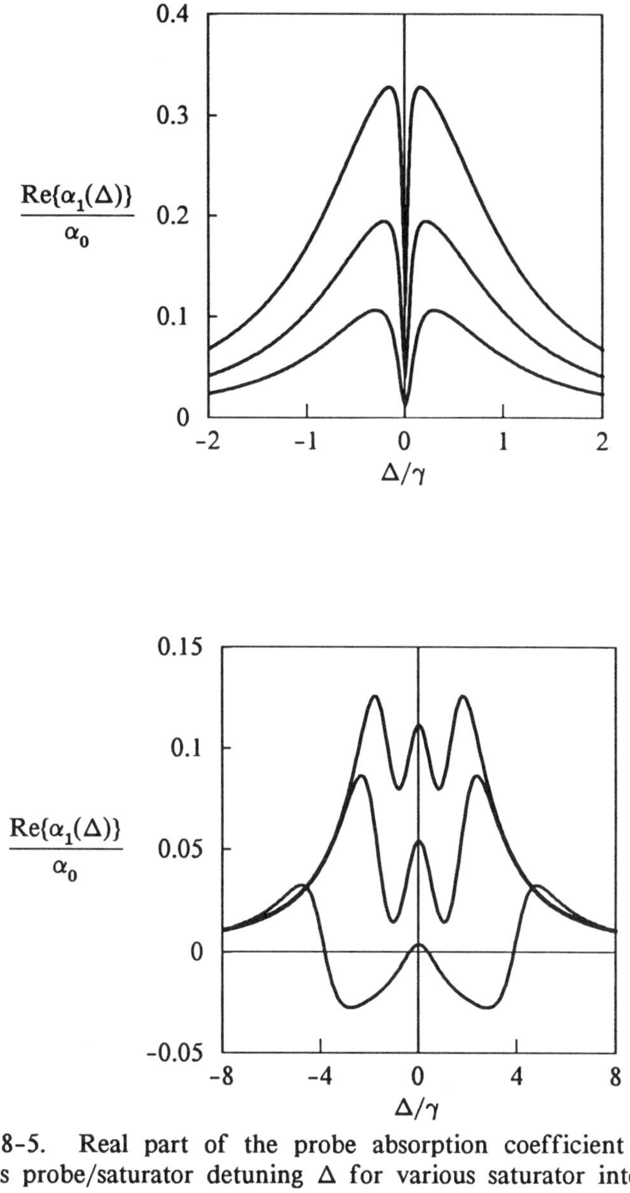

Fig. 8-5. Real part of the probe absorption coefficient (17) versus probe/saturator detuning Δ for various saturator intensities, $\nu_2 = \omega$, and the decay-constant relationships (*a*) $\Gamma = .01\gamma$ and (*b*) $\Gamma = \gamma$. In order of decreasing $\alpha_1(0)$, the curves are for the saturator intensities $I_2 = 2$, 3.3, and 16.

from Fig. 8-3, note that the equations of motion (5.9) - (5.11) for the dipole and populations form a coupled set of damped anharmonic oscillators. When subjected to an oscillating component in the electric-dipole interaction energy, both dipoles and populations can experience phase shifts for Δ values comparable to or greater than the respective power-broadened bandwidths (power-broadening factor times γ for the dipole and times Γ for the population difference). The coupled dipole-population response to the probe-saturator beat frequency yields the coherent contribution to Eq. (17). For nonzero Δ, the sum of the dipole and population phase shifts can exceed $\pi/2$, and hence cause an increase in absorption ($\alpha_{coh}>0$, whereas $\alpha_{coh}<0$ in the dip region) relative to the α_{inc} value. This results in the "shoulders" in the Fig. 8-5. Figure 8-5*a* reveals sharp power-broadened pulsation dips (produced by α_{coh}) on a broader background given by α_{inc}. In Fig. 8-5*b*, we see that the coherent dip has broadened out, changing shape into *a dynamic Stark splitting* ($I_2 = 2$ in Fig. 8-5*b*), as two sidebands appear.

Resonant Pump with Large Intensity

We can obtain a simple analytic formula valid for large I_2, i.e., $|\wp\mathcal{E}_2/\hbar|^2 \equiv \mathcal{R}_0^2 >> \gamma, \Gamma$, and resonant pump tuning ($\nu_2 = \omega$), for which $\mathcal{D}_3^* = \mathcal{D}_1$. For beat frequencies $\Delta \simeq \mp \mathcal{R}_0$, the absorption coefficient of Eq. (17) reduces to (we neglect the incoherent part)

$$\alpha_1(\Delta\simeq\mp\mathcal{R}_0) \simeq -\frac{\alpha_0\Gamma\gamma\mathcal{D}_1}{2\mathcal{R}_0^2}\,\frac{\mathcal{R}_0^2}{\Gamma + i\Delta + \mathcal{R}_0^2\mathcal{D}_1} = -\frac{\alpha_0\Gamma\gamma/2}{(\Gamma+i\Delta)(\gamma+i\Delta) + \mathcal{R}_0^2}$$

$$= -\frac{\alpha_0\Gamma\gamma/2}{(\mathcal{R}_0+\Delta)(\mathcal{R}_0-\Delta) + i\Delta(\Gamma+\gamma)} = \mp i\,\frac{\alpha_0\Gamma\gamma/4\mathcal{R}_0}{(\gamma+\Gamma)/2 \pm i(\mathcal{R}_0\pm\Delta)}$$

$$= -\frac{\alpha_0\Gamma\gamma}{4\mathcal{R}_0}\,\frac{\mathcal{R}_0 \pm \Delta \pm i(\gamma+\Gamma)/2}{(\mathcal{R}_0 \pm \Delta)^2 + (\gamma+\Gamma)^2/4}\,. \tag{24}$$

This gives a symmetrically placed pair of dispersive-like lineshape curves centered at the Rabi frequencies for the absorption (real part - see Fig. 8-5*b*, bottom curve) and corresponding Lorentzian curves for the index (imaginary part). The half width of the Lorentzian is $(\gamma+\Gamma)/2$, i.e., the average of the dipole and population-difference decay constants. This is due to the fact that the coherent term results from driving both the dipole and the populations at the frequency Δ. Similar features occur in the closely related phenomenon of resonance fluorescence (see Secs. 15-3 and 16-2).

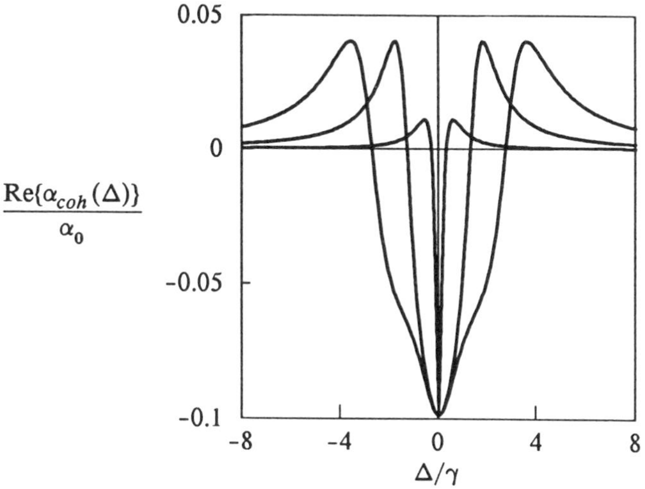

Fig. 8-6. α_{coh} *vs* probe-saturator beat frequency $\Delta = \nu_2 - \nu_1$ for a number of values of $T_1 = T_2$, $4T_2$, and $64T_2$ (in order of decreasing dip width).

Although the coherent dip and the Stark splitting behaviors appear to be quite different, plots of α_{coh} alone all resemble the twin-peak, dip structure shown in Fig. 8-6. All the curves in Fig. 8-5 are obtained by adding the shoulder-dip structure in this figure to Lorentzians. Hence the dynamic Stark splitting is an extension of the coherent dip into regions of beat frequencies as large as or larger than the homogeneous linewidth.

The dynamic Stark effect is sometimes interpreted in terms of an amplitude modulation of the dielectric polarization by Rabi flopping. When the Rabi frequency $\mathcal{R}_0$ greatly exceeds the decay constants, the atoms Rabi flop up and down many times before decaying. Clearly a pump-probe fringe that oscillates at or near the Rabi frequency can interact with such atoms in a resonant fashion. We see the results of these resonances in formulas like Eq. (24). The analytic details of the Rabi resonances are hidden by our use of a population matrix, which sums over the contributions of many atoms at different stages in their evolutions.

Pure Radiative Decay and PIER

For pure radiative decay due only to spontaneous emission, T_1 equals $\frac{1}{2}T_2$, rather than exceeding it due to phase interrupting collisions. For pump intensities sufficiently weak to be described by third-order perturbation theory, i.e., by $\mathcal{P}_1$ expanded to order $\mathcal{E}_1 I_2$, Eq. (17) reduces to

$$\alpha_1 = \alpha_0\gamma\mathscr{D}_1\left[1 - I_2\mathscr{L}_2 - I_2\mathscr{D}_1\mathscr{D}_2^*\Gamma\gamma\,\frac{2\gamma + i\Delta}{\Gamma + i\Delta}\right]. \tag{25}$$

Hence we see that for pure radiative decay ($2\gamma = \Gamma$), the pronounced Δ dependence in the coherent contribution cancels out. On the other hand, the addition of pressure produces phase-changing collisions that increase γ relative to Γ, spoiling this cancellation. This is a two-wave, two-level example of a Pressure-Induced Extra Resonance (PIER), a phenomenon first observed by Prior *et al* (1981). Larger pump fields also introduce pronounced Δ dependencies in the denominator of Eq. (17).

Nonresonant Pump Wave with Large Intensity

With a bit of algebra (Prob. 8-9), we can reduce Eq. (17) to a generalization of Eq. (24) valid for $\mathscr{R}_0^2 >> \gamma$ and Γ that includes pump detuning ($\delta \equiv \omega - \nu_2 \neq 0$). We find

$$\alpha_1(\Delta \simeq \mp\mathscr{R}) \simeq \frac{\mp i\alpha_0\gamma\mathscr{R}_0^2}{4\mathscr{R}(\gamma'^2 + \delta^2)}\,\frac{\Delta - \delta - i\gamma}{\Delta + \delta - i\gamma}\,\frac{\gamma + i\delta}{\gamma_\delta \pm i(\mathscr{R} \pm \Delta)}, \tag{26}$$

where the generalized Rabi frequency $\mathscr{R} = \sqrt{\mathscr{R}_0^2 + \delta^2}$, and the decay rate γ_δ is given by

$$\gamma_\delta = \frac{1}{2}[\gamma + \Gamma + (\gamma - \Gamma)\delta^2/\mathscr{R}^2]. \tag{27}$$

For $\delta = 0$, Eq. (26) reduces to Eq. (25). For $\delta >> \gamma$, the $\gamma + i\delta$ factor (originating from the $\mathscr{D}_2^*$ term of Eq. (17)) in Eq. (26) gives a phase shift of $\pi/2$ and the dispersive-like lineshapes of $\mathrm{Re}\{\alpha_1\}$ in Eq. (25) turn into Lorentzians of opposite sign (one gain, one absorption) and differing heights determined by the $(\Delta - \delta + i\gamma)/(\Delta + \delta + i\gamma)$ factor. Here $\Delta - \delta + i\gamma \equiv 1/\mathscr{D}_3^*$ comes from the $\mathscr{D}_3^*$ in Eq. (17) and $\Delta + \delta + i\gamma = 1/\mathscr{D}_1$ from the leading $\mathscr{D}_1$. The real part of Eq. (17) is illustrated in Fig. 8-7. The absorptive Lorentzian occurs for Δ tuned to the generalized Rabi sideband nearest to the probe resonance defined by $\mathscr{D}_1 = 1/\gamma$. The displacement of the absorption peak from the unsaturated position is sometimes called *a light shift*. The gain Lorentzian at the other generalized Rabi sideband more closely satisfies the resonance defined by $\mathscr{D}_3^* = 1/\gamma$. This is sometimes called the *Raman resonance*, since it derives from a three-photon "Raman" process involving $\mathscr{E}_2^*\mathscr{E}_1\mathscr{E}_2^*$. Equations (24) and (26) are not valid for small beat frequencies ($\Delta \simeq 0$), for which a dispersive-like lineshape occurs. This feature is sometimes called a "stimulated Rayleigh resonance"

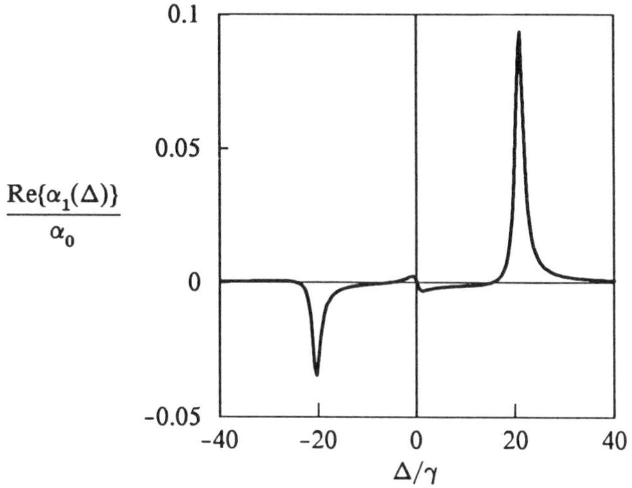

Fig. 8-7. Probe absorption coefficient α_1 of Eq. (17) versus probe/saturator beat frequency Δ in the presence of a strong pump wave of intensity I_2 = 400, detuned by $\omega - \nu_2 = -5$, and with $\Gamma = \gamma = 1$.

since it is most nearly like elastic light scattering (little frequency shift). The feature is particularly simple when $T_1 >> T_2$, as discussed above for the detuned long-T_1 spectrum in Fig. 8-4. All spectra in this section are given by Eq. (17), originally derived in an equivalent form by Mollow (1972). In particular two spectra are usually called "Mollow spectra", namely the strong field (I_2 = 16) spectrum of Fig. 8-5*b* and the spectrum in Fig. 8-7.

8-3. Inhomogeneously Broadened Media

This section treats saturation spectroscopy of inhomogeneously-broadened media and in particular of Doppler-broadened gaseous media. It was in studying this last type of spectroscopy that the subject had its origins, in a form known as Lamb-dip spectroscopy. This technique is based on the Lamb dip explained in Sec. 6-3.

One might expect a probe absorption spectrum simply to measure the inhomogeneously-broadened distribution saturated by the saturator wave. As discussed in Sec. 5-2, this saturation is not uniform as in the homogeneously-broadened case, in that it occurs only in the vicinity of the saturator-wave frequency. Specifically a Bennett hole is burnt by the saturator wave into the unsaturated distribution. The probe measures a Lorentzian (Eq. (5.51)) with a FWHM given by the sum of the probe dipole and power-broadened hole widths $2\gamma(1+I_2)^{1/2}$. The reason for this sum is that the monochromatic probe samples a Lorentzian spread of frequencies, leading to a convolution of this Lorentzian with the Bennett-hole Lorentzian. In addition a coherent interaction occurs, which is given by an average over homogeneously-broadened coherent contributions. Depending on the relative sizes of T_1 and T_2, this coherent contribution can give both coherent dips superimposed on a Bennett hole and dynamic Stark splittings. For the counterrunning Doppler case central to Lamb-dip spectroscopy, the coherent contribution tends to be smeared out, but it is far from negligible.

The nonsaturating probe-absorption coefficient for unidirectional operation in any inhomogeneously-broadened medium, and for arbitrary directions as well in stationary inhomogeneously-broadened media (i.e., not Doppler broadened) is given simply by integrating Eqs. (17) over the inhomogeneous broadening distribution $\mathcal{W}(\omega)$, that is,

$$\alpha_1 = \alpha_0 \int_{-\infty}^{\infty} d\omega' \mathcal{W}(\omega') \frac{\gamma\mathcal{D}_1}{1 + I_2\mathcal{L}_2} \left[1 - \frac{I_2\mathcal{F}(\Delta)\frac{\gamma}{2}(\mathcal{D}_1 + \mathcal{D}_2^*)}{1 + I_2\mathcal{F}(\Delta)\frac{\gamma}{2}(\mathcal{D}_1 + \mathcal{D}_3^*)} \right]$$

$$= \alpha_{\text{inc}} + \alpha_{\text{coh}} \,. \tag{28}$$

Section 5-3 evaluates the real part of the incoherent contribution to this α_1 (set $\mathcal{F} = 0$) for the case of a Lorentzian $\mathcal{W}(\omega')$. This gives a Lorentzian hole with a width equal to the sum of the homogeneous and inhomogeneous widths (see Eq. (5.52)). We generalize this result here to include the effects of the coherent contribution due to population pulsations.

We consider the short dipole-lifetime limit ($T_1 >> T_2$) first. Since the response of an inhomogeneously-broadened medium involves systems both on and off resonance, the average saturation is reduced relative to the centrally-tuned saturation of a homogeneously-broadened medium. Hence we do not expect the coherent dip to be as deep or as power-broadened as its homogeneously-broadened counterpart. As for Eq. (26), it is possible to simplify the contributions (28) by assuming $\Gamma << \gamma$ and choosing level decay constants appropriate for Eq. (25). We find

$$\alpha_1 = \frac{\alpha_0'}{(1 + I_2)^{1/2}} \left[1 - \frac{1}{2} \frac{I_2}{1+I_2 + \Delta^2/\Gamma^2} \right] , \tag{29}$$

where $\alpha_0'(\nu_1) = \pi\gamma\alpha_0\mathcal{W}(\nu_1)$. This consists of a frequency-independent term and a power broadened Lorentzian of the beat frequency. We see in comparing Eq. (29) with the homogeneous broadening expression (26) that here the power broadening enters with $(1+I_2)^{1/2}$, whereas for (26) it is proportional to $1+I_2$. Similarly the overall saturation factor is a square root for the inhomogeneous broadening case as found for the single wave case (5.39).

Another limit of interest is that for strong inhomogeneous broadening. Equation (5.52) gives $\mathrm{Re}(\alpha_{inc})$ in this limit, finding 1 minus a Lorentzian with width $\gamma+\gamma'$. This width results from using a monochromatic probe wave that samples a Lorentzian with width γ to measure a hole with width γ'. This yields a convolution of the two Lorentzians, which is a Lorentzian with the sum of the widths. The coherent contribution can also be evalu-

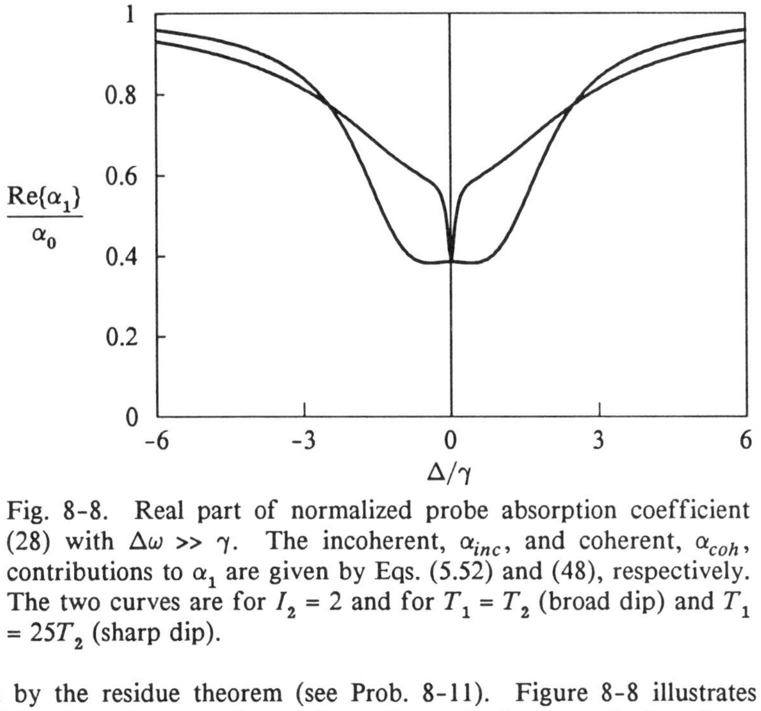

Fig. 8-8. Real part of normalized probe absorption coefficient (28) with $\Delta\omega >> \gamma$. The incoherent, α_{inc}, and coherent, α_{coh}, contributions to α_1 are given by Eqs. (5.52) and (48), respectively. The two curves are for $I_2 = 2$ and for $T_1 = T_2$ (broad dip) and $T_1 = 25T_2$ (sharp dip).

ated by the residue theorem (see Prob. 8-11). Figure 8-8 illustrates the resulting formulas, revealing a sharp dip on the bottom of a broad spectral hole for $T_1 = 25T_2$, while a deepened dip for $T_1 = T_2$.

To convert the probe absorption coefficient (28) to the bidirectional Doppler case, we Doppler shift the probe and saturator frequencies according to $\nu_1 \rightarrow \nu_1 - Kv$ and $\nu_2 \rightarrow \nu_2 + Kv$, respectively, and replace the inhomogeneous-broadening distribution $\mathcal{W}(\omega')$ by $\mathcal{W}(v)$ given by Eq. (6.34). In the Doppler limit, the incoherent contribution once again reduces to unity minus the convolution of a probe Lorentzian with a Bennett-hole Lorentzian giving (5.52). The coherent contribution is quite different from the unidirectional case and complicated by the fact that β depends on the integration variable. For simplicity, we take central tuning of both probe and saturator waves ($\nu_1=\nu_2=\omega$) and choose the level decay constants to satisfy Eq. (25). We then find (see Prob. 8-12))

$$\alpha_{coh} = \alpha_0'|\wp\mathcal{E}_2/\hbar|^2 \frac{\gamma'-\gamma}{\gamma} \frac{\gamma + 3\gamma'}{(\Gamma+2\gamma')(\gamma+3\gamma')(\gamma+\gamma') + |\wp\mathcal{E}_2/\hbar|^2(\gamma+2\gamma')} . \quad (30)$$

Two features are immediately evident: 1) α_{coh} is positive for this centrally-tuned value in contrast to all centrally-tuned cases considered earlier, and 2) as the saturator intensity goes to infinity, this expression approaches $3/(2+6\gamma/\Gamma)$, in contrast to the incoherent contribution (5.51), which bleaches to zero. The substantial size of the coherent contribution is a little surprising, since the various population pulsations that generated the effect involve gratings moving at different rates. One could imagine incorrectly that the superposition of their scatterings might well interfere destructively to zero. However the average phase shift from dipole and population responses is greater than $\pi/2$, yielding the positive contribution. Note that as $\Gamma \to 0$ (longer level lifetimes), the coherent contribution becomes small, as one would expect since the population pulsations follow less and less off-resonant behavior. For more discussion of Doppler broadened saturation spectroscopy, see Khitrova *et al.* (1988).

Lamb-dip spectroscopy

In some sense saturation spectroscopy started with the study of the Lamb dip (Sec. 6-3) in the laser output. As discussed in Prob. 6-15, the FWHM of the Lamb dip is given approximately by $2\sqrt{2}\gamma$. A limitation of studying the atomic response from the laser output is that one has to deal with an operating laser. A more generally applicable technique employs an external absorption cell, as shown in Fig. 8-9. This allows conditions such as pressure to be optimized for the study of the systems, independent of laser operation. Since the reflected wave also saturates the medium, one again gets increased saturation at line center thereby reducing the absorption. Hence, the radiation passing through the mirror exhibits an inverse Lamb dip versus tuning. By studying the width of this peak, one determines the homogeneous linewidth. A good match requires substantial numerical analysis. It is not sufficient to apply a simple formula like

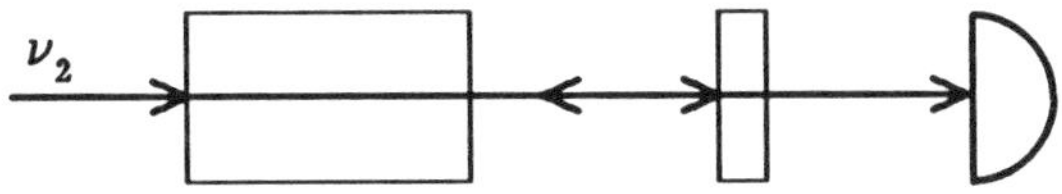

Fig. 8-9. Diagram of Lamb-dip spectroscopy technique as done by Brewer, Kelley, and Javan (1969). The medium studied was SF_6. Instead of tuning the laser through the line center, they tuned the line center through the laser frequency by Stark shifts

$$\alpha_1 = \alpha_0 \frac{e^{-(\omega-\nu_1)^2/(Ku)^2}}{1 + \mathscr{L}(\omega-\nu_1)},$$

however tempting that might be. Since a general treatment is quite complicated, the probe-saturator method has been used much more widely.

8-4. Three-Level Saturation Spectroscopy

Another level scheme often used in saturation spectroscopy involves three levels, with the saturator interacting between two of the levels and the probe between the third level and one of the first two (see Fig. 8-10). The analysis of these cases and the corresponding physical interpretation is actually easier than that for the two-level case because no polarization at the third frequency ν_3 is induced. In fact, even when probe and saturator are both arbitrarily large, the absorption coefficient can be obtained in closed form, i.e., not as a continued fraction. In this section, we derive the absorption coefficient for a nonsaturating probe wave specifically for the cascade (Fig. 8-10a) case, and then show how the corresponding coefficients for the "V" and "Λ" cases can be obtained from the cascade case.

As a fairly general model, we suppose that each of the three levels may be excited at rates λ_j and decay with rate constants γ_j. To simplify the analysis a bit, we use an interaction picture in which the appropriate $\exp(-i\nu_j t)$ factor specified by the rotating wave approximation has been removed from the Schrödinger dipole density matrix elements ρ_{32} and ρ_{21} and interaction energies $\mathcal{V}_{32}$ and $\mathcal{V}_{21}$. Together with the interaction terms given by Eq. (4.44), we find the six coupled equations of motion

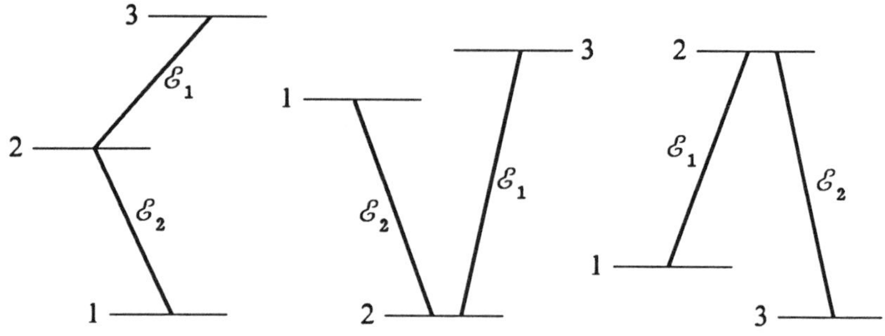

Fig. 8-10. Three three-level probe saturator configurations. (*a*) cascade, (*b*) "V", and (*c*) "Λ". With appropriate values for the detunings Δ_{21} and Δ_{32}, the derivation of the absorption coefficient in the text applies directly to the cascade and "V" cases, and the "Λ" case is given by the complex conjugate. The weak probe wave interacts only with the 2⟷3 transition, and the strong saturator wave interacts only with the 1⟷2 transition. In keeping with the notation of Secs. 8-1 through 8-3, we label the probe as mode 1 and saturator as mode 2

$$\dot{\rho}_{33} = \lambda_3 - \gamma_3\rho_{33} - [i\hbar^{-1}\mathcal{V}_{32}\rho_{23} + \text{c.c.}] \tag{31}$$

$$\dot{\rho}_{22} = \lambda_2 - \gamma_2\rho_{22} + \hbar^{-1}[i\mathcal{V}_{32}\rho_{23} - i\mathcal{V}_{21}\rho_{12} + \text{c.c.}] \tag{32}$$

$$\dot{\rho}_{11} = \lambda_1 - \gamma_1\rho_{11} + [i\hbar^{-1}\mathcal{V}_{21}\rho_{12} + \text{c.c.}] \tag{33}$$

$$\dot{\rho}_{32} = -(\gamma_{32} + i\Delta_{32})\rho_{32} - i\hbar^{-1}\mathcal{V}_{32}(\rho_{22}-\rho_{33}) + i\hbar^{-1}\rho_{31}\mathcal{V}_{12} \tag{34}$$

$$\dot{\rho}_{21} = -(\gamma_{21} + i\Delta_{21})\rho_{21} - i\hbar^{-1}\mathcal{V}_{21}(\rho_{11}-\rho_{22}) - i\hbar^{-1}\mathcal{V}_{23}\rho_{31} \tag{35}$$

$$\dot{\rho}_{31} = -(\gamma_{31} + i\Delta_{31})\rho_{31} - i\hbar^{-1}\mathcal{V}_{32}\rho_{21} + i\hbar^{-1}\rho_{32}\mathcal{V}_{21}. \tag{36}$$

For the cascade case, the detunings are given in the rotating wave approximation as $\Delta_{32} = \omega_{32}-\nu_1$ and $\Delta_{21} = \omega_{21}-\nu_2$. For the "V" case, $\Delta_{32} = \omega_{32}-\nu_1$ and $\Delta_{21} = \omega_{21}+\nu_2$, and for the "Λ" case, $\Delta_{32} = \omega_{32}+\nu_1$ and $\Delta_{21} = \omega_{21}-\nu_2$. For the cascade and "V" cases, the probe polarization $\mathcal{P}_1 = 2\wp_{23}\rho_{32}$, while for the "Λ" case, $\mathcal{P}_1 = 2\wp_{32}\rho_{23}$.

These equations of motion can be solved in closed form for arbitrarily large probe and saturator intensities, but the equations are substantially simpler for a nonsaturating probe wave. Hence we first solve for the saturator dipole moment ρ_{21} in the rate equation approximation and supposing $\mathcal{V}_{32}$=0. In terms of our interaction picture, this means setting $\dot{\rho}_{21}$=0 in Eq.

(35) above and solving for ρ_{21}. The population difference $\rho_{22}-\rho_{11}$ is found as in Sec. 5-1. We obtain

$$\rho_{21} = i\hbar^{-1}\mathcal{V}_{21}\mathcal{D}_{21}N_{21}/(1 + I_2\mathcal{L}_2), \tag{37}$$

where the complex denominator $\mathcal{D}_{21}$ is given by

$$\mathcal{D}_{ij} = 1/(\gamma_{ij} + i\Delta_{ij}), \tag{38}$$

the unsaturated population difference N_{21} is

$$N_{ij} = \lambda_i\gamma_i^{-1} - \lambda_j\gamma_j^{-1} \tag{39}$$

and the saturator dimensionless intensity is

$$I_2 = |\wp_{21}\mathcal{E}_{21}/\hbar|^2 T_1/\gamma_{21}. \tag{40}$$

Here we have introduced the $1\longleftrightarrow 2$ population-difference decay time

$$T_1 = \frac{1}{2}\left[\frac{1}{\gamma_2} + \frac{1}{\gamma_1}\right]. \tag{41}$$

Finally the saturator Lorentzian is given by

$$\mathcal{L}_2 = \gamma_{21}^2/(\gamma_{21}^2 + \Delta_{21}^2). \tag{42}$$

For the probe dipole element ρ_{32} we also need values for the populations ρ_{33} and ρ_{22}. Since the probe does not saturate, $\rho_{33}=\lambda_3/\gamma_3$. Problem 8- shows that ρ_{22} is given by

$$\rho_{22} = \frac{\lambda_2}{\gamma_2} - \frac{I_2\mathcal{L}_2}{2\gamma_2 T_1}\,\frac{N_{21}}{1 + I_2\mathcal{L}_2}\,. \tag{43}$$

We also need the "Raman" two-photon coherence ρ_{31}. Solving Eq. (36) in steady-state, we find

$$\rho_{31} = -i\hbar^{-1}\mathcal{D}_{31}[\mathcal{V}_{32}\rho_{21} - \rho_{32}\mathcal{V}_{21}]. \tag{44}$$

Setting $\dot\rho_{32}=0$ in Eq. (34) and substituting ρ_{31}, we have

$$\rho_{32} = \hbar^{-1}\mathcal{D}_{32}[i\mathcal{V}_{32}(\rho_{33}-\rho_{22}) + \hbar^{-1}\mathcal{D}_{31}(\mathcal{V}_{32}\rho_{21} - \rho_{32}\mathcal{V}_{21})\mathcal{V}_{12}]$$

$$= \frac{\hbar^{-1}\mathcal{V}_{32}\mathcal{D}_{32}[i(\rho_{33}-\rho_{22}) + \hbar^{-1}\mathcal{D}_{31}\mathcal{V}_{12}\rho_{21}]}{1 + |\mathcal{V}_{12}/\hbar|^2\mathcal{D}_{31}\mathcal{D}_{32}} .$$

Substituting the populations ρ_{22} and ρ_{33}, and using $\alpha_1 = -i(\wp_{23}K_1/\epsilon_0)\rho_{32}/\mathcal{E}_1$, we obtain the absorption coefficient

$$\alpha_1 = \alpha_0 \frac{\gamma_{32}\mathcal{D}_{32}}{1+I_2\mathcal{L}_2} \frac{N_{21}I_2(2\mathcal{L}_2/\gamma_2+\gamma_{21}\mathcal{D}_{31}\mathcal{D}_{21})/4T_1 + N_{32}(1+I_2\mathcal{L}_2)}{1 + |\mathcal{V}_{12}/\hbar|^2\mathcal{D}_{31}\mathcal{D}_{32}} , \quad (45)$$

where $\alpha_0 = K_1|\wp_{32}|^2/\hbar\epsilon_0\gamma_{21}$. Note that the "$\Lambda$" absorption coefficient is given by the complex conjugate of Eq. (45).

To illustrate Eq. (45), we excite only the lower level, i.e., $\lambda_3=\lambda_2=0$ and tune the saturator to resonance ($\nu_2=\omega_{21}$). In terms of the saturator "half" Rabi frequency $\Omega = |\mathcal{V}_{21}/\hbar|$, Eq. (45) reduces to

$$\alpha_1 = \alpha_0 \frac{\gamma_{32}N_{21}I_2}{4\gamma_2T_1(1+I_2)} \frac{2(\gamma_{31}+i\Delta_{31}) + \gamma_2}{\Omega^2 - \Delta_{32}^2 + i\Delta_{32}(\gamma_{32}+\gamma_{31}) + \gamma_{31}\gamma_{32}} . \quad (46)$$

Now consider a sufficiently intense saturator that $\Omega >> \gamma_{31}, \gamma_{32}$. This case corresponds to the two-level limit of Eq. (27). For small probe detunings in both cases, $\alpha_1 \propto 1/\Omega^2$, i.e., no central peak. For detunings near the saturator "half" Rabi frequency ($\Delta_{32} \simeq \mp\Omega$), Eq. (46) reduces to

$$\alpha_1 = \alpha_0 \frac{\gamma_{32}N_{21}}{4\gamma_2T_1} \frac{1}{(\gamma_{32}+\gamma_{31})/2 + i(\Omega \pm \Delta_{32})} . \quad (47)$$

As in the two-level case (27), this gives a symmetrical pair of complex Lorentzians with widths equal to the average of the probe dipole linewidth and that of the coherent scatterer (population pulsations or ρ_{31}). However the two coefficients differ by the factor i. Hence for the three-level case, $\mathrm{Re}(\alpha_1)$ is a pair of Lorentzians, whereas for the two-level case, one gets a symmetrical pair of Lorentzian "derivatives" (see Fig. 8-6*c*). For the three-level case, the resonance fluorescence spectrum is essentially the same as the absorption coefficient (47), while the two-level case gives a triple-peaked spectrum very different from Eq. (27), see chapter 15. The basic physical reason behind the difference between Eqs. (27) and (47) is that the three-level probe samples a single saturated level split by the dynamic Stark effect, while the two-level probe samples the difference between two such levels. A related effect is that the three-level splitting is half as large as the corresponding two-level splitting, i.e., the three-level Ω equals only half the Rabi flopping frequency.

One can also calculate the probe absorption in the two-photon, two-level model of Sec. 5-4. This involves the superposition of many three-level transitions with nonresonant intermediate states. In the simple case for which the level-shift factors k_{aa} and k_{bb} of Eq. (5.84) and (5.85) are equal, the probe absorption coefficient reduces to Eq. (17), with the replacements $\wp\mathcal{E}_n/\hbar \to k_{ab}\mathcal{E}_2\mathcal{E}_n/2\hbar$ and $\nu_n \to \nu_2 + \nu_n$.

A particularly beautiful transition occurs as Doppler broadening is introduced to the case for oppositely directed pump and probe waves. For Doppler widths small compared to the two-photon Rabi frequency $k_{ab}\mathcal{E}_2{}^2/2\hbar$, the spectra resemble those of Fig. 8-5. As the Doppler broadening is increased beyond the Rabi frequency, the coherent wiggles wash out, and in the Doppler limit, a simple Doppler-free Lorentzian of width 2γ remains [see Capron and Sargent (1986)].

References

Baklanov, E. V. and V. P. Chebotaev (1971), Sov. Phys JETP **33**, 300.

Brewer, R. G., M. J. Kelley, A. Javan (1969), Phys. Rev. Lett. **23**, 559.

Capron, B. A. and M. Sargent III (1986), Phys. Rev. A**34**, 3034.

Haroche, S., and F. Hartman (1972), Phys. Rev. A**6**, 1280.

Khitrova, G., P. Berman, M. Sargent III (1988), J. Opt. Soc. Am. B**5**, 160.

Mollow, B. R. (1972) Phys. Rev. A**5**, 2217.

Prior, Y, A. R. Bogdan, M. Dagenais, N. Bloembergen (1981), Phys. Rev. Lett. **46**, 111; A. R. Bogdan, M. Downer, N. Bloembergen (1981), Phys. Rev. A**24**, 623.

Sargent III, M. (1976), Appl. Phys. **9**, 127.

Sargent III, M. and P. E. Toschek (1976), Appl. Phys. **11**, 107.

Wu, F. Y., S. Ezekiel, M. Ducloy, and B. R. Mollow (1977), Phys. Rev. Lett. **38**, 1077.

Books and review papers include:

Boyd, R. W. and M. Sargent III (1988), J. Opt. Soc. B**5**, 99.

Demtröder, W. (1981), *Laser Spectroscopy*, SpringerVerlag, Heidelberg.

Levenson, Marc D., and S. S. Kano, *Introduction to Nonlinear Laser Spectroscopy* (1988), Revised Edition, Academic Press, New York.

Sargent III, M. (1978), Phys. Reps. **43**, 223.

Shen, Y. R. (1984), *The Principles of Nonlinear Optics*, John Wiley & Sons, New York.

B. W. Shore (1990), *Theory of Coherent Atomic Excitation*, Vol. I. *Simple Atoms and Fields*, Vol. II. *Multilevel Atoms and Incoherence*, John Wiley & Sons, New York.

Stenholm, S. (1984), *Foundations of Laser Spectroscopy*, John Wiley & Sons, New York.

Problems

8-1. Draw frequency diagrams like that in Fig. 8-2 showing the mode frequency placement corresponding to the gain and absorptive peaks in Fig. 8-4. Hint: examine Eq. (26).

8-2. Derive the population pulsation component d_{-1} using the expansion

$$\rho_{ab} = N\, e^{i(\mathbf{K}_2\cdot\mathbf{r}\, -\, \nu_2 t)} \Sigma_m\, p_m\, e^{im[(\mathbf{K}_1 - \mathbf{K}_2)\cdot\mathbf{r}\, +\, \Delta t]}\ .$$

Note that this gives the polarization components p_{-1}, p_0, and p_1, which correspond to p_3, p_2, and p_1 of Sec. 8-1.

8-3. Describe four ways of measuring T_2 and T_1.

8-4. Sketch the two-level probe absorption coefficient for an intense, centrally-tuned pump and $T_1 \simeq T_2$. How can a probe wave experience gain in such an absorbing medium? Sketch the corresponding case for $T_1 >> T_2$.

8-5. Prove Eq. (18) for the special case of $\gamma_a = \gamma_b$ and central tuning. What does this mean physically?

8-6. Show that for large pump intensity and the frequencies $\nu_2 \neq \omega$ and $\nu_1 = \omega$, that the coherent and incoherent contributions to the absorption coefficient α_1 of Eq. (17) nearly cancel one another. Hence Fig. 8-4 exhibits no special feature at the probe frequency $\nu_1 = \omega$.

8-7. Consider a medium of two-level atoms with upper-to-ground-lower-level decay with $T_1 >> T_2$, i.e., $\Gamma << \gamma$. Using the Fourier series technique of Sec. 8-1, solve for the probe absorption coefficient α_1 given by Eq. (8.21). Hint: the calculation is similar to that in Sec. 8-1, but is simpler in that the population matrix equations of motion involve the single decay constant Γ rather than γ_a and γ_b, and $\mathscr{D}_n = \mathscr{D}_2$ for all n.

8-8. Write a computer program in the language of your choice to evaluate the real part of the degenerate pump-probe absorption coefficient $\alpha_1 =$

$\alpha_0\gamma\mathscr{D}_2/(1 + I_2\mathscr{L}_2)^2$. Have the program print out (or plot) a few values of $\mathrm{Re}\{\alpha_1\}$ as the detuning is varied.

8-9. Derive the approximate probe absorption coefficient given by Eq. (26). The method is similar to the derivation of Eq. (25).

8-10. Consider a field given by Eq. (1) for two arbitrarily intense waves. Show that the Fourier components p_n and d_k satisfy the recurrence relations

$$d_k = \delta_{k0} - iT_1\wp\hbar^{-1}\mathscr{F}(-k\Delta)\,\Sigma_n\,[\mathscr{E}_n p^*_{n-k} - \mathscr{E}^*_n p_{n+k}]$$

$$p_m = -i(\wp/2\hbar)\mathscr{D}_m\,\Sigma_k\,\mathscr{E}_{m-k}d_k\,,$$

and hence that

$$d_k = \delta_{k0} + \tfrac{1}{2}T_1\mathscr{F}(-k\Delta)(\wp/\hbar)^2\,\Sigma_n\Sigma_i[\mathscr{E}_n\mathscr{E}^*_{n-k+i}\mathscr{D}^*_{n-k} + \mathscr{E}_n{}^*\mathscr{E}_{n+k-i}\mathscr{D}_{n+k}]d_i\,,$$

which can be written as $c_{-1k}d_{k-1} + c_{0k}d_k + c_{1k}d_{k+1} = 0$, where the coefficients

$$c_{jk} = \begin{cases} \dfrac{T_1\wp^2}{2\hbar^2}\displaystyle\sum_{n=1+j\geq 1}^{j+2\leq 2}\mathscr{E}_n\mathscr{E}^*_{n-j}[\mathscr{D}^*_{n-k} - \mathscr{D}_{n-j+k}] \\[2ex] \left[1 - \dfrac{\delta_{k0}}{d_0}\right]\dfrac{1}{\mathscr{F}(-k\Delta)} + \dfrac{\gamma}{2}\displaystyle\sum_{n=1}^{2} I_n[\mathscr{D}^*_{n-k} + \mathscr{D}_{n+k}] \end{cases}$$

Define the ratio $r_k = d_k/d_{k-1}$ to find the continued fraction

$$r_k = \frac{c_{-1k}}{c_{0k} + c_{1k}r_{k-1}}$$

Show that d_{-1} of Eq. (15) is given by $r_1{}^*d_0$ provided $\mathscr{E}_1$ is weak.

8-11. Write the coherent part of inhomogeneously broadened probe-absorption coefficient of Eq. (28) in the form

$$\alpha_{coh} = -\tfrac{1}{2}\alpha_0\gamma^2 I_2\mathscr{F}\int_{-\infty}^{\infty}d\omega'\,\mathscr{W}(\omega')\,\frac{\gamma + i\delta}{\gamma'^2+\delta^2}\,\frac{2\gamma+i\Delta}{\beta^2+\delta^2}\,\frac{\gamma+i\Delta-i\delta}{\gamma+i\Delta+i\delta}\,, \qquad (48)$$

where $\beta^2 = (\gamma + i\Delta)(\gamma + i\Delta + \gamma I_2\mathscr{F})$. This has two poles in the lower half plane at $\delta = -i\beta$ and $\delta = -i\gamma'$. Show using contour integration that in the inhomogeneously broadened limit, α_{coh} reduces to

$$\alpha_{coh} = -\tfrac{1}{2}\alpha_0'(\nu_2)\gamma I_2\mathscr{F}\,\frac{2\gamma+i\Delta}{\gamma'^2-\beta^2}\left[\frac{(\gamma'+\gamma)(\gamma'-\gamma-i\Delta)}{\gamma'(\gamma'+\gamma+i\Delta)} - \frac{(\beta+\gamma)(\beta-\gamma-i\Delta)}{\beta(\beta+\gamma+i\Delta)}\right] . \quad (49)$$

What does this expression reduce to for $\Delta = 0$? Hint: solve Eq. (48) for $\Delta = 0$ or note that Eq. (49) can be written in terms of a derivative with respect to β.

8-12. Write the coherent part of inhomogeneously broadened probe absorption coefficient of Eq. (28) for the counterpropagating pump and probe wave case with $\Delta = \delta = 0$ in the form

$$\alpha_{coh} = -\alpha_0\gamma^2\Gamma I_2\int_{-\infty}^{\infty} dx\,\mathscr{W}(x)\,\frac{(\gamma + 3ix)(\gamma^2 + x^2)}{(\gamma + ix)(\gamma'^2 + x^2)}$$

$$\times\,\frac{1}{(\Gamma + 2ix)(\gamma + 3ix)(\gamma + ix) + (\wp E_0/\hbar)^2(\gamma + 2ix)} .$$

Noting that this has a single pole at $x = -i\gamma'$, show that α_{coh} is given by Eq. (30).

8-13. Describe the physics illustrated in the following figure:

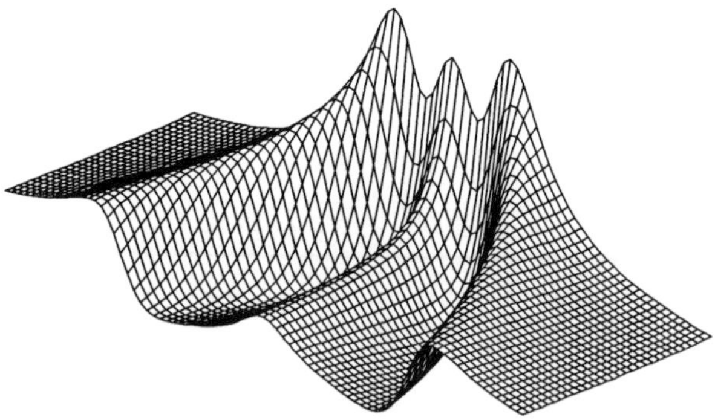

Chapter 9
THREE AND FOUR WAVE MIXING

One characteristic of nonlinear processes in quantum optics is the occurrence of products of interaction energies with complex conjugates of these energies, e.g., in the third-order product $\mathcal{V}\mathcal{V}^*\mathcal{V}$. This is built into the concept of the rotating-wave approximation, whether taken classically or quantum mechanically. Because of the $\mathcal{V}^*$, it is possible to conjugate a wave front. Section 2-4 discusses the conjugation of a plane wave front in a classical $\chi^{(3)}$ medium using four-wave mixing. More generally, imagine a point source emitting spherical waves that pass through a distorting medium. Impinging on an ordinary mirror, which obeys the rule angle of reflection equals angle of incidence, the diverging rays continue to diverge

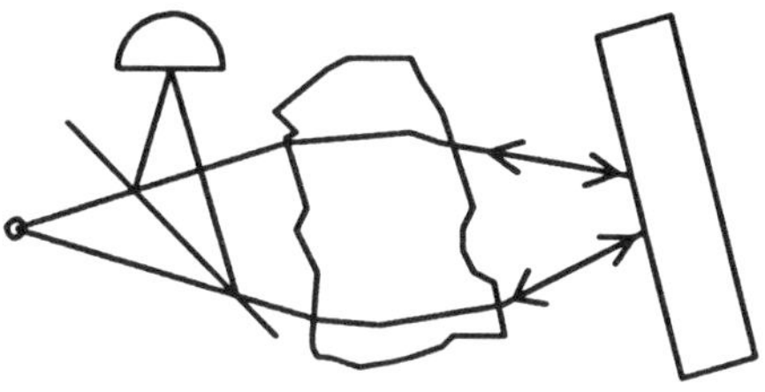

Fig. 9-1. Illustration of how a spherical wave is retroreflected by a phase conjugator.

upon reflection (Fig. 9-1). However if all the $\exp(i\mathbf{K}\cdot\mathbf{r})$ plane waves comprising the wave front could be complex conjugated, i.e., turned into the corresponding $\exp(-i\mathbf{K}\cdot\mathbf{r})$ waves, the wave front would be inverted and sent back through the distorting medium to converge on the original point source. Such a phenomenon has been demonstrated using nonlinear optics and is of substantial interest in applications in laser fusion, weapons, and compensation for bad optics in general. In addition, it is an extension of the two and three-wave interactions discussed in Chap. 8 on saturation spectroscopy, and offers useful alternative configurations to study the characteristics of matter.

This chapter continues the discussion of phase conjugation started in Sec. 2-4 specifically deriving the coupled mode coefficients for two-level atoms, for which perturbation theory typically fails to converge. The incident signal and the conjugate it generates are assumed to be weak enough that they can be treated linearly. One or two pump waves are needed, and these can be arbitrarily intense. Because the signal and conjugate are linear, a general wave front with various temporal and spatial frequencies can be treated by the principle of superposition.

Section 9-1 shows how third-order interactions appear in a quantum mechanical context. Section 9-2 derives the two-level model of Abrams and Lind (1978), for which four-waves of equal frequencies are mixed. This derivation requires a knowledge of Sec. 5-1 alone. Using the steady-state results of Sec. 8-1, the theory is then generalized to treat a pump frequency that differs from the signal and conjugate frequencies. This theory shows how phase conjugation can be used as a spectroscopic technique, for the reflectivity depends sharply on characteristics of the medium such as decay times.

9-1. Phase Conjugation in Two-Level Media

This section shows how the grating terms discussed in Sec. 2-4 show up in two-level media interacting with the three-wave field of Eq. (2.23). The induced third-order polarization contains contributions to $\mathscr{P}_1$ consisting of various permutations of the product $\mathscr{E}_1\mathscr{E}_2^*\mathscr{E}_2$. In the $\mathscr{E}_1$ polarization component, $\mathscr{E}_2$ must be accompanied by $\mathscr{E}_2^*$ to cancel out the $\mathscr{E}_2$ time and space dependence. We suppose that the order of the interactions is from the left (earliest) to the right (latest). Hence for $(\mathscr{E}_2\mathscr{E}_2^*)\mathscr{E}_1$, $\mathscr{E}_1$ mixes with a population term $(\mathscr{E}_2\mathscr{E}_2^*)$ having no space or time variations. In $(\mathscr{E}_1\mathscr{E}_2^*)\mathscr{E}_2$, $\mathscr{E}_2$ scatters off the grating induced by $\mathscr{E}_1\mathscr{E}_2^*$. If the medium can respond to the second harmonic such as for the two-photon medium of Sec. 5-4, $(\mathscr{E}_1\mathscr{E}_2)\mathscr{E}_2^*$ must also be included. This differs from $(\mathscr{E}_1\mathscr{E}_2^*)\mathscr{E}_2$ in that $\mathscr{E}_2^*$ scatters off a term with the rapid time and space dependence $\exp[i(\mathbf{K}_1+\mathbf{K}_2)\cdot\mathbf{r} - i(\nu_1+\nu_2)t]$, yielding a term with the dependence $\exp(i\mathbf{K}_1\cdot\mathbf{r} - i\nu_1 t)$, i.e., that of $\mathscr{E}_1$. These processes are diagrammed in Fig. 9-2 in terms of a kind of energy level diagram in which the horizontal axis corresponds to the time of interaction. Other time orderings occur as well, and they fall into one of these three categories.

Now let's add a second pump wave that propagates up from the bottom as shown in Fig. 2-2. We distinguish the two pump waves by subscripting an up or a down arrow to identify their directions. The pump wave $\mathscr{E}_{2\uparrow}$ also scatters off the grating induced by the interference of $\mathscr{E}_{2\downarrow}$ and $\mathscr{E}_1$, but scatters in the direction exactly *opposite* to $\mathscr{E}_1$*'s* propagation.

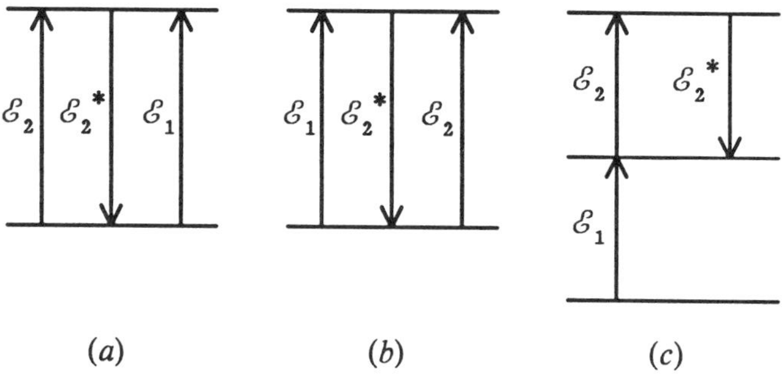

Fig. 9-2. Pseudo energy-level diagram with arrows showing three principal kinds of third-order interactions. Upward arrows correspond to positive-frequency ($e^{-i\nu t}$) fields, and the time of an interaction increases the further to the right its arrow is. (*a*) represents $(\mathcal{E}_2\mathcal{E}_2^*)\mathcal{E}_1$, where $\mathcal{E}_2\mathcal{E}_2^*$ has no spatial or time dependence. (*b*) represents $(\mathcal{E}_1\mathcal{E}_2^*)\mathcal{E}_2$, where $\mathcal{E}_1\mathcal{E}_2^*$ yields a grating varying slowly in time. (*c*) represents $(\mathcal{E}_1\mathcal{E}_2)\mathcal{E}_2^*$, where $\mathcal{E}_1\mathcal{E}_2$ yields a grating varying at twice the field frequency.

In terms of the third-order products, we have terms like $(\mathcal{E}_{2\downarrow}\mathcal{E}_1^*)\mathcal{E}_{2\uparrow}$. Here $\mathcal{E}_{2\downarrow}\mathcal{E}_1^*$ represents a grating that varies little in an optical frequency period. $\mathcal{E}_{2\uparrow}$ scatters off this grating, cancelling the $\mathcal{E}_2$ spatial dependence identically, since $\mathbf{K}_{2\uparrow} = -\mathbf{K}_{2\downarrow}$. The resulting retroreflected wave is proportional to the *conjugate* of $\mathcal{E}_1$ and hence is called the conjugate wave, which gives phase conjugation its name. More specifically, we obtain a term in the nonlinear polarization with the dependence

$$(\mathcal{E}_{2\downarrow}\mathcal{E}_1^*)\mathcal{E}_{2\uparrow}e^{-i(\mathbf{K}_1\cdot\mathbf{r}\ -\ \nu_3 t)}\ , \tag{1}$$

where the conjugate wave frequency ν_3 is given by

$$\nu_3 = \nu_2 + (\nu_2 - \nu_1) \equiv \nu_2 + \Delta. \tag{2}$$

This wave moves in the direction opposite to $\mathcal{E}_1$, which has the carrier $\exp(i\mathbf{K}_1\cdot\mathbf{r}\ -\ i\nu_1 t)$. If the medium can respond to frequencies like $2\nu_2$, then the term $(\mathcal{E}_{2\downarrow}\mathcal{E}_{2\uparrow})\mathcal{E}_1^*$ is also possible, and contributes to the conjugate wave. For this, $\mathcal{E}_1^*$ scatters off a two-photon term with no spatial variations and with the high frequency time dependence $\exp[i(\nu_1+\nu_2)t]$. In addition $\mathcal{E}_{2\downarrow}$ scatters off gratings induced by $\mathcal{E}_1$ with $\mathcal{E}_{2\uparrow}$. The two third-order phase conjugation processes are diagrammed in Fig. 9-3.

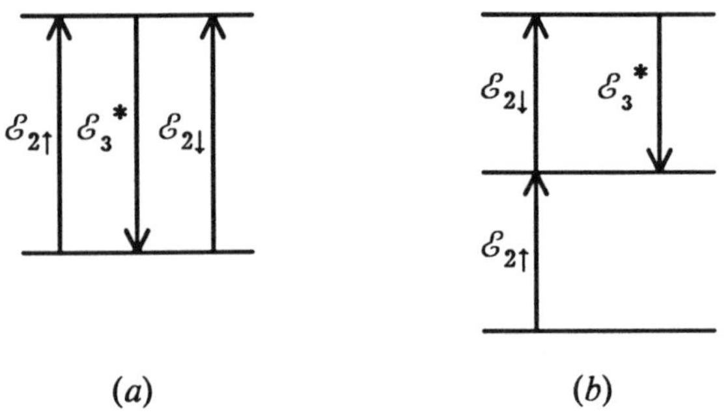

Fig. 9-3. Pseudo energy-level diagram of (*a*) "single-photon" and (*b*) "two-photon" four-wave phase conjugation processes.

If the signal and pump frequencies differ ($\nu_1 \neq \nu_2$), ν_3 is still different as given by Eq. (2) and illustrated in Fig. 2-1. This has important consequences. First consider the two-wave case of Fig. 8-1. The interference between $\mathcal{E}_1$ and $\mathcal{E}_2$ is a stationary fringe for $\nu_1 = \nu_2$. Since the product $\mathcal{E}_1 \mathcal{E}_{2\downarrow}^*$ is associated with the propagation factor $\exp[i(\mathbf{K}_1 - \mathbf{K}_{2\downarrow}) \cdot \mathbf{r} - i(\nu_1 - \nu_2)t]$, it moves in general with a velocity

$$\boxed{\mathbf{v} = \frac{\nu_1 - \nu_2}{|\mathbf{K}_1 - \mathbf{K}_{2\downarrow}|} \frac{\mathbf{K}_1 - \mathbf{K}_{2\downarrow}}{|\mathbf{K}_1 - \mathbf{K}_{2\downarrow}|}} \,. \tag{3}$$

If $\nu_1 < \nu_2$, this moves downward and to the left as indicated in Fig. 8-1, thereby Doppler downshifting ν_2 to the frequency ν_1. In the exponent, we have $\nu_2 - (\nu_2 - \nu_1) \rightarrow \nu_1$. On the other hand it Doppler upshifts the pump wave $\mathcal{E}_{2\uparrow}$ to the frequency ν_3, and contributes to the $\mathcal{E}_3^*$ amplitude. Similarly the conjugate grating is proportional to $\mathcal{E}_{2\uparrow} \mathcal{E}_1^*$ and Doppler upshifts ν_2 to ν_3 and contributes to $\mathcal{E}_3$.

Two phenomena limit the bandwidth of the reflection spectrum, 1) the finite response time of the medium [*Fu* and Sargent (1979)] and 2) the phase mismatch implied by Eq. (2) [$K_1 = \nu_1/c \neq \nu_3/c$, see Pepper and Abrams (1978)]. For simplicity, we consider the two-wave case first. In general the nonlinear response of a medium is characterized by a minimum time T_1 for which it can follow field variations. Significant changes within this time are averaged out. In particular as discussed in Sec. 8-2, if the fringe pattern moves through the medium too rapidly, the medium sees an averaged field, and no grating is induced. This fact allows one to measure the time T_1 of a medium, and is a basic feature of grating-dip spectroscopy [Sargent (1976)]. More quantitatively, if the beat frequency ν_2-ν_1 exceeds $1/T_1$, the nonlinear grating contribution to the induced polarization falls off. In two-level resonant media, the time T_1 can be quite large; in ruby, for example, it is on the order of milliseconds. However in other media, the effective T_1 can be extremely short. In principle, one can even measure femtosecond response times, since the signal can be tuned far away from the pump.

Similarly, for the four wave case of Fig. 2-2, a beat frequency large compared to $1/T_1$ fails to induce a grating, and hence does not create a conjugate wave. This leads to a narrow-band retroreflector. The conjugate generation is also limited by the phase mismatch implied by Eq. (2). In other words, the induced polarization may have a space-time dependence that cannot propagate at the speed of light in the medium. These concepts are illustrated in later sections for specific kinds of media.

Still another form of phase conjugation exists with three waves as shown in Fig. 9-4*a*. The case is described by the field product $(\mathcal{E}_2\mathcal{E}_1^*)\mathcal{E}_2$. This yields a conjugate with the frequency (2) and the spatial factor

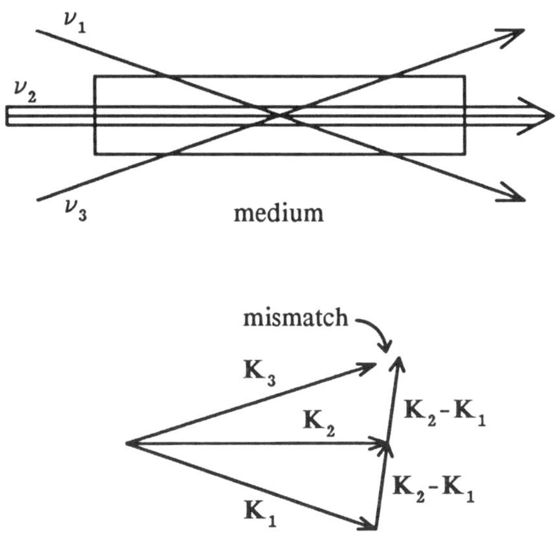

Fig. 9-4. (*a*) Three-wave phase conjugation: two incident waves generate a conjugate wave if the angle θ is small enough. (*b*) Three-wave phase mismatch.

$2\mathbf{K}_2-\mathbf{K}_1$. As shown in Fig. 9-4*b*, this results in a phase mismatch for nonzero θ. This technique is useful for transmission phase conjugation, as distinguished from the four-wave retroreflecting method described above.

9-2. Two-Level Coupled Mode Coefficients

In this section, we derive the Abrams-Lind (1978) coupled-mode absorption and coupling coefficients for two-level atoms perturbed by the superposition of three and four waves of the same frequency, but differing directions. Sections 9-3 and 9-4 deal with the more general case for different (nondegenerate) frequencies. For the degenerate case (or at least for frequency differences small compared to the atomic decay constants), we replace the electric field amplitude factor $\mathcal{E}(z)e^{iKz}$ in Eq. (5.1) by an arbitrary function of $\mathbf{r}$, $\mathcal{E}(\mathbf{r})$, that is,

$$E(\mathbf{r},t) = \frac{1}{2}\,\mathcal{E}(\mathbf{r})\,e^{-i\nu t} + \text{c.c.} \tag{4}$$

Similarly to the derivation of (5.24), this field induces the $e^{-i\nu t}$ polarization component in the two-level model

$$\rho_{ab} = -i\,\frac{(\wp/2\hbar)N\mathcal{D}\mathcal{E}(\mathbf{r})e^{-i\nu t}}{1+\dfrac{\mathcal{E}(\mathbf{r})\mathcal{E}^*(\mathbf{r})\mathcal{L}}{\mathcal{E}_s^{\,2}}}\,, \tag{5}$$

where $\mathcal{E}_s^{\,2}=(\hbar/\wp)^2/T_1T_2$. The signal and conjugate waves are then represented by a small deviation $\epsilon(\mathbf{r})$ about a possibly large pump field $\mathcal{E}_2(\mathbf{r})U_2(\mathbf{r})$, that is, $\mathcal{E}(\mathbf{r}) = \mathcal{E}_2(\mathbf{r})U_2(\mathbf{r}) + \epsilon(\mathbf{r})$, where $U_2(\mathbf{r})$ gives the pump waves rapidly varying spatial dependence. To first order in ϵ and ϵ^*, the polarization (5) is given by the truncated Taylor series (Prob. 9-7)

$$\rho_{ab}(\mathcal{E}_2U_2+\epsilon,\ \mathcal{E}_2^{\,*}U_2^{\,*}+\epsilon^*) \simeq \rho_{ab}(\mathcal{E}_2,\mathcal{E}_2^{\,*}) + \epsilon\,\frac{\partial\rho_{ab}}{\partial\mathcal{E}}\bigg|_{\mathcal{E}_2} + \epsilon^*\,\frac{\partial\rho_{ab}}{\partial\mathcal{E}^*}\bigg|_{\mathcal{E}_2^{\,*}}. \tag{6}$$

It is more physically revealing, however, to derive this result by expanding Eq. (5) as a $1/(1+x)$ series. This gives

$$\rho_{ab}(\mathcal{E}_2U_2+\epsilon,\ \mathcal{E}_2^*U_2^*+\epsilon^*) = -\frac{i(\wp/2\hbar)N\mathcal{D}(\mathcal{E}_2U_2+\epsilon)e^{-i\nu t}}{[1 + I_2|U_2(\mathbf{r})|^2\mathcal{L}_2]\left[1 + \dfrac{(\epsilon\mathcal{E}_2^*U_2^* + \text{c.c.})\mathcal{L}_2}{\mathcal{E}_s{}^2[1 + I_2|U_2(\mathbf{r})|^2\mathcal{L}_2]}\right]}$$

$$\simeq -\frac{i(\wp/2\hbar)N\mathcal{D}(\mathcal{E}_2U_2+\epsilon)e^{-i\nu t}}{1 + I_2|U_2(\mathbf{r})|^2\mathcal{L}_2}\left[1 - \frac{(\epsilon\mathcal{E}_2^*U_2^* + \text{c.c.})\mathcal{L}_2}{\mathcal{E}_s{}^2[1 + I_2|U_2(\mathbf{r})|^2\mathcal{L}_2]}\right]. \tag{7}$$

The ϵ and ϵ^* terms inside the square brackets result from pump scattering of the gratings induced by fringes between the pump, ϵ and ϵ^*. In particular, suppose $\epsilon(\mathbf{r})$ is the sum of two small fields

$$\epsilon(\mathbf{r}) = \mathcal{E}_1(\mathbf{r})e^{i\mathbf{K}_1\cdot\mathbf{r}} + \mathcal{E}_3(\mathbf{r})e^{i\mathbf{K}_3\cdot\mathbf{r}}\ . \tag{8}$$

More general wave fronts are represented by a sum over such amplitudes. We choose the z axis such that $\mathbf{K}_1\cdot\mathbf{r} = K_1z$. The polarization component $\mathcal{P}_1(z)$ contributes to a Beer's law (5.3) for $\mathcal{E}_1(z)$ given by the projection (similar to Eq. (5.55))

$$\mathcal{P}_1(z) = \frac{2\wp K_1}{2n\pi}\int_0^{2n\pi/K_1} d\zeta\ e^{-i[K_1(z+\zeta)-\nu t]}\ \rho_{ab}(\mathbf{r}+\zeta\hat{z})\ , \tag{9}$$

where ρ_{ab} is given by Eq. (6), and n is a sufficiently small integer that $\mathcal{E}_1(z)$ varies little in the distance $2n\pi/K_1$. In carrying out the projection (9), we suppose the angle between $\mathbf{K}_1$ and $\mathbf{K}_2$ is large enough that signal scattering off a standing-wave pump grating averages to 0. Two kinds of pump waves are of interest, $U_2(\mathbf{r}) = \exp(i\mathbf{K}_2\cdot\mathbf{r})$ giving three-wave mixing, and $U_2(\mathbf{r}) = 2\cos(\mathbf{K}_2\cdot\mathbf{r})$ giving four-wave mixing.

For the three-wave case of Fig. 9-4, we find

$$\mathcal{P}_1(z) = -\frac{i\wp^2N\mathcal{D}_2}{\hbar(1 + I_2\mathcal{L}_2)^2}\,[\mathcal{E}_1 - \mathcal{E}_3{}^*I_2\mathcal{L}_2e^{2i\Delta Kz}], \tag{10}$$

where for this three-wave case the phase mismatch $2\Delta Kz$ is given by $(2\mathbf{K}_2 - \mathbf{K}_1 - \mathbf{K}_3)\cdot\mathbf{r}$ and $I_2 = |\wp\mathcal{E}_2/\hbar|^2T_1T_2$. A similar expression is obtained for $\mathcal{P}_3(z)$. Substituting these polarization components into the Beer's law (5.3), we find the coupled-mode equations (2.25a) and (2.25b), where the absorption and coupling coefficients are

$$\alpha_1 = \alpha_0 \gamma \mathscr{D}_2/(1 + I_2\mathscr{L}_2)^2 \tag{11}$$

$$\chi_1 = \alpha_0 \gamma \mathscr{D}_2 I_2 \mathscr{L}_2/(1 + I_2\mathscr{L}_2)^2 \tag{12}$$

Here α_1 agrees with Eqs. (5.70) and (9.19). Corresponding coefficients for α_3^* and χ_3^* are given by complex conjugates of Eqs. (11) and (12).

For the four-wave case of Fig. 2-2, the saturation denominator in Eq. (5) has spatial holes. Assuming the signal wave is not parallel to the pump waves, the signal polarization contains an average over these holes for a combination of two reasons. First when the signal wave is nearly perpendicular to the pump waves, the width of the signal beam generates the average over the holes. Second, the projection along z in Eq. (9) typically reduces to an average over these holes by virtue of the factor $\cos^2(\mathbf{K}_2 \cdot \hat{\mathbf{z}}\zeta) \equiv \cos^2\frac{1}{2}\theta$ resulting from the $U_2(\mathbf{r})$ functions in Eq. (7). Evaluating the average as discussed in Sec. 5-3, we have

$$\mathscr{P}_1(z) = i\wp^2\hbar^{-1}N\mathscr{D}_2 \frac{1}{2\pi}\int_0^{2\pi} d\theta \left[\frac{\mathscr{E}_1}{(a+b\cos\theta)^2} - \mathscr{E}_3^* b \frac{1+\cos\theta}{(a+b\cos\theta)^2}\right], \tag{13}$$

where $b = 2I_2\mathscr{L}_2$ and $a = 1+b$. Simplifying, we have

$$\mathscr{P}_1 = -i\wp^2\hbar^{-1}N\mathscr{D}_2 \frac{\partial}{\partial a}\frac{1}{2\pi}\int_0^{2\pi} d\theta \left[\frac{\mathscr{E}_1+\mathscr{E}_3^*(a-b)}{a+b\cos\theta} - \mathscr{E}_3^* e^{i(K_3-K_1)z}\right]$$

$$= \frac{i\wp^2 N\mathscr{D}_2}{\hbar(1+2b)^{3/2}}\left[\mathscr{E}_1(1+b) - \mathscr{E}_3^* e^{i(K_3-K_1)z} b\right]. \tag{14}$$

With the corresponding formula for $\mathscr{P}_3(z)$, this yields the coupled-mode equations (2.25a) and (2.25b) with the coefficients

$$\alpha_1 = \alpha_0\gamma\mathscr{D}_2 \frac{1 + 2I_2\mathscr{L}_2}{(1 + 4I_2\mathscr{L}_2)^{3/2}} \tag{15}$$

$$\chi_1 = \frac{2\alpha_0\gamma\mathscr{D}_2 I_2\mathscr{L}_2}{(1 + 4I_2\mathscr{L}_2)^{3/2}} \tag{16}$$

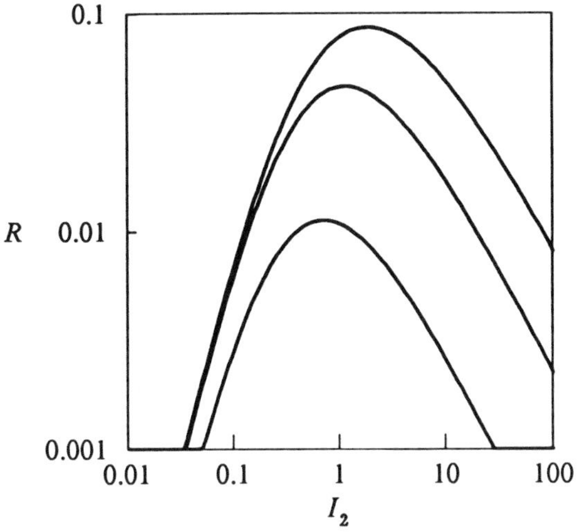

Fig. 9-5. Reflectivity $R = |r|^2$ of Eq. (2.35) using α_1 of Eq. (15) and χ_1 of Eq. (16) versus pump intensity I_2 for $\nu_2 = \omega$, and $\alpha_0 L$ = .7, 2, 4 (in order of higher values).

The $\mathcal{E}_3$ coefficients α_3 and χ_3 are given by the complex conjugates of (15) and (16), respectively. Note that although this four-wave mixing case is generally better phase matched than the three-wave case, it is only perfectly phase matched for degenerate tuning ($\nu_n = \nu_2$, for all n).

The phase-conjugate reflectivity is given by substituting these coefficients into Eq. (2.35), with results shown in Fig. 9-5. For small pump intensities, the reflectivity increases as a function of intensity. The medium then begins to saturate, and ultimately the reflectivity bleaches to zero.

9-3. Modulation Spectroscopy

Some of the most useful configurations of multiwave mixing use waves with different frequencies. These include phase conjugation by nondegenerate three- and four-wave mixing and various kinds of three-wave modulation spectroscopy. The three-wave cases are depicted in Figs. 9-4 and 9-6. For modulation spectroscopy, the saturator wave is weakly modulated, imposing sidebands at the frequencies ν_1 and ν_3. These sidebands act as a pair of coupled probe waves with the amplitudes $\mathcal{E}_1$ and $\mathcal{E}_3$. The resultant polarization of the medium includes the effects of the population pulsations induced by the fringe component between each probe wave with

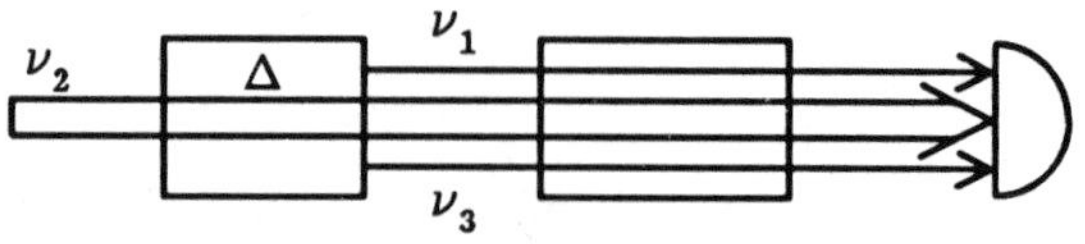

Fig. 9-6. Modulation spectroscopy configuration. The large intensity beam at frequency ν_2 passes through a modulator, producing two sidebands at the frequencies $\nu_1 = \nu_2 - \Delta$ and $\nu_3 = \nu_2 + \Delta$. The beat frequency signal is studied as a function of the beat frequency Δ or other parameters of interest.

the saturator wave. The total field including the saturator passes through the medium and impinges on the square-law detector. A special piece of electronics called *a spectrum analyzer* distills out the signal oscillating at the beat frequency Δ. This kind of measurement is called *homodyne detection*, which means that the incoming light beats with itself on a square-law detector. The beat-frequency signal is studied as a function of the beat frequency Δ or of other parameters of interest.

To quantify this method, we take the absolute value squared of the three-wave amplitude from Eq. (1) and find the signal intensity I_B oscillating at the beat frequency Δ to be

$$I_B = \frac{1}{4}\mathcal{E}_2[\mathcal{E}_1(z) + \mathcal{E}_3^*(z)]e^{i\Delta(t-z/c)} + \text{c.c.} \tag{17}$$

Here we neglect pump depletion and, without loss of generality, assume that the pump amplitude $\mathcal{E}_2$ is real. We see that I_B is the sum of two terms, each proportional to the product of $\mathcal{E}_2$ and one of the sidemode amplitudes. These products are substantially larger than a probe-wave intensity. This increase in the signal intensity is a major advantage of modulation spectroscopy over the direct detection scheme of Fig. 8-1.

Furthermore the relative phase between the three modes is important. If at some point in time all three modes are in phase with one another, then the two population pulsation sources add. This is called the *AM* case, in analogy with AM radio techniques. If the phases of the saturator and one sideband are equal and differ from the phase of the other sideband by π, the two population pulsations cancel out. This gives a constant envelope in time and is called *FM*. Both of these limiting cases have attracted substantial attention. The AM case has been used by Malcuit *et al*. (1984) to measure T_1 for cases where $T_1 >> T_2$. The FM case has been used by Bjorklund (1981) and Drewer *et al.* (1981), who use the fact that the medium may modify the phase/amplitude relationships of an FM wave thereby producing an easily detected AM component. In addition to spectroscopy, the problem is important in phase conjugation, laser instabilities, and cavity stabilization.

We can extend the two-wave treatment of Sec. 8-1 to treat three- and four-wave mixing by including a $\mathcal{E}_3$ sideband at frequency ν_3 in $E(\mathbf{r},t)$ of Eq. (9.1). This contributes an $\mathcal{E}_3$ term to the polarization coefficient p_3 of Eq. (9.13), namely,

$$p_3 = -i(\wp/2\hbar)\, \mathcal{D}_3[\mathcal{E}_3 U_3 d_0 + \mathcal{E}_2 U_2 d_1]\,, \tag{18}$$

where we include the mode functions $U_n(\mathbf{r})$ explicitly to allow for phase mismatch. Equation (18), in turn, adds a $\mathcal{E}_3^*$ contribution to d_{-1}, which becomes

$$d_{-1} = -\frac{d_0 T_1 \mathcal{F}(\Delta)\,\frac{\wp^2}{2\hbar^2}[\mathcal{E}_1 U_1 \mathcal{E}_2^* U_2^* (\mathcal{D}_1 + \mathcal{D}_2^*) + \mathcal{E}_2 U_2 \mathcal{E}_3^* U_3^* (\mathcal{D}_2 + \mathcal{D}_3^*)]}{1 + I_2 \mathcal{F}(\Delta)\,\frac{\gamma}{2}(\mathcal{D}_1 + \mathcal{D}_3^*)} \tag{19}$$

Using the three-frequency population-pulsation component (19) in Eq. (9) along with $\mathcal{P}_1 = 2\wp N p_1$ and Eq. (5.3), we find the propagation equation

$$\frac{d\mathcal{E}_1}{dz} = -\alpha_1 \mathcal{E}_1 + \chi_1 \mathcal{E}_3^* e^{2i\Delta K z}\,, \tag{20}$$

where the absorption coefficient α_1 is given by Eq. (8.17) as before and the coupling coefficient χ_1 is given by

$$\chi_1 = \frac{\alpha_0 \gamma \mathcal{D}_1}{1 + I_2 \mathcal{L}_2}\, \frac{I_2 \mathcal{F}(\Delta)\,\frac{\gamma}{2}(\mathcal{D}_2 + \mathcal{D}_3^*)}{1 + I_2 \mathcal{F}(\Delta)\,\frac{\gamma}{2}(\mathcal{D}_1 + \mathcal{D}_3^*)}\,. \tag{21}$$

Similarly $\mathcal{E}_3$ obeys the equation

$$\frac{d\mathcal{E}_3^*}{dz} = -\alpha_3^* \mathcal{E}_3^* + \chi_3^* \mathcal{E}_1 e^{-2i\Delta K z}\,, \tag{22}$$

The coefficients α_3 and χ_3 are given by α_1 and χ_1, respectively, by interchanging the subscripts $_1$ and $_3$ (note this implies replacing Δ by $-\Delta$). These equations are the same as the counterpropagating coupled-mode equations (2.25a) and (2.25b) except that they describe copropagating waves without phase mismatch. They are generalizations of our nonlinear Beer's law Eq. (5.3) for cases involving two coupled waves. The equations appear in many areas of physics and engineering in addition to the spectroscopic context here.

By finding the eigenvalues for the coupled equations (20) and (22), we can find the solutions (Prob. 9-8)

$$\mathcal{E}_1(z) = e^{-az}\,[\mathcal{E}_1(0)\cosh wz + (-\alpha\mathcal{E}_1(0) + \chi_1\mathcal{E}_3^*(0))\sinh wz/w] \qquad (23)$$

$$\mathcal{E}_3^*(z) = e^{-az}\,[\mathcal{E}_3^*(0)\cosh wz + (\alpha\mathcal{E}_3^*(0) + \chi_3^*\mathcal{E}_1(0))\sinh wz/w]\,, \qquad (24)$$

where $a = (\alpha_1 + \alpha_3^*)/2$, $\alpha = (\alpha_1 - \alpha_3^*)/2$, and $w = [\alpha^2 + \chi_1\chi_3^*]^{1/2}$. These can be substituted into Eq. (17) to find the beat-frequency intensity I_B for any initial field values.

We can find a simpler solution for central pump tuning ($\nu_2 = \omega$), for which $\alpha_1 = \alpha_3^*$ and $\chi_1 = \chi_3^* = \alpha_{coh}$, and therefore $\alpha = 0$ and $w = \chi_1$. In particular taking $\mathcal{E}_3^*(0)$ to be equal to $\mathcal{E}_1$ (AM modulation) reduces Eq. (23) to

$$\mathcal{E}_1(z) = e^{-(\alpha_{inc} + 2\alpha_{coh})z}\,\mathcal{E}_1(0)\,. \qquad (25)$$

Hence the centrally-tuned AM case has coherent contributions double the size of the two-wave case. This is because the population pulsations induced by the fringe for one sidemode with the pump add to those induced by the fringe for the other sidemode with the pump. This causes the coherent dips of Fig. 8-3 to be twice as deep, allowing the possibility of gain. The effective AM absorption coefficient corresponding to the $T_1 = T_2$ curves of Fig. 8-4 are shown in Fig. 9-7. The gain region is deepened substantially compared to the two-wave case.

In contrast for the FM case $\mathcal{E}_3^* = -\mathcal{E}_1$, Eq. (23) reduces to

$$\mathcal{E}_1(z) = e^{-\alpha_{inc}z}\,\mathcal{E}_1(0)\,. \qquad (26)$$

For FM the two sets of population pulsations cancel one another out. This isn't surprising since as Eq. (17) illustrates, a square-law detector doesn't detect an FM modulation, i.e., $I_B = 0$, and two-level atoms can be thought of as nonlinear detectors that share this property of the square-law detector. In addition to its spectroscopic value, modulation spectroscopy plays a role in some kinds of optical instabilities, as discussed in Sec. 10-2.

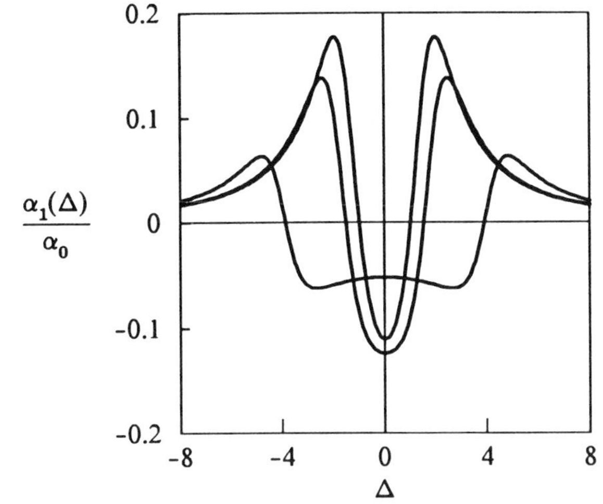

Fig. 9-7. The effective AM absorption coefficient in Eq. (25) with the α_{coh} of Eq. (8.17). The parameters are the same as for the two-wave case of Fig. 8-5*b*, that is, $T_1 = T_2$, $\nu_2 = \omega$, and $I_2 =$ 2, 3.3, and 16 (in order of decreasing maximum value).

9-4. Nondegenerate Phase Conjugation by Four-Wave Mixing

To allow for signal and conjugate frequencies different from the pump frequency, we average the nondegenerate absorption and coupling coefficients (9.17) and (21) over the pump spatial holes. The algebra is similar to that for the degenerate Eqs. (15) and (16) (see Prob. 9-9). We find

$$\alpha_1 = \frac{\alpha_0\gamma\mathscr{D}_1}{(1+4I_2\mathscr{L}_2)^{1/2}}\left[1 - \frac{2I_2\mathscr{F}\gamma(\mathscr{D}_1+\mathscr{D}_2^*)}{1+4I_2f_2+[(1+4I_2\mathscr{L}_2)(1+4I_2f_2)]^{1/2}}\right] \tag{27}$$

$$\chi_1 = \frac{\alpha_0\gamma\mathscr{D}_1}{(1+4I_2\mathscr{L}_2)^{1/2}}\frac{2I_2\mathscr{F}\gamma(\mathscr{D}_2+\mathscr{D}_3^*)}{1+4I_2f_2+[(1+4I_2\mathscr{L}_2)(1+4I_2f_2)]^{1/2}} \tag{28}$$

where $f_2 = \mathscr{F}\gamma(\mathscr{D}_1+\mathscr{D}_3^*)/2$. These coefficients reduce to the degenerate cases (15) and (16) respectively, as they should (take $\mathscr{F} = 1$, equal $\mathscr{D}$'s, and $f_2 = \mathscr{L}_2$). Figure 9-8 illustrates the conjugate reflectivity as a function of signal detuning for various values of the population difference lifetime T_1. As for the coherent dips of Figs. 8-3, the population pulsation (grating)

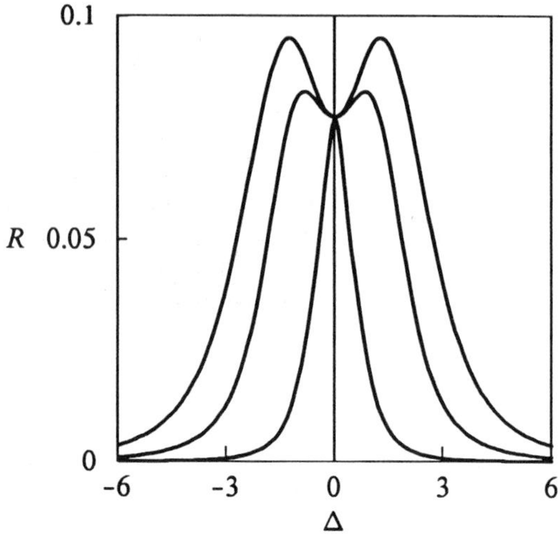

Fig. 9-8. Reflectivity $R = |r|^2$ given by Eq. (2.35) with α_1 of Eq. (27) and χ_1 of Eq. (28) versus signal-pump detuning for $\alpha_0 L = 4$, $I_2 = 1$, $\nu_2 = \omega$, and $T_1 = .5, 1, 4$ (in order of decreasing width).

contribution is reduced as a function of this tuning. The width is characterized by the grating lifetime, T_1. In practice, it is almost as easy to study the four-wave conjugate wave as the transmitted signal, and hence this offers an alternative way of determining T_1. As we see, the formulas used to analyze the results are, however, more complicated.

This "grating-washout" effect can be used to produce a narrow-band retroreflector. Alternatively note that the induced conjugate polarization has the frequency ν_3, but that its **K** vector is $-\mathbf{K}_1$, which is not equal in magnitude to ν_3/c. Hence a phase mismatch occurs. This mismatch can also be used to provide a narrow-band retroreflector.

References

Abrams, R. L. and R. C. Lind (1978), Opt. Lett. 2, 94; (1978) **3**, 205.

Boyd, R. W., M. G. Raymer, P. Narum, and D. Harter (1981), Phys. Rev. A**24**, 411.

Bjorklund, G. (1980), Opt. Lett. **5**, 15.

Fisher, R. A. (1983), Ed., *Phase Conjugate Optics*, Academic Press, NY.

Fu, T. and M. Sargent III (1979), Opt. Lett. **4**, 366.

R. W. P. Drewer, J. L. Hall, F. W. Kowalski, J. Hough, G. M. Ford, A. G. Manley, and H. Wood, Appl. Phys. **B31**, 97 (1981).

Hillman, L. W., R. W. Boyd, J. Krasinski, and C. R. Stroud (1983), Jr., Opt. Comm. **46**, 416.

Malcuit, M. S., R. W. Boyd, L. W. Hillman, J. Krasinski, and C. R. Stroud (1984), Jr., J. Opt. Soc. **B1**, 354.

McCall, S. L. (1974), Phys. Rev. **A9**, 1515.

Pepper, D., and R. L. Abrams, Opt. Lett. **3**, .

Sargent, M. III (1976), Appl. Phys. **9**, 127.

Sargent, M. III, P. E. Toschek, H. G. Danielmeyer (1976), App. Phys. **11**, 55.

Senitzky, B., G. Gould, and S. Cutler (1963), Phys. Rev. **130**, 1460 (first report of gain in an uninverted two-level system).

Relevant review papers include (see also the references to Chap. 8):

Boyd, R. W. and M. Sargent III (1988), J. Opt. Soc. **B5**, 99.

Sargent, M. III (1978), Phys. Rep. **43C**, 223.

Problems

9-1. Suppose in four-wave mixing that $\mathbf{K}_1 = K\mathrm{z}$, $\mathbf{K}_{2\downarrow} = K_2[\cos\theta\mathrm{z} + \sin\theta\mathrm{y}]$, $\mathbf{K}_{2\uparrow} = -K_2[\cos(\theta-\epsilon)\mathrm{z} + \sin(\theta-\epsilon)\mathrm{y}]$. To first order in ϵ, find the direction of the reflected wave and the phase matching coherence length.

9-2. What two mechanisms lead to narrow-band retroreflection in four-wave mixing?

9-3. Derive the Abrams-Lind degenerate four-wave mixing theory of phase conjugation for the case of unequal pump intensities.

9-4. For central pump tuning ($\nu_2 = \omega$), upper to ground lower level decay, and $I_2 >> 1$, find a simple formula for the coupling coefficient χ_1 defined by Eq. (21). *Ans*:

$$\chi_1(\Delta \simeq \pm\Omega) \simeq \mp i \, \frac{\alpha_0\gamma\Gamma/4\Omega}{(\gamma+\Gamma)/2 \mp i(\Delta\mp\Omega)} \, .$$

$$\chi_1(\Delta\simeq 0) \simeq \frac{\alpha_0(1 + \gamma\mathscr{D}_1)}{2I_2} \, .$$

9-5. Using the two-photon single-mode absorption coefficient of Eq. (5.99), derive the probe absorption coefficient α_1 and the mode coupling coefficient χ_1 for this model by applying the Abrams-Lind method.

Assume $\mathcal{E} = \mathcal{E}_2\exp(i\mathbf{K}_2\cdot\mathbf{r}) + \epsilon$, where ϵ is given by Eq. (8), $|\mathcal{E}_1|$ and $|\mathcal{E}_3|$ $<< |\mathcal{E}_2|$, and that the fields are copropagating.

9-6. Using the definition $\mathcal{E}_n = |\mathcal{E}_n|e^{-i\phi_n}$, derive the real coupled amplitude equations of motion

$$\frac{d|\mathcal{E}_1|}{dz} = -\mathrm{Re}\{\alpha_1\}|\mathcal{E}_1| + \mathrm{Re}\{\chi_1 e^{i\Psi}\}|\mathcal{E}_3|$$

$$\frac{d|\mathcal{E}_3|}{dz} = -\mathrm{Re}\{\alpha_3\}|\mathcal{E}_3| + \mathrm{Re}\{\chi_3 e^{i\Psi}\}|\mathcal{E}_1|,$$

where $\Psi = \phi_1+\phi_3$. What are the corresponding equations for $d\phi_n/dz$?

9-7. Derive $\mathcal{P}_1$ of Eq. (10) using the Taylor-series expansion (6).

9-8. Using Laplace transforms or the substitution $\mathcal{E}_1 = e^{\mu z}$, find the solutions (23) and (24) to the coupled-mode Eqs. (20) and (22).

9-9. Derive the nondegenerate four-wave mixing coefficients (27) and (28) by averaging Eqs. (8.17) and (21), respectively, over pump spatial holes. Hint: use partial fractions and the technique in Eq. (5.65).

9-10. Describe the physics illustrated by the following figure:

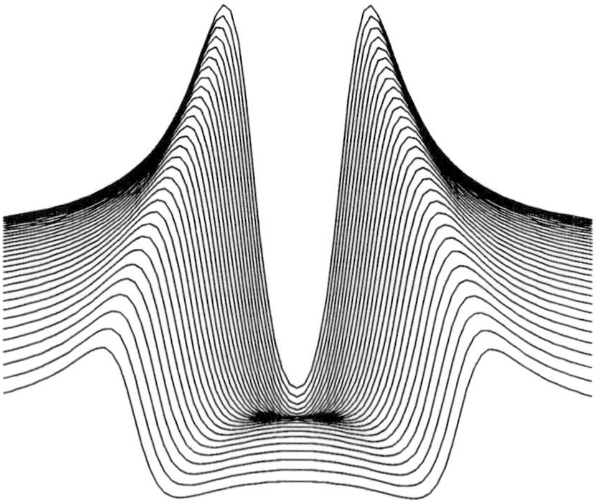

Chapter 10
TIME-VARYING PHENOMENA IN CAVITIES

Sections 6-4 and 6-5 consider the bidirectional ring laser, which is described by two oppositely-running waves of possibly different frequencies. This is a simple and useful example of multimode phenomena that can produce time variations in the laser. The present chapter considers a number of time-varying laser processes such as the relaxation oscillations in the ruby or semiconductor laser output, the build-up of multimode operation in lasers, the generation of steady-state pulse trains, and chaotic operation. Some of these time-varying processes result when you might expect single-mode steady-state operation such as in a homogeneously-broadened unidirectional ring laser. As such their multimode character can be thought of as an instability of single-mode operation. Other cases of multimode operation are expected intuitively, since different cavity modes interact to a considerable extent with different atoms, such as in a Doppler-broadened gaseous medium. While these can also be thought of as optical instabilities, we prefer to refer to them in the traditional way simply as multimode operation. In particular, this kind of operation can produce a periodic train of short pulses.

The relaxation oscillation phenomenon is treated in Sec. 10-1 and is analogous in some ways to the car turn-signal blinker. In the old days, such a blinker consisted of a bimetalic strip with the property that the two metal pieces expanded different amounts when heated, causing the strip to bend. By using the strip to switch an electric current that heated it, it disconnected its heat source when it got hot and reconnected it when it got cold again. This kind of process is called a *limit cycle* and occurs in a wide range of situations including in your heart beat. The laser relaxation oscillation works in the rate equation limit so that when the population difference is pumped high enough to compensate for the cavity losses, the field turns on, builds up rapidly, thereby driving the population difference down and removing its gain source. The simplest laser relaxation oscillations differ from the blinker and hopefully your heart in that the oscillations damp out to a steady-state value. Although a transient phenomenon, laser spiking can be treated nicely as a single-mode phenomenon.

Section 10-2 discusses the build-up of more than one mode in lasers and in absorptive optical bistability. It uses the results of Chaps. 8 and 9, which predict when probe waves experience gain in the presence of a saturating wave. When a resonant laser sidemode experiences enough gain to exceed its losses, it builds up. This buildup relies on spontaneous emission to occur and Sec. 17-3 shows how this takes place. Section 10-3 discusses the results of having a number of such modes in the cavity. The resulting time variations can give a frequency-modulated output, pulse trains useful for picosecond spectroscopy, and chaos. Section 10-4 shows how various cavity instabilities can lead to dynamical chaos, which occurs in many coupled nonlinear problems and might be mistaken for noise. In particular it gives the isomorphism derived by Haken (1975) between the Bloch equations and the Lorenz equations.

10-1. Relaxation Oscillations in Lasers

To lend analytic clothes to the intuitive explanation above of spiking in a ruby laser, we derive a coupled set of equations of motion for the single-mode laser intensity and the population difference. We use the upper to ground-state lower level model of Fig. 4-5 and include a pump term $\Lambda\rho_{bb}$ pumping the upper level a from the ground level b. This model is appropriate for the active transition in the ruby laser. An analysis of semiconductor laser relaxation oscillations can be made using the model in Sec. 6-6. The derivation of the field intensity equation differs from that in Sec. 6-2, where the atoms were adiabatically eliminated, since in the present case the population difference varies little in the cavity lifetime Q/ν. In fact, we assume that $T_2 \ll Q/\nu \ll T_1$, which allows us to use the rate equation approximation to determine the induced polarization in terms of the field amplitude and the population difference. This in turn yields two equations of motion that couple the intensity to the population difference. From Eq. (5.14) and (6.8) for central tuning and with $U_n(z) = \exp[iK_n(z)]$, i.e., a unidirectional ring laser for simplicity, we obtain the rate-equation polarization

$$\mathscr{P}_n(t) = -i(\wp^2/\gamma\hbar)E_n D\ , \tag{1}$$

where D is the population difference $\rho_{aa}-\rho_{bb}$ of Eq. (5.33), N' is the sum of the populations/volume (5.31), and $\mathscr{D}$ is the complex Lorentzian (5.25). Evaluating ρ_{ab} in the rate equation approximation, we find that Eq. (5.34) reduces to (Prob. 10-1)

$$\dot{D} = (\Lambda - \Gamma)N' - (\Lambda + \Gamma)(1 + I_n)D = -\frac{(1 + I_n)D - D_0}{T_1}, \tag{2}$$

where Λ is the rate constant for pumping the upper level from the lower level, $T_1 = (\Lambda+\Gamma)^{-1}$ for this case, and D_0 is the steady-state population difference in the absence of electric-dipole interactions

$$D_0 = N'(\Lambda - \Gamma)T_1 . \tag{3}$$

Further substituting Eq. (1) into the field self-consistency equation (6.5) and multiplying through by $2(\wp/\hbar)^2 T_1 T_2 E_n^*$, we find the dimensionless-intensity equation of motion

$$\boxed{\dot{I}_n = \frac{\nu}{Q}\left[\frac{\mathcal{N}D}{D_0} - 1\right]I_n}\ , \tag{4}$$

where the relative excitation $\mathcal{N}$ is given by

$$\mathcal{N} = \frac{\alpha_0}{\nu/2Q} \tag{5}$$

and $\alpha_0 = \nu\wp^2 D_0/\hbar\epsilon_0\gamma$.

Let us repeat the relaxation oscillation description armed with the coupled equations of motion (2) and (4). Initially we suppose that the intensity I_n is zero because the population difference D is too small to compensate for the cavity losses ν/Q in Eq. (4). If the pumping rate Λ is greater than the population-difference decay rate Γ, Eq. (2) shows that D builds up linearly in time. This build-up is slow since it is characterized by the population-difference lifetime $T_1 = (\Lambda+\Gamma)^{-1}$. When D gets sufficiently large that the gain $2\alpha_0\mathcal{L}$ exceeds the losses ν/Q, the intensity builds up rapidly according to Eq. (4). This makes the damping term $(\Lambda+\Gamma)I_n\mathcal{L}$ in Eq. (2) large, forcing the population difference down and killing the gain. The intensity decays away in the cavity lifetime $Q/\nu \ll T_1$, allowing D to build up again, repeating the cycle.

We can gain an analytic feel for these equations by analyzing the stability of the steady-state solutions ($\dot{I} = \dot{D} = 0$) of Eqs. (2) and (4) using a linear stability analysis. Two steady states exist, namely

$$I^{(s)} = 0\ , \quad D^{(s)} = D_0\ , \tag{6}$$

and

$$I^{(s)} = \mathcal{N} - 1\ , \quad D^{(s)} = \frac{D_0}{\mathcal{N}}\ . \tag{7}$$

The second solution is physical only if it predicts that $I^{(s)} > 0$. With the intensity and population difference written as $I_n(t) = I^{(s)} + \epsilon(t)$ and $D(t) = D^{(s)} + d(t)$, respectively, Eqs. (2) and (4) give

$$\frac{d}{dt}\begin{bmatrix} \epsilon(t) \\ d(t) \end{bmatrix} = \begin{bmatrix} 2\alpha_0 D^{(s)}/D_0 - \nu/Q & 2\alpha_0 I^{(s)}/D_0 \\ -D^{(s)}/T_1 & -(1+I^{(s)})/T_1 \end{bmatrix} \begin{bmatrix} \epsilon(t) \\ d(t) \end{bmatrix} . \tag{8}$$

Note that since the electric field amplitude is complex, two equations are needed in general to analyze the field stability. Here we assume central tuning of a single mode to allow only an intensity equation to be used. If either eigenvalue of Eq. (8) has a positive real part for a given steady-state solution, that solution is unstable. Problem 10-1 shows that the solution (6) has the eigenvalues

$$\lambda = 2\alpha_0 - \nu/Q, \quad -1/T_1 . \tag{9}$$

For a medium with sufficient gain to lase, the first eigenvalue is positive, which leads to a buildup of the mode, i.e., the solution (6) is unstable. Problem 10-2 also shows that the solution (7) has the eigenvalues

$$\lambda = -\mathcal{N}/2T_1 \pm i\sqrt{(\mathcal{N}-1)\nu/QT_1 - (\mathcal{N}/2T_1)^2}\,, \tag{10}$$

which have negative real parts for $\mathcal{N} > 1$. Hence the solution (7) is stable under lasing conditions. Depending on the sign of the argument of the square root, the solution may or may not show oscillations as it converges to the steady state. Problem 10-3 shows that oscillations occur if the relative excitation $\mathcal{N}$ satisfies

$$\mathcal{N} > 1 + \frac{1}{4T_1\nu/Q} . \tag{11}$$

In situations where T_1 is large compared to the cavity lifetime, such as e.g. in Ruby lasers, this condition is almost always obeyed if the laser oscillates at all. Note however that it is *not* obeyed in many other lasers, such as the He-Ne laser, for which relaxation oscillations do not occur.

During the evolution of the laser intensity between the unstable solution of Eq. (6) and the final solution of Eq. (7), damped relaxation oscillations occur, as illustrated by the numerical integrations graphed in Fig. 10-1. Note that a relation crucial to the stability of the solution (7) is that the dipole decay time T_2 is short compared to the level and cavity decay times, and hence that the rate equation approximation can be made. As discussed in Sec. 10-4, if the decay times are comparable, the rate equation approximation cannot be made and chaotic operation may ensue.

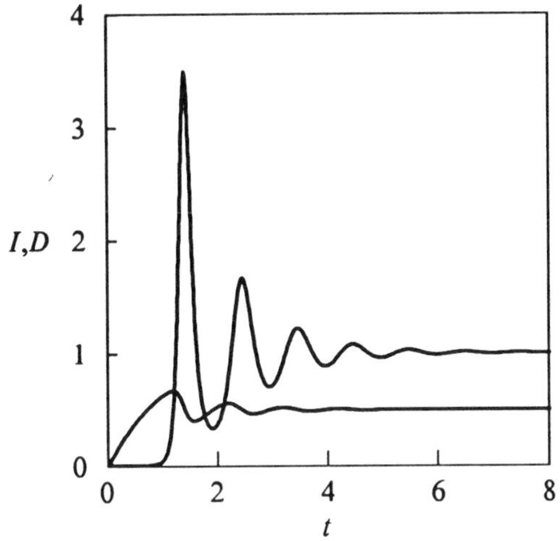

Fig. 10-1. Intensity I_n and population difference D/D_0 (curve with smaller final value) versus time as given by the coupled relaxation-oscillation equations (2) and (4) for $\mathcal{N} = 2$, $\nu/Q = 40$, and $T_1 = 1$.

In this model, we take the population-difference lifetime T_1 to be long compared to the cavity and dipole lifetimes. Hence the eigenvalues of Eq. (10) are small, causing the laser to converge slowly with low frequency oscillations to the steady state Eq. (7). As such the laser operates very near an instability that would occur if the real part of the eigenvalues could actually go through zero. This convergence can be prevented by periodically varying some laser parameter such as the loss or pump. In fact if the magnitude and frequency of the variation are comparable to the real and imaginary parts, respectively, of the eigenvalues in Eq. (10), the laser may frequency lock to the variation. Other variations can lead to chaotic behavior. This is an important problem in the semiconductor laser, which is characterized by a long T_1. For further discussion of these effects, see Erneux *et al.* (1987) and Boyd *et al.* (1986) and references therein.

The long-T_1 (Rate Equation Approximation) limit considered here is the same as that considered in Sec. 8-2. The damped oscillations that occur are due to the decay of sidemodes that correspond to the same cavity resonance (wavelength) as the main mode. The sidemode frequencies differ from the main mode frequency by Im{λ} given by Eq. (10). With a longer T_2 and/or with inhomogeneous broadening, such sidemodes can grow instead of decaying. This phenomenon is sometimes called *mode splitting*, and is discussed in greater detail in the next section.

10-2. Stability of Single-Mode Laser Operation

A basic question in laser physics is, can sidemodes build up in the presence of a single *cw* oscillating mode, that is, is single-mode operation stable? Typically all modes compete for the gain provided by the amplifying systems. As we have seen in Sec. 6-4 on the bidirectional ring laser, a mode not only saturates its own gain (self saturation), but it may also saturate the gain of other modes (cross saturation). Multimode operation occurs relatively easily in lasers with inhomogeneous broadening and/or standing waves, since different cavity modes interact at least in part with different groups of atoms. Similarly multiple transverse modes can often coexist, since they also interact in part with different atoms. To suppress such multiple transverse mode operation, experimentalists put apertures in the laser cavity, thereby greatly increasing the higher-transverse-mode losses without appreciably affecting the fundamental mode. To suppress multiple longitudinal mode operation, a second etalon can be introduced, requiring an oscillating mode to be simultaneously resonant for two cavities.

It is less obvious that a sufficiently intense single mode of a homogeneously broadened unidirectional ring laser or optical bistability cavity can also lead to multimode operation. Intuitively one would think that since all modes interact with precisely the same atoms, only that mode with the highest net gain would oscillate. Nevertheless both for the laser and for absorptive optical bistability, theory predicts that sidemodes may grow, each with a distinct wavelength (multiwavelength instability). In addition, for the laser multimode operation may occur under conditions such that the induced anomalous dispersion allows three frequencies to correspond to the same wavelength. This single-wavelength instability may become chaotic as discussed in Sec. 10-4.

The origin of these homogeneously-broadened single-mode instabilities is population pulsations, which are generated by the response of the medium to the mode beat frequencies or equivalently to the oscillating field fringe components. These pulsations can scatter energy from the oscillating mode into the sidemode(s), and may amplify them even in an otherwise absorbing medium as discussed in Sec. 8-1. We also saw this kind of coherent mode interaction in a degenerate frequency case in the homogeneously broadened bidirectional ring laser of Sec. 6-4 and in the Ikeda instability of Sec. 7-3. In the ring laser, the cross-saturation coefficient θ_{+-} of Eq. (6.44) is twice as big as the self-saturation coefficient β_{+}. For this case the response of the medium to the stationary field fringe pattern is just as large as the ordinary gain saturation, and thereby doubles the cross saturation and leads to bistable single-mode operation. Similarly the degenerate probe absorption coefficient given by $\alpha_1 = \alpha_0\mathscr{D}_2/(1 + I_2\mathscr{L}_2)^2$ of

Eq. (9.9) is significantly smaller than the single-mode absorption coefficient $\alpha_2 = \alpha_0\mathscr{D}_2/(1 + I_2\mathscr{L}_2)$ of Eq. (5.27). Hence coherent modes compete with one another in a more complicated fashion than by simple gain saturation, since the medium may respond to the mode interference patterns.

Multimode operation in homogeneously-broadened unidirectional configurations is difficult to obtain experimentally. However with the addition of some inhomogeneous broadening, the single-wavelength laser case is readily observed, e.g., in the He-Xe laser (see Casperson (1978) and Abraham (1985)).

In general to determine whether a sidemode can build up in the presence of a single strong mode, we ask the two questions, "Does the sidemode experience more gain than losses, and is it resonant?" For the single-wavelength instability, the nonlinear index contributions are just as critical as the gain contributions since more than one frequency must be made resonant for the same wavelength. Sections 8-1 and 8-2 allow us to answer these questions by providing us with the complex probe polarizations $\mathscr{P}_n$ to insert into the multimode self-consistency equations (6.5) and (6.6). Chapter 8 uses the complex notation $\mathscr{E}_n$ instead of Chap. 6's $E_n \exp(i\phi_n)$. Switching to the latter, we find $\mathscr{P}_1$ by comparing the coupled mode equation (9.20) with the slowly-varying Maxwell Eq. (5.3) using the absorption coefficient α_1 of Eq. (8.28) and the coupling coefficient χ_1 of Eq. (9.21) (integrated over ω'). This gives

$$\mathscr{P}_1 = \frac{\wp^2}{\hbar}\int_{-\infty}^{\infty} d\omega' \mathscr{W}(\omega')\,\frac{\gamma\mathscr{D}_1}{1 + I_2\mathscr{L}_2}$$

$$\times \left[E_1 - \frac{I_2\mathscr{F}(\Delta)\frac{\gamma}{2}[(\mathscr{D}_1 + \mathscr{D}_2^*)E_1 + (\mathscr{D}_2 + \mathscr{D}_3^*)E_3 e^{i\Psi}]}{1 + I_2\mathscr{F}(\Delta)\frac{\gamma}{2}(\mathscr{D}_1 + \mathscr{D}_3^*)}\right], \qquad (12)$$

where the relative phase angle Ψ is given by

$$\Psi = (\phi_2 - \phi_1) - (\phi_3 - \phi_2). \qquad (13)$$

$\mathscr{P}_3$ is given by this formula with the interchange $_1 \longleftrightarrow {}_3$, which implies $\Delta \rightarrow -\Delta$. Equation (12) is valid for unidirectional plane waves, although by including appropriate spatial mode factors $U_n(x,y,z)$, it can be generalized to treat standing waves and transverse mode interactions as well. Note that the relative phase angle Ψ must be determined in a self-consistent fashion. For the symmetric tuning implied by $\nu_2 = \omega$ and equal mode losses, Ψ vanishes and $E_1 = E_3$, which is the AM operation described at the end of Sec.

8-1. For this case, the sidemode fields E_1 and E_3 grow together if the sidemode gain $-(\nu/2\epsilon_0)\text{Im}\{\mathscr{P}_1\}$ from Eq. (12) exceeds the loss $(\nu/Q_1)E_1$. Examples of the corresponding gain/absorption spectra are given in Fig. 9-7. For a lasing medium, the gain is given by positive values in Fig. 9-7, while for an absorbing medium it is given by negative values.

More generally, Eqs. (6.5) and (6.6) with the polarization (12) provide a theory equivalent to a linear stability analysis, provided the mode frequencies include an imaginary part interpreted as the mode net gain or loss. Hence the beat frequency Δ acquires an imaginary part, so that it contains the complete linear time response, oscillatory and growth/decay, of the sidemodes. These equations treat the strong mode to all orders in perturbation theory, although the incipient modes are treated only to first order. The corresponding quantized sidemode theory given in Secs. 15-4 and 16-2 allows one to study how sidemodes build up from quantum noise. We have already discussed how when different cavity modes interact at least in part with different groups of atoms, the mode competition is reduced and more than one mode may oscillate. For more detailed discussions of such multimode operation, see Sargent, Scully, and Lamb (1977). Here we restrict our discussion to two situations: 1) the *multiwavelength instabilities* in a homogeneously-broadened unidirectional ring laser, and 2) *single-wavelength instabilities*.

By multiwavelength instabilities, we mean that each mode frequency corresponds to a different passive cavity mode. Hence the instability problem reduces to finding cavity sidemodes with gain that can grow. If the oscillating mode is centrally tuned, pairs of modes grow together. If the oscillating mode is detuned, a single mode may grow alone, or an asymmetric pair may grow. The role of the frequency equations (6.3) in these cases is to determine the relative phase Ψ between the modes, which can significantly influence the growth of a mode pair. For example, if $\Psi = -\pi$ and $E_1 = E_3$, the population pulsation part of Eq. (12) (that with the $\mathscr{F}$) cancels out completely. In general the pair problem reduces to solving the coupled-mode equations.

In addition to the multiwavelength case, operation on a single wavelength may also be unstable. In this case, population pulsations create a region of anomalous dispersion that allows three frequencies to correspond to one wavelength as shown in Fig. 10-2 . If these frequencies also experience net gain (gain exceeding the cavity losses), they can build up. To have net gain, the cavity linewidth ν/Q has to be comparable to or broader than the atomic linewidth. This kind of instability is marginal unless inhomogeneous broadening is added. Such broadening both reduces the mode competition and increases the anomalous dispersion by contributing a term something like the derivative of the hole burned into the gain line.

To quantify the single-wavelength instability, consider central tuning of the oscillating mode ($\nu_2 = \omega$), for which the sidemode polarizations $\mathscr{P}_1$

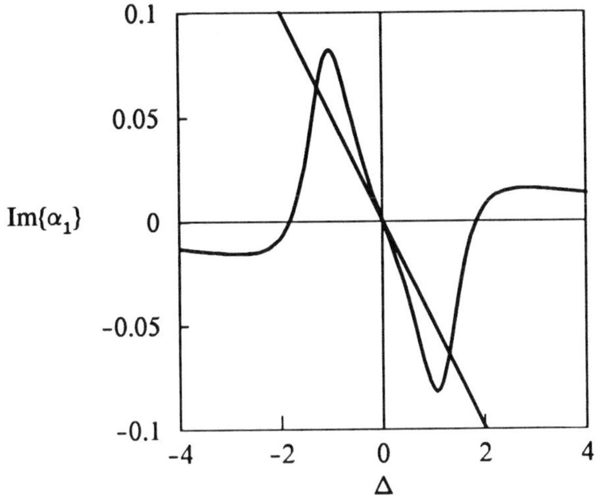

Fig. 10-2. Nonlinear anomalous dispersion $\eta(\nu_n) - 1 \propto \text{Im}\{\alpha_1\}$ versus intermode beat frequency. The index of refraction $\eta(\nu_n)$ also has to obey the relation $\eta(\nu_n) = c/\lambda\nu_n$, where the wavelength λ is fixed by the cavity boundary conditions. The diagonal line depicts this second η dependence on frequency. The allowed oscillation frequencies are given by the intersections between the curve and the straight line, here yielding three frequencies resonant for a single wavelength. Equation (8.17) with a doubled α_{coh} contribution (for *two* sideband operation) was used for $\text{Im}\{\alpha_1\}$ with $\alpha_0 = -1$ (gain), $I_2 = 16$, $T_1 = 10$, $T_2 = 1$, and $\nu_2 = \omega$. The slope of the diagonal line depends on ν/Q and the relative excitation.

and $\mathcal{P}_3$ given by Eq. (12) are complex conjugates of one another. We can use these polarizations to find the eigenvalues of the instability. By symmetry, we conclude that $E_1 = E_3$ and that the relative phase angle $\Psi = 0$ or π. In terms of the complex amplitude $\mathcal{E}_n$, the field self consistency equations (6.5) and and (6.6) can be combined into the single complex equation

$$\frac{d\mathcal{E}_n}{dt} = -[\nu/2Q_n + i(\Omega_n - \nu_n)]\mathcal{E}_n - i(\nu/2\epsilon_0)\mathcal{P}_n \, . \tag{14}$$

In steady state, we find the dispersion relation

$$\kappa + i(\Omega_n - \nu_n) = -i(\nu/2\epsilon_0)\mathcal{P}_n/\mathcal{E}_n \, , \tag{15}$$

where $\kappa \equiv \nu/2Q_n$. Taking $n = 1$, substituting $\delta = \Omega_2 - \Omega_1$, $\Delta = \nu_2 - \nu_1 + ig$ (to study the possible growth of instabilities, we include an imaginary part giving the sidemode net gain or loss in the Δ in $\mathcal{P}_1$), and setting $\Omega_2 = \omega = \nu_2$, we find

$$\Delta^3 - i(\kappa + \gamma + T_1^{-1} + i\delta)\Delta^2 - [(1 + I_2)\gamma T_1^{-1} + \kappa T_1^{-1} + i\delta(\gamma + T_1^{-1})]\Delta$$
$$+ 2iI_2\kappa\gamma T_1^{-1} - (1 + I_2)\gamma T_1^{-1}\delta = 0 \,.$$

Here we have set $g = 0$, which gives the instability edge of the laser. In particular setting $\delta = 0$ (single cavity wavelength) and substituting $\lambda = i\Delta$, we find the eigenvalue equation

$$\lambda^3 + (\kappa + \gamma + T_1^{-1})\lambda^2 + [(1 + I_2)\gamma T_1^{-1} + \kappa T_1^{-1}]\lambda + 2I_2\kappa\gamma T_1^{-1} = 0. \quad (16)$$

Problem 10-4 applies the Hurwitz criterion to this equation to predict that the steady-state single-mode intensity $I_2 = \mathcal{N} - 1$ is unstable if

$$\kappa > \gamma + T_1^{-1} \text{ (bad cavity)}$$
$$I_2 > \frac{(\gamma + T_1^{-1} + \kappa)(\gamma + \kappa)}{\gamma(\kappa - \gamma - T_1^{-1})} \,. \quad (17)$$

Haken (1975) showed that this instability condition corresponds to chaos in the Lorenz model.

Our discussion outlines various ways in which a single mode may become unstable. It does not predict what may happen once one or more sidemodes become strong enough to saturate the medium. If the mode intensities are much less than the saturation intensity, a multimode third-order theory can be used (see Sargent, Scully, and Lamb (1977)). In general the coupled atom-field equations can be integrated numerically. In the remainder of this chapter, we consider various cases of multimode operation, first periodic in nature, and then chaotic.

10-3. Multimode Mode Locking

Section 6-5 introduces the idea of mode locking and considers the frequency locking of the oppositely directed running waves in a bidirectional ring laser. Three or more longitudinal modes can also lock: in such cases all mode frequencies are separated by integral multiples of a common adjacent mode separation (in essence the adjacent mode beat frequencies lock to the same value), and the relative phases of the modes are fixed. In particular, three-mode mode locking can be described directly by the analysis of Sec. 6-5. In this section, we discuss this problem, and make some observations about multimode mode locking in general.

When three adjacent modes are oscillating, a detector ordinarily registers two distinct low-frequency beat notes at frequencies $\nu_2 - \nu_1$ and $\nu_3 - \nu_2$, which are approximately equal to the cavity mode separation $c/2L$. In addition there is a very low-frequency (even audio) tone resulting from the beat between these beats, namely, a signal at frequency $(\nu_3 - \nu_2) - (\nu_2 - \nu_1)$ which is generated by the nonlinearities in the medium. The last of these is represented in the theory by the very slowly varying relative phase angle

$$\Psi = (2\nu_2 - \nu_1 - \nu_3)t - (2\phi_2 - \phi_1 - \phi_3)\,, \tag{18}$$

which is a generalization of the Ψ of Eq. (13) to include the possibility that the mode frequencies may not be evenly spaced. In fact, the ν's are close to the passive-cavity frequencies Ω_n, which satisfy $2\Omega_2 - \Omega_1 - \Omega_3 = 0$. The Ψ terms arise from the beating between modes in the nonlinear medium to produce *combination tones* with frequencies very nearly equal to the mode frequencies themselves. Such combination tones are a multimode generalization of the coupling terms considered in multiwave mixing in Chaps. 8 and 9. For example, the mode at frequency ν_2 beats with that at ν_3 forming the population pulsation with frequency $\nu_2 - \nu_3$ which, in turn, interacts with mode 2 to give the tone $\nu_1' = 2\nu_2 - \nu_3$. This is very nearly equal to ν_1 and contributes in third-order to the complex polarization for mode 1. This is illustrated in Fig. 10-3, which shows four such combination tones. The difference $\nu_1' - \nu_1$ is $\dot{\Psi}$, and we see that ν_1' acts very much like an injected signal that tempts mode one to oscillate at the frequency ν_1', that is, to frequency lock such that $\nu_1 = \nu_1' = \nu_2 - (\nu_3 - \nu_2)$. This condition can be rewritten as $\nu_2 - \nu_1 = \nu_3 - \nu_2$, i.e., the beat notes between adjacent modes are equal. This is multimode mode locking. It differs from other kinds of mode locking such as that for the ring laser in Sec. 6-5 in that it is the beat frequencies between the modes that lock to one another rather than the mode frequencies themselves.

Taking the time rate change of Eq. (18) and using Eq. (6.4), we find the equation of motion

$$\dot{\Psi} = \frac{\nu}{2\epsilon_0}\,\mathrm{Re}\left\{\frac{2\mathscr{P}_2}{E_2} - \frac{\mathscr{P}_1}{E_1} - \frac{\mathscr{P}_3}{E_3}\right\}. \tag{19}$$

The polarization $\mathscr{P}_1$ of Eq. (12) gives the lowest-order values of the polarization for sidemodes subject to an arbitrarily intense pump. For the mode-locking problem, we need higher-order values. It is beyond the

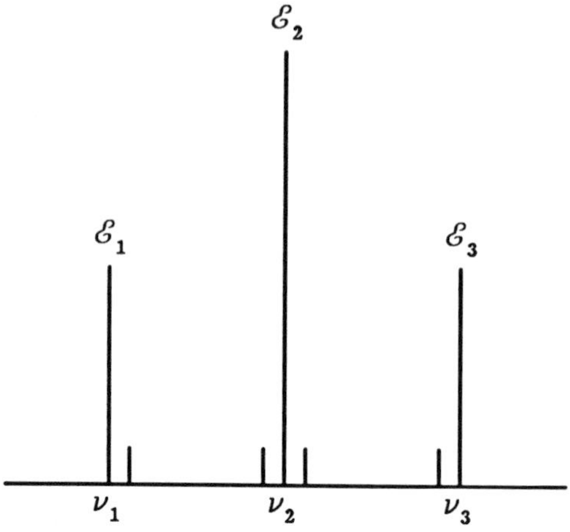

Fig. 10-3. Diagram of three-mode operation showing the amplitudes $\mathcal{E}_1$, $\mathcal{E}_2$, and $\mathcal{E}_3$ and oscillation frequencies ν_1, ν_2, and ν_3. The small lines drawn next to the mode intensities represent combination tones. Their placement here is suggestive only.

scope of this book to derive such values (see Sargent *et al* (1977) for a thorough derivation of third-order values). Suffice it to say that Eq. (19) yields an equation of the same form as Eq. (6.69) for the ring laser Ψ. For the three-mode case, the a factor consists of differences between the net mode pulling coefficients generated in the nonlinear medium, and the b term consists of a superposition of the combination-tone terms.

As the second mode is tuned toward line center, it sees less and less mode pulling. Furthermore the mode pulling for modes one and three become equal and opposite. Consequently for nearly central tuning of mode two, the a in Eq. (6.69) becomes smaller in magnitude than b, and mode locking takes place. In fact, one can perform a simple experiment that illustrates this with a cheap He-Ne laser that runs in three modes. Upon being turned on, this laser heats up, changing its length over a period of several hours. This causes the laser to slowly scan its frequencies. By hooking up the output of a detector that looks at the laser light to an audio amplifier and speaker, one can hear the beat frequency between the adjacent -mode beat notes ($\dot{\Psi}$) as it tunes down from the upper to the lower limits of one's hearing, then hear nothing for a while (mode locked) and then hear the frequency tune in the opposite direction. A scanning interferometer reveals that the mode amplitudes dance about during the unlocked intervals, while they become quite steady in the "DC" intervals.

Taking $E_1 = E_3$, we find $\Psi_0 \to 0$ or π for $b < 0$ or $b > 0$, respectively. The value $\dot{\Psi} = 0$ requires $\phi_2 - \phi_1 = \phi_3 - \phi_2$. Choosing the time origin so that $\phi_2 = \phi_1$, we have $\phi_1 = \phi_2 = \phi_3$, that is, at points periodic in time spaced by the interval $2\pi/\Delta$, the field phasors $E_n \exp[-i(\nu_n t+\phi_n)]$ add constructively. At other times they tend to cancel one another. This is three-mode pulsing (or AM operation). The time widths of the pulses are inversely proportional to the mode spacing (as is the interval between pulses), but the degree to which there is destructive interference between pulses is limited by the small number of modes. The second value $\Psi = \pi$ leads to the phase relation $\phi_1 = \phi_2 = \pi + \phi_3$, which corresponds to smaller variations (even nearly no variations) in the field intensity (called FM operation in analogy to the corresponding phase relation in FM radio). Both cases occur in three-mode locking, depending on the placement of the medium in the cavity. For further discussion of the three-mode case, see Sargent (1976).

The multimode case for more than three modes has the same general physical phenomena as the three mode, namely population pulsations that act as modulators in generating sidebands that, in turn, act as injected signals attempting to lock the laser. The most important special case is that for AM (pulsed) operation with its potentially very sharp, subpicosecond pulses. One useful relation is that between the time and frequency domains. Specifically the mode domain consists of a product of a Dirac comb (delta functions separated by the adjacent mode spacing Δ) and an amplitude envelope. Hence the time domain is given by a convolution of another Dirac comb, this one with spacing $2\pi/\Delta$, and an envelope equal to the Fourier transform of the amplitude envelope in the mode domain. This convolution is just a train of pulses separated in time by $2\pi/\Delta$ and having widths roughly (for Gaussian envelopes, exactly) inversely proportional to the inverse of the width in the mode domain.

The medium may mode lock spontaneously due to the combination tones, but typically special methods are used to force locking. One way is to include a saturable absorber within the cavity. Perhaps the easiest way to see how this works is directly in the time domain. Imagine that the noise in the initial laser radiation consists of random fluctuations or "minipulses." A group of these concentrating sufficient energy within the short relaxation time of the saturable absorber effectively "opens" the absorber, allowing radiation to pass unattenuated for a short time. Groups with smaller energy are absorbed. The most energetic packet builds up at the expense of the others and bounces back and forth in the cavity, producing a train of pulses. For further discussion of this and other multimode mode locking techniques, see Smith, Duguay, and Ippen (1974).

10-4. Single-Mode Laser and the Lorenz Model

So far we have discussed the stability edge of nonlinear optical systems using the concept of linear side-mode gain. Although instructive in that they teach us the range of parameters for which the system is unstable, linear stability analyses do not give any indication of the dynamics of the system once the side-modes build-up significantly.

An unstable nonlinear system can evolve toward another fixed point or toward a limit cycle (undamped periodic oscillations). Alternatively, it can exhibit an irregular behavior reminiscent of a system driven by random forces. This noisy-looking dynamics occurring in a purely deterministic system is called *deterministic chaos*. It occurs commonly in nonlinear systems, although only recently has its detailed study been made possible by powerful modern computers. Chaos can occur in just about any nonlinear difference equation such as in the Ikeda's instability (Sec. 7-3), and in any coupled system of at least three autonomous ordinary differential equations, a consequence of the Poincaré-Bendixon theorem. A detailed discussion of chaos, its various scenarios, universality aspects, etc. is beyond the scope of this book; a partial list of references relevant to quantum optics is given at the end of this Chapter. Here we briefly discuss the Lorenz model/single-mode laser, which illustrates one of the powerful aspects of the study of chaos: generic features that are learned from one field - here hydrodynamics - can readily be adapted to another - laser physics.

The original purpose of the Lorenz equations was to provide a simple model of the Bénard instability, which occurs when a fluid is heated from below and kept at a constant temperature from above. For small temperature gradients, the fluid remains quiescent, but it starts a macroscopic motion as the gradient approaches a critical value. The heated parts of the fluid expand, move up by buoyancy, cool and fall back to the bottom: a regular spatial pattern appears out of a completely homogeneous state.

The fluid motion is described by the nonlinear, partial differential Navier-Stokes equations of hydrodynamics. By introducing a spatial Fourier decomposition of the velocity and temperature fields in the fluid, Lorenz derived a set of truncated dimensionless equations coupling just one component x of the velocity field to two components y and z of the temperature field (for more details, see Haken (1978)). These equations are

$$\frac{dx}{dt} = \sigma y - \sigma x ,$$

$$\frac{dy}{dt} = -xz + rx - y \tag{20}$$

$$\frac{dz}{dt} = xy - bz .$$

Here σ is the Prandtl number and $r = R/R_c$, where R is the Rayleigh number and R_c the critical Rayleigh number.

Haken (1975) showed that Eqs. (20) are mathematically equivalent to the equations governing the dynamics of a homogeneously broadened, single-mode unidirectional ring laser operating at line center. This is best seen in the U, V, W Bloch vector notation of Sec. 4-3. For central tuning we can assume $U = 0$, and the remaining Bloch equations (4.49) and (4.50) are

$$\frac{dV}{dt} = - V/T_2 + (\wp E/\hbar)\, W \,, \tag{21a}$$

$$\frac{dW}{dt} = - (W + 1)/T_1 - (\wp E/\hbar)\, V \,, \tag{21b}$$

coupled to the slowly-varying amplitude equation (6.5)

$$\frac{dE}{dt} = - \frac{\nu}{2Q} E - \frac{\wp\nu N'}{2\epsilon_0} V \,, \tag{21c}$$

where we consider upper to ground state transitions. We can prove the equivalence of the single-model laser and Lorenz equations by rescaling (21) to the fixed-point values V_s, W_s and E_s obtained by setting $d/dt = 0$. With the new variables $x = \alpha E/E_s$, $y = \alpha V/V_s$, $z = -(W+1)/W_s$, $\alpha = (-(1+W_s)/W_s\, T_2/T_1)^{1/2}$ we obtain precisely Eqs. (20), except that now $\sigma = T_2\nu/2Q$, $r = -1/W_s$, $b = T_2/T_1$ and time is now in units of T_2.

The stability of the fixed points of (20) can be determined by linear stability analysis. The origin (0,0,0) is a fixed point, and it is the only one for $r < 1$. For $r > 1$ there are two additional fixed points $x^* = y^* = \pm[b(r-1)]^{1/2}$, $z^* = r - 1$, which correspond to the laser threshold. These fixed points in turn become unstable for $\sigma > b+1$ and $r > \sigma(\sigma+b+3)/(\sigma-b-1)$. For $\sigma = 3$ and $b = 1$, for instance, this criterion is satisfied whenever $r > 21$. At this point, the real parts of the complex eigenvalues of the linearized problem cross the real axis to the right. This is called a Hopf bifurcation.

Computer simulations are necessary to go much beyond this stage. Figure 4 plots the time evolution of the scaled polarization $y(t)$ for $\sigma = 3$, $b = 1$ and various values of the order parameter r. In each case the initial conditions are given by $x(0) = y(0) = z(0) = 1$. For $r = 15$, the origin (0,0,0) is unstable, but the fixed point ($\sqrt{42}$, $\sqrt{42}$, 14) is stable. For $r = 21$, $y(t)$ begins as an orderly but growing oscillation, but then breaks into chaos, see Fig. 10-4*b*. A similar behavior is found for $r = 22$. Figure 10-5 shows plots of x *vs.* y for the same case as Fig. 10-4*b*. Although trapped in a finite region of phase space, the system's trajectory does not evolve toward a fixed point or a limit cycle. On the other hand, the laser is a dis-

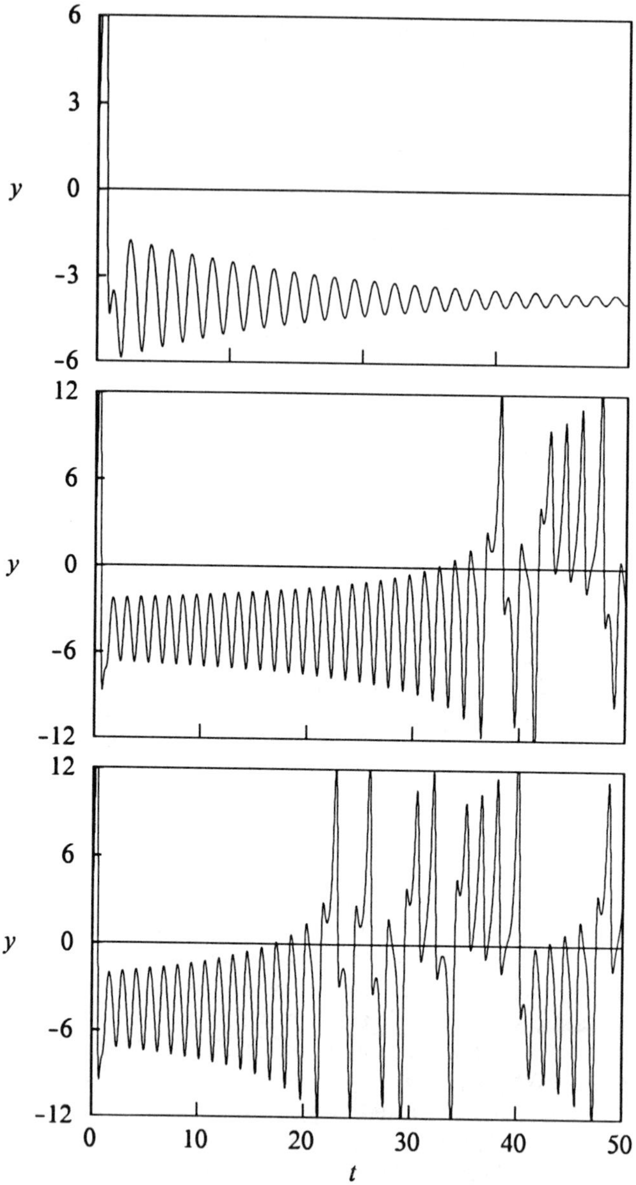

Fig. 10-4. Scaled polarization $y(t)$ of Eq. (20) versus t for (from top to bottom) $r = 15$, 21, and 22. Other parameters are $x(0) = y(0) = z(0) = 1$, $\sigma = 3$, $b = 1$.

sipative system and hence the volume of phase space region it asymptotically covers must have zero volume. Trajectories such as shown in Fig. 10-5 are thus said to cover *a strange attractor*. characterized by a zero volume, by one or more *fractal dimensions* which help distinguish them from points or segments. (One of the more well-known fractals in the Cantor set.)

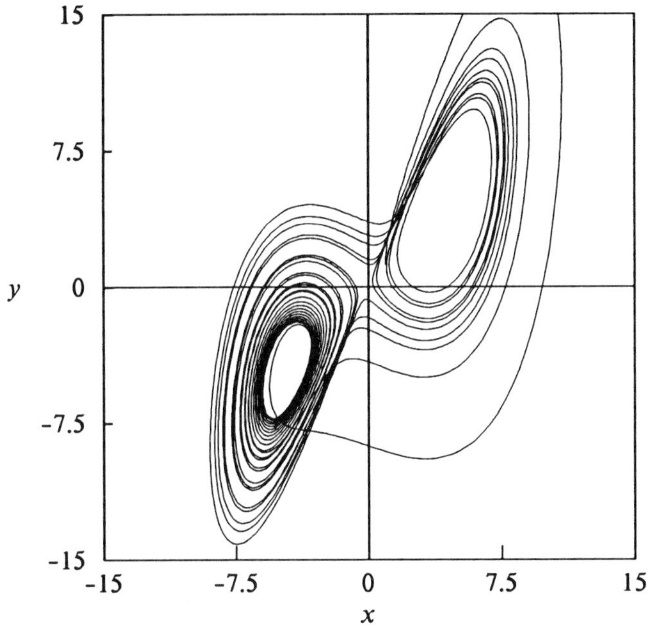

Fig. 10-5. x of Eq. (20) versus y for the same parameters as in Fig. 10-4 with $r = 21$.

These results generate a series of questions of mathematical as well as physical significance. Ideally one would like to be able to predict the behavior of trajectories for any set of initial conditions and parameter values. This is very much an open question in general. The Lorenz equations play a central role in this task, and have encouraged the development of techniques to study more and more complicated and higher dimensional models.

References

For a review of instabilities and chaos in optical systems, see Milonni, P. W., M. L. Shih, J. R. Ackerhalt (1987), *Chaos in Laser-Matter Interactions*, World Scientific Publishing Co., Singapore.

For an overview of laser instabilities including many references to the early literature, see N. B. Abraham, L. A. Lugiato, and L. M. Narducci (1985), J. Opt. Soc. B2, January issue. The semiclassical treatment most closely corresponding to Chaps. 8 - 10 is given in this same issue by S. Hendow and M. Sargent III starting on *p*. 84.

Abraham, N. B., L. Narducci, and P. Mandel (1988), *Progress in Optics.*

Boyd, R. W., M. G. Raymer, and L. M. Narducci, Eds. (1986), *Optical Instabilities*, Cambridge University Press, Cambridge.

Important early references to instabilities in lasers with homogeneously broadened media include:

Grazyuk, A. Z. and A. N. Oraevski (1964), in *Quantum Electronics and Coherent Light*, P. A. Miles, *ed*., Academic Press, New York.
Haken, H. (1966), Z. Physik **190**, 327.
Risken, H., C. Schmidt, and W. Weidlich (1966), Z. Physik **194**, 337.
Graham, R. and H. Haken (1966), Z. Physik **213**, 420.
Risken, H. and K. Nummedal (1968), J. Appl. Phys. **39**, 4662.

The corresponding theory for optical bistability was given by:

Bonifacio, R. and L. A. Lugiato (1978), Lett. Nuovo Cimento **21**, 505. The relationship to the three-mode semiclassical theory is explained in Ref. 8 and M. Gronchi, V. Benza, L. A. Lugiato, P. Meystre, and M. Sargent III (1981), Phys. Rev. A**24**, 1419.

Other references made in this Chapter are:

Casperson, L. W. (1978), IEEE J. Quant. Elect. **QE-14**, 756. This paper shows how more than one laser frequency can have gain and be resonant for the *same* wavelength.

Erneux, T., S. M. Baer, and P. Mandel (1987), Phys. Rev. A**35**, 1165.

Haken, H. (1975), Phys. Lett. **53A**, 77. This paper gives the isomorphism between the Maxwell-Bloch and Lorenz equations.

Haken, H. (1985), *Light, Vol. 2, Laser Light Dynamics*, North-Holland Physics Publishing, Amsterdam.

Haken, H. (1978), *Synergetics, An Introduction*, Springer-Verlag, Heidelberg, *p*. 123.

Sargent, M. III (1976), in *Applications of Lasers to Atomic and Molecular Physics, Proc. Les Houches Summer School*, Ed. by R. Balian, S. Haroche, and S. Lieberman, North-Holland, Amsterdam.

Sargent, M. III, M. O. Scully, and W. E. Lamb, Jr. (1977), *Laser Physics*, Addison-Wesley, Reading, MA.

Smith, P. W., M. A. Duguay, and E. P. Ippen (1974), *Mode Locking of Lasers*, Pergamon Press, Oxford.

Problems

10-1. Show that the rate equation (2) for $\dot{D}$ follows from the coherent version Eq. (5.34).

10-2. Show that the solution (7) to the linear-stability equation (8) has the eigenvalues given by Eq. (10).

10-3. Show that relaxation oscillations occur near threshold ($\mathcal{N} \simeq 1$) if the relative excitation $\mathcal{N}$ satisfies Eq. (11). Derive a more general condition for oscillations to occur.

10-4. Consider the polynomial

$$f(\lambda) = c_0\lambda^n + c_1\lambda^{n-1} + \ldots + c_n = 0.$$

where the coefficients c_n are real. The Hurwitz criterion states that all zeros of this polynomial have negative real parts if and only if the following conditions are fulfilled:

(*a*) $\frac{c_1}{c_0} > 0,\ \frac{c_2}{c_0} > 0, \ldots.\ \frac{c_n}{c_0} > 0$

(*b*) The principal subdeterminants H_j of the quadratic scheme

$$\begin{matrix} c_1 & c_0 & 0 & 0 & \cdots\cdots & 0 & 0 \\ c_3 & c_2 & c_1 & c_0 & \cdots\cdots & 0 & 0 \\ c_5 & c_4 & c_3 & c_2 & \cdots\cdots & 0 & 0 \\ \cdots & \cdots & \cdots & \cdots & \cdots & \cdots & \cdots \\ 0 & 0 & 0 & 0 & & c_{n-1} & c_{n-2} \\ 0 & 0 & 0 & 0 & & 0 & c_n \end{matrix}$$

that is, $H_1 = c_1$, $H_2 = c_1c_2 - c_0c_3$, ..., $H_n = c_n H_{n-1}$, are all positive. Apply this criterion to the eigenvalue equation (16) to show the stability conditions of Eq. (17).

10-5. Given the dispersion relation (16), derive the eigenvalue Eq. (16).

10-6. Derive an eigenvalue equation like Eq. (25) but for an arbitrary α in Eq. (18) instead of the Kerr nonlinearity of Eq. (7.11). Express your answer in terms of $\partial\alpha_2/\partial\mathcal{E}_2$ and $\partial\alpha_2/\partial\mathcal{E}_2^*$.

10-7. For a single-mode unidirectional ring laser containing a homogeneously-broadened medium, discuss the possibility that another mode can build up.

Chapter 11
COHERENT TRANSIENTS

Chapter 8 on saturation spectroscopy treats the interaction of two and three-level systems with simple time varying fields consisting of the Fourier superposition of two or three *cw* modes. The response of the medium depends heavily on the decay times of the medium and thereby provides ways of measuring these times. Similarly Chap. 10 considers time-varying fields in cavities, typically given by the superposition of modes. In this chapter we consider the response of atoms to time varying fields consisting of either short pulses or step functions that prepare the medium in special ways. In such situations, we find it convenient to follow the time-varying atomic (or molecular) response using the coupled Maxwell-Bloch equations developed in Secs. 4-3 and 5-1 without a modal decomposition of the field. Since the induced atomic polarization is initially coherent, the situations we consider are categorized as "coherent transients." As the induced polarization decays due to various line broadening mechanisms, the coherent response is lost. The time taken for this decay is typically directly related to the dipole (T_2) and population (T_1) decay times as well as to the width of the inhomogeneous broadening. Hence coherent transients provide alternative ways to measure these basic decay times. While saturation spectroscopy works in the frequency domain, coherent transients work in the time domain. Some media are equally easily studied in either domain, but often one domain is more convenient than the other. Most of the coherent transient methods correspond to techniques developed in studies of nuclear magnetic resonance (NMR), which uses microwave frequencies. In the optical domain Stark switching techniques pioneered by Brewer and Shoemaker (1972) for gaseous media are the most common. For gases, the final polarization must be averaged over the Doppler velocity distribution. As in our discussions of saturation spectroscopy, we usually assume that the radiated field amplitudes are small compared to the incident fields, which we can then take to be unaffected by the medium. In particular, this approximation is valid for "thin samples". Sections 11-1 through 11-3 discuss three phenomena in this category known as optical nutation, free-induction decay, and photon echo, respectively. Section 11-4 treats optical Ramsey fringes, which involve interactions with spatially separated light beams.

These fringes are described by a formalism very similar to photon echo, although the physics is different since homogeneously broadened systems are used. Section 11-5 discusses both simple pulse propagation and a coherent interaction called self-induced transparency, which applies the principle of Rabi flopping to achieve lossless propagation of light through absorbers of arbitrary length.

11-1. Optical Nutation

Optical nutation is the name given to the transient effect that occurs when a molecular ensemble is suddenly exposed to intense *resonant* laser light. The name was chosen in analogy to the corresponding NMR phenomenon called "spin nutation". The molecules undergo alternate absorption and emission of the light as they are driven coherently between the upper and lower levels. This Rabi flopping is destroyed both by homogeneous and inhomogeneous broadening, and hence provides a way to measure the associated widths.

In optical nutation and free induction decay we study the total field striking the detector. It consists of sum of the incident field E_0 and the field E_s emitted by the induced polarization in the sample. This total field after passing through a sample of length L is given by

$$E(L,t) = \frac{1}{2}\,[\mathcal{E}_s(L,t) + E_0]\,e^{i(KL - \nu t)} + \text{c.c.} \tag{1}$$

The intensity measured by a detector is $|E(L,t)|^2$, namely

$$|E(L,t)|^2 \simeq E_0^2 + E_0\mathrm{Re}[\mathcal{E}_s(L,t)]\,, \tag{2}$$

which is a DC signal plus a homodyne term containing the signal response of Eq. (1). Here and in the Bloch equations we neglect the small term proportional to $|\mathcal{E}_s|^2$ and take the incident field amplitude E_0 to be real without loss of generality since an overall field phase factor cancels out.

We first find the field emitted by a thin sample of length L containing two-level systems with no decay ($T_1 = T_2 = 0$), but irradiated by an incident field of arbitrary intensity and tuning. We assume that the sample is thin enough that the transit time through the sample is negligible compared to atomic response times, and that the incident field is unaffected by the signal field radiated by the induced polarization. The signal field is then given by integrating the propagation equation

$$\frac{d\mathcal{E}_s(z)}{dz} = i(K/2\epsilon)\,\mathcal{P} \tag{5.3}$$

from 0 to L with the initial signal field amplitude $\mathcal{E}(0) = 0$. Since we assume that the incident field is not affected by thin samples, the Rabi frequency for the induced polarization $\mathcal{P}$ is constant, yielding the real part of the signal field envelope

$$\mathrm{Re}\{\mathcal{E}_s(L,t)\} = -(KL/2\epsilon)\,\mathrm{Im}\{\mathcal{P}\}\ . \tag{3}$$

Here we have included a time dependence which is slow compared to the transit time through the sample and yields the coherent transients. In general we consider media with inhomogeneous broadening, so that the polarization $\mathcal{P}$ is given in terms of the induced dipole density matrix element ρ_{ab} by averaging Eq. (5.8) over the inhomogeneous broadening distribution $\mathcal{W}(\omega')$

$$\mathrm{Im}\{\mathcal{P}\} = 2\wp\int d\omega'\,\mathcal{W}(\omega')\,\mathrm{Im}\{e^{i(\nu t\,-\,Kz)}\,\rho_{ab}(z,t)\} = -\wp N\int d\omega'\,\mathcal{W}(\omega')\,V\ , \tag{4}$$

where N is the number of systems per volume and the Bloch vector component V is defined in terms of ρ_{ab} by Eq. (4.43). Substituting the appropriate solution (4.65) for V and using the Doppler inhomogeneous broadening distribution

$$\mathcal{W}(\omega'-\omega) = (Ku\sqrt{\pi})^{-1}\,e^{-(\omega-\omega')^2/(Ku)^2}\ , \tag{5}$$

we find

$$\begin{aligned}\mathrm{Im}[\mathcal{P}] &= -\wp\mathrm{N}\int_{-\infty}^{\infty} d\omega'\,\mathcal{W}(\omega-\omega')V(\omega'-\nu)\\ &\simeq -2\wp\mathcal{R}_0 NW(0)\mathcal{W}(\omega-\nu)\int_0^{\infty} d\delta\,\frac{\sin\sqrt{\delta^2+\mathcal{R}_0^{\,2}}\,t}{\sqrt{\delta^2+\mathcal{R}_0^{\,2}}}\\ &= -\,\wp N\mathcal{R}_0 W(0)\mathcal{W}(\omega-\nu)J_0(\mathcal{R}_0 t)\ ,\end{aligned} \tag{6}$$

where $J_0(\mathcal{R}_0 t)$ is the zeroth-order Bessel function. Here we assume that the Doppler Gaussian is very broad compared to the width of V, so that it may be evaluated at the peak of V ($\omega'=\nu$) and factored outside the integral.

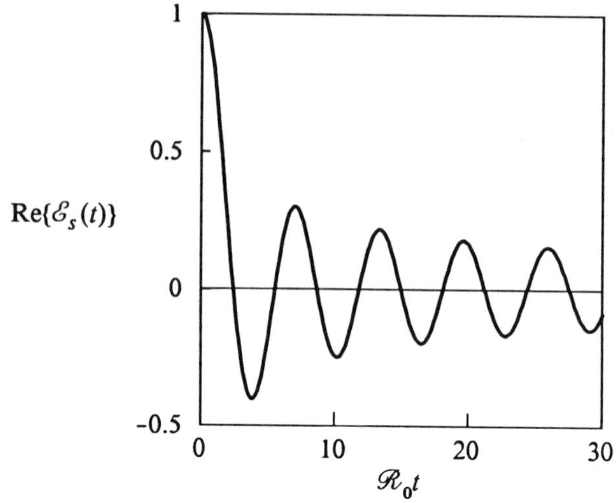

Fig. 11-1. Normalized optical nutation signal versus $\mathcal{R}_0 t$ as given by Eq. (7).

The Bessel function gives the "damped" Rabi flopping response of Fig. 11-1. The decay there is not actually due to damping at all, but rather to the dynamical interference between Rabi flopping precessions at different rates $(\delta^2 + \mathcal{R}_0^2)^{1/2}$. Unlike irreversible damping, this dynamical decay due to the inhomogeneous broadening can be reversed, as demonstrated by the photon-echo phenomenon discussed in Sec. 11-3. The "decay" time so measured gives the Doppler width Ku.

The V component of Eqs. (4.63) and (4.65) can be generalized to include equal dipole and population-difference decay times $T_2 = T_1$ by multiplying by the damping factor $\exp(-t/T_2)$. Combining this generalization, which neglects terms of order $(\mathcal{R}_0 T_2)^{-1}$, with Eqs. (3) and (6), we find the field envelope

$$\boxed{\mathrm{Re}\{\mathcal{E}_s(L,t)\} = \frac{\wp N K L \mathcal{W}(\omega-\nu)}{2\epsilon}\mathcal{R}_0 e^{-t/T_2}\, W(0) J_0(\mathcal{R}_0 t)} \,. \qquad (7)$$

11-2. Free Induction Decay

To observe this phenomenon, we first allow the response of a sample to a *cw* incident field to reach steady state. Such a field pumps up the medium in analogy with the way a current charges up an inductor. For a medium of two-level atoms, the steady-state Bloch vector is given by Eqs. (4.75), (4.74), and (4.71). Following the experimental method of Brewer and Shoemaker (1972), we then Stark switch the molecules or frequency switch the laser out of resonance so that further interaction with the field is negligible. Similarly to an inductor attempting to maintain a current when the driving current is switched off, the steady-state induced polarization continues to radiate after the incident field is switched off. This radiation decays away in the dipole lifetime. The process is called free induction decay after its NMR analog, which consists of spins in magnetic fields and acts very similarly to the free decay of an inductor.

While in steady-state equilibrium with the incident field, the inhomogeneously broadened atoms are characterized by the detuning $\delta = \omega' - \nu$. After the Stark switching, the atoms evolve freely with the detuning $\delta' = \omega' - \nu + \Delta_s = \delta + \Delta_s$. The corresponding transient solutions for large T_1 are given by Eq. (4.68) with $\gamma = 0$ and δ replaced by δ'. In particular, to evaluate $\mathrm{Im}(\mathscr{P})$ we need the V component

$$V(t) = [\sin\delta' t\; U(0) + \cos\delta' t\; V(0)]\, e^{-t/T_2}\,. \tag{8}$$

Substituting this into Eq. (4) with the steady state values given by Eqs. (4.75) and (4.74) as the initial values $U(0)$ and $V(0)$, evaluating $\mathscr{W}(\omega'-\omega)$ at the center of the $U(0)$ and $V(0)$ values ($\omega' = \nu$), and neglecting terms odd in δ, we find

$$\mathrm{Im}[\mathscr{P}] = -\wp N\, \mathscr{W}(\omega-\nu) \int_{-\infty}^{\infty} d\delta\, [\sin\delta' t\; U(0) + \cos\delta' t\; V(0)]$$

$$\simeq 2\wp^2\hbar^{-1}T_2E_0NW(0)\mathscr{W}(\omega-\nu)\, e^{-t/T_2}\cos\Delta_s t \int_0^{\infty} d\delta\, \frac{\delta T_2\sin\delta t - \cos\delta t}{1 + I + \delta^2T_2^{\,2}}$$

$$\simeq -\pi\wp^2\hbar^{-1}T_2E_0NW(0)\mathscr{W}(\omega-\nu)\, e^{-(1+\sqrt{I+1})t/T_2}\cos\Delta_s t\, [(1+I)^{-1/2}-1]\,. \tag{9}$$

Substituting Eqs. (3) and (9) into Eq. (2), we find a total field containing the beat-frequency component $E_0Q\cos\Delta_s t$, where

$$Q(t) = \frac{\pi N K L \wp^2 E_0}{4\hbar\epsilon} W(0)\mathscr{W}(\omega-\nu)\, e^{-(1+\sqrt{I+1})t/T_2}\, [(1+I)^{-1/2}-1] \quad . \qquad (10)$$

The beat-note exponential decay factor $\exp[-t(1+\sqrt{I+1})/T_2]$ arises from two sources, 1) the ordinary dipole decay $\exp(-t/T_2)$ appearing in Eq. (8), and 2) the $\exp[-t\sqrt{I+1}/T_2]$ factor which results from the Doppler dephasing of the molecules in the power broadened hole excited by the incident field. At moderate incident intensities (few watts/cm²), the dimensionless intensity $I = \mathscr{R}_0^2 T_1 T_2$ can be much greater than 1 causing the dephasing contribution to dominate and the free induction to decay in about one Rabi flopping period.

In actual Stark shift experiments the Stark shift is only strong enough to shift the initial velocity group out of resonance and another group into resonance. This leads to a free induction decay signal for the initially resonant group superimposed on an optical nutation signal for the subsequently resonant group. The large I ($\mathscr{R}_0^2 \gg (T_1T_2)^{-1}$) case helps to separate the two phenomena, since the free induction decay takes place within one Rabi flopping cycle, while the optical nutation decays in the time T_2.

11-3. Photon Echo

The inhomogeneous-broadening dephasing of the free induction decay is a dynamical process governed by the two-level Hamiltonian. As such it is *a reversible process*, in contrast to the homogeneous broadening decay, which results from Markoffian interactions with one or more reservoirs (see Chap. 14). The reversible character of this dynamical dephasing can be observed in experiments such as photon echoes. Specifically we can apply one pulse preparing the system someway, let it evolve freely for a time, apply a second pulse, and then watch the response. If the time between pulses is less than T_2, the polarization will rephase to some extent after additional free evolution. This rephased polarization emits an "echo" field even in the absence of applied fields. This echo was named "photon echo" by Abella, Kurnit, and Hartmann (1966), in analogy with the NMR "spin echo" discovered by Hahn (1950).

We first give a simple Bloch vector discussion of this phenomenon for the most pronounced pulse combination, namely a sharp $\pi/2$ pulse followed a time τ later by a sharp π pulse. Provided these pulses are sharp compared to the inhomogeneous broadening dephasing time, and provided this time is, in turn, short compared to T_1 and T_2, we see a nice echo at time 2τ. We then derive a more general solution for two pulses of arbitrary area but still short compared to the inhomogeneous dephasing time.

Before the first pulse, the Bloch vectors for all molecules regardless of velocity point down, namely $\mathbf{U}(0-) = -\mathbf{e}_3$, where we use $t = 0-$ to indicate the time just before the first pulse, and 0+ for the time just afterward. *A* $\pi/2$ pulse sharp compared to the inhomogeneous broadening dephasing time rotates all Bloch vectors up to $\mathbf{e}_2$, i.e., $\mathbf{U}(0+) = \mathbf{e}_2$. Effectively the bandwidth of the sharp pulse is assumed to be so broad that all molecules are resonant and their Bloch vectors rotate about the $\mathbf{e}_1$ axis up to e_2. We then let the Bloch vectors precess freely in the rotating frame at their rates ω-ν. Due to the distribution in ω they spread out in the UV plane (Fig.

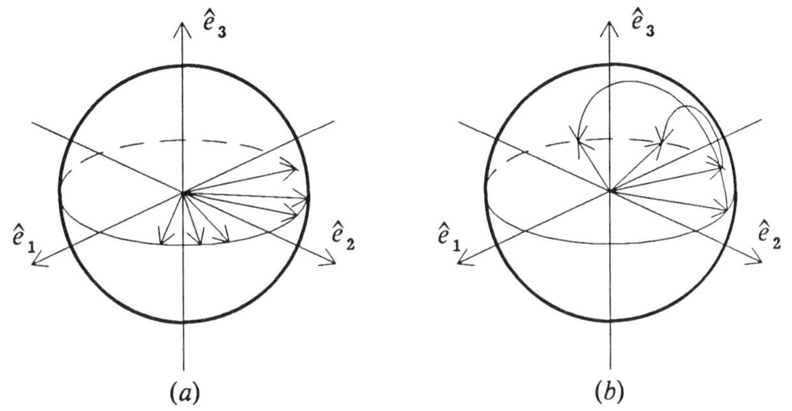

Fig. 11-2. Evolution of the Bloch vectors in an ideal photon echo experiment. (*a*) **U** vectors for molecules with varying resonant frequencies spread out in the UV plane after an initial $\pi/2$ pulse rotates them up from $=\mathbf{e}_3$. (*b*) *A* sharp π pulse rotates all vectors π radians about the $\mathbf{e}_1$ axis, putting the more slowly precessing vectors ahead of the more rapid ones

11-2*a*), with more detuned systems precessing faster than more resonant ones. After a time τ we then subject the molecules to a sharp π pulse, which rotates all vectors π radians about the $\mathbf{e}_1$ axis (Fig. 11-2*b*). This simply sets $V(\tau-) \to -V(\tau+)$. The nearly resonant Bloch vectors haven't precessed much, so the π pulse gives them a big shove towaıd $-\mathrm{e}_2$. After another time τ, they precess the same amounts and end up at $-\mathrm{e}_2$. The less resonant vectors have precessed further, so the π pulse leaves them relatively far from $-\mathbf{e}_2$. After another τ, they also end up at $-\mathbf{e}_2$, since they precess faster. The net effect is that all Bloch vectors end up at $-\mathrm{e}_2$ at time 2τ, regardless of their individual precession rates. Thus they add constructively, yielding a macroscopic polarization that radiates an echo field.

Hahn's analogy with race horses helps to make this clearer. At time t = 0, all horses start at the starting gate (like the vectors along $\mathbf{e}_2$) running at different speeds and spreading out around the race track. At the time t = τ, someone fires a gun, signaling all horses to about face and run back to the starting gate. In the second time interval of τ each horse travels the same distance it did in the first time interval of τ. Hence at the time t = 2τ, all horses end up back at the starting gate. Crash! The analogy is good, except that the horses end up together where they started, while the Bloch vectors end up together on the other side of the *UV* circle from where they started.

We can derive the echo signal for two pulses of arbitrary areas θ_1 and θ_2, provided they are short compared to T_1 and T_2, as follows. The evolution for a given Bloch vector is obtained by multiplying the initial value of $-\mathbf{e}_3$ by four matrices, first an optical-nutation matrix (4.63) for the time 0 to t_1 of the first square pulse, second a free precession matrix (4.68) for t = t_1 to t_2, third another optical nutation matrix (4.63) for $t = t_2$ to t_3 of the second square pulse, and fourth, another free precession matrix (4.68) for t = t_3 to t_4. In the final matrix we only need to calculate the Δ-symmetrical parts of the $U(t)$ and $V(t)$ components, since they're all that survive the integral over the inhomogeneous broadening. If we consider only a single decay rate, we can solve this problem substantially quicker using the wave function U-matrix of Eq. (3.106). Hence we leave the Bloch vector solution to Prob. 11-3 and carry out the wave function solution here.

In terms of the two-level U-matrix of Eq. (3.106), we find the wave function after the two pulses of areas θ_1 and θ_2 and corresponding free evolutions for times τ_1 and τ_2 to be

$$\begin{pmatrix} C_a(\tau_1+\tau_2) \\ C_b(\tau_1+\tau_2) \end{pmatrix} = \begin{pmatrix} e_2^* & 0 \\ 0 & e_2 \end{pmatrix} \begin{pmatrix} c_2 - i\delta\mathcal{R}^{-1}s_2 & i\mathcal{R}_0\mathcal{R}^{-1}s_2 \\ i\mathcal{R}_0^*\mathcal{R}^{-1}s_2 & c_2+i\delta\mathcal{R}^{-1}s_2 \end{pmatrix} \begin{pmatrix} e_1^* & 0 \\ 0 & e_1 \end{pmatrix} \begin{pmatrix} i\mathcal{R}_{01}\mathcal{R}_1^{-1}s_1 \\ c_1+i\delta\mathcal{R}_1^{-1}s_1 \end{pmatrix} \tag{11}$$

where $\mathcal{R}_{0i} = \wp E_i/\hbar$ for i = 1 and 2, $\delta = \omega - \nu$, $\mathcal{R}_i^2 = \mathcal{R}_{0i}^2 + \delta^2$. For the sake of brevity we have used the notation

$$e_i = \exp(i\delta\tau_i),\ s_i = \sin(\theta_i/2),\ c_i = \cos(\theta_i/2). \tag{12}$$

We have already multiplied the U-matrix for the first pulse by the initial ground state column vector. Multiplying the diagonal free evolution matrices next and then performing the final multiplication, we obtain

$$\begin{bmatrix} C_a(\tau_1+\tau_2) \\ C_b(\tau_1+\tau_2) \end{bmatrix} = \begin{bmatrix} (c_2-i\delta\mathcal{R}_2^{-1}s_2)e_2^* & i\mathcal{R}_{02}\mathcal{R}_2^{-1}s_2e_2^* \\ i\mathcal{R}_{02}^*\mathcal{R}_2^{-1}s_2e_2 & (c_2+i\delta\mathcal{R}_2^{-1}s_2)e_2 \end{bmatrix} \begin{bmatrix} i\mathcal{R}_{01}\mathcal{R}_1^{-1}s_1e_1^* \\ (c_1+i\delta\mathcal{R}_1^{-1}s_1)e_1 \end{bmatrix}$$

$$= \begin{bmatrix} i\mathcal{R}_{01}\mathcal{R}_1^{-1}s_1(c_2-i\delta\mathcal{R}_2^{-1}s_2)e_1^*e_2^* + i\mathcal{R}_{02}\mathcal{R}_2^{-1}s_2(c_1+i\delta\mathcal{R}_1^{-1}s_1)e_1e_2^* \\ -\mathcal{R}_{02}^*\mathcal{R}_{01}(\mathcal{R}_1\mathcal{R}_2)^{-1}s_1s_2e_1^*e_2 + (c_2+i\delta\mathcal{R}_2^{-1}s_2)(c_1+i\delta\mathcal{R}_1^{-1}s_1)e_1e_2 \end{bmatrix} . \quad (13)$$

We now average $C_a(\tau_1+\tau_2)C_b^*(\tau_1+\tau_2)$ over the inhomogeneous broadening distribution. In this average, only the cross term proportional to $\mathcal{R}_{01}^*\mathcal{R}_{02}^2$ contains free evolution phase factors that cancel when $\tau_1 = \tau_2$ and give an appreciable polarization. The δs_1 contribution to this term averages to zero over the even velocity integral, leaving us with the slowly-varying complex polarization

$$\wp\langle C_a(\tau_1+\tau_2)C_b^*(\tau_1+\tau_2)\rangle = -\frac{i}{2}\,\wp\mathcal{R}_{01}^*\mathcal{R}_{02}^2\langle e^{2i\delta(\tau_1-\tau_2)}(\mathcal{R}_1\mathcal{R}_2^2)^{-1}\sin\theta_1\sin^2(\theta_2/2)\rangle \quad (14)$$

The observed photon-echo signal is proportional to the absolute value of this slowly-varying polarization. The functions $\mathcal{R}$, θ_1, and θ_2 are all functions of δ, the variable of integration in the average over the inhomogeneous broadening distribution. Hence in general the integral must be performed numerically, but in the limit when the pulses are short compared to the inhomogeneous dephasing time, we can ignore these dependencies. Then $\mathcal{R} \to \mathcal{R}_0$, and we see that the maximum echo occurs for a first pulse of area $\theta_1 = \pi/2$ and a second pulse of area $\theta_2 = \pi$, the choices we used in our intuitive discussion at the start of this section. The maximum signal given by Eq. (14) occurs for equal free evolution times, $\tau_1 = \tau_2$, since otherwise the $\exp[2i\delta(\tau_1-\tau_2)]$ factor oscillates rapidly, canceling out the integral over δ. To include simple effects of decay, multiply the answer of Eq. (14) by $\exp[-\gamma(\tau_1+\tau_2)]$. The $\mathcal{R}_{01}^*\mathcal{R}_{02}^2$ factor reveals that photon echo can be interpreted as a superposition of four-wave mixing processes in which the Fourier components of the second pulse scatter off the gratings induced by the interference between the Fourier components of both pulses. At the echo time these processes add constructively.

11-4. Ramsey Fringes

A coherent interaction very similar in some ways to photon echo is the phenomenon of Ramsey fringes, which are generated by the scheme of Fig. 11-3. Initially unexcited atoms pass through two uniform parallel beams of light separated by a gap. We suppose that the areas of the Rabi precession in the beams are given by θ_1 and θ_2, as for the pulses in photon

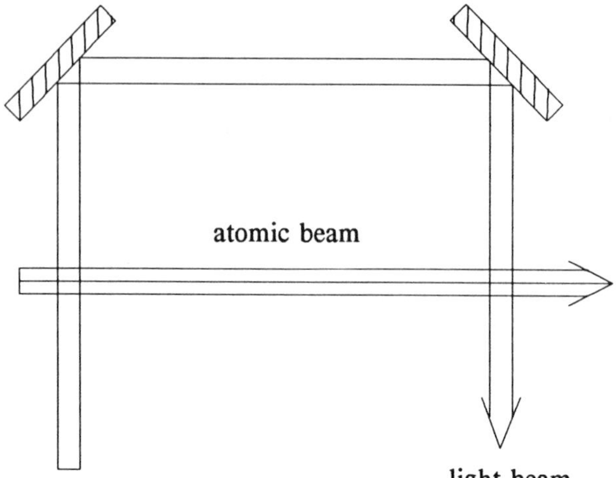

Fig. 11-3. Configuration yielding Ramsey fringes

echo. We further suppose that the atoms move between the beams in a time T. The wave function at the exit of the second beam is then given by the product of three 2×2 matrices which are precisely the same as the first three matrices for photon echo. Hence the wave function is given by Eqs. (11) and (13) with the factor $e_2 = 1$ and without the subscripts 1 and 2. Unlike photon echoes, Ramsey fringes experiments typically detect the spontaneous emission from the upper level rather than the radiation from a coherent dipole. Hence they effectively measures the upper-level probability $|C_a|^2$ for a homogeneously broadened system, rather than an average over an inhomogeneous distribution of dipoles.

Calculating $|C_a|^2$ from Eq. (13) with $e_2 = 1$, we have

$$|C_a|^2 = |a_1|^2 + |a_2|^2 + 2\mathrm{Re}\{a_1^{*}a_2\exp[i(\omega-\nu)T]\}\ , \tag{15}$$

where a_1 and a_2 are given by

$$a_1 = \mathcal{R}_0\mathcal{R}^{-1}s_1(c_2 - i\delta\mathcal{R}_2{}^{-1}s_2) \tag{16}$$

$$a_2 = \mathcal{R}_0\mathcal{R}^{-1}s_2(c_1 + i\delta\mathcal{R}^{-1}s_1)\ , \tag{17}$$

and s_i and c_i are sines and cosines of the Rabi precession angles as given by Eq. (12). The Ramsey fringes technique is used in high-resolution spectroscopy to reduce the effects of transit time broadening resulting from the finite light-atom interaction time.

11-5. Pulse Propagation and Area Theorem

Up to this point, we have approximated the slowly-varying Maxwell propagation equation (1.43) by keeping only the time derivative in cavity problems and only the space derivative in propagation problems. For pulse propagation in general and self-induced transparency in particular, we must keep both derivatives. To this end, this section derives the polarization $\mathcal{P}(z,t)$ in Eq. (1.43) as a convolution of the field envelope and a susceptibility that includes the effects of inhomogeneous broadening. We solve the resulting equations for the pulse area, finding the McCall-Hahn (1967) *area theorem*, which provides the basis for an explanation of self-induced transparency.

The interaction energy $\mathcal{V}_{ab}$ for the field (1.41) is

$$\mathcal{V}_{ab} = -\frac{\wp}{2}\,\mathcal{E}(z,t)e^{i(Kz - \nu t)} . \tag{18}$$

Substituting this into the population matrix equations of motion (5.9) and (5.34), we have

$$\dot{\rho}_{ab} = -(i\omega + \gamma)\rho_{ab} - \frac{i\wp}{2\hbar}\mathcal{E}(z,t)e^{i(Kz - \nu t)}D , \tag{19}$$

$$\frac{dD}{dt} = -\Gamma(D + N') + [i\hbar^{-1}\mathcal{E}(z,t)e^{i(Kz - \nu t)}\rho_{ab}{}^* + \text{c.c.}] , \tag{20}$$

in which we drop the pump (Λ) terms for the present problems. A formal integral of Eq. (19) is

$$\rho_{ab}(z,\omega,t) = -\frac{i\wp}{2\hbar}e^{i(Kz - \nu t)}\int_{-\infty}^{t} dt'\mathcal{E}(z,t')D(z,\omega,t')e^{-(i\omega - i\nu + \gamma)(t - t')} . \tag{21}$$

Substituting this into Eq. (20), we find the integro-differential equation

$$\frac{dD}{dt} = -\Gamma(D + N') - \frac{1}{2}\left(\frac{\wp}{\hbar}\right)^2\int_{-\infty}^{t} dt'D(z,\omega,t')$$
$$\times[e^{(i\omega - i\nu - \gamma)(t - t')}\mathcal{E}(z,t)\mathcal{E}^*(z,t') + \text{c.c.}] . \tag{22}$$

Substituting Eq. (21) into Eq. (5.8), we obtain the complex polarization

$$\mathscr{P}(z,t) = -\frac{i\wp^2}{\hbar}\int_{-\infty}^{t} dt'\mathscr{E}(z,t')D(z,\omega,t')e^{-(i\omega - i\nu + \gamma)(t - t')} \tag{23}$$

Combining this with the slowly-varying Maxwell propagation equation (1.43), we have

$$\left[\frac{\partial}{\partial z} + \frac{1}{c}\frac{\partial}{\partial t} + \kappa\right]\mathscr{E} = \frac{\alpha'}{N'}\int_{-\infty}^{t} dt'\mathscr{E}(z,t')D(z,\omega,t')e^{-(i\omega - i\nu + \gamma)(t - t')}\ , \tag{24}$$

where $\alpha' = K\wp^2N'/2\hbar\epsilon$ (note that α' does not have the dimensions of an absorption coefficient). The coupled atom-field Eqs. (22) and (24) allow us to predict the evolution of a pulse as it propagates down a homogeneously broadened two-level medium.

For propagation in inhomogeneously broadened media, we sum Eq. (23) times the distribution $\mathscr{W}(\omega)$ over ω to find the convolution

$$\mathscr{P}(z,t) = -\frac{i\wp^2N'}{\hbar}\int_{-\infty}^{t} dt'\mathscr{E}(z,t')e^{-\gamma(t - t')}\chi(z,t-t',t')\ , \tag{25}$$

where the dimensionless complex susceptibility

$$\chi(z,T,t) = N'^{-1}\int_{-\infty}^{\infty} d\omega\mathscr{W}(\omega)e^{-i(\omega-\nu)T}D(z,\omega,t)\ . \tag{26}$$

Like Eq. (2.37), the convolution in Eq. (25) generalizes the usual susceptibility relation (1.33) to include the effects of the medium's memory. Equation (25) describes both the irreversible decay in memory due to finite T_2 $(1/\gamma)$ as well as the reversible decay resulting from the interference of population differences at different ω's in the integral of Eq. (26).

Inserting Eq. (25) into the propagation Eq. (1.43), we find

$$\left[\frac{\partial}{\partial z} + \frac{1}{c}\frac{\partial}{\partial t} + \kappa\right]\mathscr{E} = -\alpha'\int_{-\infty}^{t} dt'\mathscr{E}(z,t')e^{-\gamma(t - t')}\chi(z,t-t',t')\ . \tag{27}$$

The equation of motion for χ is given by differentiating Eq. (26) with respect to t. Using Eq. (24) for $\dot{D}$, we find

$$\frac{\partial}{\partial t}\chi(z,T,t) = -\Gamma[\tilde{W}(T) + \chi(z,T,t)] - \frac{1}{2}\left(\frac{\wp}{\hbar}\right)^2 \int_{-\infty}^{t} dt' e^{-\gamma(t - t')}$$

$$\times[\mathscr{E}(z,t)\mathscr{E}^*(z,t')\chi(z,T-t+t',t') + \text{c.c.}] , \tag{28}$$

where

$$\tilde{W}(T) = \int_{-\infty}^{\infty} d\omega \mathscr{W}(\omega) e^{-i(\omega-\nu)T} \tag{29}$$

is the Fourier transform of the inhomogeneous broadening function $\mathscr{W}(\omega)$. The coupled Eqs. (27) and (28) determine the development of a plane-wave pulse propagating through a two-level medium with the inhomogeneous broadening distribution $\mathscr{W}(\omega)$.

Pulse Area Theorem

In particular we are interested to see how the pulse area defined by

$$\theta(z) \equiv \int_{-\infty}^{\infty} dt \frac{\wp\mathscr{E}(z,t)}{\hbar} \tag{30}$$

changes along z. In a rate-equation Beer's law limit, we would expect that $\theta(z)$ obeys the propagation equation

$$\frac{d}{dz}\theta(z) = -\alpha\theta(z). \tag{31}$$

In contrast, McCall and Hahn (1969) found that under conditions of *coherent propagation*, i.e. for light pulses short compared to the atomic decay times T_1 and T_2, $\theta(z)$ obeys the more general propagation equation

$$\boxed{\frac{d}{dz}\theta(z) = -\alpha\sin[\theta(z)]} \tag{32}$$

known as the *area theorem*. This theorem is remarkable in several ways, one of which is that it works for an arbitrary amount of inhomogeneous broadening, and another of which is that it predicts multiple stable solutions for the values $\theta = 2q\pi$, where $n = 0, 1, \ldots$. Note that a stable area does not necessarily imply a stable pulse shape! We discuss conditions under which the pulse shape itself is stable in Section 11-6. In this subsection, we solve Eqs. (27) and (28) with $\gamma \simeq \Gamma \simeq 0$ and a real envelope $\mathcal{E}(z,t)$ to find the area theorem.

Integrating the propagation Eq. (27) with $\gamma = 0$ over t, we find

$$\frac{\partial\theta}{\partial z} + \kappa\theta = -\alpha'\int_{-\infty}^{\infty} dt \int_{-\infty}^{t} dt' \frac{\wp\mathcal{E}(z,t')}{\hbar}\chi(z,t-t',t') ,$$

where the integral over the $\partial\mathcal{E}/\partial t$ vanishes since $\mathcal{E}$ vanishes at $t = \pm\infty$. Interchanging the integrals with corresponding changes in the integration limits we find

$$\begin{aligned}\frac{\partial\theta}{\partial z} + \kappa\theta &= -\alpha'\int_{-\infty}^{\infty} \frac{dt'\wp\mathcal{E}(z,t')}{\hbar}\int_{t'}^{\infty} dt\chi(z,t-t',t') \\ &= -\alpha'\int_{-\infty}^{\infty} dt' \frac{\wp\mathcal{E}(z,t')}{\hbar}\int_{0}^{\infty} dT\chi(z,T,t') \\ &= -\frac{\alpha'}{2}\int_{-\infty}^{\infty} dt \frac{\wp\mathcal{E}(z,t)}{\hbar}\int_{-\infty}^{\infty} dT\chi(z,T,t) ,\end{aligned} \tag{33}$$

where the last step follows since the reality of χ implies that $\chi(z,-T,t) = \chi(z,T,t)$. Integrating Eq. (28) over T with $\gamma = \Gamma = 0$, we find

$$\frac{\partial}{\partial t}\int_{-\infty}^{\infty} dT\chi(z,T,t) = -\left(\frac{\wp}{\hbar}\right)^2\int_{-\infty}^{t} dt'\mathcal{E}(z,t)\mathcal{E}(z,t')\int_{-\infty}^{\infty} dT\chi(z,T,t') , \tag{34}$$

where we have taken $\mathcal{E}$ real and used the fact that χ is real so that the c.c. in Eq. (28) yields a multiplicative factor of 2. At $t = -\infty$, i.e., before the pulse arrives, the medium is in its ground state $[D(z,\omega,-\infty) = -N']$, so that Eq. (26) gives

$$\int_{-\infty}^{\infty} dT \chi(z,T,-\infty) = -\int_{-\infty}^{\infty} dT \mathscr{W}(T) = -2\pi \mathscr{W}(\nu) . \tag{35}$$

Problem 11-5 solves a differential equation that has the same form as Eq. (34). Writing the corresponding solution for Eq. (34) with the initial condition (35) gives

$$\int_{-\infty}^{\infty} dT \chi(z,T,t) = 2\pi \mathscr{W}(\nu) \cos\left[\int_{-\infty}^{t} dt' \frac{\wp \mathscr{E}(z,t')}{\hbar}\right]$$

Substituting this into the area equation (33), we find

$$\frac{\partial \theta}{\partial z} + \kappa\theta = -\alpha \int_{-\infty}^{\infty} dt \frac{\wp \mathscr{E}(z,t)}{\hbar} \cos\left[\int_{-\infty}^{t} dt' \frac{\wp \mathscr{E}(z,t')}{\hbar}\right], \tag{36}$$

where the absorption coefficient $\alpha = \pi \mathscr{W}(\nu)\alpha'$. Finally, changing the variable of integration t' to the partial area

$$\vartheta(z,t) = \int_{-\infty}^{t} dt' \frac{\wp \mathscr{E}(z,t')}{\hbar}, \tag{37}$$

we obtain the area theorem

$$\frac{\partial \theta}{\partial z} + \kappa\theta = -\alpha \int_{0}^{\theta(z)} d\vartheta \cos\vartheta = -\alpha \sin[\theta(z)] . \tag{38}$$

11-6. Self-Induced Transparency

For no losses ($\kappa = 0$), the area theorem shows that under the influence of coherent propagation through absorbing media, the area of optical pulses evolves toward $2q\pi$, where q is an integer (see Prob. 11-6). This very general result still does not predict the shape of the pulse. This Section addresses this question, and more precisely discusses pulses that propagate through the absorber without reshaping. This effect, also discovered by McCall and Hahn (1967), is called *self-induced transparency.*

Consider the Bloch Eqs. (4.48), (4.49), and (4.50) without decay and with an electric field that varies in space and time

$$dU/dt = -\delta V \tag{39}$$

$$dV/dt = \delta U + \mathcal{R}_0(z,t)\, W \tag{40}$$

$$dW/dt = -\mathcal{R}_0(z,t)\, V\ , \tag{41}$$

where $\mathcal{R}_0 = \wp\mathcal{E}(z,t)/\hbar$ and $\delta = \omega - \nu$. In the resonant case $\delta = 0$, the solution to these equations is

$$V(z,t;0) = -\sin[\vartheta\,(z,t)] \tag{42}$$

$$W(z,t;0) = -\cos[\vartheta\,(z,t)]\ , \tag{43}$$

where we have used the initial condition $U(z,-\infty) = V(z,-\infty) = 0$, $W(z,-\infty) = -1$, and $\vartheta\,(z,t)$ is the pulse area of Eq. (37). We consider propagation through an inhomogeneous medium and hence need also to find the solution of the Bloch equations for $\delta \neq 0$. We proceed by introducing the factorization Ansatz

$$V(z,t;\delta) = F(\delta)V(z,t;0) = -F(\delta)\sin[\vartheta\,(z,t)]\ . \tag{44}$$

Physically this means that we assume the detuned atoms have essentially the same response as the resonant atoms, except perhaps with a reduced amplitude. Substituting this Ansatz in Eq. (41) leads to a form that can be integrated exactly to give

$$W(z,t;\delta) = -F(\delta)[\cos\vartheta\,(z,t) - 1] - 1\ . \tag{45}$$

Taking the second derivative of Eq. (40) and substituting for dU/dt and W, we find the equation for the pulse partial area

$$\frac{d^2\vartheta}{dt^2} = \frac{\delta^2 F(\delta)}{1-F(\delta)}\sin\vartheta\ . \tag{46}$$

Note that since ϑ is independent of the detuning δ, the same must hold for $F(\delta)$. Identifying

$$\frac{\delta^2 F(\delta)}{1-F(\delta)} = \frac{1}{\tau^2} \tag{47}$$

or

$$F(\delta) = \frac{1}{1 + (\delta\tau)^2}, \tag{48}$$

which, when substituted in Eq. (46), yields the pendulum equation

$$\frac{d^2\vartheta\,(z,t)}{dt^2} - \frac{1}{\tau^2} = 0\,. \tag{49}$$

In general this equation can be solved in terms of elliptic functions. For the initial condition $E(-\infty) = dE(-\infty)/dt = 0$, however, the solution reduces to

$$\vartheta\,(z,t) = 4\tan^{-1}[\exp(t-t_0)/\tau]\,, \tag{50}$$

or

$$\boxed{\mathcal{E}(z,t) = \frac{2\hbar}{\tau\wp}\,\mathrm{sech}[(t-t_0)/\tau]}\,, \tag{51}$$

so that τ can be interpreted as the pulse duration. It is easily seen that for $t \to \infty$ this pulse has the area 2π. This result indicates that despite the fact that the medium is inhomogeneously broadened, it can still sustain 2π pulses, provided their shape is a hyperbolic secant. Even more remarkable perhaps is the fact that the Bloch equations alone are sufficient to determine the temporal evolution of the pulse: so far no mention has been made of Maxwell's wave equation. Of course the spatial dependence of the pulse is still implicit and hidden in t_0. What Maxwell's equations provide is the pulse velocity. Introducing $t_0 = z/v_p$, it can be shown that the 2π-hyperbolic secant solutions propagate unchanged through the absorber at velocity

$$v_p = \frac{c}{1 + \alpha c\tau}, \tag{52}$$

which can be substantially slower than the speed of light. The 2π-sech pulses are an optical example of *solitons*, and have a number of fascinating properties that have been studied in considerable detail over the years [see Lamb (1971)].

Equation (52) shows that the pulse velocity is inversely proportional to its length. Shorter solitons move faster than broader ones. Thus solitons can "collide" as the faster one passes through the slow one, but at the end of the collision the shapes of the two solitons remain as they were initially. Although the area theorem indicates that 4π, 6π ... pulses are also asymptotic solutions to coherent propagation through absorbers, such pulses are not stable and break up into 2, 3 etc... 2π-sech pulses. This break-up has been studied in detail for example by Lamb (1971).

References

Allen, L. and J. H. Eberly (1975), *Optical Resonance and Two-Level Atoms*, John Wiley & Sons, New York, reprinted (1987) with corrections by Dover, New York. This book gives a more detailed tutorial discussion of the coherent transients discussed in this Chapter.

Abella, I. D., N. A. Kurnit, and S. R. Hartmann (1966), Phys. Rev. **141**, 391.

Brewer, R. G. and R. L. Shoemaker (1971), Phys. Rev. Lett. 27, 631.

Hahn, E. L. (1950), Phys. Rev. **80**, 580.

McCall, S. L. and E. L. Hahn (1967), Phys. Rev. Lett. **18**, 908

McCall S. L. and E. L. Hahn (1969), Phys. Rev. **183**, 457.

Detailed reviews of coherent transient spectroscopy are given in

Brewer, R. G. (1977), in *Nonlinear spectroscopy, ed.* by N. Bloembergen, North-Holland, Amsterdam.

Brewer, R. G. (1977), in *Frontiers in Laser Spectroscopy, ed.* by R. Balian,, S. Haroche and S. Liberman, North-Holland, Amsterdam.

Lamb, G. L. (1971), Rev. Mod. Phys. **43**, 99.

Levenson, Marc D., and S. S. Kano, *Introduction to Nonlinear Laser Spectroscopy* (1988), Revised Edition, Academic Press, New York. This is an authoritative modern monograph on laser spectroscopy emphasizing experimental aspects. Its Chap. 6 is on optical coherent transients.

Sargent, M. III, M. O. Scully, and W. E. Lamb, Jr. (1977), *Laser Physics*, Addison-Wesley, Reading, MA.

Shoemaker, R. L. (1978), in *Laser and Coherence Spectroscopy, ed.* by J. I. Steinfeld, Plenum, New York gives another exhaustive review of coherent transient spectroscopy.

Stenholm, S. (1984), *Foundations of Laser Spectroscopy*, John Wiley & Sons, New York.

Problems

11-1. Compare and contrast self-induced transparency and bleaching.

11-2. Name two ways to measure the homogeneous linewidth in the presence of large inhomogeneous broadening.

11-3. Calculate the photon echo polarization using the Bloch vector formalism. Specifically the result of the first optical nutation from the ground-state initial condition is given by Eqs. (4.63) with $W(0) = -1$, $U(0) = V(0) = 0$, Ωt replaced by θ_1, and Ω approximated by $\mathcal{R}_0$, which for simplicity we take to be big compared to the magnitudes of the relevant Δ's. Hence the desired complex polarization is given by calculating

$$\begin{bmatrix} U(\tau_1+\tau_2) \\ V(\tau_1+\tau_2) \\ W(\tau_1+\tau_2) \end{bmatrix} = \begin{bmatrix} \cos\Delta\tau_2 & -\sin\Delta\tau_2 & 0 \\ \sin\Delta\tau_2 & \cos\Delta\tau_2 & 0 \\ 0 & 0 & 1 \end{bmatrix}$$

$$\times \begin{bmatrix} \Delta^2\Omega^{-2}\cos\theta_2+1 & -\Delta\Omega^{-1}\sin\theta_2 & \Delta\Omega^{-1}(\cos\theta_2-1) \\ \Delta\Omega^{-1}\sin\theta_2 & \cos\theta_2 & \sin\theta_2 \\ \Delta\Omega^{-1}(\cos\theta_2-1) & -\sin\theta_2 & \Delta^2\Omega^{-2}+\cos\theta_2 \end{bmatrix}$$

$$\times \begin{bmatrix} \cos\Delta\tau_1 & -\sin\Delta\tau_1 & 0 \\ \sin\Delta\tau_1 & \cos\Delta\tau_1 & 0 \\ 0 & 0 & 1 \end{bmatrix} \begin{bmatrix} \Delta\Omega^{-1}(\cos\theta_1-1) \\ \sin\theta_1 \\ \Delta^2\Omega^{-2}+\cos\theta_1 \end{bmatrix}$$

and integrating the resulting $U+iV$ over the inhomogeneous distribution. The algebra is substantially more complex than that for the two-level U-matrix technique used in the text, in spite of the simplifying assumption that Ω can be approximated by $\mathcal{R}_0$ (Eqs. (11) and (13) do not make this assumption). This is one more example of how the wave function can involve less algebra in cases with no or uniform decay.

11-5. Show that the equations

$$\dot{C}_a = \frac{i\wp\mathcal{E}(t)}{2\hbar}C_b, \quad \dot{C}_b = \frac{i\wp\mathcal{E}(t)}{2\hbar}C_a,$$

have the solution

$$C_a(t) = A\cos(\vartheta/2) + B\sin(\vartheta/2)$$

with a corresponding expression for $C_b(t)$, where the partial area

$$\vartheta = \frac{\wp}{\hbar}\int_{-\infty}^{t} dt'\,\mathcal{E}(t') \; .$$

11-6. Show that the stable solutions of Eq. (38) are given by $\theta = 2q\pi$, where q is an integer.

Chapter 12
FIELD QUANTIZATION

Up to now, we have treated many problems in light-matter interactions and have obtained results in excellent agreement with experiments without having to quantize the electromagnetic field. Such a semiclassical description is sufficient to describe most problems in quantum optics. However, there are a few notable exceptions where a classical description of the field leads to the wrong answer. These include spontaneous emission, the Lamb shift, resonance fluorescence, the anomalous gyromagnetic moment of the electron, and "non-classical" states of light such as squeezed states. The remainder of this book deals with selected problems in light-matter interactions that require field quantization. The present chapter treats the quantization of the electromagnetic field in free space. Those familiar with this subject might want to glance at our notation and then proceed directly to Chap. 13.

Section 12-1 quantizes a single-mode electromagnetic field using the results of Sec. 3-4 for the harmonic oscillator. Section 12-2 generalizes these results to multimode fields. In Sec. 12-3, we find that an electromagnetic field in thermal equilibrium is described by a density matrix leading to a Maxwell-Boltzmann photon statistics. In Sec. 12-4, we review briefly the properties of the coherent states of the electromagnetic field, finding in particular their photon statistics. Section 12-5 discusses the coherence properties of quantum fields in generalization of Sec. 1-4, and Section 12-6 discusses the $P(\alpha)$ and other quasi-distributions which allow to cast certain quantum optics problems into a classical-looking formalism.

12-1. Single-Mode Field Quantization

To quantize the electromagnetic field, we consider a cavity of volume V, closed by perfectly reflecting mirrors as diagrammed in Fig. 12-1. For problems in free space, we take this volume to be infinite at the end of the calculation. This need not be the case and tailored electromagnetic environments can be experimentally realized as discussed in Chap. 15.

A classical monochromatic, single-mode electromagnetic field polarized in the $\hat{x}$-direction has the form

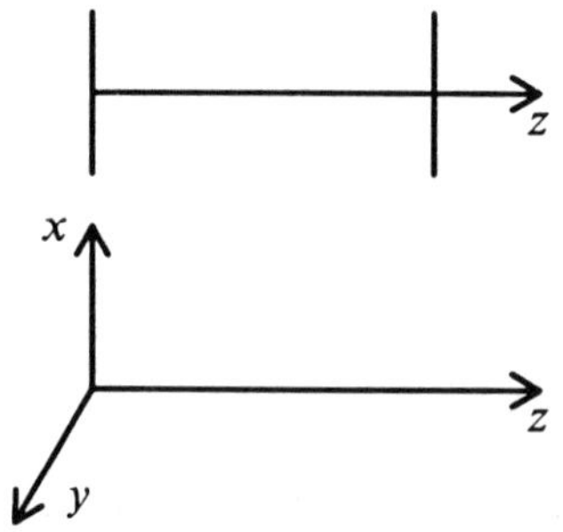

Fig. 12-1. One dimensional cavity used to quantize electromagnetic field.

$$\mathbf{E}(z,t) = \hat{x}q(t)[2\Omega^2/\epsilon_0 V]^{1/2}\sin Kz \ , \tag{1}$$

Here Ω is the single-mode field oscillation frequency, K is the wave number Ω/c, and $q(t)$ is a measure of the field amplitude.

The electromagnetic field satisfies Maxwell's equations in a vacuum, namely Eqs. (1.1) through (1.4). Substituting Eq. (1) into Eq. (1.3) yields the magnetic field

$$\mathbf{B}(z,t) = \frac{\hat{y}}{c^2K}\,\dot{q}(t)\left[\frac{2\Omega^2}{\epsilon_0 V}\right]^{1/2} \cos Kz \ . \tag{2}$$

The classical electromagnetic energy density is given by

$$\mathcal{U} = \frac{1}{2}\,[\epsilon_0 E^2 + B^2/\mu_0] \ , \tag{3}$$

with the corresponding Hamiltonian

$$\mathcal{H} = \int_V dV\,[\epsilon_0 E^2 + B^2/\mu_0] \ , \tag{4}$$

where dV is a volume element and E and B are the magnitudes of $\mathbf{E}$ and $\mathbf{B}$, respectively. Inserting Eqs. (1) and (2) into Eq. (4), we find

$$\mathcal{H} = \tfrac{1}{2}(\Omega^2 q^2 + p^2), \tag{5}$$

which is formally identical with the Hamiltonian for a simple harmonic oscillator with unit mass. We can therefore immediately quantize a single mode of the electromagnetic field by applying the results of Sec. 3-4 on the quantization of the simple harmonic oscillator. We find that in terms of the annihilation and creation operators a and $a^\dagger$, the single-mode electromagnetic field Hamiltonian is given by

$$\boxed{\mathcal{H} = \hbar\Omega(a^\dagger a + 1/2)}\ . \tag{6}$$

The corresponding eigenstates $|n\rangle$ of the field satisfy

$$\mathcal{H}|n\rangle = \hbar\Omega(n + 1/2)|n\rangle\ ,\ n = 0, 1, 2,..., \tag{7}$$

where n may be loosely interpreted as the "number of photons" in the state $|n\rangle$. The corresponding state vector is a linear superposition of these energy eigenstates

$$|\psi\rangle = \Sigma_n\ c_n\ |n\rangle\ . \tag{8}$$

More general descriptions of the single-mode field are given in terms of density operators in the next Sections.

Substituting the position operator of Eq. (3.120) into Eq. (1), we find the electric field operator

$$\boxed{E(z,t) = \mathcal{E}_\Omega(a + a^\dagger)\sin Kz}\ , \tag{9}$$

where the "electric field per photon"

$$\mathcal{E}_\Omega \equiv [\hbar\Omega/\epsilon_0 V]^{1/2}\ . \tag{10}$$

The Photon

Although the word "photon" is ubiquitous in quantum optics, it is commonly used in several different ways. One common use interprets Eqs. (3.135) and (3.136) as creating or annihilating a photon

$$a|n\rangle = \sqrt{n}|n-1\rangle \tag{3.135}$$

$$a^\dagger|n\rangle = \sqrt{n+1}|n+1\rangle\ . \tag{3.136}$$

According to this interpretation, the photon is a quantum of a single mode of the electromagnetic field. As such, it fills the cavity of quantization.

One might argue that although precise, such a definition of the photon is not particularly useful from an experimental point of view. In this context, "photon" is often used to describe a relatively localized (and therefore multimode) packet of radiation with an average energy of $\hbar\Omega$. In such situations, photons are always multimode objects since Fourier analysis tells us that the localization of any wavepacket requires a superposition of modes. This "particle" interpretation has intuitive appeal, but presents the drawback that each source emits its own kind of wave packet and hence such "photons" have a wide variety of analytic forms or worse, no analytic form at all. Section 13-4 on quantum beats gives an explicit example of a wave packet with energy $\hbar\Omega$. The most common use of the word photon is as a unit of electromagnetic radiation with the energy $\hbar\Omega$. This use is found in chemistry, in some parts of atomic physics dealing with "multi-photon processes," and often in semiconductor physics and in engineering optics. In these and similar cases, the field is classical, with absolutely no quantum character. "Photon" is used as a catchy synonym for "light" as in "photon echo" and photonics. Because the word photon is used in so many ways, it is a source of much confusion. The reader always has to figure out what the writer has in mind.

12-2. Multimode Field Quantization

Section 12-1 describes the quantization of a single-mode of the electromagnetic field. Many problems of interest, such as spontaneous emission, require many modes of the field with various frequencies and wave vectors. This section generalizes our analysis to treat many modes quantum mechanically. Modes are characterized by frequency, wave vector, polarization, and transverse variation. We consider linearly polarized plane-wave modes for simplicity.

Generalizing Eq. (1), we find the multimode electric field

$$\mathbf{E}(z,t) = \hat{x} \sum_s q_s(t)[2\Omega_s{}^2/\epsilon_0 V]^{1/2} \sin K_s z \,, \tag{11}$$

where $\Omega_s = cK_s$, and $K_s = s\pi/L$, $s = 1,2,3,\ldots$, and L is the length of the cavity in the $\hat{z}$ direction. The Hamiltonian $\mathcal{H}$ of such a multimode field is the sum of the Hamiltonians $\mathcal{H}_s$ of the single modes

$$\mathcal{H} = \Sigma_s\, \mathcal{H}_s \,, \tag{12}$$

where in $\mathcal{H}_s$ is given in terms of the single mode annihilation and creation operators a_s and $a_s^\dagger$ by Eq. (6) as

$$\mathcal{H}_s = \hbar\Omega_s(a_s^\dagger a_s + 1/2) . \tag{13}$$

The annihilation and creation operators satisfy the commutation relations

$$\boxed{[a_s, a_{s'}{}^\dagger] = \delta_{s,s'}} . \tag{14}$$

The electric field operator of Eq. (9) generalizes to

$$E(z,t) = \sum_s \mathcal{E}_s(a_s + a_s^\dagger) \sin K_s z . \tag{15}$$

where $\mathcal{E}_s = [\hbar\Omega_s/\epsilon_0 V]^{1/2}$ is the electric-field per photon.

The eigenstates of the multimode Hamiltonian (12) are given by products of the single-mode eigenstates

$$|n_1 n_2 \ldots n_s \ldots\rangle \equiv |\{n\}\rangle , \tag{16}$$

which have eigenvalues given by the eigenvalue equation

$$\mathcal{H}|\{n\}\rangle = \hbar \sum_s \Omega_s(n_s + \tfrac{1}{2})|\{n\}\rangle . \tag{17}$$

When acting on $|\{n\}\rangle$, the creation and annihilation operators $a_p{}^\dagger$ and a_p of the p-th mode give

$$a_p{}^\dagger|\{n\}\rangle = \sqrt{n_p+1}\,|n_1 n_2 \ldots n_p+1 \ldots\rangle \tag{18}$$

$$a_p|\{n\}\rangle = \sqrt{n_p}\,|n_1 n_2 \ldots n_p-1 \ldots\rangle . \tag{19}$$

The general state vector of a multimode field is a linear superposition of states such as $|\{n\}\rangle$, namely,

$$|\psi\rangle = \sum_{n_1}\sum_{n_2} \cdots \sum_{\{n_s\}} c_{n_1 n_2 \ldots n_s \ldots} |n_1 n_2 \ldots n_s \ldots\rangle \equiv \sum_{\{n\}} c_{\{n\}}\, |\{n\}\rangle . \tag{20}$$

Note that this is more general than

$$|\psi\rangle = |\psi_1\rangle|\psi_2\rangle \ldots |\psi_s\rangle \ldots , \tag{21}$$

where the $|\psi_s\rangle$ are state vectors for individual modes. Equation (20) includes state vectors of the type (21) as well as more general states with correlations between different modes resulting from interactions, for example, with atoms.

Quantization of Standing Waves versus Traveling Waves

Both standing and traveling waves are routinely used as basis fields for quantum theories of electrodynamics. To appreciate the difference between them, consider the positive frequency field operator

$$E^+(z,t) = \mathcal{E}_r \, [a_1 e^{iKz} + a_2 e^{-iKz}] \, e^{-i\nu t} \; . \tag{22}$$

Here $\mathcal{E}_r$ is the running-wave electric field per photon and a_i and $a_i{}^\dagger$ are the annihilation operators for two oppositely running wave modes, obeying the boson commutation relations $[\hat{a}_i, \hat{a}^\dagger{}_j] = \delta_{ij}$. Their action on the state $|n_1\, n_2\rangle_r$ describing the two running waves is

$$a_1|n_1\, n_2\rangle_r = \sqrt{n_1}\,|n_1{-}1\; n_2\rangle_r \; ; \qquad a_2|n_1\, n_2\rangle_r = \sqrt{n_2}\,|n_1\; n_2{-}1\rangle_r \; . \tag{23}$$

Alternatively, in terms of the operators $a_c = (a_1 + a_2)/\sqrt{2}$ and $a_s = (a_1 - a_2)/\sqrt{2}$, Eq. (22) is given by

$$E^+(z,t) = \sqrt{2}\, \mathcal{E}_r \, [a_c \cos(Kz) + i a_s \sin(Kz)] \, e^{-i\nu t} \; , \tag{24}$$

where the operators a_c and a_s also obey boson commutation relations, but act on standing-wave rather than traveling-wave modes. For example,

$$a_s|n_s\, n_c\rangle_s = \sqrt{n_s}\,|n_s{-}1\; n_c\rangle_s \; ; \qquad a_c|n_s\, n_c\rangle = \sqrt{n_c}\,|n_s\; n_c{-}1\rangle_s \; . \tag{25}$$

Although the choice of standing or running waves is typically one of mathematical convenience, there are cases in which the two are not equivalent. Standing-wave modes are the natural choice when a field is contained within a two-mirror cavity. We can then use a single-mode description as in Eq. (15), with the standing-wave electric field per photon related to the running-wave electric field per photon by $\mathcal{E}_s = \sqrt{2}\,\mathcal{E}_r$. Conversely, traveling waves are the natural choice when the field consists of counterpropagating waves in a three-mirror ring cavity.

Both approaches predict the same single-photon transition rates, such as those for spontaneous emission and photoionization. However, the situation is different for the effects of light forces on atomic motion, discussed semiclassically in Sec. 5-6. For small photon numbers, atoms are diffracted differently by a true standing wave than by a superposition of two counter-propagating waves of equal amplitude and opposite direction. We can understand this result intuitively from the following argument: with running waves, it is possible in principle to know which running wave exchanges a unit of momentum with the atom. In contrast, a standing wave is an inseparable quantum unit with zero average momentum. This unity is imposed by the fixed mirrors that establish the standing wave and that act as infinite sinks and sources of momentum. Quantum mechanics forbids one *even in principle* to determine, via a field measurement, "which traveling wave" exchanges momentum with the atom, and hence one expects interference phenomena. Indeed the atomic diffraction patterns reflect this fundamental difference between a "true" standing wave and a superposition of two running waves [see Shore *et al.* (1991)].

12-3. Single-Mode Field in Thermal Equilibrium

Consider again a single mode electromagnetic field with the Hamiltonian (6) and state vector (8), or more generally a density operator ρ. As such, a quantum mechanical single-mode electromagnetic field is characterized by an infinite number of complex numbers $c_n(t)$, while in the absence of classical fluctuations, its classical counterpart is completely determined by a single complex number, or equivalently by the real coefficient $q(t)$ in Eq. (1) and a phase. This fact alone illustrates the wealth of the quantum world as opposed to its classical counterpart. It is true that some kinds of quantal fluctuations can be modeled by classical statistics, but not all (see Sec. 12-6 on distributions). In the next two sections, we discuss two of the most important states of a single mode quantized field, namely a thermal "state" and a coherent state.

A thermal single-mode field is a field from which we only know the average energy $\langle\mathcal{H}\rangle$

$$\langle\mathcal{H}\rangle = \mathrm{tr}\{\rho\mathcal{H}\}, \tag{26}$$

Our goal here is to determine the density matrix ρ describing this field. We can find ρ much as in classical physics by using Lagrange multipliers and maximizing the entropy of the mode subject to constraints like Eq. (26). The classical entropy of a system is given by

$$S = -k_B\,\Sigma_\ell\, P_\ell \ln P_\ell, \tag{27}$$

where k_B is Boltzmann's constant and P_ℓ is the probability of finding the system in the state ℓ. In quantum mechanics, the concept of entropy still holds, but P_ℓ is replaced by the density matrix ρ, and the sum is replaced by the trace

$$S = -k_B\,\mathrm{tr}\{\rho\ln\rho\}. \tag{28}$$

The quantum equivalent of the normalization condition $\Sigma_\ell P_\ell = 1$ is

$$\mathrm{tr}\{\rho\} = 1. \tag{29}$$

Equations (29) and (26) are the two constraints under which we wish to maximize the entropy.

We proceed by finding the change δS in entropy corresponding to a small variation $\delta\rho$ of the density matrix. Without the constraints (26) and (29), we find from Eq. (28)

$$\delta S \simeq -\, k_B\,\mathrm{tr}\left\{\frac{\partial}{\partial\rho}(\rho\ln\rho)\delta\rho\right\} = -k_B\,\mathrm{tr}\{(1 + \ln\rho)\delta\rho\}\ . \tag{30}$$

The constraint (29) leads to an extra contribution $\mathrm{tr}\{\delta\rho\} = 0$, while Eq. (26) gives $\mathrm{tr}\{\mathcal{H}\delta\rho\} = 0$. Inserting these values into Eq. (30), we find

$$\delta S = -k_B\ \mathrm{tr}\{(1 + \ln\rho + \lambda + \beta\mathcal{H})\delta\rho\}\ . \tag{31}$$

To maximize the entropy, we must have $\delta S = 0$ for any $\delta\rho$, which yields

$$1 + \ln\rho + \lambda + \beta\mathcal{H} = 0,$$

i.e.,

$$\rho = e^{-(1+\lambda)}\exp(-\beta\mathcal{H})\ . \tag{32}$$

We still have to determine the two Langrange multipliers λ and β from the constraints (26) and (29). Substituting Eq. (32) into (29), we find

$$e^{1+\lambda} = \mathrm{tr}\{\exp(-\beta\mathcal{H})\} \equiv Z\ , \tag{33}$$

where Z is the so-called partition function of the system. Substituting this into Eq. (32), we have

$$\boxed{\rho = \frac{\exp(-\beta\mathcal{H})}{\mathrm{tr}\{\exp(-\beta\mathcal{H})\}} = \frac{\exp(-\beta\mathcal{H})}{Z}}\ . \tag{34}$$

From classical statistical mechanics, we recognize

$$\beta \equiv 1/k_B T \tag{35}$$

as the Boltzmann coefficient, which we use as a definition of the temperature T. So far, our discussion is quite general. The density operator of Eq. (34) describes any system with Hamiltonian $\mathcal{H}$ in thermal equilibrium. We use this fact explicitly in the reservoir theory of Chap. 14.

We now specialize our discussion to the case of the simple harmonic oscillator Hamiltonian (6). We proceed by redefining the zero of the energy scale by removing the "zero-point energy" $\hbar\Omega/2$ from $\mathcal{H}$. The density operator (34) becomes then

$$\rho = \frac{e^{-\beta\hbar\Omega a^\dagger a}}{\mathrm{tr}\{e^{-\beta\hbar\Omega a^\dagger a}\}} . \tag{36}$$

In general we can expand the field density operator in any complete set of states such as

$$\rho = \sum_n \sum_m |n\rangle\langle n|\rho|m\rangle\langle m| = \sum_n \sum_m \rho_{nm}\ |n\rangle\langle m| . \tag{37}$$

Noting that $\langle n|a^\dagger a|m\rangle = n\delta_{nm}$, we obtain the photon number expansion

$$\rho_{nm} = e^{-n\beta\hbar\Omega}\,[\Sigma_n\ e^{-n\beta\hbar\Omega}]^{-1}\,\delta_{nm} = e^{-n\beta\hbar\Omega}\,[1 - e^{-\beta\hbar\Omega}]\,\delta_{nm} . \tag{38}$$

Hence we see that the thermal distribution has a diagonal expansion in terms of the photon number states. This diagonality causes the electric field expectation value to vanish in thermal equilibrium.

From this we can calculate the probability p_n $(=\rho_{nn})$ that the field has n photons, the so-called *"photon statistics."* From Eq. (38), we find

$$p_n = [1-e^{-\beta\hbar\Omega}]\ e^{-n\beta\hbar\Omega} . \tag{39}$$

This is a Maxwell-Boltzmann distribution illustrated in Fig. 12-2 and justifies calling a field with such statistical properties a thermal field.

The density matrix (36) permits to compute any observable of interest, such as e.g. the mean energy in the field. We have

$$\langle\mathcal{H}\rangle = \mathrm{tr}\{\rho\mathcal{H}\} = \frac{1}{Z}\,\mathrm{tr}\{\hbar\Omega a^\dagger a\ e^{-\beta\hbar\Omega a^\dagger a}\} = \frac{1}{Z}\,\Sigma_n\ \hbar\Omega n\ \exp(-\hbar\Omega n/k_B T) \tag{40}$$

Differentiating both sides of the partition function Z gives

$$\frac{dZ}{dT} = \frac{1}{k_B T^2}\sum_n n\hbar\Omega\ \exp(-\hbar\Omega n/k_B T) . \tag{41}$$

Comparing this result with Eq. (40), we find

$$\langle\mathcal{H}\rangle = k_B T^2\,\frac{1}{Z}\,\frac{dZ}{dT} = \frac{\hbar\Omega}{\exp(\hbar\Omega/k_B T) - 1} , \tag{42}$$

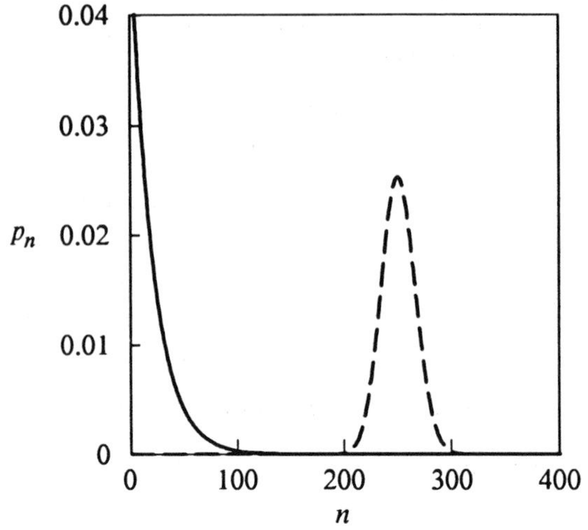

Fig. 12-2. Thermal distribution (solid line) given by photon statistics formula (39) with $\beta\hbar\Omega$ = .05; Poisson distribution (dashed line) of Eq. (54) for the coherent state with $\alpha^*\alpha = 250$.

or, reintroducing the "zero-point energy" $\hbar\Omega/2$,

$$\langle\mathscr{H}\rangle = \frac{\hbar\Omega}{2} + \hbar\Omega/(\exp(\hbar\Omega/k_B T - 1) \ . \tag{43}$$

For $T = 0$, $\langle\mathscr{H}\rangle = \hbar\Omega/2$. At absolute zero, the oscillator is in its ground state, with energy $\hbar\Omega/2$. In contrast, in thermal equilibrium the classical oscillator energy is $2\cdot\frac{1}{2}k_B T$, which vanishes as $T \to 0$. Note that for high temperatures, $\hbar\Omega << k_B T$, we find from Eq. (42) that $\langle\mathscr{H}\rangle \to k_B T$, that is, the quantum and classical oscillators have the same mean energy.

In Eq. (43), it was important to keep the zero-point energy of the oscillator since we wish to compare the quantum to the classical energy and therefore need to use the same reference point in both cases. In most of this book, however, we redefine the energy of the quantum oscillator such that its zero-point energy is zero, a considerable typographical simplification. The mean energy calculation also permits us to find the mean number of photons $\bar{n}$ in the mode, namely

$$\bar{n} = \Sigma_n \, n\rho_{nn} = \frac{1}{e^{\beta\hbar\Omega} - 1} \, . \tag{44}$$

12-4. Coherent States

In this section, we discuss a class of states of the simple harmonic oscillator that play a central role in the quantum theory of radiation discussed in Chaps. 12 - 18. These states are defined as those that minimize the Heisenberg uncertainty relation $\Delta p \Delta q \geq 1/2 \, |\langle [p,q] \rangle| = \hbar/2$, where $\Delta p = (\langle p^2 \rangle - \langle p \rangle^2)^{1/2}$ is the root-mean-square deviation of p and similarly for Δq. One subclass, the so-called "coherent" states, simultaneously minimizes the spread in both the position and momentum operators. More generally we can minimize the uncertainty product with a smaller uncertainty for one conjugate variable at the expense of increasing that for the other. Such states are called "squeezed coherent states" (see Sec. 16-1).

There are several ways to introduce coherent states. Here we choose a method that emphasizes their nearly classical character. Specifically we seek pure states of the harmonic oscillator with mean energy equal to the classical energy. We proceed by applying Ehrenfest's theorem, which states that in the case of free particles, of particles in uniform fields, as well as of particles in quadratic potentials (harmonic oscillators), the motion of the center of the quantum wave packet obeys precisely the laws of classical mechanics

$$\begin{aligned} \langle\psi| q(t) |\psi\rangle &= q_c(t) , \\ \langle\psi| p(t) |\psi\rangle &= p_c(t) , \end{aligned} \tag{45}$$

where we use the index c to label the classical variables. Inserting these into Eq. (3.113), we obtain the classical energy

$$\mathcal{H}_c = \frac{p_c^{\,2}}{2} + \frac{\Omega^2 q_c^{\,2}}{2} = \frac{1}{2} \left[\langle\psi| p(t) |\psi\rangle^2 + \Omega^2 \langle\psi| q(t) |\psi\rangle^2 \right] . \tag{46}$$

With the substitution of Eqs. (3.120) and (3.121), $\mathcal{H}_c$ becomes

$$\mathcal{H}_c = \hbar\Omega \langle\psi| a^\dagger |\psi\rangle \langle\psi| a |\psi\rangle . \tag{47}$$

The corresponding quantum-mechanical oscillator has the energy

$$\langle\mathcal{H}\rangle = \langle\psi| \mathcal{H} |\psi\rangle = \hbar\Omega \langle\psi| a^\dagger a |\psi\rangle , \tag{48}$$

where we have shifted the zero of energy by the zero-point energy $\hbar\Omega/2$. Our requirement that the classical energy be equal to the quantum-mechanical energy for the coherent state $|\psi\rangle$ leads therefore to the factorization condition

$$\langle\psi|a^\dagger|\psi\rangle\langle\psi|a|\psi\rangle = \langle\psi|a^\dagger a|\psi\rangle \,. \tag{49}$$

From our discussion of coherence in Sec. 1-4, it is natural to follow Glauber (1963, 1964) and call states satisfying this condition coherent states. The right-hand side of Eq. (49) has the same form as the first-order correlation $G^{(1)}(t;t)$ (see Eq. (1.80)), except that the classical ensemble average is replaced by a quantum-mechanical average. Furthermore, $\langle\psi|a|\psi\rangle$ is the average of the operator a in the state $|\psi\rangle$, and is some complex number $\mathcal{E}(t)$, since a is nonhermitian. Thus Eq. (49) may be rewritten formally as

$$G^{(1)}(t;\, t) = \mathcal{E}^*(t)\mathcal{E}(t) \,, \tag{50}$$

which is formally the same as the coherence condition (1.88) for $\mathbf{r}_1 t_1 = \mathbf{r}_2 t_2$.

We can show that the definition of the coherent state (49) implies that coherent states are eigenstates of the annihilation operator

$$\boxed{a|\alpha\rangle = \alpha|\alpha\rangle} \,. \tag{51}$$

It is immediate to see by direct substitution that Eq. (51) does satisfy Eq. (49). To show that Eq. (49) implies (51), we note that Eq. (49) may be written as

$$|\langle\psi|a^\dagger|\psi\rangle|^2 = \langle\psi|a^\dagger a|\psi\rangle \,. \tag{52}$$

The Gram-Schmidt orthogonalization procedure tells us that starting with $|\psi\rangle$, we can construct a complete orthonormal basis set consisting of $|\psi\rangle$ and an infinite complementary set of vectors $\{|R\rangle\}$. Writing this statement in terms of the identity operator, we have

$$I = |\psi\rangle\langle\psi| + \Sigma_R\; |R\rangle\langle R| \,.$$

Inserting this operator in between the $a^\dagger$ and the a on the right-hand side of Eq. (52), we have

$$\begin{aligned}\langle\psi|a^\dagger a|\psi\rangle &= \langle\psi|a^\dagger|\psi\rangle\langle\psi|a|\psi\rangle + \Sigma_R\; \langle\psi|a^\dagger|R\rangle\langle R|a|\psi\rangle \\ &= |\langle\psi|a^\dagger|\psi\rangle|^2 + \Sigma_R\; |\langle R|a|\psi\rangle|^2 \,.\end{aligned}$$

Equating this result with the LHS of Eq. (52), we find

$$\Sigma_R\; |\langle R|a|\psi\rangle|^2 = 0.$$

Because every term in this sum is positive definite, we must have $\langle R|a|\psi\rangle = 0$ for all $|R\rangle$, which implies that $a|\psi\rangle$ must be orthogonal to any $|R\rangle$, i.e., proportional to $|\psi\rangle$, namely

$$a|\psi\rangle = \lambda_\psi|\psi\rangle \; .$$

This concludes the proof that all states satisfying the factorization property (49) must be eigenstates of the annihilation operator. To agree with the usual notation, we write these states as $|\alpha\rangle$ with eigenvalue α as in Eq. (51).

To obtain an explicit form for $|\alpha\rangle$ in terms of the number states $|n\rangle$, we write

$$|\alpha\rangle = \sum_n |n\rangle\langle n|\alpha\rangle = \sum_n |n\rangle\langle 0|a^n/\sqrt{n!}\,|\alpha\rangle = \langle 0|\alpha\rangle \sum_n \frac{\alpha^n}{\sqrt{n!}}|n\rangle,$$

where we have used Eq. (3.137) for $|n\rangle$. Using the normalization condition $\langle\alpha|\alpha\rangle = 1$, we find that $|\langle 0|\alpha\rangle|^2 = e^{-|\alpha|^2}$. Choosing a unit phase factor, we obtain

$$\boxed{|\alpha\rangle = e^{-|\alpha|^2/2} \sum_n \frac{\alpha^n}{\sqrt{n!}}|n\rangle} \; . \tag{53}$$

This form immediately gives the probability of finding n photons (the photon statistics) in the coherent state as

$$\boxed{p_n = |\langle n|\alpha\rangle|^2 = e^{-|\alpha|^2}\frac{|\alpha|^{2n}}{n!}} \; . \tag{54}$$

This is called a Poisson distribution and is illustrated in Fig. 12-2. This is centered at $n \simeq |\alpha|^2$ with width $|\alpha|$ (see Prob. 12-8).

It is also useful to express the coherent states in terms of the vacuum state $|0\rangle$. Substituting Eq. (3.137) for $|n\rangle$ into Eq. (53), we find

$$\begin{aligned} |\alpha\rangle &= e^{-|\alpha|^2/2} \sum_n \frac{(\alpha a^\dagger)^n}{n!}|0\rangle = e^{-|\alpha|^2/2}\, e^{\alpha a^\dagger}|0\rangle \\ &= e^{-|\alpha|^2/2}\, e^{\alpha a^\dagger}\, e^{-\alpha^* a}|0\rangle \; , \end{aligned} \tag{55}$$

where we have used the fact that $a|0\rangle = 0$ to perform the last step. Using the Baker-Hausdorff relation (Prob. 3-19)

$$e^{A+B} = e^A e^B e^{-[A,B]/2} , \tag{56}$$

which holds if A and B are operators that commute with their commutator, $[A,[A,B]] = [B,[A,B]] = 0$, we can write Eq. (55) as

$$|\alpha\rangle = D_\alpha |0\rangle . \tag{57}$$

Here

$$\boxed{D_\alpha \equiv e^{\alpha a^\dagger - \alpha^* a}} \tag{58}$$

is the "displacement" operator, and we can call the coherent states $|\alpha\rangle$ displaced states of the vacuum.

To get a physical picture of this displacement, we write $|\alpha\rangle$ in a two-dimensional coordinate representation, one dimension for q and one for p. Equation (3.121)

$$p = -i\sqrt{\hbar\Omega/2}\,(a - a^\dagger) \tag{3.121}$$

has the coordinate representation $-i\hbar d/dq$. Hence $a^\dagger - a$ has the coordinate representation $-d/d\xi$, where the dimensionless position coordinate $\xi \equiv (\Omega/2\hbar)^{1/2}q$. In terms of ξ, the ground-state eigenfunction of Eq. (3.139) is given by

$$u_0(\xi) = (2/\pi)^{1/4}\, e^{-\xi^2} .$$

Similarly the position operator q has the momentum-space representative $i\hbar d/dp$. Combining this with Eq. (3.120), we find that $a^\dagger + a$ has the momentum representation $-d/d\eta$, where the dimensionless momentum coordinate $\eta \equiv (2\hbar\Omega)^{-1/2}p$. Using both coordinate axes, the coordinate representation of $D(\alpha)$ of Eq. (58) is

$$D_\alpha(\xi,\eta) = e^{-\alpha_r \partial/\partial\xi \,-\, \alpha_i \partial/\partial\eta} , \tag{59}$$

where $\alpha = \alpha_r + i\alpha_i$. When operating on the function $f(\xi,\eta)$, $D_\alpha(\xi,\eta)$ defines a two-dimensional Taylor series that displaces the function to the value $f(\xi - \alpha_r, \eta - \alpha_i)$. Projecting Eq. (57) onto the two-dimensional coordinate eigenvector $|\xi\eta\rangle$, we have

$$\langle\xi\eta|\alpha\rangle = \langle\xi\eta|D_\alpha|0\rangle = \iint d\xi' d\eta' \langle\xi\eta|D_\alpha|\xi'\eta'\rangle\langle\xi'\eta'|0\rangle \ . \qquad (60)$$

Noting that D_α is a differential operator and therefore local (see Sec. 3-1), we have $\langle\xi\eta|D_\alpha|\xi'\eta'\rangle = D_\alpha(\xi,\eta)\delta(\xi - \xi')\delta(\eta - \eta')$, which gives

$$\begin{aligned}\langle\xi\eta|\alpha\rangle &= D_\alpha(\xi,\eta)u_0(\xi)u_0(\eta) = e^{-\alpha_r \partial/\partial\xi \ - \ \alpha_i \partial/\partial\eta} u_0(\xi)u_0(\eta) \\ &= u_0(\xi - \alpha_r)u_0(\eta - \alpha_i) \ ,\end{aligned} \qquad (61)$$

The function $u_0(\xi - \alpha)$ is the displaced ground state of the harmonic oscillator.

Furthermore this displaced state oscillates back and forth across the ξ origin at the frequency Ω. To see this, note that the time-dependent coherent state $|\alpha(t)\rangle$ is given by Eq. (53) with the replacement $\alpha \rightarrow \alpha\exp(-i\Omega t)$. Hence Eq. (61) becomes

$$\langle\xi\eta|\alpha(t)\rangle = \phi_0(\xi - |\alpha|\cos(\Omega t+\theta))\phi_0(\eta - |\alpha|\sin(\Omega t+\theta)) \ , \qquad (62)$$

where $\alpha = |\alpha|e^{i\theta}$. Since this wave function oscillates with constant width, we say that it "coheres" in time, which is another reason to call it a coherent state. Chapter 18 deals with similar wave functions with time-varying widths, called squeezed states.

More generally, D_α displaces the two-dimensional ground-state simple-harmonic-oscillator wave function to a point α in the complex plane. In time this displacement is given by $\alpha\exp(-i\Omega t)$, which rotates about the origin at the optical frequency Ω. A projection on any particular axis through the origin looks like Eq. (62) with α replaced by $|\alpha|$, aside from a phase shift in time.

As next section shows, the coherent states are very useful in describing the electromagnetic field, although they have the complication of being overcomplete (Prob. 12-1)

$$\int d(\mathrm{Re}\alpha)d(\mathrm{Im}\alpha)\ |\alpha\rangle\langle\alpha| = \pi \qquad (63)$$

and nonorthogonal (Prob. 12-2)

$$\langle\alpha|\beta\rangle = \exp[-\tfrac{1}{2}(|\alpha|^2 + |\beta|^2 - 2\alpha^*\beta)] \ , \qquad (64)$$

which does not vanish for $\alpha \neq \beta$. Squaring Eq. (64)

$$|\langle\alpha|\beta\rangle|^2 = \exp(-|\alpha-\beta|^2) \ , \qquad (65)$$

we see that the states becomes increasingly orthogonal if α differs sufficiently from β.

12-5. Coherence of Quantum Fields

Section 1-4 discusses nth-order coherence classically. An important result is that first-order coherence and monochromaticity are related simply only for stationary fields. This section generalizes the treatment of coherence to quantized fields. This is easy to do; all we need to do is to reinterpret the averages of Sec. 1-4 as quantum averages.

In particular, in the quantum case Eq. (1.80) becomes

$$\boxed{G^{(1)}(x_1,x_2) \equiv \mathrm{tr}\{\rho E^-(x_1)E^+(x_2)\}}\ , \tag{66}$$

where $x_n \equiv (r_n, t_n)$, E^- and E^+ are the negative and positive frequency parts of the electric field operator (see Eq. (1.9)). Using Eq. (9) and remembering the free evolution (3.125) and (3.126) of the annihilation and creation operators

$$a(t) = a(0)e^{-i\Omega t} \tag{3.125}$$

$$a^\dagger(t) = a^\dagger(0)e^{i\Omega t}\ , \tag{3.126}$$

we find that for a single-mode field

$$E^+(x) = \mathcal{E}_\Omega a e^{-i\Omega t}\sin Kz \tag{67}$$

$$E^-(x) = (E^+(x))^\dagger = \mathcal{E}_\Omega a^\dagger e^{i\Omega t}\sin Kz\ . \tag{68}$$

With Eq. (66), this yields the single-point correlation function

$$G^{(1)}(x_1,x_1) = \mathcal{E}_\Omega{}^2\sin^2 Kz\,\mathrm{tr}\{\rho a^\dagger a\}\ . \tag{69}$$

Here $\mathrm{tr}\{\rho a^\dagger a\} = \langle n\rangle$ is the average number of photons in the field. For the monochromatic field, the two-point correlation function of Eq. (66) is given by

$$G^{(1)}(x_1,x_2) = \mathcal{E}_\Omega{}^2\sin Kz_1\sin Kz_2 e^{i\Omega(t_1-t_2)}\mathrm{tr}\{\rho a^\dagger a\}\ . \tag{70}$$

This shows that the first-order correlation function is insensitive to the photon statistics, since Eq. (70) only depends on the average photon number. We see the same first-order intensity interference pattern using filtered blackbody radiation with the photon statistics of Eq. (38) as we see using coherent light with the statistics of Eq. (54). Higher-order correlation functions do allow us to distinguish between such sources. The quantum-mechanical nth-order coherence function is given by

$$G^{(n)}(x_1 \ldots x_1) \propto \mathrm{tr}\{\rho a^\dagger a^\dagger \cdots a^\dagger aa \cdots a\} . \tag{71}$$

We wish to show how these correlation functions enter into the quantum theory of photon detection. We suppose that the response of our atomic detector can be described by first-order atom-field interactions, in the spirit of the Fermi Golden rule. Classically the functions E^+ and E^- play equally important roles, because in the limit $\hbar\Omega \to 0$ test charges absorb radiation as readily as they emit it. However for optical frequency transitions in atoms in thermal equilibrium, the probability that atoms are initially excited is negligibly small [$\propto \exp(-\hbar\Omega/k_BT) \simeq \exp(-40)$ for room temperature]. As such, the atoms start in their lower state and can only absorb radiation. Mathematically, this means that only the annihilation operator plays a significant role.

To get a quantitative description of how such an atomic detector works at optical frequencies, we calculate the transition probability for absorbing a photon at the position $\mathbf{r}$ at time t, i.e, at $x = (\mathbf{r}, t)$. This is proportional to

$$w_{i\to f} = |\langle f|E^+(x)|i\rangle|^2, \tag{72}$$

where $|i\rangle$ is the initial state of the coupled atom-field system before interacting with the field, and $|f\rangle$ is its final state. We are only interested in the photon counting rate and not in the final state. Hence our counting rate w is proportional to the sum over all possible final states, that is,

$$w = \sum_f |\langle f|E^+(x)|i\rangle|^2 = \langle i|E^-(x)E^+(x)|i\rangle , \tag{73}$$

where we have used the completeness relation $\Sigma_f |f\rangle\langle f| = 1$. Furthermore, although we do know that the atom starts in the ground state, we typically do not know the initial state of the field precisely. To allow for the corresponding statistical variations, we average the rate (73) over $|i\rangle$ using the field density operator $\rho = [|i\rangle\langle i|]_{av.\ over\ i}$. Inserting this into Eq. (73), we obtain

$$w = \mathrm{tr}[\rho E^-(x)E^+(x)] . \tag{74}$$

Comparing this to Eq. (66), we see that the counting rate (74) of our ideal photodetector is in fact just the first-order correlation function.

The fact that the coherent states are eigenstates of the annihilation operator [Eq. (51)] is very important in quantum optics. Because the quantized single-mode complex electric field is proportional to the annihilation operator [Eq. (67)], the coherent states are eigenstates of this field. In calculations involving fully quantum mechanical matter-field interactions, this eigenstate property often reduces operator algebra to algebra of complex functions. This simplification is however somewhat offset by the fact that the coherent states are overcomplete [see Eq. (63)].

Higher-order interference experiments such as that of Hanbury-Brown and Twiss require the use of higher-order correlation functions like

$$\boxed{G^{(n)}(x_1...x_n, y_1...y_n) = \mathrm{tr}\{\rho E^-(x_1) \cdots E^-(x_n)\, E^+(y_1) \cdots E^+(y_n)\}} \,. \tag{75}$$

We could have chosen a larger class of correlation functions including unequal numbers of creation and annihilation operators, but they do not correspond to quantities measured in typical photon-counting experiments.

At this point, we have achieved to develop a very close analogy between the quantum mechanical and classical theories of coherence. We can use the results of Sec. 1-4 with a straightforward reinterpretation of the averaging process, keeping in mind that the electric field is now an operator. In particular the field is said to exhibit nth-order coherence if all of its mth-order correlation functions for $m \le n$ satisfy

$$G^{(m)}(x_1...x_m, y_1...y_m) = \mathcal{E}^*(x_1) \cdots \mathcal{E}^*(x_m)\, \mathcal{E}(y_1) \cdots \mathcal{E}(y_m) \,, \tag{76}$$

where $\mathcal{E}$ is a complex function.

As a first example, consider a single mode of the electromagnetic field in an eigenstate $|n\rangle$, i.e., $\rho = |n\rangle\langle n|$. From Eq. (76), a field possessing second-order coherence satisfies

$$G^{(2)}(x_1x_1, x_1x_1) = |\mathcal{E}(x_1)|^4 = |G^{(1)}(x_1)|^2 \,. \tag{77}$$

However directly calculating $G^{(1)}$ and $G^{(2)}$ from Eq. (75), we find

$$|G^{(1)}(x_1)|^2 = |\mathcal{E}_\Omega \sin Kz|^4 n^2 \,, \tag{78}$$

$$G^{(2)}(x_1x_1, x_1x_1) = |\mathcal{E}_\Omega \sin Kz|^4 n(n-1) \,. \tag{79}$$

These values do not satisfy Eq. (77), and hence an n-photon state does not possess second-order coherence. It is left as an exercise to show that the thermal field of Sec. 12-3 does not exhibit second-order coherence either.

Section 1-4 briefly mentions the nonclassical phenomenon of antibunching, which is characterized by a normalized second-order correlation function (Eq. (1.97) at τ=0)

$$g^{(2)}(0) = \frac{G^{(2)}(0)}{|G^{(1)}(0)|^2} \tag{80}$$

that becomes smaller than 1. From Eqs. (78) and (79) we see that for the n-photon state,

$$g^{(2)}(0) = 1 - \frac{1}{n} . \tag{81}$$

As such the n-photon state exhibits antibunching, although as $n \to \infty$, the classical lower limit of unity is attained. In contrast, Prob. 12-3 shows that the thermal field exhibits bunching, rather than antibunching.

We conclude this section by determining which states of the field are coherent to all orders. In Sec. 12-4, we showed that the only pure states $|\psi\rangle$ of the field that factorize as in Eq. (49) are the coherent states $|\alpha\rangle$, which satisfy Eq. (51), that is, these are the only pure states exhibiting second-order coherence. As we readily see, they also satisfy the general coherence condition (76), since for all m we have

$$\begin{aligned}\langle\alpha| a^\dagger(x_1)a^\dagger(x_2) \cdots a^\dagger(x_m)\, a(y_m) \cdots a(y_1)|\alpha\rangle \\ = \mathcal{E}^*(x_1)\mathcal{E}^*(x_2) \cdots \mathcal{E}^*(x_m)\, \mathcal{E}(y_m) \cdots \mathcal{E}(y_1) ,\end{aligned} \tag{82}$$

where the field amplitudes

$$\mathcal{E}^*(x_i) \equiv \alpha \exp(-i\Omega t_i) . \tag{83}$$

The generalization to many-mode fields is straightforward.

12-6. $P(\alpha)$ Representation

In many problems, it is useful to describe the state of the field in terms of coherent states, rather than with photon number states. This presents some surprises and minor difficulties since as discussed in Sec. 3-4, the coherent states are not orthogonal and are overcomplete. On the other hand, as we see in this section, this overcompleteness allows us to obtain a useful diagonal expansion of the density operator with complex matrix element $P(\alpha)$. This contrasts with the number state representation, for which ρ has the general nondiagonal expansion (35). We use this $P(\alpha)$ representation in the quantum Fokker-Planck formalism of Sec. 14-2.

Specifically we expand ρ as

$$\boxed{\rho = \int d^2\alpha \, P(\alpha)|\alpha\rangle\langle\alpha|} \, . \tag{84}$$

In terms of this expansion, the expectation value of an operator $\mathcal{O}$ is given by

$$\begin{aligned}\langle\mathcal{O}\rangle = \mathrm{tr}\{\rho\mathcal{O}\} &= \sum_n \langle n| \int d^2\alpha \, P(\alpha)|\alpha\rangle\langle\alpha|\mathcal{O}|n\rangle \\ &= \int d^2\alpha P(\alpha) \sum_n \langle\alpha|\mathcal{O}|n\rangle\langle n|\alpha\rangle \\ &= \int d^2\alpha P(\alpha)\mathcal{O}(\alpha) \, ,\end{aligned} \tag{85}$$

where $\mathcal{O}(\alpha) = \langle\alpha|\mathcal{O}|\alpha\rangle$. This leads to simple calculations involving only c-numbers provided that the operator $\mathcal{O}$ is expressed in *normal order*, that is, so that creation operators stand to the left of annihilation operators in any product appearing in $\mathcal{O}$.

As an example, we derive $P(\alpha)$ for the thermal field. This is most easily accomplished by using Fourier transforms and the $Q(\alpha)$ distribution

$$Q(\alpha) \equiv \langle\alpha|\rho|\alpha\rangle = \int d^2\beta P(\beta)\langle\alpha|\beta\rangle\langle\beta|\alpha\rangle$$

or

$$\boxed{Q(\alpha) = \int d^2\beta P(\beta) e^{-|\alpha-\beta|^2}} \, , \tag{86}$$

which is the convolution of $P(\alpha)$ with a Gaussian. We can determine $P(\alpha)$ by taking the two-dimensional Fourier transform of Eq. (86), which by the inversion theorem for convolutions is the product

$$\mathcal{F}[Q(\alpha)] = \mathcal{F}[P(\beta)]\mathcal{F}[\exp(-|\alpha|^2)] \, . \tag{87}$$

Dividing through by $\mathcal{F}[\exp(-|\alpha|^2)]$ and taking the inverse transform, we find

$$P(\beta) = \mathcal{F}^{-1} \frac{\mathcal{F}[Q(\alpha)]}{\mathcal{F}[\exp(-|\alpha|^2)]} \; . \tag{88}$$

In fact using Eq. (38), we have (setting $x = \hbar\Omega/k_B T$)

$$Q(\alpha) = \sum_n \sum_m \langle\alpha|n\rangle\rho_{nm}\langle m|\alpha\rangle = (1 - e^{-x}) \sum_n e^{-nx} \langle\alpha|n\rangle\langle n|\alpha\rangle$$

$$= (1 - e^{-x}) \sum_n e^{-nx} \frac{(\alpha^*\alpha)^n}{n!} e^{-\alpha^*\alpha} = (1 - e^{-x})\, e^{-|\alpha|^2} e^{|\alpha|^2 e^{-x}}$$

$$= (1 - e^{-x}) \exp[-|\alpha|^2(1 - e^{-x})] \; . \tag{89}$$

Inverting Eq. (44) for the average photon number $\bar{n}$, we find $1 - e^{-x} = 1/(\bar{n} + 1)$, which gives for Eq. (89)

$$Q(\alpha) = (\bar{n} + 1)^{-1} e^{-|\alpha|^2/(\bar{n} + 1)} \; . \tag{90}$$

Combining Eqs. (87), (88), and (90), we find

$$\mathcal{F}[Q(\alpha)] = (\bar{n} + 1)^{-1} \mathcal{F}[e^{-x^2/(\bar{n} + 1)}]\, \mathcal{F}[e^{-y^2/(\bar{n} + 1)}]$$

$$= \pi\, e^{-k^2(\bar{n} + 1)/4} \; , \tag{91}$$

where we have set $|\alpha|^2 = \mathrm{Re}(\alpha)^2 + \mathrm{Im}(\alpha)^2 = x^2 + y^2$. Similarly, we have

$$\mathcal{F}[\exp(-|\alpha|^2)] = \pi\, e^{-k^2/4} \; .$$

Substituting this and Eq. (91) into (88), we have

$$P(\alpha) = \mathcal{F}^{-1}[e^{-k^2/4\bar{n}}] = (\pi\bar{n})^{-1} e^{-|\alpha|^2/\bar{n}} \; . \tag{92}$$

It is interesting to note that, in the classical limit of large $\bar{n}$, $Q(\alpha)$ given by Eq. (90) and $P(\alpha)$ coincide. Also in the classical limit, distinctions depending on the ordering of operators vanish. This point is discussed further in Glauber's Les Houches Lectures (1965).

The probability for finding n photons is given by the photon statistical distribution ρ_{nn} from Eq. (38). This exponentially decaying distribution contrasts with the Poisson distribution for the coherent state. The difference between $P(\alpha)$ for the two cases is perhaps even more striking, since the thermal light $P(\alpha)$ has a Gaussian dependence (92) on the magnitude of α, while the coherent state $|\alpha\rangle$ case is the Dirac delta function $P(\alpha) = \delta(\alpha - \alpha_0)$. We have interpreted $P(\alpha)$ as a probability distribution for finding the coherent state $|\alpha\rangle$. This is not always correct, however, for $P(\alpha)$ may take on negative values or singularities worse than the delta function. In particular, whenever the photon distribution becomes narrower than the Poisson distribution for the coherent state, $P(\alpha)$ becomes badly behaved. Such a case is the photon number state $|n\rangle$ as discussed in Prob. 12-4.

Sometimes fields that can be described by a positive definite $P(\alpha)$ are called "classical" fields, since they can be simulated by a classical stochastic process (see Chap. 14). This does not mean that such "classical" fields have vanishing quantum mechanical uncertainties. For example, the coherent state itself is described by a positive definite Dirac delta function, but has minimum, not vanishing, quantum mechanical uncertainties. As such the coherent state is the most nearly classical state, but it is not classical.

In the same fashion that $P(\alpha)$ is convenient to evaluate the expectation values of normally ordered operators, a number of other quasi-distributions have been introduced to evaluate operators in different orderings. Two such quasi-distributions worth mentioning are the $Q(\alpha)$ distribution encountered in Eq. (86) and the Wigner distribution

$$W(\alpha_1,\alpha_2) = \frac{2}{\pi}\int d^2\beta P(\beta) e^{-2|\beta - \alpha_1 - i\alpha_2|^2} . \tag{93}$$

The $Q(\alpha)$ representation is well adapted to evaluate the expectation value of operators expressed in *antinormal order*, while the Wigner distribution is particularly suited in the case of operators expressed in terms of symmetrized products of $a_1 = (a + a^\dagger)/\sqrt{2}$ and $a_2 = i(a^\dagger - a)/\sqrt{2}$ (Haken 1975). It is possible to define a single distribution function that varies continuously between the P, W, and Q distributions [see, for example, Vogel and Risken (1989)]. While $P(\alpha)$ is particularly suited to describing square-law detection, the Wigner distribution is ideal for treating homodyne detection, used in the detection of squeezed states [see Lai and Haus (1990)].

References

Cohen-Tannoudji, C., J. Dupont-Roc, and G. Grynberg (1989), *Photons and Atoms, Introduction to Quantum Electrodynamics*, John-Wiley & Sons, New York.

Glauber, R. J. (1963), Phys. Rev. **130**, 2529, and **131**, 2766. These are the classic and still unsurpassed papers on coherent states.

Glauber, R. J. (1965), in: *Quantum Optics and Electronics*, Les Houches, Ed. by C. DeWitt, A. Blandin, and C. Cohen-Tannoudji, Gordon and Breach, New York. *pp.* 331-381, gives a pedagogical review of quantum coherence, coherent states and detection theory.

Haken, H. (1975), Rev. Mod. Phys. **47**, 67.

Klauder, J. R. and E. C. G. Sudarshan (1968), *Fundamentals of Quantum Optics*, W. A. Benjamin, New York.

Lai, Y. and H. Haus (1990), Quant. Opt. **1**, 99.

Louisell, W. H. (1990), *Quantum Statistical Properties of Radiation*, John Wiley & Sons, New York. A classic reference book on boson operator algebra, quantized fields and their applications in quantum optics.

Nussenzveig, H. M. (1974), *Introduction to Quantum Optics*, Gordon and Breach, New York.

Risken, H. (1984), *The Fokker-Planck Equation*, Springer-Verlag, Heidelberg.

Sargent, M. III, M. O. Scully, and W. E. Lamb, Jr. (1977), *Laser Physics*, Addison-Wesley Publishing Co., Reading, MA. See especially Chaps. 14 and 15.

Shore, B. J., P. Meystre and S. Stenholm (1990), J. Opt. Soc. Am., in press.

Vogel, K. and H. Risken (1989), Phys. Rev. A**39**, 4675.

Problems

12-1. By converting to polar coordinates, prove the coherent state over-completeness relation (63).

12-2. Prove that the coherent states satisfy Eq. (64), i.e., that they are not orthogonal.

12-3. Show that the thermal field described by the density operator of Eq. (34) exhibits bunching, rather than antibunching.

12-4. Show that $P(\alpha)$ given by

$$P(\alpha) = \frac{n!}{2\pi r(2n)!} e^{r^2}(-\partial/\partial r)^{2n}\,\delta(r)\ , \tag{93}$$

where $\alpha = re^{i\theta}$ yields the photon number state $|n\rangle\langle n|$. Hint: transform to (r, θ) coordinates and use integration by parts.

12-5. Write $\int d^2\alpha\ |\alpha\rangle\langle\alpha|$ in terms of $|n\rangle\langle m|$.

12-6. Calculate the variance of the single-mode electric-field operator in the vacuum state. Write $(aa^\dagger)^3$ in normal order.

12-7. Given an average of one photon, what is the probability of having n photons for *a*) a Poisson distribution, and *b*) a thermal distribution? Calculate the variance of a Poisson distribution.

12-8. Compute the photon number fluctuations $\langle(a^\dagger a)^2 - \langle a^\dagger a\rangle^2\rangle$ for a coherent state $|\alpha\rangle$.

12-9. Write the operator $a^2 a^{\dagger 3} a$ in normal order.

12-10. Show for the multimode annihilation/creation operator

$$d = \sum_s (\alpha_s a_s + \beta_s a_s^\dagger)\ ,$$

that $\langle d^\dagger d\rangle \geq 0$ for any state vector by proving that

$$\langle\{n_s\}|d^\dagger d|\{n_s\}\rangle = \sum_s [|\alpha_s|^2 n_s + |\beta_s|^2(n_s + 1)]\ .$$

Chapter 13

INTERACTION BETWEEN ATOMS AND QUANTIZED FIELDS

In Chaps. 1 - 11, we consider the interactions of atoms with optical fields in the semiclassical approximation, a treatment which gives good agreement with many experiments. Nevertheless several significant experiments show strong disagreements with such semiclassical theories. The present Chapter treats the atom-field interaction fully quantum mechanically, providing a basic understanding of spontaneous emission and laying the foundations for the treatment of more advanced problems in quantum optics such as resonance fluorescence, squeezed states and the laser linewidth that are developed in the remaining chapters in the book. Basically all we need to do in laying the foundations for these treatments is to combine the knowledge we have gained from the semiclassical theory of the atom-field interactions with the quantized field treatment of Chap. 12.

It would be misleading to say that this marriage is easy. There are some nasty infinities floating around that one has to argue away. In principle this is hard to do. Quantum electrodynamics (QED), which handles electron-photon interactions for arbitrary electron velocities and field frequencies, has developed techniques such as renormalization to solve these problems. There are numerous excellent books on the subject, such as e.g. Itzykson and Zuber (1980). There are also subtle difficulties associated with gauge invariance which are of direct relevance to a number of quantum optics applications and have been the subject of heated debate in the recent past. Cohen-Tannoudji *et al.* (1989) discuss these problems in great details in the framework on nonrelativistic QED. Our ambition in this book is much more limited. We restrict our considerations to nonrelativisitic velocities and low (optical) frequency photons. In these limits, we further make plausible assumptions that sidestep the real problems with QED.

We start with the relatively simple problem of an atom coupled to a single quantized mode of the field. This problem is called the Jaynes-Cummings model, which is perhaps ironic, since after working on the problem, Jaynes has steadfastly championed atom-field theories that use classical fields. We introduce the "dressed-atom" picture, which can be very useful in explaining how elementary phenomena take place. In particular, we gain a simple understanding of the "light shift" and set up the understanding of the three peaks of strong field resonance fluorescence. Section 13-2 discusses the dynamics of the atom-field model for various states of the field. We obtain an elementary picture of spontaneous and stimulated emission and absorption, and briefly discuss the "Cummings Collapse" and revivals due to the quantum granularity of the field. Section 13-3 derives the spontaneous emission decay rate both using the Fermi Golden Rule and using the more generally accurate Weisskopf-Wigner theory. This problem is important in its own right and provides an excellent example of system-reservoir interactions, discussed in greater detail in Chap. 14.

13-1. Dressed States

In Chap. 3 we show that the interaction Hamiltonian between an atom and a classical field is given in the dipole approximation by

$$\mathcal{V} = -e\mathbf{r}\cdot\mathbf{E}\ , \tag{1}$$

where $\mathbf{E}$ is the electric field and $e\mathbf{r}$ is the atomic dipole moment operator. The form of the interaction energy remains the same for quantized fields, but $\mathbf{E}$ becomes the electric field operator discussed in Chap. 12. For the single mode field (12.13), the interaction Hamiltonian (1) becomes

$$\mathcal{V} = \hbar(a + a^\dagger)(g\sigma_+ + g^*\sigma_-), \tag{2}$$

where σ_+ and σ_- are the Pauli spin-flip matrices (3.110) and the electric-dipole matrix element

$$g = -\frac{\wp\mathcal{E}_\Omega}{2\hbar}\sin Kz\ . \tag{3}$$

With the two-level unperturbed Hamiltonian $\frac{1}{2}\hbar\omega\sigma_z$ (see Eq. (3.110) and the free-field Hamiltonian (12.6) (without the zero point energy), we obtain the total atom-field Hamiltonian

$$\mathcal{H} = \tfrac{1}{2}\hbar\omega\sigma_z + \hbar\Omega a^\dagger a + \hbar(a + a^\dagger)(g\sigma_+ + g^*\sigma_-)\ . \tag{4}$$

Without loss of generality for two level systems, we can choose the atomic quantization axis such that the matrix element g is real. One of the basic approximations in the theory of two-level atoms is the rotating wave approximation. We can understand how this approximation works with quantized fields by considering the various terms in the interaction energy Eq. (2). $a\sigma_+$ corresponds to the absorption of a photon and the excitation of the atom from the lower state $|b\rangle$ to the upper state $|a\rangle$. Conversely, $a^\dagger\sigma_-$ describes the emission of a photon and the deexcitation of the atom. These combinations correspond to those kept in the rotating wave approximation. To see how the remaining two pairs, $a\sigma_-$ and $a^\dagger\sigma_+$ are dropped in this approximation, consider the free evolution (g=0) of these operators in the Heisenberg picture. The annihilation and creation operators have the time dependence of Eqs. (3.120) and (3.121). Similarly the spin-flip operators have the Heisenberg time dependence

$$\sigma_\pm(t) = \sigma_\pm(0)\, e^{\pm i\omega t} \ . \tag{5}$$

Combining these dependencies we have

$$a(t)\sigma_-(t) = a(0)\sigma_-(0)\, e^{-i(\Omega+\omega)t} \tag{6}$$

$$a^\dagger(t)\sigma_+(t) = a^\dagger(0)\sigma_+(0)\, e^{i(\Omega+\omega)t} \ . \tag{7}$$

Hence these pairs evolve at optical frequencies. In contrast $a\sigma_+$ and $a^\dagger\sigma_-$ have the dependencies

$$a(t)\sigma_+(t) = a(0)\sigma_+(0)\, e^{-i(\Omega-\omega)t} \tag{8}$$

$$a^\dagger(t)\sigma_-(t) = a^\dagger(0)\sigma_-(0)\, e^{i(\Omega-\omega)t} \ , \tag{9}$$

which vary slowly near resonance. In the course of a few optical periods, the "antiresonant" combinations (6) and (7) tend to average to zero compared to the combinations (8) and (9). This amounts to the same physics and mathematics that we used in discussing the rotating-wave approximation in Sec. 3-2.

Consequently dropping the combinations (6) and (7), we obtain the total atom-field Hamiltonian

$$\boxed{\mathcal{H} = \mathcal{H}_0 + \mathcal{V} = \tfrac{1}{2}\hbar\omega\sigma_z + \hbar\Omega a^\dagger a + \hbar(ga\sigma_+ + \text{adj})} \ . \tag{10}$$

The unperturbed Hamiltonian satisfies the eigenvalue equations

$$\begin{aligned} \mathcal{H}_0|an\rangle &= \hbar(\tfrac{1}{2}\omega + n\Omega)|an\rangle \ , \\ \mathcal{H}_0|bn\rangle &= \hbar(-\tfrac{1}{2}\omega + n\Omega)|bn\rangle \ . \end{aligned} \tag{11}$$

The interaction energy $\mathcal{V}$ couples the atom-field states $|an\rangle$ and $|bn{+}1\rangle$ for each value of n, but does not couple other states such as $|an\rangle$ and $|bn{-}1\rangle$

($a\sigma_-$ would, but is dropped in the RWA). Hence we can consider the atom-field interaction for each manifold $\mathcal{E}_n = \{|an\rangle, |bn{+}1\rangle\}$ independently and write $\mathcal{H}$ as the sum

$$\mathcal{H} = \sum_n \mathcal{H}_n\,, \tag{12}$$

where $\mathcal{H}_n$ acts only on the manifold $\mathcal{E}_n$ and is given in the $\{|an\rangle\ |bn{+}1\rangle\}$ basis by the matrix

$$\mathcal{H}_n = \hbar(n + \tfrac{1}{2})\Omega \begin{pmatrix} 1 & 0 \\ 0 & 1 \end{pmatrix} + \frac{\hbar}{2}\begin{pmatrix} \delta & 2g\sqrt{n+1} \\ 2g\sqrt{n+1} & -\delta \end{pmatrix}, \tag{13}$$

where the detuning $\delta = \omega - \Omega$.

The second matrix in Eq. (13) has the same form as in the semiclassical case of Eq. (3.104), but with the Rabi frequency $\mathcal{R}_0$ replaced by $-2g\sqrt{n+1}$ and an overall minus sign. Hence we can diagonalize this matrix just as the semiclassical version is diagonalized in Prob. 3-14. We find the same eigenvalues (interchanged) and eigenvectors [Eqs. (3.105), (3.142) and (3.143)] with the semiclassical Rabi frequency $\mathcal{R}_0$ replaced by $-2g\sqrt{n+1}$. This gives the energy eigenvalues

$$\begin{aligned} E_{2n} &= \hbar(n + \tfrac{1}{2})\Omega - \tfrac{1}{2}\hbar\mathcal{R}_n = \hbar[\tfrac{1}{2}\omega + n\Omega - \tfrac{1}{2}(\mathcal{R}_n + \delta)]\,, \\ E_{1n} &= \hbar(n + \tfrac{1}{2})\Omega + \tfrac{1}{2}\hbar\mathcal{R}_n = \hbar[-\tfrac{1}{2}\omega + (n+1)\Omega + \tfrac{1}{2}(\mathcal{R}_n + \delta)]\,, \end{aligned} \tag{14}$$

where we have introduced the quantized generalized Rabi flopping frequency

$$\boxed{\mathcal{R}_n = \sqrt{\delta^2 + 4g^2(n+1)}}\,. \tag{15}$$

The energy eigenvectors are

$$\boxed{\begin{aligned} |2n\rangle &= \cos\theta_n\,|an\rangle - \sin\theta_n\,|bn{+}1\rangle \\ |1n\rangle &= \sin\theta_n\,|an\rangle + \cos\theta_n\,|bn{+}1\rangle \end{aligned}} \tag{16}$$

where

$$\cos\theta_n = \frac{\mathcal{R}_n - \delta}{\sqrt{(\mathcal{R}_n - \delta)^2 + 4g^2(n+1)}}$$
$$\sin\theta_n = \frac{2g\sqrt{n+1}}{\sqrt{(\mathcal{R}_n - \delta)^2 + 4g^2(n+1)}} \ . \tag{17}$$

Note in particular that

$$\cos 2\theta_n = -\frac{\delta}{\mathcal{R}_n} \quad \text{and} \quad \sin 2\theta_n = \frac{2g\sqrt{n+1}}{\mathcal{R}_n} \ , \tag{18}$$

which are useful for calculating expectation values in the dressed-state basis.

We can gain some insight into the Hamiltonian (13) by plotting its eigenvalues as a function of the atomic transition frequency ω as shown in Figure 13-1. There the dashed lines correspond to the unperturbed eigenvalues of the atom-field system of Eq. (11) and the solid lines correspond

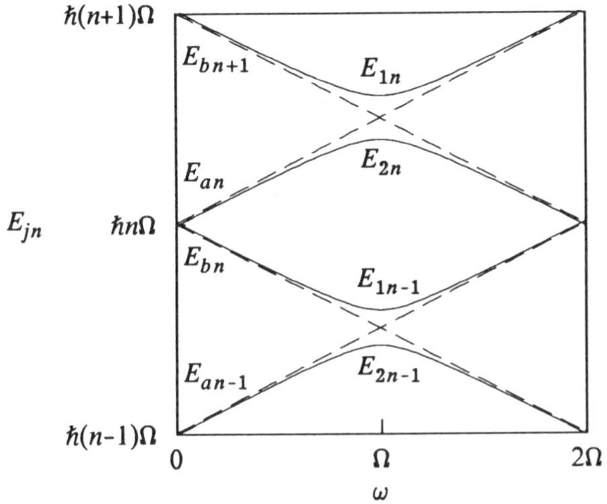

Fig. 13-1. Dressed atom energy level diagram. The dashed lines are the energy eigenvalues (11) for atom-field system with no interaction energy. Solid lines include atom-field interaction as in Eq. (14).

to the perturbed eigenvalues of Eq. (14). We say that the atomic levels are "dressed" by the radiation field and the levels $|an\rangle$ and $|bn+1\rangle$ of the composite system cross for $\omega = \Omega$. In contrast the atom-field interaction causes the eigenstates $|1n\rangle$ and $|2n\rangle$ of the total Hamiltonian to repel one another. This repulsion phenomenon is called "anticrossing." Combining Eqs. (14)

and (15), we find the general level separation $E_{1n} - E_{2n} = \mathscr{R}_n$. The minimum separation occurs for $\omega = \Omega$, and is given by $|2\hbar g\sqrt{n+1}|$. The states given by Eq. (16) are called *dressed states*, namely the eigenstates of the Hamiltonian describing the two-level atom interacting with a single-mode field. We refer to the eigenstates of the unperturbed Hamiltonian, i.e., not including the atom-field interaction as *bare states*.

At resonance ($\omega = \Omega$), the dressed states of Eq. (16) reduce to

$$\begin{aligned} |2n\rangle &= [|an\rangle - |bn{+}1\rangle]/\sqrt{2} \\ |1n\rangle &= [|an\rangle + |bn{+}1\rangle]/\sqrt{2} \end{aligned} \tag{19}$$

with eigenvalues

$$\begin{aligned} E_{2n} &= \hbar(n{+}\tfrac{1}{2})\Omega - \hbar g\sqrt{n+1} \\ E_{1n} &= \hbar(n{+}\tfrac{1}{2})\Omega + \hbar g\sqrt{n+1}\ . \end{aligned} \tag{20}$$

Thus at resonance, the $|an\rangle$ and $|bn{+}1\rangle$ levels have equal contributions of the upper and lower atomic levels, while for $\omega << \Omega$, $\cos\theta_n \simeq 1$, giving $|2n\rangle$ mostly $|an\rangle$ character and for $\omega >> \Omega$, $\sin\theta_n \simeq 1$, which gives $|2n\rangle$ mostly $|bn{+}1\rangle$ character. Note that for very large $|\delta|$, the rotating wave approximation used in Eq. (10) breaks down and the diagonalization procedure used in Eqs. (14) and (16) is invalid. This limitation is not important for most of our work, which is close to resonance.

The eigenstates given in Eqs. (16) define a rotation matrix

$$T = \begin{pmatrix} \cos\theta_n & -\sin\theta_n \\ \sin\theta_n & \cos\theta_n \end{pmatrix} \tag{21}$$

that diagonalizes the Hamiltonian of Eq. (13) by the matrix product $T\mathscr{H}_n T^{-1}$. In general it allows us to transform operators and state vectors between the bare-atom and dressed-atom bases. In particular it relates the dressed-atom and bare-atom probability amplitudes by

$$\begin{pmatrix} c_{2n}(t) \\ c_{1n}(t) \end{pmatrix} = T \begin{pmatrix} c_{an}(t) \\ c_{bn+1}(t) \end{pmatrix}, \tag{22}$$

where a general Schrödinger state vector is expanded in terms of bare states as

$$|\psi\rangle = \sum_n [c_{an}(t)\ |an\rangle + c_{bn+1}\ |bn+1\rangle]$$

and in terms of dressed states as

$$|\psi\rangle = \sum_n [c_{1n}(t)\ |1n\rangle + c_{2n}\ |2n\rangle]\ .$$

We use an interaction picture version of this relation in the next section to derive Rabi flopping using dressed states. In general in using dressed states, substantial effort is saved by using the abbreviations $c \equiv \cos\theta_n$ and $s \equiv \sin\theta_n$. The transformation matrix (21) and its inverse are simply

$$T = \begin{pmatrix} c & -s \\ s & c \end{pmatrix} \qquad T^{-1} = \begin{pmatrix} c & s \\ -s & c \end{pmatrix}\ .$$

Similarly the notations $s_2 \equiv \sin 2\theta_n = 2cs$ and $c_2 \equiv \cos 2\theta_n = c^2 - s^2$ are also useful.

13-2. Jaynes-Cummings Model

The dressed-atom picture developed in Sec. 13-1 provides us with very useful physical insight into the dynamics of a two-level atom interacting with a quantized field mode. In this section, we use this picture to treat Rabi flopping in a fully quantized way and compare the results to the corresponding semiclassical treatment of Sec. 3-3. We see a simple example of spontaneous emission and the phenomenon known as the "Cummings collapse" of Rabi flopping induced by a coherent state. We also discuss the revivals of the Rabi flopping that result from the discrete nature of the photon field.

Quantum Rabi Flopping

Since the dressed atom states are the eigenstates of the two-level atom interacting with a single mode of the radiation field, we can use them to obtain the state vector of the combined system as a function of time. Writing the Schrödinger equation (3.43) in the integral form

$$|\psi(t)\rangle = \exp(-i\mathcal{H}t/\hbar)\,|\psi(0)\rangle\ , \tag{23}$$

we insert the identity operator expressed in terms of the dressed-atom states $|jn\rangle$ to find

$$|\psi(t)\rangle = \sum_{n=0}^{\infty}\sum_{j=1}^{2} \exp(-iE_{jn}t/\hbar)\,|jn\rangle\langle jn|\psi(0)\rangle\ , \tag{24}$$

where E_{jm} is given by Eq. (14). We have seen in the preceding section that the various manifolds $\mathcal{E}_n$ are uncoupled. In matrix form and in an interaction picture rotating at the frequency $(n+\frac{1}{2})\Omega$, the dressed state amplitude coefficients inside one such manifold $\mathcal{E}_n$ read

$$\begin{bmatrix} C_{2n}(t) \\ C_{1n}(t) \end{bmatrix} = \begin{bmatrix} \exp(\frac{1}{2}i\mathcal{R}_n t) & 0 \\ 0 & \exp(-\frac{1}{2}i\mathcal{R}_n t) \end{bmatrix} \begin{bmatrix} C_{2n}(0) \\ C_{1n}(0) \end{bmatrix}, \tag{25}$$

as readily seen from Eq. (14). We can use this result together with the transformation relation (22) and its inverse to derive the quantized field version of the two-level state vector of Eq. (3.106). We find

$$\begin{bmatrix} C_{an}(t) \\ C_{bn+1}(t) \end{bmatrix} = T^{-1} \begin{bmatrix} \exp(\frac{1}{2}i\mathcal{R}_n t) & 0 \\ 0 & \exp(-\frac{1}{2}i\mathcal{R}_n t) \end{bmatrix} T \begin{bmatrix} C_{an}(0) \\ C_{bn+1}(0) \end{bmatrix}$$

$$= \begin{bmatrix} \cos\frac{1}{2}\mathcal{R}_n t - i\delta\mathcal{R}_n^{-1}\sin\frac{1}{2}\mathcal{R}_n t & -2ig\sqrt{n+1}\,\mathcal{R}_n^{-1}\sin\frac{1}{2}\mathcal{R}_n t \\ -2ig\sqrt{n+1}\,\mathcal{R}_n^{-1}\sin\frac{1}{2}\mathcal{R}_n t & \cos\frac{1}{2}\mathcal{R}_n t + i\delta\mathcal{R}_n^{-1}\sin\frac{1}{2}\mathcal{R}_n t \end{bmatrix} \begin{bmatrix} C_{an}(0) \\ C_{bn+1}(0) \end{bmatrix}, \tag{26}$$

which is the same as Eq. (3.106) with $\mathcal{R}_0$ replaced by $-2g\sqrt{n+1}$. In particular for a resonant atom initially in the upper level, we obtain the probabilities

$$|C_{an}(t)|^2 = \cos^2(g\sqrt{n+1}\,t) \tag{27}$$

$$|C_{bn+1}(t)|^2 = \sin^2(g\sqrt{n+1}\,t) \tag{28}$$

which show how the atom Rabi flops between the upper and lower levels within a manifold $\mathcal{E}_n$.

As in the semiclassical case of Sec. 3-3, we can obtain these results directly using the bare-atom state vector

$$|\psi(t)\rangle = \sum_n [C_{an}(t)|an\rangle + C_{bn+1}(t)|bn+1\rangle]e^{-i(n+\frac{1}{2})\Omega t} , \tag{29}$$

corresponding to the semiclassical wave function of Eq. (3.98). Substituting this into the Schrödinger equation (3.43) and projecting onto the $|an\rangle$ and $|bn+1\rangle$ states, we find the equations of motion for the atom-field probability amplitudes in the manifold $\mathcal{E}_n$ to be

$$\dot{C}_{an} = -\tfrac{1}{2}i\delta C_{an} - ig\sqrt{n+1}\,C_{bn+1} \tag{30}$$

$$\dot{C}_{bn+1} = \tfrac{1}{2}i\delta C_{bn+1} - ig\sqrt{n+1}\,C_{an} . \tag{31}$$

These have the same form as the semiclassical equations (3.99) and (3.100), respectively, with the semiclassical Rabi frequency $\mathcal{R}_0 \equiv \wp E_0/\hbar$ replaced by the quantum mechanical value $-2g\sqrt{n+1}$. Hence the solution is the same [Eq. (3.106)] with this substitution, which gives Eq. (26).

We see that both the dressed-atom and bare-atom approaches lead to Rabi flopping as they must, and in general anything you can study with one basis set you can study with the other. The insights gained with one typically differ from those gained with the other, however, and each method has its advantages. In the Rabi flopping problem, we see that the dressed-state solution is very easy to get once the dressed states and eigenvalues have been found. To find them, we calculate the eigenvalues and eigenvectors of a 2x2 matrix, while the bare-atom state method used in Sec. 4-1 derives the eigenvalues alone. For this particular problem, the bare-atom approach requires somewhat less algebra.

Another advantage of the bare-atom approach is that it can easily include the effects of level decay, yielding the solution of Eq. (4.10) with $\mathcal{R}_0$ replaced by $-2g\sqrt{n+1}$. While the dressed-atom approach simplifies the Hamiltonian by diagonalization, it *un*diagonalizes the decay rates, making their inclusion relatively cumbersome. Hence the dressed-state approach is usually used either when decay can be ignored, or when the Rabi-flopping frequency greatly exceeds the decay constants.

Single-Mode Spontaneous Emission

One intriguing difference between the semiclassical and fully quantum Rabi flopping problems is that in the quantum case (29), an initially excited atom Rabi flops in the absence of an applied field, i.e., even for $n = 0$ ($C_{a0}(0) = 1$). This is because the quantum Rabi-flopping frequency is

$-2g\sqrt{n+1}$, while the semiclassical is $\mathcal{R}_0 \equiv \wp E_0/\hbar$. Equivalently we see from Eq. (3) that the quantized field has the amplitude $\mathcal{E}_\Omega \sqrt{n+1}$, which reduces to the electric field per photon, $\mathcal{E}_\Omega$, for $n = 0$. The reason for this lies in the fact that even though the vacuum expectation value for the field amplitude vanishes

$$\langle E \rangle = \mathcal{E}_\Omega \langle 0 | a + a^\dagger | 0 \rangle = 0 \,, \tag{32}$$

that for the intensity does not:

$$\langle E^2 \rangle = \mathcal{E}_\Omega^2 \langle 0 | (a + a^\dagger)^2 | 0 \rangle = \mathcal{E}_\Omega^2 \,. \tag{33}$$

Stated another way, there are vacuum fluctuations in the electromagnetic field. These vacuum fluctuations effectively stimulate an excited atom to emit, a process called spontaneous emission. The weak Rabi flopping that occurs for $n = 0$ in Eq. (29) is due to spontaneous emission alone, which is neglected in the semiclassical approximation. If the atom is initially unexcited ($C_{b0}(0) = 1$), however, no flopping occurs, since spontaneous absorption doesn't conserve energy.

This model of spontaneous emission is unrealistic except for special cavities, since it involves only a single mode of the field. As such it yields vacuum Rabi flopping instead of the well-known exponential decay. Usually we have a continuous spectrum of vacuum fluctuations that all attempt to make an excited atom Rabi flop. The resulting lower-level probability amplitudes interfere with one another, giving the exponential decay. Section 13-3 derives a very successful theory of this multimode spontaneous emission.

Collapse and Revival

The quantum Rabi flopping frequency $-2g\sqrt{n+1}$ explicitly shows that different photon number states have different quantum Rabi flopping frequencies. In particular, consider an initially excited atom interacting with a field initially in a coherent state. Combining the coherent state photon number probability (12.54) with the single-photon state result (27), we have the probability for an excited atom regardless of the field state

$$p_a = \sum_n p_n \, |C_{an}(t)|^2 = e^{-|\alpha|^2} \sum_n \frac{|\alpha|^{2n}}{n!} \cos^2[g\sqrt{(n+1)}t] \,. \tag{34}$$

For a sufficiently intense field and short enough times $t \ll |\alpha|/g$, this sum can be shown (Prob. 13-3) to reduce to

$$\rho_{aa} \simeq \frac{1}{2} + \frac{1}{2}\cos(2|\alpha|gt)\, e^{-g^2t^2} . \tag{35}$$

Intuitively this result can be understood by noting that the range of dominant Rabi frequencies in (34) is from $\Omega = g[\bar{n} + \Delta n]^{1/2}$ to $g[\bar{n} - \Delta n]^{1/2}$ and the probability (34) dephases in a time

$$t_c^{-1} = g[\bar{n} + \Delta n]^{1/2} - g[\bar{n} - \Delta n]^{1/2} \simeq g,$$

which is independent of $\bar{n}$. Here we have used the property of a Poisson distribution $\Delta n = \bar{n}^{1/2}$. Hence the Rabi oscillations are damped with a Gaussian envelope independent of the photon number $\bar{n} = |\alpha|^2$, a result sometimes called the "Cummings collapse." This collapse is due to the interference of Rabi floppings at different frequencies. For still longer times, the system exhibits a series of "revivals" and "collapses" discussed in detail by Eberly, Narozhny, and Sanchez-Mondragon (1980). Because the photon numbers n are discrete in the quantum sum (34), the oscillations rephase in the revival time

$$t_r \simeq 4\pi\alpha/g = 4\pi\bar{n}^{1/2}\, t_c \quad ,$$

as illustrated in Fig. 13-2. This revival property is a much more unambiguous signature of quantum electrodynamics than the collapse: any spread in

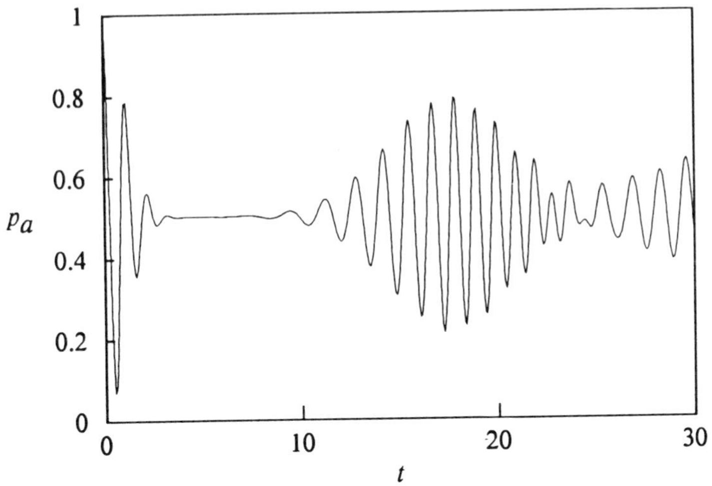

Fig. 13-2. Collapse and revival in the interaction of a quantized single-mode field with an atom.

field strengths will dephase Rabi oscillations, but the revivals are entirely due to the grainy nature of the field, so that the atomic evolution is determined by the individual field quanta. Eventually, the revivals (which are never complete and get broader and broader) overlap and give a quasi-random time evolution. The Jaynes-Cummings model thus exhibits interesting nontrivial quantum features, despite its conceptual simplicity. The remarkable fact is that these have recently started to be observed experimentally [see Rempe and Walther (1987)].

It is rather surprising that while the coherent state is the most classical state allowed by the uncertainty principle, it leads to a result qualitatively different from the classical Rabi flopping formula of Eq. (3.100). In contrast, the very quantum mechanical number state $|n\rangle$ has the nice correspondence given following Eq. (31). The number state and the classical field share the property of a definite intensity, which is needed to avoid the interferences leading to a collapse. The indeterminacy in the field phase associated with the number state (but not with the classical field), is not important for Rabi flopping since the atom and field maintain a precise relative phase in the absence of decay processes. In contrast the coherent state field features a minimum uncertainty phase, but its minimum uncertainty intensity causes the atom-field relative phase to "diffuse" away.

13-3. Spontaneous Emission in Free Space

The last section shows that an excited atom interacting with a single-mode field can make a transition to the ground state even in the absence of cavity photons. Specifically, in a single-mode cavity the atom Rabi flops at a slow but nonzero rate purely due to vacuum fluctuations. In free space the atom interacts with a continuum of modes, which leads to an exponential decay of the excited state probability. We can get an idea of how this works by applying the Fermi Golden Rule developed in Sec. 3-2. We consider the general state vector

$$|\psi(t)\rangle = \sum_n [C_{an}(t)|an\rangle e^{-i(\omega_a+n\Omega)t} + C_{bn+1}(t)|bn{+}1\rangle e^{-i(\omega_b+(n+1)\Omega)t}] , \quad (36)$$

where the C's are interaction picture probability amplitudes. We start with an initially excited atom and a vacuum field, that is,

$$C_{a0}(0) = 1 \quad (37)$$

and all other probability amplitudes vanish. We solve to first-order in perturbation theory for the probability that the atom emits into one mode of the field and then sum this probability over all possible modes. The time derivative of this total transition probability gives the transition rate according to the Fermi Golden Rule. The relevant single-mode equations of motion are

$$\dot{C}_{a0} = -ig\, e^{i\delta t}\, C_{b1} \tag{38}$$

$$\dot{C}_{b1} = -ig\, e^{-i\delta t}\, C_{a0}\ . \tag{39}$$

Setting $C_{a0} = 1$ in Eq. (39) and integrating from 0 to t, we find the first order contribution to $C_{b1}(t)$ to be

$$C_{b1}^{(1)}(t) = g\, \frac{e^{-i\delta t} - 1}{\delta}\ .$$

This gives the first-order single-mode transition probability

$$|C_{b1}^{(1)}(t)|^2 = g^2\, \frac{\sin^2\frac{1}{2}\delta t}{\frac{1}{4}\delta^2}\ . \tag{40}$$

As such we may think of spontaneous emission as emission stimulated by the continuum of vacuum modes. The total upper to lower state transition probability is given by summing this probability over all modes of the electromagnetic field. Neglecting for now the effects of field polarization reduces this task to a simple sum over the density of modes $\mathcal{D}(\Omega)$ (derived in the next section), we write this sum as the integral

$$P_T = \frac{1}{4}\int_0^\infty d\Omega t^2 g(\Omega)^2 \mathcal{D}(\Omega)\, \frac{\sin^2[(\Omega-\omega)t/2]}{(\Omega-\omega)^2 t^2/4}\ . \tag{41}$$

The density of states $\mathcal{D}(\Omega)$ given by Eq. (46) varies as Ω^2 which is a slowly-varying, very broad function. Similarly $g^2(\Omega)$ is a slowly-varying function of Ω. Hence except for extremely short times, the $\sin^2 x/x^2$ function in Eq. (41) is significant only for values of Ω near ω. Evaluating $g^2(\Omega)\mathcal{D}(\Omega)$ at the peak of the $\sin^2 x/x^2$ function (at $\Omega = \omega$), we factor the product out of the integral and obtain

$$P_T = \Gamma t\ , \tag{41}$$

where the spontaneous emission decay rate $\Gamma = dP_T/dt$ is given by

$$\Gamma = 2\pi g^2(\omega)\mathcal{D}(\omega)\ . \tag{42}$$

Although important features of spontaneous emission are explained by this simple derivation, it is missing two crucial pieces: 1) the vacuum modes belong to three dimensional space, and hence the dot product in Eq. (1), which involves explicitly the field polarization, cannot be ignored; and 2) the Fermi Golden Rule logic assumes that the initial state probability remains equal to unity rather than decaying exponentially. Hence it is valid only for times short enough that the upper state population is not significantly depleted. The three dimensional piece can be incorporated in a straightforward fashion using the integrals over space that appear as in the Weisskopf-Wigner discussion below. More fundamentally, the Weisskopf-Wigner theory predicts that the initial state decays exponentially. As such it derives a very important result in the interaction of radiation with atoms and it shows one way in which the Fermi Golden Rule can be generalized to treat long time responses. Another feature of the Weisskopf-Wigner theory is that predicts a frequency shift in addition to a decay.

Free Space Density of States

To give a complete discussion of spontaneous emission, we need to have a precise formulation of the density of electromagnetic states in free space. Both for the simple Fermi Golden Rule theory above and for the more complete Weisskopf-Wigner theory below, this density is used to transform from a discrete summation over field modes to a continuous one.

Consider the three dimensional cubic cavity with side length L shown in Fig. 13-3. Along the $\hat{x}$ direction, the cavity can sustain running modes with wave numbers $K_x = 2\pi n_x/L$, $n_x = 1, 2, \ldots$. Taking differentials of this expression, we find the number of modes between K_x and $K_x + dK_x$ to be $dn_x = dK_x\, L/2\pi$. Performing the same calculation for the $\hat{y}$ and $\hat{z}$ directions, we find the number of modes in the volume element $dK_x\, dK_y\, dK_z$ to be

$$dn = d^3K\, L^3/(2\pi)^3 \,. \tag{43}$$

For large L, a summation over K can be written as the integral

$$\frac{1}{V}\sum_K f(\mathbf{K}) \rightarrow \frac{1}{V}\int dn\, f(\mathbf{K}) = \frac{1}{(2\pi)^3}\int d^3K\, f(\mathbf{K}) \,.$$

where $V = L^3$ is the volume of the cavity. Using spherical coordinates $d^3K = dKK\sin\theta d\theta K d\phi$ and transforming to frequency Ω using $K = c\Omega$, we find

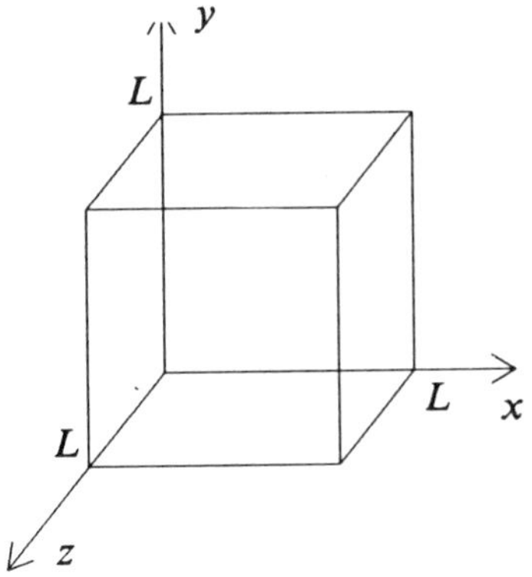

Fig. 13-3. Three dimensional cavity used in deriving density of states formulas (44) and (46).

$$\frac{1}{V}\sum_{K} f(\mathbf{K}) \to \frac{1}{(2\pi)^3}\int d\Omega\, \frac{\Omega^2}{c^3}\int_0^{\pi} d\theta\, \sin\theta \int_0^{2\pi} d\phi\, f(\mathbf{K})\ , \tag{44}$$

In addition we have to sum over the two polarizations of the transverse electromagnetic field.

In performing any particular sum over states, we should insert the desired function $f(\mathbf{K})$ into Eq. (44) and carry out the three integrals taking into account the two possible field polarizations. However if $f(\mathbf{K})$ is independent of field polarization we multiply by 2 and if $f(\mathbf{K})$ doesn't depend on θ and ϕ, we obtain 4π for the angular integrations. These simplifications give the correspondence

$$\frac{1}{V}\sum_{K} f(K) \to \int d\Omega\, \mathscr{D}(\Omega)\, f(\Omega)\ , \tag{45}$$

where the density of states between Ω and $\Omega+d\Omega$ is given by

$$\boxed{\mathscr{D}(\Omega) = \frac{\Omega^2}{\pi^2 c^3}}\ . \tag{46}$$

This simple formula is not directly applicable for electric-dipole interactions, since $-e\mathbf{r}\cdot\mathbf{E}$ depends on the angle between the electric field polarization and the atomic quantization axis. Hence we use Eq. (44) in evaluating the spontaneous emission coefficient. Equations (44) and (46) are derived for free space. In a cavity with finite dimensions, the density of states is substantially modified for frequencies close to the cavity cutoff as shown in Fig. 13-4.

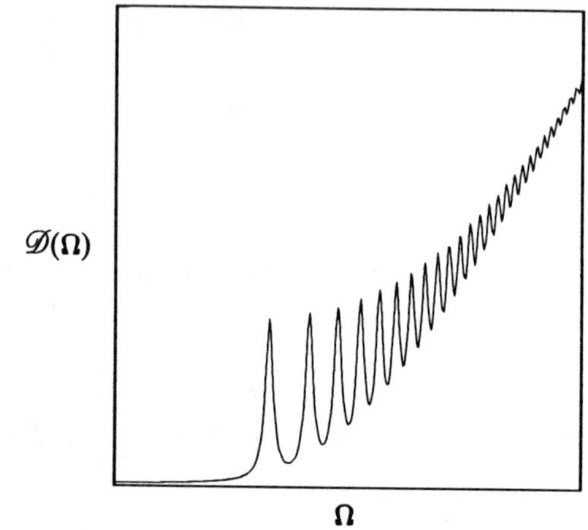

Fig. 13-4. Artist's drawing of the density of states for a finite cavity volume. Below cutoff the cavity sustains no modes at all, and near but above cutoff, the density of states can be increased relative to a continuum case.

Weisskopf-Wigner Theory of Spontaneous Emission

Armed with a proper three-dimensional density of states formalism and the multimode quantum theory of radiation of Sec. 12-2, we can treat the spontaneous emission of an excited state atom into the electromagnetic vacuum. The initial condition is given by Eq. (37), where by C_{a0} we mean the probability amplitude that the atom is excited and that all modes of the field are empty. To the Schrödinger picture Hamiltonian (12.13), we add the two-level atom energy and the multimode rotating-wave-approximation interaction energy

$$\mathcal{V} = \hbar\Sigma_s g_s a_s \sigma_+ + \text{adj} . \tag{47}$$

This gives

$$\mathscr{H} = \hbar \sum_s \Omega_s a_s^\dagger a_s + \hbar\omega_a \sigma_a + \hbar\omega_b \sigma_b + \hbar \sum_s (g_s a_s \sigma_+ + \text{adj.}) \,. \tag{48}$$

The interaction energy (47) can only connect the state $|a\{0\}\rangle$ to the state $|b\{1_s\}\rangle$, which describes an atom in the lower state with one photon in the sth mode and no photons in any other mode. This reduces the very general state vector (12.20) to

$$|\psi(t)\rangle = C_{a0}(t)\, e^{-i\omega_a t}\, |a\{0\}\rangle + \sum_s C_{b\{1_s\}}(t) e^{-i(\omega_b+\Omega_s)t}\, |b\{1_s\}\rangle \,. \tag{49}$$

Substituting Eqs. (48) and (49) into the Schrödinger equation (3.43) and projecting onto the states $|a\{0\}\rangle$ and $|b\{1_s\}\rangle$, we find the probability amplitude equations of motion

$$\dot{C}_{a0}(t) = -i \sum_s g_s e^{-i(\Omega_s-\omega)t}\, C_{b\{1_s\}}(t) \tag{50}$$

$$\dot{C}_{b\{1_s\}}(t) = -i g_s^{*}\, e^{i(\Omega_s-\omega)t}\, C_{a0}(t) \,. \tag{51}$$

Inserting the formal time integral of Eq. (51) into Eq. (50), we find an integro-differential equation for C_{a0} alone:

$$\dot{C}_{a0}(t) = -\sum_s |g_s|^2 \int_{t_0}^{t} dt'\, e^{-i(\Omega_s-\omega)(t-t')}\, C_{a0}(t') \,. \tag{52}$$

This equation is nonlocal in time since $C_{a0}(t)$ is a function of the earlier values $C_{a0}(t')$. In other respects it resembles what we obtained with the Fermi Golden Rule. To solve it, we use the concept of coarse graining, which is valid provided $C_{a0}(t')$ varies little in the time interval $t - t_0$ over which the remaining part of the integrand has nonzero value. Section 14-1 shows that this time interval is related to the correlation time of the vacuum field.

We are interested in the spontaneous decay of an excited atom in *free space*, hence we can convert the sum over states to a three dimensional integral using Eq. (44). This gives

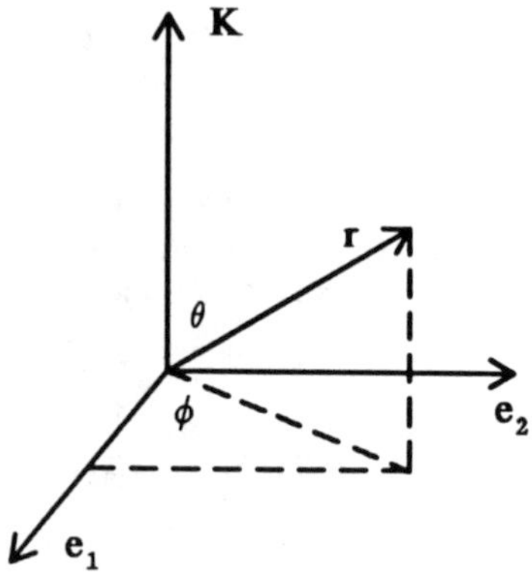

Fig. 13-5. Diagram of coordinate system for plane running wave with wave vector **K** and two transverse polarizations along the directions $\mathbf{e}_1$ and $\mathbf{e}_2$. The atomic dipole points in a direction at angle θ with respect to the propagation direction **K** and ϕ with respect to $\mathbf{e}_1$.

$$\dot{C}_{a0}(t) = -\frac{V}{(2\pi c)^3}\int d\Omega\,\Omega^2 \int_0^{\pi} d\theta\,\sin\theta \int_0^{2\pi} d\phi$$
$$\times\,|g(\Omega,\theta)|^2 \int_{t_0}^{t} dt'\, e^{-i(\Omega-\omega)(t-t')}\, C_{a0}(t')\,. \tag{53}$$

The interaction constant g depends on the dot product in Eq. (1). Figure 13-5 shows the coordinate system for one running wave with two possible polarizations interacting with the atomic dipole. To calculate $|g(\Omega,\theta)|^2$, we evaluate

$$\begin{aligned}|g(\Omega,\theta)|^2 &= \hbar^{-2}\sum_{\sigma=1}^{2} |\langle a|e\mathbf{r}\cdot\mathbf{e}_\sigma|b\rangle \mathcal{E}_\Omega\, U_\Omega|^2\\ &= |\mathcal{E}_\Omega\wp/\hbar|^2 \sin^2\theta\,|\cos^2\phi + \sin^2\phi|\\ &= |\mathcal{E}_\Omega \wp\sin\theta/\hbar|^2\,,\end{aligned} \tag{54}$$

which is independent of the azimuthal coordinate ϕ. For running waves, the square of the "electric field per photon" is given by

$$\mathcal{E}_\Omega{}^2 = \hbar\Omega/2\epsilon_0 V\ , \tag{55}$$

as seen by combining Eqs. (12.10) and (12.24). In substituting Eq. (54) into Eq. (53), we encounter the integral

$$\int_0^\pi d\theta\, \sin^3\theta = \int_{-1}^1 d(\cos\theta)\,(1 - \cos^2\theta) = \frac{4}{3}\ . \tag{56}$$

Substituting Eqs. (54) and (55) into (53) and using (56), we find

$$\dot{C}_{a0}(t) = -\frac{1}{6\epsilon_0\pi^2\hbar c^3}\int d\Omega\,\Omega^3|\wp|^2\int_{t_0}^t dt'\, e^{-i(\Omega-\omega)(t-t')}\, C_{a0}(t')\ . \tag{57}$$

As for the Fermi Golden Rule, we note that $\Omega^3|\wp|^2$ varies little in the frequency interval for which the integral over t' has appreciable value. In addition we perform a coarse-grained integration, i.e., we suppose that $C_{a0}(t')$ varies sufficiently slowly that it can be evaluated at $t' = t$ and factored outside the integrals. The remaining time integral has the highly peaked value

$$\lim_{t\to\infty}\int_{t_0}^t dt'\, e^{-i(\Omega-\omega)(t-t')} = \pi\delta(\Omega-\omega) - \mathcal{P}\left[\frac{i}{\Omega-\omega}\right]\ , \tag{58}$$

which allows us to evaluate the product $\Omega^3|\wp|^2$ at the peak value ω. The $\mathcal{P}[i/(\Omega-\omega)]$ term leads to a frequency shift related to the Lamb shift. The $\pi\delta(\Omega-\omega)$ term gives

$$\dot{C}_{a0}(t) = -\frac{\Gamma}{2}\, C_{a0}(t)\ , \tag{59}$$

where Γ is the Weisskopf-Wigner *spontaneous emission decay rate*

$$\boxed{\Gamma = \frac{\omega^3|\wp|^2}{3\pi\epsilon_0\hbar c^3} = \frac{1}{4\pi\epsilon_0}\,\frac{4\omega^3|\wp|^2}{3\hbar c^3}}\ . \tag{60}$$

To obtain the cgs expression for Γ, remove the $1/4\pi\epsilon_0$ factor.

The Weisskopf-Wigner theory predicts an *irreversible* exponential decay of the upper state population with no revivals, in contrast to the Jaynes-Cummings problem of Sec. 13-2. Whereas in the latter a quasiperiodicity results from the interaction with a single mode and the discrete nature of the possible photon numbers, in ordinary free-space spontaneous emission, the atom interacts with a continuum of modes. Although under the action of each individual mode the atom would have a finite probability to return to the upper state in a way similar to the Jaynes-Cummings revival, the probability amplitudes for such events interfere destructively when summed over the continuum of free space modes. In cavities with volumes comparable to the interaction wavelength, the density of states differs appreciably from the result of Eqs. (44) and (46) as shown in Fig. 13-4. For wavelengths below the cavity cutoff, the density of states vanishes altogether, while for wavelengths somewhat above the cutoff, the density of states may be substantially larger or smaller than the free space value of Eq. (46). Accordingly, in such cavities the spontaneous emission decay rate can differ substantially from the Weisskopf-Wigner value and be either enhanced or inhibited. For further discussion see O'Brien *et al.* (1985).

13-4. Quantum Beats

We just mentioned an interference phenomenon leading to irreversible exponential spontaneous decay in free space. Interference phenomena are ubiquitous in physics. They occur each time there are two or more possible paths in the evolution of a system and one does not know which of these it actually follows. To put it simply, if the question "which way ?" cannot be answered, then interferences are to be expected.

Such a situation can occur in the process of free space spontaneous emission, and is known as quantum beats. Suppose we have a three level system with levels $|a\rangle$, $|b\rangle$ and $|c\rangle$, with transitions allowed between levels $|a\rangle$ - $|c\rangle$ and $|b\rangle$ - $|c\rangle$ only. The Hamiltonian describing this system in interaction with the electromagnetic field is

$$\mathcal{H} = \hbar\omega_a |a\rangle\langle a| + \hbar\omega_b |b\rangle\langle b| + \hbar\omega_c |c\rangle\langle c| + \sum_{\mathbf{K}} \hbar\Omega_k a_{\mathbf{K}}{}^\dagger a_{\mathbf{K}}$$

$$+ \hbar \sum_{\mathbf{K}} [g_{a\mathbf{K}} a_{\mathbf{K}}{}^\dagger |c\rangle\langle a| + g_{b\mathbf{K}} a_{\mathbf{K}}{}^\dagger |c\rangle\langle b| + \text{h.c.}] , \tag{61}$$

where $\hbar g_{a\mathbf{K}}$ and $\hbar g_{b\mathbf{K}}$ are the dipole coupling constants between the running-wave mode of the field of wave vector $\mathbf{K}$ and the $|a\rangle$-$|c\rangle$ and $|b\rangle$-$|c\rangle$ transitions, respectively.

We consider the situation where the system is prepared initially in a coherent superposition of the two upper states

$$|\psi(0)\rangle = C_{a0}(0)|a\{0\}\rangle + C_{b0}(0)|b\{0\}\rangle \,, \tag{62}$$

and want to observe the fluorescence light emitted by this system with a broadband detector sensitive to both frequencies $\omega_{ac} \equiv \omega_a - \omega_c$ and $\omega_{bc} \equiv \omega_b - \omega_c$. In this experimental arrangement, we have no way of knowing if the light received at the detector originates from $|a\rangle \rightarrow |c\rangle$ or from a $|b\rangle \rightarrow |c\rangle$ transition, and hence should expect some kind of interference phenomenon. Indeed, we find that the fluorescence light is modulated at the difference frequency $\omega_{ab} \equiv \omega_a - \omega_b$. To see how this happens, we note that at time t, the state of the system is of the form

$$|\psi(t)\rangle = C_{a0}(t)e^{-i\omega_a t}\,|a\{0\}\rangle + C_{b0}(t)e^{-i\omega_b t}\,|b\{0\}\rangle + \sum_{\mathbf{K}} C_{c\mathbf{K}}(t)e^{-i\omega_k t}\,|c\{1_{\mathbf{K}}\}\rangle, \tag{63}$$

where $\omega_k = \omega_c + \Omega_{\mathbf{K}}$.

The equations of motion for the time-dependent coefficients in Eq. (63) are readily obtained from the Schrödinger equation. Substituting Eq. (63) into Eq. (3.43), and projecting onto $|a\{0\}\rangle$, $|b\{0\}\rangle$, $|c\{1_{\mathbf{K}}\}\rangle$, we find

$$\frac{dC_{a0}}{dt} = -i \sum_{\mathbf{K}} g^*_{a\mathbf{K}}\, C_{c\mathbf{K}}\, e^{i\delta_{a\mathbf{K}} t} \tag{64a}$$

$$\frac{dC_{b0}}{dt} = -i \sum_{\mathbf{K}} g^*_{b\mathbf{K}}\, C_{c\mathbf{K}}\, e^{i\delta_{b\mathbf{K}} t} \tag{64b}$$

$$\frac{dC_{c\mathbf{K}}}{dt} = -ig_{a\mathbf{K}}\, C_{a0}\, e^{-i\delta_{a\mathbf{K}} t} - ig_{b\mathbf{K}} C_{b0}\, e^{-i\delta_{b\mathbf{K}} t} \,, \tag{64c}$$

where $\delta_{a\mathbf{K}} = \omega_a - \omega_{\mathbf{K}}$ and $\delta_{b\mathbf{K}} = \omega_b - \omega_{\mathbf{K}}$. Substituting the formal integral of Eq. (64c)

$$C_{c\mathbf{K}}(t) = -ig_{a\mathbf{K}} \int_0^t dt'\, C_{a0}(t')\, e^{-i\delta_{a\mathbf{K}} t'} - ig_{b\mathbf{K}} \int_0^t dt' C_{b0}(t') e^{-i\delta_{b\mathbf{K}} t'} \tag{65}$$

into Eq. (64a), we obtain

$$\frac{dC_{a0}}{dt} = -\sum_{\mathbf{K}} |g_{a\mathbf{K}}|^2 \int_0^t dt' C_{a0}(t') e^{i\delta_{a\mathbf{K}}(t-t')}$$

$$-\sum_{\mathbf{K}} g_{a\mathbf{K}}{}^* g_{b\mathbf{K}} \int_0^t dt' C_{b0}(t') e^{i\delta_{a\mathbf{K}}t - \delta_{b\mathbf{K}}t'} . \qquad (66)$$

We recognize that the first term in this expression is precisely the same as that obtained in the Weisskopf-Wigner theory of spontaneous emission. The integral over the continuum of modes leads to an exponential decay. However Eq. (66) has a second term, which corresponds to multiple scattering events: this contribution describes processes in which a quantum is first emitted in the $|b\rangle \rightarrow |c\rangle$ transition, and then reabsorbed in a $|c\rangle \rightarrow |a\rangle$ transition. We ignore such processes in the following.

Having realized that this step involves a supplementary assumption, we perform the Weisskopf-Wigner approximation and obtain

$$\frac{dC_{a0}}{dt} = -\frac{\gamma_a}{2} C_{a0} \qquad (67a)$$

$$\frac{dC_{b0}}{dt} = -\frac{\gamma_b}{2} C_{b0} . \qquad (67b)$$

Integrating Eq. (67) and substituting the result into Eq. (65) yields

$$C_{c\mathbf{K}}(t) = C_{a0}(0) \frac{ig_{a\mathbf{K}}}{\frac{1}{2}\gamma_a + i\delta_{a\mathbf{K}}} [e^{(-\frac{1}{2}\gamma_a - i\delta_{a\mathbf{K}})t} - 1]$$

$$+ C_{b0}(0) \frac{ig_{b\mathbf{K}}}{\frac{1}{2}\gamma_b + i\delta_{b\mathbf{K}}} [e^{(-\frac{1}{2}\gamma_b - i\delta_{b\mathbf{K}})t} - 1] . \qquad (68)$$

According to Eq. (12.74), a detector registers a signal proportional to

$$\mathcal{S} = \langle\psi(t)|E^-(\mathbf{r})E^+(\mathbf{r})|\psi(t)\rangle , \qquad (69)$$

where $\mathbf{r}$ is the position of the detector relative to the emitting atom,

$$E^+(\mathbf{r}) = \mathcal{E}_\Omega \sum_k a_K e^{i\mathbf{K}\cdot\mathbf{r}} , \qquad (70)$$

and $\mathcal{E}_\Omega$ is given by Eq. (55). With the form (63) of the state vector and the observation that $\langle 1_\mathbf{K} | 1_{\mathbf{K}'} \rangle = \delta_{\mathbf{KK}'}$, Eq. (69) yields readily

$$\mathcal{I} = \left| \mathcal{E}_\Omega \sum_\mathbf{K} C_{c\mathbf{K}}(t) \exp[i(\mathbf{K}\cdot\mathbf{r} - \omega_\mathbf{K} t)] \right|^2 , \tag{71}$$

so that knowing $C_{c\mathbf{K}}(t)$ suffices to determine the measured field intensity.

We thus have to evaluate the sum

$$\mathcal{S} = \sum_\mathbf{K} C_{c\mathbf{K}}(t) \exp[i(\mathbf{K}\cdot\mathbf{r} - \omega_\mathbf{K} t)] . \tag{72}$$

As in Eq. (53), we replace the $\Sigma_\mathbf{K}$ by an integral, but here we simplify the algebra by neglecting the effects of light polarization, e.g., we take $g_{a\mathbf{K}} = g_a = \wp_{ac}\mathcal{E}_\Omega/\hbar$. In spherical coordinates $\exp(i\mathbf{K}\cdot\mathbf{r})$ is given by $\exp(iKr\cos\theta)$, which has the property

$$-iKr\sin\theta \exp(iKr\cos\theta) = \frac{d}{d\theta} \exp(iKr\cos\theta) .$$

Combining these observations, we find

$$\mathcal{S} = C_{a0}(0) \frac{\wp_{ac}\omega_{ac}\mathcal{E}_\Omega V}{(2\pi c)^2 \hbar r} \int d\Omega \frac{1}{\frac{1}{2}\gamma_a + i\delta_{a\mathbf{K}}} [e^{-(\frac{1}{2}\gamma_a + i\omega_a)t} - e^{-i\omega_k t}][e^{iKr} - e^{-iKr}]$$

$$+ \text{ same with } a \to b . \tag{73}$$

Here the first term originates from the $|a\rangle - |c\rangle$ transition and the second from the $|b\rangle - |c\rangle$ transition. Remembering that $\delta_{a\mathbf{K}} = \omega_{ac} - \Omega$, we note that the first term in $\mathcal{S}$ has a single pole in lower half-plane at $\Omega = \omega_{ac} - i\gamma_a/2$. The integral is thus easily performed by contour integration.

At this point, the astute reader will notice that in the four terms contributing to each transition, one is zero, one is a retarded term going as $\exp(...)(t-r/c)$, and two are *advanced* contributions proportional to $\exp(...)(t+r/c)$. Problem 13-4 shows that these last two terms cancel exactly, so that the system is causal and well-behaved. This observation also applies to the single-transition spontaneous emission of Sec. 13-3. Performing the integrals, we find

$$\mathcal{S}(t) = iC_{a0}(0)\,\frac{\wp_{ac}\omega_{ac}\mathcal{E}_\Omega V}{2\pi c^2\hbar r}\,e^{-i\omega_c t}\,e^{-(\gamma_a/2\,-\,i\omega_{ac})(t-r/c)}\,\Theta(t-r/c)$$

$$+\ \text{same with } a \to b\,, \tag{74}$$

where Θ is a Heaviside function, $\Theta(x) = 1$ for $x > 0$ and $\Theta(x) = 0$ for $x < 0$.

When substituting this form into (71), we observe that the contributions of the two possible transitions give a temporal interference term. We obtain interferences, because if we detect a signal we have no way of knowing whether the light received at the detector comes from the $|a\rangle$ - $|c\rangle$ or from the $|b\rangle$ - $|c\rangle$ transition. This is the phenomenon of quantum beats. Carrying out the last trivial step gives finally

$$\mathcal{I}(t) = \Theta(t-r/c)\left[\frac{\Omega}{4\pi\epsilon_0 c^2 r}\right]^2\Big\{|\wp_{ac}\omega_{ac}C_{a0}(0)|^2 e^{-\gamma_a(t-r/c)}$$

$$+\ \wp_{ac}\wp_{cb}\omega_{ac}\omega_{bc}\,e^{-(\gamma_a+\gamma_b)(t-r/c)/2}\,C_{a0}(0)C^*_{b0}(0)\,e^{-i\omega_{ab}(t-r/c)}$$

$$+\ \text{same with } a \to b\Big\}. \tag{75}$$

If either $C_{a0}(0)$ or $C_{b0}(0)$ vanishes, we recover the usual exponential decay of the spontaneous transition between the other two levels. But if the system is initially in a coherent superposition of levels $|a\rangle$ and $|b\rangle$, the fluorescence signal has a component modulated at the difference frequency $\omega_a-\omega_b$. This result is at the origin of a spectroscopic technique used to determine the difference in frequency between two levels. Problem 13-5 shows that if a single upper level can decay to two lower levels, no quantum beat signal is observed. Why?

References

The major references for this chapter are the same as those for Chap. 12. Further references are:

Barnett, S. M., P. Filipowicz, J. Javanainen, P. L. Knight, and P. Meystre, "The Jaynes-Cummings model and beyond," in *Frontiers in Quantum Optics*, E. R. Pike and S. Sarkar, eds. (Adam Hilger, Bristol 1986), p. 485. This gives a detailed review of the Jaynes-Cummings model.

Cohen-Tannoudji, C., J. Dupont-Roc and G. Grynberg (1989), *Photons and Atoms, Introduction to Quantum Electrodynamics*, John-Wiley & Sons, New York. An excellent advanced book on atom-light interactions.

Eberly, J. H., N. B. Narozhny, and J. J. Sanchez-Mondragon (1980), Phys. Rev. Lett. **44**, 1323.

Itzykson, C. and J. B. Zuber (1980), *Quantum Field Theory*, McGraw-Hill, is an excellent monograph on this topic.

O'Brien, C., P. Meystre, and H. Walther (1985), in *Advances in Atomic and Molecular Physics*, Vol. 21, Ed. by Sir David Bates and B. Bederson, Academic Press, Orlando, FL.

Rempe, G. and H. Walther (1987), Phys. Rev. Lett. **58**, 353.

Stenholm, S. (1973), Phys. Reps. **6C**, 1.

Weisskopf, V. and E. Wigner (1930), Z. Phys. **63**, 54.

Problems

13-1. Solve the dressed-atom eigenvalue problem obtaining Eqs. (14) through (18).

13-2. Complete the steps in Eq. (26) for the quantized two-level state vector.

13-3. Show that the sum in Eq. (34) reduces to Eq. (36) for a sufficiently intense field and short enough times ($t << |\alpha|/g$).

13-4. Show that the two advanced contributions in Eq. (73) cancel identically, leading to a causal result.

13-5. Show that if a single upper level can decay to two lower levels, no quantum beat signal is observed.

13-6. In what ways do approximations of this chapter fail outside of the optical frequency regime?

13-7. Calculate the expectation value of the dipole moment operator in terms of dressed-state probability amplitudes. Ans:

$$\langle er \rangle = \wp \sum_n \{\sin 2\theta_n [|C_{1n}|^2 - |C_{2n}|^2] + \cos 2\theta_n [C_{1n} C_{2n}{}^* + \text{c.c.}]\} .$$

where $\sin 2\theta_n$ and $\cos 2\theta_n$ are given by Eq. (18).

13-8. An interaction energy operator for all field modes *except* for the dressing mode can be written as

$$\mathcal{V} = \hbar G[|a\rangle\langle b| + |b\rangle\langle a|] .$$

for some constant G. This operator allows the atom to change state without affecting the dressing mode (some other mode or modes provide or get the atomic energy). Using Eq. (16), calculate the matrix elements of $\mathcal{V}$ between all combinations of the dressed states $|2n\rangle$, $|1n\rangle$, $|2n-1\rangle$, $|1n-1\rangle$.

13-9. Derive the operator form of Eq. (26) as follows. Split the Hamiltonian of Eq. (4) as

$$\mathcal{H} = \mathcal{H}_1 + \mathcal{H}_2 ,$$

where

$$\mathcal{H}_1 = \hbar\Omega a^\dagger a + \tfrac{1}{2}\hbar\Omega\sigma_z , \tag{76}$$

$$\mathcal{H}_2 = \tfrac{1}{2}\hbar\delta\sigma_z + \hbar(ga\sigma_+ + \text{adj}) . \tag{77}$$

Show that the commutator $[\mathcal{H}_1, \mathcal{H}_2]$ vanishes, allowing the time development matrix $U(t, 0)$ to be factored as

$$U(t,0) = \exp(-i\mathcal{H}_1 t/\hbar)\exp(-i\mathcal{H}_2 t/\hbar) .$$

Writing $\mathcal{H}_1$ in 2x2 matrix form, show that

$$U_1(t,0) = e^{-i\Omega a^\dagger a t} \begin{pmatrix} \exp(-i\Omega t/2) & 0 \\ 0 & \exp(i\Omega t/2) \end{pmatrix} .$$

A simple 2x2 form for $U_2(t,0)$ requires proving first that

$$\begin{pmatrix} \tfrac{1}{2}\delta & ga \\ ga^\dagger & -\tfrac{1}{2}\delta \end{pmatrix}^{2\ell} = \begin{pmatrix} (\varphi + g^2)^\ell & 0 \\ 0 & \varphi^\ell \end{pmatrix}$$

$$\begin{pmatrix} \frac{1}{2}\delta & ga \\ ga^\dagger & -\frac{1}{2}\delta \end{pmatrix}^{2\ell+1} = \begin{pmatrix} \frac{1}{2}\delta(\varphi + g^2)^\ell & g(\varphi + g^2)^\ell a \\ g\varphi^\ell a^\dagger & -\frac{1}{2}\delta\varphi^\ell \end{pmatrix},$$

where the operator $\varphi = g^2a^\dagger a + \frac{1}{4}\delta^2$. Expanding $U_2(t,0)$ in a Taylor series, using these relations and resumming the matrix elements, show that

$$U_2(t,0) = \begin{pmatrix} \cos(t\sqrt{\varphi+g^2}) - i\delta\dfrac{\sin(t\sqrt{\varphi+g^2})}{2\sqrt{\varphi+g^2}} & -ig\dfrac{\sin(t\sqrt{\varphi+g^2})}{\sqrt{\varphi+g^2}}\,a \\ \dfrac{-ig\sin(t\sqrt{\varphi})}{\sqrt{\varphi}}\,a^\dagger & \cos(t\sqrt{\varphi}) + i\delta\dfrac{\sin(t\sqrt{\varphi})}{2\sqrt{\varphi}} \end{pmatrix}. \tag{78}$$

The projection onto $|an\rangle$ and $|bn{+}1\rangle$ gives the time development matrix in Eq. (26).

13-10. The phenomenological decay rates $-\frac{1}{2}\gamma_a C_{an}$ and $-\frac{1}{2}\gamma_a C_{bn+1}$ can be added to the equations of motion (30) and (31), respectively. Those equations then have the same form as the semiclassical Eqs. (4.1) and (4.2), with $\mathscr{R}_0$ replaced by $-2g\sqrt{n+1}$ and hence have the solution (4.10) with this substitution.

Calculate the corresponding decay rates for the dressed-atom probability amplitudes $C_{1n}(t)$ and $C_{2n}(t)$. Solve the resulting equations of motion to find a generalization of Eq. (25). As in Eq. (26), use this solution to find $C_{an}(t)$ and $C_{bn+1}(t)$ with decay. Note that although this is a good exercise in working with dressed states, it is also the more cumbersome way to solve the problem.

13-11. Show that the electric-field-per-photon $\mathscr{E}_\Omega$ for a running wave is given by Eq. (55).

13-12. Solve the first-order perturbation result corresponding to Eq. (40) using $|\psi(t)\rangle$ of Eq. (29) with the equation of motion (30) and (31). Remember that $C_a{}^{(0)}(t)$ is *not* constant.

Chapter 14
SYSTEM-RESERVOIR INTERACTIONS

The Weisskopf-Wigner theory of spontaneous emission of Chap. 13 is an example of a general class of problems involving the coupling of a small system to a large system. In that case the small system is the atom and the large system is the continuum of modes of the electromagnetic field. When computing the atomic decay rate, we were not interested in the field itself, but only on its effect on the atomic dynamics. Thus we never explicitly computed the field dynamics. Our theory leads to an irreversible decay of the upper state population. At first this should come as a surprise, since our starting equation (the Schrödinger equation) is reversible. Irreversibility results from two main approximations: 1) the assumption that the probability amplitude $C_{a0}(t)$ varies little during the time interval defined by the inverse bandwidth of the continuum of modes of the electromagnetic field, and 2) the replacement of the remaining non-local time integration in Eq. (13.58) by *a* δ-function. These choices comprise the Weisskopf-Wigner approximation.

Similar situations occur repeatedly in physics in general, and in quantum optics in particular, and always lead to an irreversible decay of the small system. In fact, each time one wishes to describe properly irreversible damping in quantum mechanics, one does so by coupling the small system under study to a large, broad-band system which typically remains in thermal equilibrium. For this reason, the large system is usually called a bath, or a reservoir. It is of considerable importance to develop a general formalism to handle this problem.

Just as there are two fundamental ways to treat a general quantum mechanics problem, the Heisenberg and the Schrödinger pictures, there are also two basic ways to tackle system-reservoir interactions. The first one is based on the Schrödinger picture and leads to the so-called *master equation*. We discuss it in Sec. 14-1 and use it in Sec. 15-4 on resonance fluorescence and in Sec. 16-2 on the quantum theory of multiwave mixing. In Sec. 14-2, we show how an expansion of the system density operator in a basis of coherent states permits us to transform the master equation into a Fokker-Planck equation. The Heisenberg approach is presented in Sec. 14-3 and leads to the introduction of *quantum noise operators* giving the description of the problem a flavor reminiscent of the Langevin approach to stochastic problems in classical physics. Section 14-4 explains the quantum regression theorem and applies it to the evaluation of two-time correlation functions such as appear in spectrum calculations in resonance fluorescence and in the generation of squeezed states.

The material of this chapter is rather technical, and our experience has been that a general presentation tends to mask the physics involved in deriving the results. We prefer therefore to sacrifice generality and concentrate on an illustration of the theory for the case of a simple harmonic oscillator coupled to a bath of harmonic oscillators. This model system is described by the Hamiltonian

$$\mathcal{H} = \mathcal{H}_s + \mathcal{H}_r + \mathcal{V} , \tag{1}$$

where

$$\mathcal{H}_s = \hbar\Omega\, a^\dagger a \tag{2}$$

is the unperturbed Hamiltonian of the small system,

$$\mathcal{H}_r = \sum_j \hbar\omega_j\, b_j{}^\dagger b_j \tag{3}$$

the unperturbed Hamiltonian of the reservoir, consisting of a very large number of harmonic oscillators, and

$$\mathcal{V} = \hbar \sum_j (g_j a^\dagger b_j + g_j{}^* b_j{}^\dagger a) \tag{4}$$

is a model for the system-reservoir interaction. The elementary exchange of energy between system and bath is thus assumed to consist of the simultaneous creation of a quantum of excitation of the system with annihilation of a quantum in the jth mode of the bath, or the reverse process.

Different problems require of course different model Hamiltonians. For instance, section 14-1 concludes with a corresponding discussion for a resonant bath of two-level atoms and the remaining chapters in the book deal with related problems. The general behavior of the system is however not very sensitive to the explicit form of $\mathcal{H}_r$, provided that it meets some general requirements, the most important one being that it has a broad-band spectrum, and hence a very short correlation time. The specific model defined by Eqs. (1)-(4) is a good model of coupling to the continuum of modes of the electromagnetic field, of phonons in a crystal, etc., which is one reason why we consider it here. The other reason is that harmonic oscillators are the simplest quantum systems, and make our lives particularly easy.

The Hamiltonian (1), together with initial conditions, completely defines our problem. We suppose that at the initial time t_0 the small system is described by a density operator $\rho_s(t_0)$, where the subscript $_s$ indicates that this is the system's density operator. Of course, $\text{tr}_s\,\{\rho_s(t_0)\} = 1$, where tr_s means "trace over the system". In contrast we suppose that the reservoir is a very large system with an immense number of degrees of freedom. Usually (but not always) it is described by a time-independent density operator ρ_r in thermal equilibrium at the temperature T. According to Eq. (12.34), such a density operator is given by

$$\rho_r(\mathcal{H}_r) = \frac{e^{-\beta\mathcal{H}_r}}{\text{tr}_r\{e^{-\beta\mathcal{H}_r}\}} . \tag{5}$$

Note that $\text{tr}_r\rho_r(\mathcal{H}_r) = 1$, where tr_r means "trace over the reservoir". Assuming that the system and reservoir are brought into contact at time $t = t_0$, they initially don't exhibit any correlations and thus the initial state of the system is described by the factorized density operator

$$\rho_{sr}(t_0) = \rho_s(t_0)\rho_r(\mathcal{H}_r) . \tag{6}$$

In the next sections, we solve the problem defined by the Hamiltonian (1) and the initial density operator (6) under the conditions that we are only interested in the system's dynamics and that the reservoir has a very broad band spectrum.

14-1. Master equation

We analyze the problem first in the Schrödinger picture. We might try to achieve this goal by solving the system-reservoir Schrödinger equation

$$\dot{\rho}_{sr} = -\frac{i}{\hbar}[\mathcal{H}, \rho_{sr}] \tag{7}$$

for all times. In general this is a hopeless goal. Fortunately all we are interested in is the system's evolution; we do not care what the reservoir is doing. Thus we need an object that allows us to compute the expectation value of system operators like $\mathcal{O}$. We know from Eq. (4.37) that

$$\langle \mathcal{O}(t)\rangle = \mathrm{tr}_{sr}\,\{\mathcal{O}\rho_{sr}(t)\}\,, \tag{8}$$

where we write tr_{sr} to remind ourselves that we must trace over the system *and* the reservoir. Since $\mathcal{O}$ is a system operator alone, we may rewrite this as

$$\langle \mathcal{O}(t)\rangle = \mathrm{tr}_s\,\{\mathcal{O}\ \mathrm{tr}_r\ \rho_{sr}(t)\} \equiv \mathrm{tr}_s\,\{\mathcal{O}\ \rho_s(t)\}\,. \tag{9}$$

The operator $\rho_s(t)$, which is the trace over the reservoir of the total density operator, is called the *reduced density operator* of the system. All we need to know to determine the expectation value of system operators is $\rho_s(t)$ at all times. The equation of motion for $\rho_s(t)$ is called *a master equation.*

Our strategy to derive the master equation is quite simple: we solve the problem to second order in perturbation theory, trace over the reservoir, take into account that it has a very broad bandwidth to perform the Markoff approximation, and obtain directly an equation for $\rho_s(t)$ that is valid for times long compared to the inverse bandwidth τ_c of the reservoir. While following this program, we have to be careful not to confuse the fast free evolution of the system with τ_c. To make sure that things don't get mixed up, we therefore first go into an interaction picture where all free evolutions are eliminated.

The interaction picture was formally introduced in Sec. 3-1, where we restricted ourselves to systems described by a state vector. Things are essentially the same when dealing with density operators. Rewriting the Hamiltonian (1) as

$$\mathcal{H} = \mathcal{H}_0 + \mathcal{V}\,, \tag{10}$$

where

$$\mathcal{H}_0 = \mathcal{H}_s + \mathcal{H}_r\,, \tag{11}$$

the density operator $P_{sr}(t)$ in the interaction picture is obtained from the Schrödinger picture density operator via the unitary transformation

$$\rho_{sr}(t) = e^{-i\mathcal{H}_0(t-t_0)/\hbar}\,P_{sr}(t)\,e^{i\mathcal{H}_0(t-t_0)/\hbar}\,. \tag{12}$$

Compare this with the state vector transformation Eq. (3.51). Differentiating this equation with respect to time and making use of (4.38) and (12), we obtain the equation of motion of the density operator in the interaction picture:

$$\frac{\partial P_{sr}}{\partial t} = -\frac{i}{\hbar}\,[\mathcal{V}_I(t-t_0), P_{sr}]\,, \tag{13}$$

where

$$\mathcal{V}_I(t-t_0) = e^{-i\mathcal{H}_0(t-t_0)/\hbar}\mathcal{V}e^{i\mathcal{H}_0(t-t_0)/\hbar} , \tag{14}$$

is the interaction Hamiltonian in the interaction picture, see Eq. (3.54). Using Eqs. (2), (3) and (4) for the various Hamiltonians we obtain immediately

$$\mathcal{V}_I(t) = \hbar \sum_j g_j\, a^\dagger e^{i\Omega(t-t_0)}\, b_j e^{-i\omega_j(t-t_0)} + \text{adj}. \tag{15}$$

Here we made use of the free evolution (3.125) and (3.126) of the annihilation and creation operators and remembered that system and reservoir operators commute at equal times.

We finally introduce the interaction-picture reduced density operator for the system

$$\rho(t) \equiv \text{tr}_r\, \{P_{sr}(t)\}\, . \tag{16}$$

This is the operator that typically plays the central role in the master equation approach. It is related to the Schrödinger picture reduced density operator $\rho_s(t)$ by the unitary transformation

$$\rho_s(t) = e^{-i\mathcal{H}_s(t-t_0)/\hbar}\, \rho(t)\, e^{i\mathcal{H}_s(t-t_0)/\hbar} , \tag{17}$$

Note that in Eq. (17), $\mathcal{H}_s$ appears rather than $\mathcal{H}_0$, as is discussed in Prob. 14-1. Differentiating Eq. (17) with respect to time, we obtain

$$\frac{\partial \rho_s}{\partial t} = -\frac{i}{\hbar}\, e^{-i\mathcal{H}_s(t-t_0)/\hbar} \left[[\mathcal{H}_s, \rho(t)] + \frac{\partial \rho}{\partial t} \right] e^{i\mathcal{H}_s(t-t_0)/\hbar} , \tag{18}$$

which relates the equations of motion for the reduced density operator in the Schrödinger and interaction pictures.

Having now all the required formalism at our disposal, we can proceed with the solution of the problem. In general system-reservoir coupling problems are not amenable to an exact solution. We use the iterative method outlined in Sec. 3-2 to solve the problem to second-order in perturbation theory. Specifically we integrate Eq. (13) from t_0 to t, taking $P_{sr}(t) \simeq P_{sr}(t_0)$ in the commutator to obtain a first-order solution for $P_{sr}(t)$. We then use this improved value in the commutator in integrating again to obtain a value of $P_{sr}(t)$ accurate to second order. We find

$$P_{sr}(t) = P_{sr}(t_0) - \frac{i}{\hbar}\int_{t_0}^{t} dt'\,[\mathcal{V}_I(t'-t_0), P_{sr}(t_0)]$$

$$-\frac{1}{\hbar^2}\int_{t_0}^{t} dt'\int_{t_0}^{t'} dt''\,[\mathcal{V}_I(t'-t_0), [\mathcal{V}_I(t''-t_0), P_{s-r}(t_0)]] + \ldots\,. \tag{19}$$

Tracing over the reservoir yields the evolution of the reduced density operator $\rho(t)$ of Eq. (16). Performing this trace, we can define *a coarse-grained equation of motion* for $\rho(t)$ by

$$\dot{\rho}(t) \simeq \frac{\rho(t) - \rho(t-\tau)}{\tau}, \tag{20}$$

where the time interval $\tau = t - t_0$ is a time long compared to the reservoir memory time τ_c, but short compared to times yielding significant changes in the system variables. In explicit calculations, it is convenient to shift our time origin by τ, i.e., to write

$$\dot{\rho}(t+\tau) \simeq \frac{\rho(t+\tau) - \rho(t)}{\tau}.$$

Since we assume that $\rho(t)$ does not vary significantly in the time τ, we suppose that $\dot{\rho}(t)$ itself is given by this expression. We further note that the double commutator in Eq. (19) simplifies somewhat since

$$[\mathcal{V}', [\mathcal{V}'', P_{sr}]] = \mathcal{V}'\mathcal{V}''P_{sr} - \mathcal{V}'P_{sr}\mathcal{V}'' + \text{adj}.$$

This is easily shown since $(ABC)^\dagger = C^\dagger B^\dagger A^\dagger$ and all of these operators are self-adjoint (Hermitian). Combining these observations, we find the coarse-grained system-density-operator equation of motion

$$\dot{\rho}(t) \simeq -\frac{i}{\hbar\tau}\int_0^{\tau} d\tau'\,\mathrm{tr}_r\{\mathcal{V}_I(\tau')P_{sr}(t)\} - \frac{1}{\hbar^2\tau}\int_0^{\tau} d\tau'\int_0^{\tau'} d\tau''\,\mathrm{tr}_r\{\mathcal{V}_I(\tau')\mathcal{V}_I(\tau'')P_{sr}(t) - \mathcal{V}_I(\tau')P_{sr}(t)\mathcal{V}_I(\tau'')\} + \text{adj}. \tag{21}$$

To proceed further we have to use the explicit form of the interaction Hamiltonian $\mathcal{V}_I$. We can simplify the algebra for the $\mathcal{V}_I$ of Eq. (15) by writing that equation in the more compact form

$$\mathcal{V}_I(\tau) = \hbar a^\dagger F(\tau) + \hbar a F^\dagger(\tau) , \tag{22}$$

where the operator

$$F(\tau) = -i \sum_j g_j b_j e^{i(\Omega-\omega_j)\tau} \tag{23}$$

acts only in the reservoir Hilbert space. In the Heisenberg approach of section 14-3, operators such as $F(\tau)$ will be identified as noise operators.

When tracing over the reservoir, we encounter terms of the type

$$\mathrm{tr}_r\{a^\dagger F(\tau) P_{sr}(t)\} = a^\dagger \rho(t) \mathrm{tr}_r\{F(\tau)\rho_r(\mathcal{H}_r)\} ,$$

where we have used Eq. (6) and the fact that at time t the interaction picture and Schrödinger density matrices are identical. The second trace in this equation is readily identified as the expectation value $\langle F\rangle_r$ of the reservoir operator $F(\tau)$. This vanishes provided the reservoir density operator ρ_r is diagonal as is the case for Eq. (5). We also note that we can cyclically permute reservoir operators under a reservoir trace so that

$$\langle F^\dagger(\tau')\rho_r F(\tau'')\rangle_r = \langle F(\tau'')F^\dagger(\tau')\rho_r\rangle_r \equiv \langle F(\tau'')F^\dagger(\tau')\rangle_r .$$

Substituting Eq. (22) into (21), we find

$$\begin{aligned}\dot{\rho}(t) = -\frac{1}{\hbar^2 \tau}\int_0^\tau d\tau' \int_{t_0}^{\tau'} d\tau'' \Big[& a^\dagger a\rho(t)\langle F(\tau')F^\dagger(\tau'')\rangle_r - a\rho(t)a^\dagger \langle F(\tau'')F^\dagger(\tau')\rangle_r \\ & + aa^\dagger \rho(t)\langle F^\dagger(\tau')F(\tau'')\rangle_r - a^\dagger\rho(t)a\langle F^\dagger(\tau'')F(\tau')\rangle_r \\ & + aa\rho(t)\langle F^\dagger(\tau')F^\dagger(\tau'')\rangle_r - a\rho(t)a\langle F^\dagger(\tau'')F^\dagger(\tau')\rangle_r \\ & + a^\dagger a^\dagger \rho(t)\langle F(\tau')F(\tau'')\rangle_r - a^\dagger\rho(t)a^\dagger \langle F(\tau'')F(\tau')\rangle_r \Big] + \text{adj.}\end{aligned} \tag{24}$$

Using Eq. (23), we find that the reservoir average terms have values like

$$\langle F(\tau')F^\dagger(\tau'')\rangle_r = \sum_{i,j} g_i g_j^* \langle b_i b_j^\dagger\rangle_r \, e^{i\Omega(\tau'-\tau'')} \, e^{i(\omega_j\tau''-\omega_i\tau')}$$

$$= \sum_i |g_i|^2 \langle b_i b_i^\dagger \rangle_r \, e^{i(\Omega-\omega_i)(\tau'-\tau'')} \,, \tag{25}$$

where the reservoir density matrix must be diagonal to obtain the last equality. This simplification is relaxed when we consider "squeezed reservoirs" in Sec. 16-4. Averages like $\langle F(\tau')F^\dagger(\tau'')\rangle_r$ are readily identified as first-order correlation functions of the bath ,see section 12-5. They depend only on the time difference $T = \tau' - \tau''$, i.e. the bath is stationary, a direct consequence of the thermal equilibrium assumption (5). Hence $\langle F(\tau')F^\dagger(\tau'')\rangle_r = \langle F(\tau'')F^\dagger(\tau')\rangle_r^*$. Equation (25) tells us how fast the bath correlations decay away. In the following, we will perform the *Markoff approximation*, which assumes that this correlation time is infinitely short compared to all times of interest for the system. Using T, we have, for example

$$\int_0^\tau d\tau' \int_0^{\tau'} d\tau'' \langle F(\tau')F^\dagger(\tau'')\rangle_r = \int_0^\tau d\tau' \sum_i |g_i|^2 \langle b_i b_i^\dagger\rangle_r \int_0^{\tau'} dT e^{i(\Omega-\omega_i)T} \,. \tag{26}$$

This kind of expression appears in the Weisskopf-Wigner theory of spontaneous emission of Sec. 13-3. There we replaced the sum over modes by an integral [Eq. (16.44)] and interpreted the integral over the exponential as a delta-function [Eq. (16.58)]. Similarly calling $\mathcal{D}(\omega)$ the density of modes in the reservoir, and assuming that the reservoir has sufficient bandwidth to justify the δ-function approximation of the integral, we find that Eq. (26) becomes

$$\int_0^\tau d\tau' \int_0^{\tau'} d\tau'' \, \langle F(\tau')F^\dagger(\tau'')\rangle_r = \frac{\gamma\tau}{2} \langle b(\Omega) b^\dagger(\Omega)\rangle_r \,, \tag{27}$$

where as in Eq. (16.59) we introduce the decay rate

$$\gamma = 2\pi \, \mathcal{D}(\Omega) |g(\Omega)|^2 \,. \tag{28}$$

Here we neglect the small shift due to the principal part in Eq. (16.58) (this is the equivalent of the Lamb shift for our simple harmonic oscillator). The factor $|g(\omega)|^2 \langle b(\omega) b^\dagger(\omega)\rangle_r$ is a measure of the strength of the coupling of the simple harmonic oscillator with the mode of the reservoir of frequency ω.

When extending the upper limit of the T-integration to infinity in Eq. (26), we implicitly assume that the reservoir correlation time τ_c is sufficiently small that the integrand vanishes already after times such that second-order perturbation theory remains valid. (Remember, we are doing perturbation theory, and this is valid only for times so short that the population of the different levels of the system remain practically unchanged!) Thus our approximate solution (27) is valid for times *short* compared to the decay of the system, but *long* compared to the correlation time of the reservoir. This is the essence of the Markoff approximation. In Sec. 14-3, we show explicitly using the Heisenberg picture that this is equivalent to assuming that the correlation functions of the bath are δ-correlated, that is, that the reservoir loses its memory instantaneously.

According to Eq. (15.36), the average of the number operator $b^\dagger b$ over a thermal distribution is given by

$$\langle b^\dagger(\Omega)b(\Omega)\rangle_r = \bar{n} = \frac{1}{e^{\beta\hbar\Omega} - 1} . \tag{29}$$

Similarly

$$\langle b(\Omega)b^\dagger(\Omega)\rangle = \bar{n} + 1 . \tag{30}$$

Substituting Eqs. (27) and (30) into Eq. (24) along with corresponding expressions for the terms with $\langle F^\dagger(\tau')F(\tau'')\rangle_r$, we find the master equation in the interaction picture

$$\boxed{\dot{\rho}(t) = -\frac{\gamma}{2}(\bar{n}+1)\left[a^\dagger a\rho(t) - a\rho(t)a^\dagger\right] - \frac{\gamma}{2}\bar{n}\left[\rho(t)aa^\dagger - a^\dagger\rho(t)a\right] + \text{adj.}} \tag{31}$$

Here we have neglected terms containing averages like $\langle F(\tau')F(\tau'')\rangle$ and $\langle F^\dagger(\tau')F^\dagger(\tau'')\rangle$, an approximation valid if the reservoir is thermal and its density operator is diagonal in the energy representation, see Eq. (5). This assumption is relaxed in Sec. 16-4, where the squeezed vacuum is considered. If the reservoir is at zero temperature, $\bar{n} = 0$, and the remaining terms are due to vacuum fluctuations.

We can use the master equation (31) to determine the evolution of the expectation of system operators. For example, the average photon number $\langle a^\dagger a\rangle_s = \mathrm{tr}_s\,(a^\dagger a\,\rho(t))$ becomes (Prob. 14-3)

$$\frac{d\langle a^\dagger a\rangle_s}{dt} = -\gamma\langle a^\dagger a\rangle_s + \gamma\bar{n} , \tag{32}$$

where we have used the cyclicity property of the trace and the boson commutation relation $[a, a^\dagger] = 1$.

A useful way to interpret the result (32) is to reexpress it as

$$\frac{d\langle a^\dagger a\rangle_s}{dt} = -\gamma\langle a^\dagger a\rangle_s(\bar{n} + 1) + \gamma\bar{n}\,(\langle a^\dagger a\rangle_s + 1)\,.$$

The rate of change of the mean number $\langle a^\dagger a\rangle_s$ of system photons is seen to result from the balance between emission from the system into the bath and from the bath into the system. In both terms, the "+1" is the contribution from spontaneous emission and has no classical equivalent, i.e. it would vanish if the boson creation and annihilation operators commuted. The other term is (classical) stimulated emission. For a reservoir at zero temperature, $\bar{n} = 0$ and all that is left is spontaneous decay from the system to the reservoir.

Equation (32) can be solved readily to give

$$\langle a^\dagger a\rangle_s(t) = \langle a^\dagger a\rangle_s(0)\,e^{-\gamma t} + \bar{n}[1 - e^{-\gamma t}]\,. \tag{33}$$

For large times, the average number of photons in the simple harmonic oscillator equilibrates to that of the bath oscillator of same frequency Ω. Similarly we find that the expectation value of the complex electric field operator obeys the interaction-picture equation of motion

$$\frac{d\langle a\rangle_s}{dt} = -\frac{\gamma}{2}\langle a\rangle_s\,. \tag{34}$$

It is instructive to calculate the equation of motion for the diagonal matrix elements $p_n \equiv \rho_{nn}$ in the number-state representation. From Eq. (31), we find

$$\dot{p}_n = -\gamma(\bar{n} + 1)[np_n - (n+1)p_{n+1}] - \gamma\bar{n}\,[(n+1)p_n - np_{n-1}]\,. \tag{35}$$

Figure 14-1 shows how the four terms in this equation represent flows of number probability up and down the system simple harmonic oscillator energy level diagram. Here we see very clearly the absorptive and emissive roles of the first and second bracketed expressions of Eq. (31), respectively. A steady state occurs when *a detailed balance* takes place between the absorptive and emissive processes, that is, when

$$\gamma(\bar{n}+1)\,np_n = \gamma\bar{n}\,np_{n-1}\,. \tag{36}$$

Note that steady state implies this detailed balance: Eq. (36) must be true for the lowest probability p_0 to be constant, which implies it must be true for p_1 to be constant, and so on up the ladder. This gives the thermal number distribution

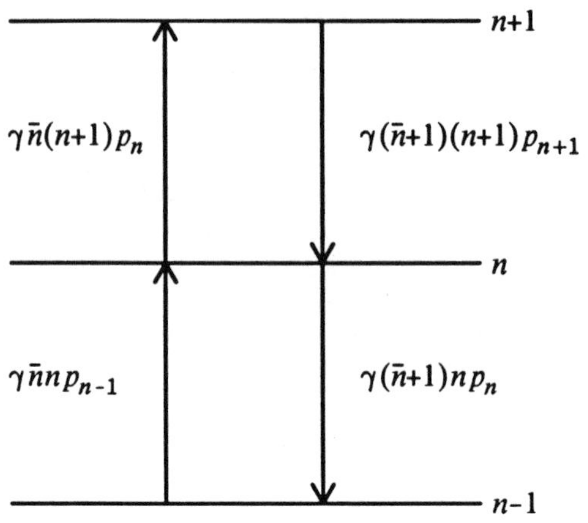

Fig. 14-1. Simple harmonic oscillator energy level diagram showing number probability flows in Eq. (35)

$$p_n = [\bar{n}/(\bar{n}+1)]\, p_{n-1} = [\bar{n}/(\bar{n}+1)]^n\, p_0 \\ = e^{-n\hbar\Omega/k_BT}\,[1 - e^{-\hbar\Omega/k_BT}]\,. \tag{37}$$

Other Master Equations

It is straightforward to modify the derivation of the master equation (31) to describe a two-level atom damped by a reservoir of simple harmonic oscillators. In this case, the interaction Hamiltonian (15) is replaced by

$$\mathcal{V}_I(\tau) = \hbar\sigma_+F(\tau) + \hbar\sigma_-F^\dagger(\tau)\,, \tag{38}$$

where $F(\tau)$ is still given by Eq. (23). The derivation of the master equation follows along exactly the same lines as that for Eq. (31), with the replacement of a by σ_- and $a^\dagger$ by σ_+. Hence we find

$$\dot{\rho}(t) = -\frac{\gamma}{2}(\bar{n}+1)\,[\sigma_+\sigma_-\rho(t) - \sigma_-\rho(t)\sigma_+] - \frac{\gamma}{2}\,\bar{n}\,[\rho(t)\sigma_-\sigma_+ - \sigma_+\rho(t)\sigma_-] + \text{adj.} \tag{39}$$

Another example of a reservoir, sometimes used in laser theory to model pump mechanisms, consists of a beam of two-level atoms traversing a single-mode cavity, as shown in Fig. 14-2. Here the system is the cavity

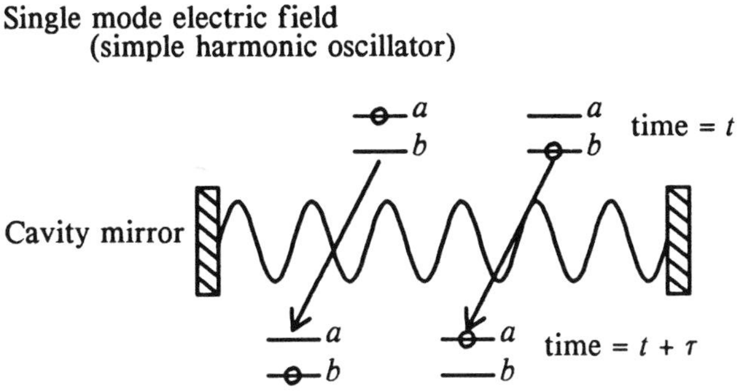

Fig. 14-2. Atomic beam of two-level atoms acts as reservoir for a simple harmonic oscillator.

mode, a simple harmonic oscillator. We assume that the atoms are injected inside the cavity in their upper state $|a\rangle$ and lower state $|b\rangle$ at rates r_a and r_b, respectively, where r_a and r_b satisfy the thermal-equilibrium Boltzmann distribution

$$\frac{r_a}{r_b} = e^{-\beta\hbar\omega} , \tag{40}$$

where as usual $\beta = 1/k_B T$. Thus the reservoir reduced density operator is given by

$$\rho_{atom}(t) = \frac{1}{Z}\begin{pmatrix} e^{-\beta\hbar\omega_a} & 0 \\ 0 & e^{-\beta\hbar\omega_b} \end{pmatrix} = \begin{pmatrix} \rho_{aa} & 0 \\ 0 & \rho_{bb} \end{pmatrix} . \tag{41}$$

We suppose the system field frequency Ω is resonant with the atomic transition frequency, $\Omega = \omega = \omega_a - \omega_b$. We proceed by considering first the interaction between the cavity mode and a single atom, and then multiply the result by the number of atoms present in the cavity at a time. This is a proper procedure provided that the atoms act incoherently on the field i.e. that collective atomic effects can be ignored. This important point is further discussed in Chap. 17. The interaction-picture interaction energy for the coupled atom-field system is given by the time-independent expression

$$\mathcal{V} = \hbar g \sigma_- a^\dagger + \text{adjoint} = \hbar g \begin{bmatrix} 0 & a \\ a^\dagger & 0 \end{bmatrix} . \tag{42}$$

We substitute $\mathcal{V}$ into the second-order system-reservoir density operator of Eq. (21) with the initial factorized value

$$P_{sr}(t) = \rho(t) \otimes \begin{bmatrix} \rho_{aa} & 0 \\ 0 & \rho_{bb} \end{bmatrix} . \tag{43}$$

The first-order term is

$$\mathcal{V} P_{sr} = \hbar g \begin{bmatrix} 0 & a\rho_{bb}\rho \\ a^\dagger \rho_{aa} \rho & 0 \end{bmatrix} . \tag{44}$$

Using Eqs. (42) and (44), we find the second-order terms

$$\mathcal{V}\mathcal{V}P_{sr} - \mathcal{V}P_{sr}\mathcal{V} = \hbar^2 g^2 \begin{bmatrix} \rho_{aa} a a^\dagger \rho - a\rho_{bb}\rho a^\dagger & 0 \\ 0 & a^\dagger a \rho_{bb}\rho - a^\dagger \rho_{aa}\rho a \end{bmatrix} . \tag{45}$$

Substituting Eqs. (44) and (45) into Eq. (21) and noting that $(\rho a a^\dagger)^\dagger = a a^\dagger \rho$, we obtain the coarse-grained equation of motion per atom

$$\dot{\rho}(t) = -\tfrac{1}{2} g^2 \tau [(a^\dagger a \rho - a \rho a^\dagger)\rho_{bb} + (\rho a a^\dagger - a^\dagger \rho a)\rho_{aa}] + \text{adj.}, \tag{46}$$

which is very similar to Eq. (31) for the reservoir of simple harmonic oscillators. Note however that the time interval τ is the *transit* time of an atom passing through the cavity. Multiplying this result by $r\tau$, the number of atoms passing through the cavity in this transit time, we find the system master equation

$$\dot{\rho}(t) = -\mathcal{R}_b\,[a^\dagger a \rho(t) - a\rho(t)a^\dagger] - \mathcal{R}_a\,[\rho(t) a a^\dagger - a^\dagger \rho(t) a] + \text{adj.}\,, \tag{47}$$

where the absorption rate constant $\mathcal{R}_b = r_b g^2 \tau^2 = r \rho_{bb} g^2 \tau^2$. Similarly, $\mathcal{R}_a = r_a g^2 \tau^2$.

The master equation (47) for a simple harmonic oscillator damped by an atomic-beam reservoir has the same form as that of Eq. (31) for a simple-harmonic-oscillator reservoir. With Eq. (47), it is particularly obvious that the contribution containing $a^\dagger a \rho(t) - a\rho(t)a^\dagger$ + adjoint corresponds to *absorption* of system photons, while the contribution containing $\rho(t) a a^\dagger - a^\dagger \rho(t) a$ + adjoint corresponds to *emission* of system photons.

14-2. Fokker-Planck Equation

The master equation of Sec. 14-1 allows us to readily compute equations of motion for the expectation values of various system observables. It is however an operator equation that by itself doesn't provide much physical intuition into the evolution of the system. We can use the master equation to derive a more classical equation with complementary insights by using quasi-probability distributions. If the system is a harmonic oscillator, we can expand the density operator in a coherent states basis as given by Eq. (15.80)

$$\rho = \int d^2\alpha\, P(\alpha)\, |\alpha\rangle\langle\alpha| \,. \tag{15.80}$$

Substituting this into Eq. (31) gives

$$\int d^2\alpha\, \dot{P}(\alpha,t)|\alpha\rangle\langle\alpha| = -\frac{\gamma}{2}(\bar{n}+1)\int d^2\alpha\, P(\alpha,t)\,[a^\dagger a|\alpha\rangle\langle\alpha| - a|\alpha\rangle\langle\alpha|a^\dagger]$$
$$-\frac{\gamma}{2}\bar{n}\int d^2\alpha\, P(\alpha,t)\,[aa^\dagger|\alpha\rangle\langle\alpha| - a^\dagger|\alpha\rangle\langle\alpha|a] + \text{c.c.} \tag{48}$$

From Eq. (15.51), we have the representation of a coherent state

$$|\alpha\rangle = e^{-|\alpha|^2/2}\, e^{\alpha a^\dagger}\, |0\rangle \,, \tag{49}$$

which gives

$$|\alpha\rangle\langle\alpha| = e^{-|\alpha|^2}\, e^{\alpha a^\dagger}|0\rangle\langle 0|e^{\alpha^* a} \,. \tag{50}$$

Hence $|\alpha\rangle\langle\alpha|a$ may be written as

$$|\alpha\rangle\langle\alpha|a = e^{-|\alpha|^2}\,\frac{\partial}{\partial\alpha^*}[e^{\alpha a^\dagger}|0\rangle\langle 0|e^{\alpha^* a}] = \left(\frac{\partial}{\partial\alpha^*} + \alpha\right)|\alpha\rangle\langle\alpha| \,. \tag{51}$$

Similarly,

$$a^\dagger|\alpha\rangle\langle\alpha| = \left(\frac{\partial}{\partial\alpha} + \alpha^*\right)|\alpha\rangle\langle\alpha| \,. \tag{52}$$

Substituting Eqs. (51) and (52) into Eq. (48) and using the definition $a|\alpha\rangle = \alpha|\alpha\rangle$ of the coherent state, we find

$$\int d^2\alpha\, \dot{P}(\alpha,t)|\alpha\rangle\langle\alpha| = -\frac{\gamma}{2}(\bar{n}+1)\int d^2\alpha\, P(\alpha,t)\,\alpha\frac{\partial}{\partial\alpha}|\alpha\rangle\langle\alpha|$$

$$-\frac{\gamma}{2}\bar{n}\int d^2\alpha\, P(\alpha,t)\left(\alpha\frac{\partial}{\partial\alpha}+\frac{\partial^2}{\partial\alpha\partial\alpha^*}\right)|\alpha\rangle\langle\alpha| + \text{adj}. \qquad (53)$$

We can integrate the RHS by parts and drop the constants of integration since $P(\alpha,t)$ vanishes for $|\alpha| \to \infty$. Thus, for example,

$$\int d^2\alpha P(\alpha,t)\alpha\,\frac{\partial}{\partial\alpha}|\alpha\rangle\langle\alpha| = -\int d^2\alpha\left[\frac{\partial}{\partial\alpha}[\alpha P(\alpha,t)]\right]|\alpha\rangle\langle\alpha|\,. \qquad (54)$$

Equation (53) becomes, after equating the coefficients of $|\alpha\rangle\langle\alpha|$ in the integrand

$$\dot{P}(\alpha,t) = \frac{\gamma}{2}\left[\frac{\partial}{\partial\alpha}[\alpha P(\alpha,t)] + \text{c.c.}\right] + \gamma\bar{n}\,\frac{\partial^2}{\partial\alpha\partial\alpha^*}P(\alpha,t)\,, \qquad (55)$$

which is in the form of a *Fokker-Planck equation* for the quasi-probability $P(\alpha,t)$ of finding the harmonic oscillator in the coherent state $|\alpha\rangle$ at time t. For reasons explained shortly, the coefficients of the first derivatives on the RHS of Eq. (55) are the elements of the *drift matrix* and the coefficients of the second derivatives compose the *diffusion matrix*. Problem (17-7) solves this equation at steady-state and finds

$$P(\alpha) = \frac{1}{\pi\bar{n}}\,e^{-|\alpha|^2/\bar{n}}\,, \qquad (56)$$

which is a thermal distribution with the average value $\bar{n}$.

Although it is hard in general to find the time-dependent solution of $P(\alpha,t)$, Eq. (55) can readily be used to obtain the rate of change of the expectation value of observables of interest. For instance,

$$\frac{d}{dt}\langle a\rangle = \int d^2\alpha\,\alpha\dot{P}(\alpha,t) = -\frac{\gamma}{2}\langle a\rangle\,, \qquad (57)$$

in agreement with Eq. (34).

It is important to realize that the derivation of a Fokker-Planck equation via quasi-probability distributions does not always lead to well-behaved results. We have seen in chapter 12 that $P(\alpha)$ needs not be positive and does not lend itself to a simple interpretation as a probability distribution. In situations where $P(\alpha)$ becomes negative or singular, which are typical if truly non-classical effects are important, we often find that the resulting Fokker-Planck equation has a non-positive diffusion matrix and hence is not mathematically well-behaved. In such situations, one can take advantage of the over-completeness of the coherent states to introduce generalizations of the $P(\alpha)$ distribution that eliminate this difficulty, typically at the expanse of doubling the phase-space dimensions. These techniques are discussed in detail by Gardiner (1980).

Returning to the problem of the damped oscillator, we can also find an approximate Fokker-Planck equation for the diagonal elements of the density matrix from the photon number probability equation of motion p_n of Eq. (35). In this heuristic approach we replace the discrete variable n by a continuous variable x, a step through which we immediately lose effects such as the Jaynes-Cummings revivals of chapter 12. Nonetheless, this is a useful approximation in a number of situations where the discrete nature of the field plays an insignificant role. We proceed by writing Eq. (35) in the form

$$\dot{p}_n = -\gamma(\bar{n}+1)[np_n - (n+1)p_{n+1}] - \gamma\bar{n}\,[np_n - (n-1)p_{n-1} + p_n - p_{n-1}] \,. \tag{58}$$

We then write xp in place of np_n, and $(x\pm1)p(x\pm1)$ in place of $(n\pm1)p_{n\pm1}$ and use a second-order Taylor series with $\Delta x = \pm1$ to find

$$(n\pm1)p_{n\pm1} \rightarrow xp \pm \frac{\partial}{\partial x}(xp) + \frac{1}{2}\left(\frac{\partial}{\partial x}\right)^2 (xp) \,.$$

Substituting this along with a second-order Taylor series for the lone p_{n-1} into Eq. (58), we find

$$\dot{p} = \gamma \frac{\partial}{\partial x}[(x - \bar{n})p] + \gamma\bar{n}\left(\frac{\partial}{\partial x}\right)^2 (xp) \,. \tag{59}$$

In steady state ($\dot{p} = 0$), we can equate the argument of $\partial/\partial x$ to zero, whereupon we find the differential equation $p' = -1/\bar{n}$. This has the solution

$$p(x) = p(0)\, e^{-x/\bar{n}} = \bar{n}^{-1}\, e^{-x/\bar{n}} \,, \tag{60}$$

where we set the initial value $p(0) = \bar{n}^{-1}$ since $p(x)$ has to be normalized. This answer agrees with value of Eq. (56) derived using the full coherent state expansion since $|\alpha|^2 = \bar{n}$. One reason this simple approach agrees with the more general one is that the thermal distribution is diagonal in the number state representation.

We can get an intuitive understanding of how a Fokker-Planck equation works by considering the general one-dimensional form

$$\dot{p}(x,t) = -\frac{\partial}{\partial x}(M_1 p) + \frac{\partial^2}{\partial x^2}(M_2 p)\ , \tag{61}$$

where $M_1(x)$ and $M_2(x)$ are called the first- and second-order moments of the distribution $p(x)$, or alternatively the *drift* and *diffusion* coefficients of the Fokker-Planck equation. In the more general multidimensional case,

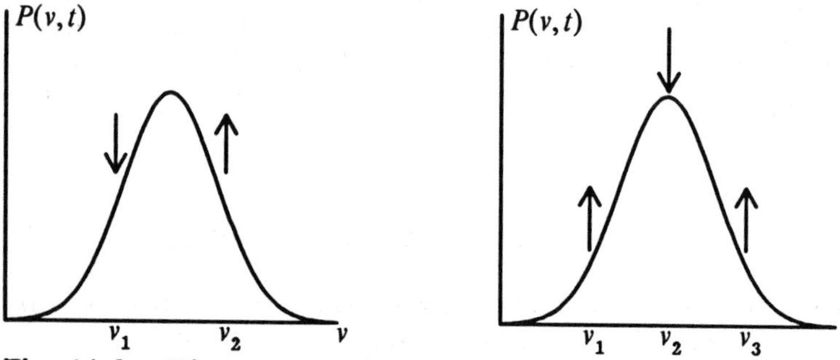

Fig. 14-3. Diagrams showing that the $M_1 p$ term in Eq. (61) causes the distribution $p(x)$ to "drift" along the x axis, while the $M_2 p$ term causes $p(x)$ to diffuse.

one speaks of drift and diffusion matrices. As shown in Fig. 14-3, for x values to the left of the peak of $M_1 p$, the slope of $M_1 p$ is positive which according to Eq. (61) causes a decrease of $p(x)$ in that region, while for x values to the right of the peak of $M_1 p$, the slope is negative, causing an increase of $p(x)$. This results in a movement of the peak toward larger values of x provided M_1 is itself positive. The second derivative at the peak of $M_2 p(x)$ is negative, which according to Eq. (61) causes a decrease in $p(x)$, while on either side of the peak of $M_2 p$, the second derivative is positive causing an increase of $p(x)$. This results in a diffusion of $p(x)$. For these reasons the M_1 term is called the drift term and the M_2 term is called the diffusion term.

14-3. Langevin Equations

In order to gain more insight into the system-reservoir interaction, we now solve the same problem in the Heisenberg picture. We show that the reservoir operators can then be interpreted similarly to Langevin forces in classical statistical mechanics. These *quantum noise operators* are the source of both fluctuations and irreversible dissipation of energy from the system to the reservoir.

From the Hamiltonians of Eq. (1), we readily obtain the equations of motion for the annihilation operators $a(t)$ and $b_j(t)$

$$\dot{a}(t) = -i\Omega a(t) - i\sum_j g_j b_j(t) \,, \tag{62}$$

$$\dot{b}_j(t) = -i\omega_j b_j(t) - ig_j{}^* a(t) \,. \tag{63}$$

We can integrate Eq. (63) formally to get

$$\begin{aligned} b_j(t) &= b_j(t_0)e^{-i\omega_j(t-t_0)} - ig_j{}^* \int_{t_0}^{t} dt' a(t')e^{-i\omega_j(t-t')} \\ &\equiv b_{free}(t) + b_{radiated}(t) \,. \end{aligned} \tag{64}$$

The first term b_{free} on the RHS of Eq. (64) is the homogeneous solution of Eq. (63), and describes the free evolution of b_j in the absence of interaction with the system. The second term $b_{radiated}$ gives the modification of this free evolution due to the coupling with the system, and shows that $a(t)$ is the source of $b_j(t)$. If the small system were, say, a two-level system instead of a harmonic oscillator, the appropriately modified Eq. (64) would show that the atomic polarization is the source of the electromagnetic field. Inserting Eq. (64) into Eq. (62), we find

$$\dot{a}(t) = -i\Omega a(t) - i\sum_j g_j b_j(t_0)e^{-i\omega_j(t-t_0)} - \sum_j |g_j|^2 \int_{t_0}^{t} dt' a(t')e^{-i\omega_j(t-t')} \,. \tag{65}$$

Here the first summation gives fluctuations and the second gives the radiation reaction.

We now change to an interaction picture in order to separate the rapid free evolution of $a(t)$ at the frequency Ω from the fast evolution due to the large bandwidth of the bath. This is done by introducing the slowly varying operator

$$A(t) = a(t)e^{i\Omega t}\ , \qquad \text{with } [A(t), A^\dagger(t)] = 1\ . \tag{66}$$

From Eq. (65), the evolution of $A(t)$ is given by

$$\dot{A}(t) = -\sum_j |g_j|^2 \int_{t_0}^{t} dt' A(t')e^{-i(\omega_j-\Omega)(t-t')} + F(t)\ , \tag{67}$$

where $F(t)$ is the *noise operator*

$$F(t) = -i\sum_j g_j b_j(t_0)e^{i(\Omega-\omega_j)(t-t_0)}\ , \tag{68}$$

which is the same as Eq. (23) aside from a shift in time origin. Note that this operator varies rapidly in time due to the presence of all the reservoir frequencies. Furthermore since the reservoir is described by a density operator diagonal in energy representation, $\langle F(t')\rangle_r$ vanishes. We have already encountered integrals similar to the first term on the RHS of Eq. (67) both when studying the Weisskopf-Wigner theory of spontaneous emission, and in Sec. 14-1. We handle it in the same fashion here. We replace the sum over modes of the bath by an integral and make the Markoff approximation by claiming that $A(t)$ varies little over the inverse reservoir bandwidth. This allows us to extend the limit of integration to infinity. Using the representation (16.58) of the delta-function, we obtain

$$\boxed{\dot{A}(t) = -\frac{\gamma}{2}A(t) + F(t)}\ . \tag{69}$$

Except for the presence of the noise operator $F(t)$, this is the same as Eq. (34) for the expectation value of $\langle a\rangle_s(t)$ obtained from the master equation. The *nature* of these equations is however completely different. One of them gives the evolution of an expectation value of an operator, while the second gives the evolution of the operator itself. One could not expect an operator to have an evolution as simple as that given by Eq. (34). If that were the case, its value at time t would be

$$A(t) = A(0)e^{-\gamma t/2}$$

and for times long compared to γ, we would have

$$[A(t), A^\dagger(t)] \to 0 \; .$$

This would be a terrible violation of the laws of quantum mechanics, since commutation relations must be valid at all times! The rapidly fluctuating operator $F(t)$ in Eq. (69) guarantees *by construction* that the commutation relations of the system remain valid at all times. With Eq. (67) and (69) we have

$$\langle \dot{A} \rangle = - \frac{\gamma}{2} \langle A \rangle \; . \tag{70}$$

The noise operator $F(t)$ therefore plays a role similar to that of Langevin forces in the theory of Brownian motion. In both cases, a random force of zero average value leads to dissipation. The only difference here is that this force has the character of an operator. Because of this analogy, Eq. (69) is sometimes called *a quantum Langevin equation* and the operator $F(t)$ *a quantum noise operator*. In principle, one can write down a quantum Langevin equation for any system operator, but of course each equation has a different noise operator. For instance, the adjoint of Eq. (69) is

$$\dot{A}^\dagger(t) = -\tfrac{1}{2}\gamma \; A^\dagger(t) + F^\dagger(t) \; . \tag{71}$$

The Heisenberg picture has the appealing feature that the equations of motion look like the corresponding classical equations of motion, but it does present a few pitfalls. The most important has to do with the ordering of operators. System operators commute with reservoir operators at equal times, but not necessarily at different times. This is illustrated in Eq. (63), which shows that as time evolves, $b_j(t)$ acquires some of the character of $a(t)$. More importantly, the homogeneous (free field) part of $b_j(t)$ alone doesn't commute with $a(t)$ even at equal times, although $[a(t), b_j(t)] = 0$. For this reason, after separating $b_j(t)$ into b_{free} and $b_{radiated}$ we can no longer interchange the order of system and reservoir operators without taking the chance of committing serious errors. Therefore once we chose the order in which to write system and reservoir operators in the initial Hamiltonian, we must *stick to it*! Here we always put all operators with a "†" to the left, which is called the "normal ordering". Any other ordering will do, provided that it is used consistently throughout the calculation. Although final answers do not depend on the choice of ordering, the physical interpretation of the results is different in different orderings [see

Milonni and Smith (1975)]. For instance, the normal ordering attributes spontaneous emission to radiation reaction, since $\langle F(t)\rangle_r = 0$, while vacuum fluctuations give the Langevin force $F(t)$. In contrast, Dalibard *et al* (1982) advocate the use of symmetrical ordering, which presents the advantage of making the contributions of the free and radiated fields to the system evolution separately hermitian. With this choice of ordering, radiation reaction and vacuum fluctuations give contributions of equal magnitude to spontaneous emission. These contributions have equal phase and add for the upper level of a two-level atom, but have opposite phase and cancel exactly for the lower level.

To illustrate one kind of problem one may get into when not being consistent in the choice of ordering, we now find the quantum Langevin equation for the number operator $a^\dagger a(t) = A^\dagger A(t)$. Its Heisenberg equation of motion gives readily

$$\frac{d}{dt}(A^\dagger A) = -i\sum_j g_j A^\dagger b_j e^{i\Omega t} + \text{adjoint}\,, \tag{72}$$

or, by substitution of Eq. (63) and its adjoint

$$\frac{d}{dt}(A^\dagger A) = -\sum_j |g_j|^2 A^\dagger(t)\int_{t_0}^{t} dt' A(t')e^{i(\Omega-\omega_j)(t-t')} - i\sum_j g_j\, A^\dagger(t)b_j(t_0)e^{i(\Omega-\omega_j)(t-t_0)} + \text{adjoint}\,. \tag{73}$$

Note that if we are not careful with ordering, we might have $AA^\dagger$ in place of $A^\dagger A$. After performing the Markoff approximation as before, we obtain the "Langevin" equation for the number operator

$$\frac{d}{dt}(A^\dagger A) = -\gamma A^\dagger A + \mathcal{G}_{A^\dagger A}(t)\,, \tag{74}$$

with

$$\mathcal{G}_{A^\dagger A}(t) = i\sum_j g_j^* b_j^\dagger(t_0)A(t)e^{-i(\Omega-\omega_j)(t-t_0)} + \text{adjoint}\,. \tag{75}$$

Here we put "Langevin" in quotation marks, because we like to think that in such equations the fluctuating force should have zero average. This is not the case for $\langle \mathcal{G}_{A^\dagger A} \rangle_r$. Problem (17-8) shows that its mean value is equal to $\gamma \bar{n}$. To obtain a proper Langevin equation, we introduce a new quantum noise operator

$$G_{A^\dagger A}(t) = \mathcal{G}_{A^\dagger A}(t) - \langle \mathcal{G}_{A^\dagger A}(t) \rangle_r = \mathcal{G}_{A^\dagger A}(t) - \gamma \bar{n} . \tag{76}$$

This noise operator does have a vanishing reservoir average and Eq. (74) becomes

$$\frac{d}{dt}(A^\dagger A) = -\gamma A^\dagger A + \gamma \bar{n} + G_{A^\dagger A}(t) . \tag{77}$$

This equation gives the same evolution as Eq. (32) for the mean excitation number, as it should.

Let us reconsider the correlation function (25) of the noise operators $F(t)$ and $F^\dagger(t)$. As for Eq. (26), we convert the sum over modes to an integral and then use Eq. (30) for the reservoir average $\langle b(\Omega) b^\dagger(\Omega) \rangle$. This gives

$$\langle F(t') F^\dagger(t'') \rangle_r = \int d\omega \mathcal{D}(\omega)\, |g(\omega)|^2\, [\bar{n}(\omega)+1]\, e^{i(\Omega-\omega_j)(t'-t'')} . \tag{78}$$

Assuming that the coupling factor $\mathcal{D}(\omega)|g(\omega)|^2 \bar{n}(\omega)$ varies little over regions for which the exponential oscillates slowly, we can evaluate it at Ω and remove it from the integral. This gives

$$\langle F(t') F^\dagger(t'') \rangle_r = \mathcal{D}(\Omega)\, |g(\Omega)|^2\, [\bar{n}(\Omega)+1] \int d\omega e^{i(\Omega-\omega_j)(t'-t'')} . \tag{79}$$

For a broad-band reservoir, we can extend the limits of integration to infinity. Using Eq. (28) for γ and the integral representation of the δ-function

$$\int_{-\infty}^{\infty} d\omega e^{i(\Omega-\omega)(t'-t'')} = 2\pi \delta(t'-t'') , \tag{80}$$

we find

$$\boxed{\langle F(t')F^\dagger(t'')\rangle_r = \gamma(\bar{n}+1)\,\delta(t'-t'')}\,. \tag{81}$$

Here $\frac{1}{2}\gamma(\bar{n}+1)$ is the diffusion coefficient for this noise operator correlation function. Similarly, we can find the correlation function

$$\langle F^\dagger(t')F(t'')\rangle_r = \gamma\bar{n}\,\delta(t'-t'')\,. \tag{82}$$

In this approximation, the noise operator correlation functions depend on the operator ordering, but not on the time ordering. Equations (81) and (82) justify the statement that in the limit of a reservoir of very large bandwidth, its correlation functions are δ-correlated in time. This is the Markoff approximation, for which the noise operators are assumed to have no memory. Integrating both sides of Eq. (82) over $\tau = t'-t''$ yields a simple example of the *fluctuation-dissipation* theorem

$$\gamma = \bar{n}^{-1}\int_{-\infty}^{\infty} d\tau\,\langle F(\tau)^\dagger F(0)\rangle_r\,, \tag{83}$$

which relates the rate of dissipation of the system to the correlations of the fluctuations of the reservoir.

Chapter 15 uses the Hamiltonian of Eq. (18.4) to find the Bloch-Langevin equations (18.9) for a two-level system interacting with a classical field and coupled to a reservoir of harmonic oscillators. These equations are very useful for deriving resonance-fluorescence and squeezing spectra.

14-4. Quantum Regression Theorem and Noise Spectra

Having seen a simple example of system-reservoir interaction, let us generalize the treatment to consider an abstract set of system operators $\{A_\mu\}$. Examples include $\{A, A^\dagger\}$ of Sec. 14-3, $\{\sigma_-, \sigma_+, \sigma_z\}$ for a two-level system, and $\{a_1, a_1^\dagger, a_3, a_3^\dagger\}$ for signal and conjugate fields in the quantum theory of four-wave mixing (see Chap. 18). This section has three major objectives: It gives a generalization of the fluctuation-dissipation theorem known as the generalized Einstein relations for the diffusion coefficients; it introduces the quantum regression theorem, which finds applications in the evaluation of expectation values of two-time correlation functions; finally, it finds the fluctuation spectrum of operator-product expectation values.

Einstein Relations

The system operators have the Langevin equations of motion

$$\dot{A}_\mu = D_\mu(t) + F_\mu(t) , \tag{84}$$

where D_μ is the drift term and F_μ is the noise operator, which has the correlation functions

$$\langle F_\mu(t')F_\nu(t'')\rangle = 2\langle D_{\mu\nu}\rangle\delta(t'-t'') . \tag{85}$$

From the identity

$$A_\mu(t) = A_\mu(t-\Delta t) + \int_{t-\Delta t}^{t} dt' \, \dot{A}_\mu(t') , \tag{86}$$

we obtain the system-operator, noise-operator correlation function

$$\langle A_\mu(t)F_\nu(t)\rangle = \langle A_\mu(t-\Delta t)F_\nu(t)\rangle + \int_{t-\Delta t}^{t} dt'\langle(D_\mu(t') + F_\mu(t'))F_\nu(t)\rangle . \tag{87}$$

Because the operator $A_\mu(t')$ at time t' cannot be affected by a fluctuation at a later time t, the first term on the RHS of Eq. (87) is zero. Similarly the correlation $\langle D_\mu(t')F_\nu(t)\rangle$ is zero except at the point $t'=t$, but the interval of integration is zero (set of measure zero). All that remains is

$$\langle A_\mu(t)F_\nu(t)\rangle = \int_{t-\Delta t}^{t} dt'\langle F_\mu(t')F_\nu(t)\rangle = \frac{1}{2}\int_{-\infty}^{\infty} dt'\langle F_\mu(t')F_\nu(t)\rangle , \tag{88}$$

where in the last step we have assumed the noise to be stationary. Substituting Eq. (85) into (88) we have

$$\langle A_\mu(t)F_\nu(t)\rangle = \langle D_{\mu\nu}\rangle . \tag{89}$$

In an analogous manner we find

$$\langle F_\mu(t)A_\nu(t)\rangle = \langle D_{\mu\nu}\rangle . \tag{90}$$

We now may use Eqs. (89) and (90) to determine the equation of motion for the average $\langle A_\mu A_\nu \rangle$. From Eq. (84), we have

$$\frac{d}{dt}\langle A_\mu A_\nu \rangle = \langle \dot{A}_\mu A_\nu \rangle + \langle A_\mu \dot{A}_\nu \rangle$$
$$= \langle D_\mu A_\nu \rangle + \langle F_\mu A_\nu \rangle + \langle A_\mu D_\nu \rangle + \langle A_\mu F_\nu \rangle \, .$$

Substituting Eqs. (89) and (90) and rearranging gives

$$\boxed{2\langle D_{\mu\nu} \rangle = - \langle A_\mu D_\nu \rangle - \langle D_\mu A_\nu \rangle + \frac{d}{dt}\langle A_\mu A_\nu \rangle} \, . \tag{91}$$

This equation is called the *generalized Einstein relation.* It shows that the diffusion coefficients $\langle D_{\mu\nu} \rangle$ are directly related to the drift coefficients D_μ and D_ν and thus comprises a quantum fluctuation-dissipation theorem. This equation makes it possible to calculate the diffusion coefficients immediately from the drift coefficients, provided one can independently calculate the equation of motion for $\langle A_\mu A_\nu \rangle$. In particular, we can use the master equation to calculate these equations as well as the drift terms. Hence once we know the master equation, we can find the Langevin equations of motion and the corresponding diffusion coefficients.

Quantum Regression Theorem

We can also use Eq. (84) to calculate the expectation values of two-time correlation functions. Specifically, we multiply Eq. (84) on the right by $A_\nu(t')$, where $t' < t$, and average to find

$$\frac{d}{dt}\langle A_\mu(t) A_\nu(t') \rangle = \langle D_\mu(t) A_\nu(t') \rangle + \langle F_\mu(t) A_\nu(t') \rangle \, . \tag{92}$$

Since in the Markoff approximation, the system operator $A_\nu(t')$ cannot know about the future noise source $F_\mu(t)$, the average $\langle F_\mu(t) A_\nu(t') \rangle$ vanishes. This gives the *quantum regression theorem*

$$\boxed{\frac{d}{dt}\langle A_\mu(t) A_\nu(t') \rangle = \langle D_\mu(t) A_\nu(t') \rangle} \, , \tag{93}$$

which simply states that the expectation value of the two-time correlation function $\langle A_\mu(t) A_\nu(t') \rangle$ satisfies the same equation of motion as the single-time $\langle A_\mu(t) \rangle$ does.

The quantum regression theorem can also be formulated in terms of reduced density operators. To see this we consider the evolution of the system-reservoir density operator $\rho_{sr}(t)$ starting at $t = 0$, when the Schröd-inger and Heisenberg pictures coincide. At this time, we suppose that $\rho_{sr}(0)$ factors as

$$\rho_{sr}(0) = \rho_s(0)\rho_r(0) , \tag{94}$$

where the system and reservoir reduced density operators are defined by

$$\rho_s(t) = \mathrm{tr}_r\{\rho_{sr}(t)\} \quad \text{and} \quad \rho_r(t) = \mathrm{tr}_s\{\rho_{sr}(t)\} , \tag{95}$$

respectively. Equation (94) states that at the initial time $t = 0$, correlations between the system and reservoir vanish. In terms of $\rho_{sr}(t)$, the expectation value $\langle B(t)\rangle$ of the system operator B is given by

$$\begin{aligned}\langle B(t)\rangle &= \mathrm{tr}_s\{\mathrm{tr}_r\{B(0)\rho_{sr}(t)\}\}\\ &= \mathrm{tr}_s\{B(0)\mathrm{tr}_r\{U(t)\rho_{sr}(0)U^{-1}(t)\}\}\\ &= \mathrm{tr}_s\{B(0)\rho_s(t)\} ,\end{aligned} \tag{96}$$

where the evolution operator $U(t)$ is given by Eq. (3.45). The two-time correlation function $\langle A(t)B(t')\rangle$ is given by

$$\begin{aligned}\langle A(t)B(t')\rangle &= \mathrm{tr}_s\{\mathrm{tr}_r\{A(t)B(t')\rho_{sr}(0)\}\}\\ &= \mathrm{tr}_s\{\mathrm{tr}_r\{U^{-1}(t)A(0)U(t)U^{-1}(t')B(0)U(t')\rho_{sr}(0)\}\}\\ &= \mathrm{tr}_s\{\mathrm{tr}_r\{A(0)U(t-t')B(0)U(t')\rho_{sr}(0)U^{-1}(t')U^{-1}(t-t')\}\}\\ &= \mathrm{tr}_s\{A(0)\mathrm{tr}_r\{U(t-t')B(0)\rho_{sr}(t')U^{-1}(t-t')\}\} .\end{aligned} \tag{97}$$

No approximations have been made up to this point. Comparing Eqs. (96) and (97), we see that both $\rho_s(t)$ and the function

$$\Omega(t,t') = \mathrm{tr}_r\{U(t-t')B(0)\rho_{sr}(t')U^{-1}(t-t')\} \tag{98}$$

appear to evolve in the same way, i.e., with the same U matrix. However the initial time for the evolution of $\rho_s(t)$ is $t = 0$ when the system-reservoir correlations vanish, while that for $\Omega(t,t')$ is t' when such correlations may exist. Such correlations couple the evolution of $\Omega(t,t')$ to previous system-reservoir interactions thereby causing $\Omega(t,t')$ to evolve differently than $\rho_s(t)$. These correlations are destroyed provided $\rho_{sr}(t')$ factorizes for all times, that is

$$\rho_{sr}(t') = \rho_s(t')\rho_r(0) \,. \tag{99}$$

This is another way of stating the Markoff approximation, which we make also in the Langevin formulation of the quantum regression theorem. When Eq. (99) is satisfied, two-time expectation values like Eq. (97) evolve the same way as the single-time expectation values like Eq. (96), that is, the quantum regression theorem holds. For more detailed discussion of this approach, see Swain (1981).

Noise Spectra

In noise problems like resonance fluorescence (Chap. 15) and the generation of squeezed states of the electromagnetic field (Chap. 16), we consider cases of Eq. (84) for which the drift term D_μ is a linear function of the system operators. In fact in general, the Langevin approach is most valuable for linear problems, or for linear variations about some nonlinear operating point. For such problems, Eq. (84) becomes

$$\dot{A}_\mu = -\sum_\nu \Lambda_{\mu\nu} A_\nu(t) + F_\mu(t) \,, \tag{100}$$

where the $\Lambda_{\mu\nu}$ is a matrix of scalar coefficients. It is convenient to arrange the operators in the vector form

$$\boldsymbol{\alpha} = \begin{pmatrix} A_1 \\ \cdot \\ \cdot \\ A_\mu \\ \cdot \\ \cdot \end{pmatrix} \qquad \boldsymbol{\alpha}^T = (A_1 \,.\,.\, A_\mu \,.\,.) \tag{101}$$

In this notation, we obtain the matrix of products of operators

$$\boldsymbol{\alpha} \times \boldsymbol{\alpha}^{\mathbf{T}} \equiv \boldsymbol{\alpha}\boldsymbol{\alpha} \,, \tag{102}$$

where $\times$ denotes the direct product. From Eq. (100), we find the equation of motion for $\boldsymbol{\alpha}(t)$ as

$$\frac{d}{dt}\boldsymbol{\alpha}(t) = -\boldsymbol{\Lambda}\,\boldsymbol{\alpha}(t) + \mathbf{F}(t) \,, \tag{103}$$

which gives

$$\frac{d}{dt}\langle\boldsymbol{\alpha}(t)\rangle = -\mathbf{A}\,\langle\boldsymbol{\alpha}(t)\rangle\,. \tag{104}$$

The quantum regression theorem of Eq. (93) yields

$$\frac{d}{dt}\langle\boldsymbol{\alpha}(t)\boldsymbol{\alpha}(0)\rangle = -\mathbf{A}\,\langle\boldsymbol{\alpha}(t)\boldsymbol{\alpha}(0)\rangle\,. \tag{105}$$

where we take $t' = 0$ and assume that the noise process is stationary. This has the formal solution

$$\langle\boldsymbol{\alpha}(t)\boldsymbol{\alpha}(0)\rangle = e^{-\mathbf{A}t}\,\langle\boldsymbol{\alpha}(0)\boldsymbol{\alpha}(0)\rangle\,, \tag{106}$$

where $e^{-\mathbf{A}t}$ is defined from its power series. Similarly

$$\langle\boldsymbol{\alpha}(0)\boldsymbol{\alpha}(t)\rangle = \langle\boldsymbol{\alpha}(0)\boldsymbol{\alpha}(0)\rangle\,e^{-\mathbf{A}^T t}\,. \tag{107}$$

To obtain the low-frequency spectra of the various mode correlations, we need to calculate the Fourier transform of the corresponding two-time correlations, a consequence of the Wiener-Khintchine theorem which is valid for ergodic, stationary processes. We thus define the spectral matrix

$$\mathcal{S}(\delta) = \int_{-\infty}^{\infty} e^{-i\delta t}\,\langle\boldsymbol{\alpha}(t)\boldsymbol{\alpha}(0)\rangle\,dt\,, \tag{108}$$

where $\boldsymbol{\alpha}(0)$ means that $\boldsymbol{\alpha}$ is evaluated at a time $t = 0$ such that all transients have died away and the system has reached stationarity. Breaking Eq. (108) into two time domains $0 \to \infty$ and $-\infty \to 0$ and using stationarity in the latter, we find

$$\mathcal{S}(\delta) = \int_{0}^{\infty} e^{-i\delta t}\,\langle\boldsymbol{\alpha}(t)\boldsymbol{\alpha}(0)\rangle dt + \int_{-\infty}^{0} e^{-i\delta t}\,\langle\boldsymbol{\alpha}(0)\boldsymbol{\alpha}(-t)\rangle dt\,. \tag{109}$$

Substituting Eqs. (106) and (107) into Eq. (109) and changing $-t \to t$ in the second integral yields

$$\mathcal{S}(\delta) = \int_0^\infty e^{-i\delta t}\, e^{-\mathbf{A}t}\, \langle\boldsymbol{\alpha}\boldsymbol{\alpha}\rangle dt + \int_0^\infty e^{i\delta t}\, \langle\boldsymbol{\alpha}\boldsymbol{\alpha}\rangle e^{-\mathbf{A}^T t}\, dt$$

$$= (\mathbf{A} + i\delta)^{-1} \langle\boldsymbol{\alpha}\boldsymbol{\alpha}\rangle + \langle\boldsymbol{\alpha}\boldsymbol{\alpha}\rangle (\mathbf{A}^T - i\delta)^{-1} , \qquad (110)$$

where $\langle\boldsymbol{\alpha}\boldsymbol{\alpha}\rangle \equiv \langle\boldsymbol{\alpha}(0)\boldsymbol{\alpha}(0)\rangle$ and where $\pm i\delta$ is multiplied by the identity matrix.

We could determine $\mathcal{S}(\delta)$ by finding the matrix $\langle\boldsymbol{\alpha}\boldsymbol{\alpha}\rangle$ from steady-state solutions of the coupled equations of motion for the operator products and carrying out the matrix multiplications in Eq. (110). It is simple to use the generalized Einstein relation for the linear case. Specifically, Eq. (91) has the form

$$\frac{d}{dt} \langle\boldsymbol{\alpha}\boldsymbol{\alpha}\rangle = -\mathbf{A} \langle\boldsymbol{\alpha}\boldsymbol{\alpha}\rangle - \langle\boldsymbol{\alpha}\boldsymbol{\alpha}\rangle\mathbf{A}^T + 2\mathbf{D} , \qquad (111)$$

where $2\mathbf{D}$ is a matrix of diffusion coefficients. In steady state this becomes

$$2\mathbf{D} = \mathbf{A}\langle\boldsymbol{\alpha}\boldsymbol{\alpha}\rangle + \langle\boldsymbol{\alpha}\boldsymbol{\alpha}\rangle\mathbf{A}^T$$

$$= (\mathbf{A} + i\delta)\langle\boldsymbol{\alpha}\boldsymbol{\alpha}\rangle + \langle\boldsymbol{\alpha}\boldsymbol{\alpha}\rangle(\mathbf{A}^T - i\delta) . \qquad (112)$$

Multiplying this equation on the left by $(\mathbf{A} + i\delta)^{-1}$ and on the right by $(\mathbf{A}^T - i\delta)^{-1}$ gives

$$(\mathbf{A} + i\delta)^{-1} 2\mathbf{D}(\mathbf{A} - i\delta)^{-1} = \langle\boldsymbol{\alpha}\boldsymbol{\alpha}\rangle(\mathbf{A}^T - i\delta)^{-1} + (\mathbf{A} + i\delta)^{-1}\langle\boldsymbol{\alpha}\boldsymbol{\alpha}\rangle. \qquad (113)$$

The right-hand side of Eq. (113) is precisely the same as the RHS of Eq. (110), so

$$\boxed{\mathcal{S}(\delta) = (\mathbf{A} + i\delta)^{-1} 2\mathbf{D}(\mathbf{A}^T - i\delta)^{-1}} . \qquad (114)$$

This general form for the spectral matrix can be used to calculate the spectrum of resonance fluorescence or the expectation values needed in the calculation of variances in squeezed states. We use the appropriate Langevin equations (105) directly for resonance fluorescence and for the quantum theory of multiwave mixing in Chaps. 15 and 16.

References

Dalibard, J., J. Dupont-Roc and C. Cohen-Tannoudji (1982), J. Physique **43**, 1617.

Gardiner, C. (1980), *Handbook of Stochastic Methods*, Springer, Berlin.

Lax, M. (1968), *Brandeis University Summer Institute Lectures (1966)*, Vol II, ed. by M. Chretien, E. P. Gross and S. Deser, Gordon and Breach, New York.

Louisell, W. H. (1973), *Quantum Statistical Properties of Radiation*, John Wiley & Sons, New York.

Milonni, P. W. and W. A. Smith (1975), Phys. Rev. A**11**, 814.

Risken, H. (1984), *The Fokker-Planck Equation*, Springer Verlag, Heidelberg.

Sargent, M. III, M. O. Scully, and W. E. Lamb, Jr. (1974), *Laser Physics*, Addison-Wesley Publishing Co., Reading, MA. See especially Chaps. 15 and 16.

Swain, S. (1981), J. Phys. *B*: Math. Gen. **14**, 2577.

Problems

14-1. Explain why $\mathcal{H}_S$ appears in Eq. (17) rather than $\mathcal{H}_0$.

14-2. Show that $\rho(t) = \exp(-i\mathcal{H}t/\hbar)\rho(0)\exp(i\mathcal{H}t/\hbar)$ satisfies the Schrödinger equation of motion (4.38). Writing the corresponding form for the Heisenberg operator $a(t)$, show that the following two expressions for $\langle a(t)\rangle$ are equivalent:

$$\langle a(t)\rangle = \mathrm{tr}\{a(t)\rho(0)\} = \mathrm{tr}\{a(0)\rho(t)\}\ .$$

Similarly find $\Omega(t,t')$ in the two-time correlation function

$$\langle a(t)b(t')\rangle = \mathrm{tr}\{a(t)b(t')\rho(0)\} = \mathrm{tr}\{a(0)\Omega(t,t')\}\ .$$

What equation of motion does $\Omega(t,t')$ satisfy?

14-3. Given the master equation

$$\dot{\rho}(t) = - B[a^\dagger a\rho(t) - a\rho(t)a^\dagger] - A[\rho(t)aa^\dagger - a^\dagger \rho(t)a] + \text{adj.} \tag{115}$$

derive $d\langle a^\dagger a\rangle/dt$ and interpret your result.

14-4. Write the Fokker-Planck equation for Eq. (115).

14-5. Using the quantum regression theorem and the Langevin Eq. (69), find the equation of motion for $\langle A(t)A^\dagger(0)\rangle$.

14-6. Calculate $d\langle n^2\rangle/dt$ given the equation of motion (35).

14-7. Show that Eq. (56) is the steady-state solution of the Fokker-Planck Eq. (55).

14-8. Show that $\langle \mathscr{G}_{A^\dagger A}(t)\rangle$ of Eq. (75) is given by $\gamma\bar{n}$.

14-9. Find the Langevin equations for the Pauli spin operators. Ans: Eq. (18.9).

14-10. Show that $a^\dagger|\alpha\rangle = (\partial/\partial\alpha + \frac{1}{2}\alpha^*)|\alpha\rangle$.

14-11. Using the Langevin Eq. (69), show that the commutator $[A(t), A^\dagger(t)]$ remains unity in time (due to the presence of the noise operators).

14-12. Prove Eq. (82) for $\langle F^\dagger(t')F(t'')\rangle_r$ along the lines outlined for Eq. (81).

14-13. Given the Langevin equation

$$\dot{A}(t) = D(t) + F(t) ,$$

where the noise operator $F(t)$ has the Markoffian two-time correlation

$$\langle F^\dagger(t)F(t')\rangle = 2D_{A^\dagger A}\,\delta(t - t') ,$$

show that $\langle A^\dagger(t)F(t)\rangle = D_{A^\dagger A}$. Hint: this is a special case of the general theory of Sec. 14-4.

14-14. Consider a spin-oscillator system described by the state $|\psi\rangle = 2^{-1/2}[|a,n\rangle + |b,n+1\rangle]$. Compute the reduced density matrix for the field alone.

14-15. Given the master equation

$$\frac{d\rho}{dt} = -\frac{i}{\hbar}[\mathcal{H}, \rho] - \frac{\Gamma}{2}[\sigma_+\sigma_-\rho + \rho\sigma_+\sigma_-] + \Gamma\sigma_-\rho\sigma_+$$

for a two-level atom coupled to a bath of harmonic oscillator and $\sigma_\pm$ are the usual Pauli spin-flip operators. What is the corresponding Hamiltonian? Derive the equations of motion for $\langle\sigma_z\rangle$ and $\langle\sigma_+\rangle$.

14-16. Show for a thermal reservoir density operator $\rho_r(\mathcal{H}_r)$ and a reservoir operator $F(\tau)$ that

$$\langle F(\tau)\rangle_r = \mathrm{tr}_r\{F(\tau)\rho_r(\mathcal{H}_r)\} = \mathrm{tr}_r\{F(0)\rho_r(\mathcal{H}_r)\} = \langle F(0)\rangle_r \ ,$$

where the Schrödinger and Heisenberg pictures coincide at $t = 0$.

14-17. Usually tracing the density operator over a subsystem produces a mixture. Given the state vector

$$|\psi(t)\rangle = \cos(\sqrt{n+1}\,t)\,|an\rangle + \sin(\sqrt{n+1}\,t\,|b\; n+1\rangle \ ,$$

when does the trace over the atom yield a pure state?

Chapter 15
RESONANCE FLUORESCENCE

Sections 13-3 and 14-4 analyze spontaneous emission from an atom interacting with the vacuum electromagnetic field. The present chapter studies the spontaneous emission of an atom irradiated by a continuous, monochromatic field. This emission is called *resonance fluorescence*. We compute its spectrum, which is given in steady state by the Fourier transform of the first-order correlation function of the field. We also discuss the phenomenon of photon antibunching, a purely quantum-mechanical effect described by the intensity correlation function of the emitted light. This chapter is an application of the general methods of Chap. 14 and illustrates the use of the quantum regression theorem in a central problem of quantum optics. It also establishes the connection between resonance fluorescence and the semiclassical probe absorption studies of Chap. 8, and lays the foundations for studying the generation of squeezed states by resonance fluorescence and four-wave mixing in Chap. 18.

Section 15-1 presents in general terms the phenomenology of resonance fluorescence and uses the bare and dressed-atom pictures to motivate the three-peak structure of the fluorescence spectrum. Section 15-2 introduces the quantum Langevin equations of an atom driven by a continuous monochromatic field. Section 15-3 calculates the resonance fluorescence intensity and spectrum by using the Wiener-Khintchine theorem and applying the quantum regression theorem to the Langevin-Bloch equations. Section 15-4 uses the techniques of Sec. 14-1 to derive a master equation for a weak sidemode of the field. Examination of this equation establishes the close connection between probe absorption of Sec. 8-1 and resonance fluorescence. In particular, we see that probe stimulated emission and resonance fluorescence are described by the same function. We also show that the resonance fluorescence spectrum is given by one of four coefficients in the side-mode master equation. A generalization to two sidemodes is used in Sec. 16-2 to study the generation of squeezed states. Section 15-5 computes the intensity correlation function of the scattered light and photon antibunching. Section 15-6 discusses bichromatic photon correlations for off-resonant excitation. These temporal correlations correspond to the frequency-domain correlations that can produce squeezing. As explained in

Chap. 16, the squeezing produced in resonance fluorescence is minimal, because the fluorescence tends to swamp the squeezing. However by using four-wave mixing (see Sec. 16-2 and 16-3) instead of the two-wave mixing of resonance fluorescence, tunings with little fluorescence can be used and substantial squeezing can be obtained.

Prerequisites for this chapter are a knowledge of the Heisenberg picture (Sec. 3-1), Secs. 12-1 and 12-2, as well as Chaps. 13 and 14. The general techniques of this chapter are utilized further in Sec. 16-2.

15-1. Phenomenology

One of the simplest quantum optics problems involving more than just vacuum fluctuations is that of determining the spectrum of the fluorescence light radiated by a single two-level atom strongly driven by a continuous monochromatic field. This situation is achieved experimentally by

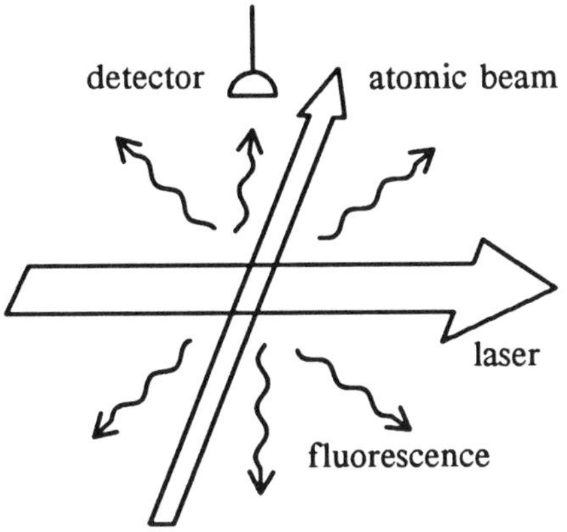

Fig. 15-1. Diagram showing interaction between atomic beam and laser leading to resonance fluorescence.

scattering a laser off a collimated atomic beam, as illustrated in Fig. 15-1. The direction of the atomic beam and the axes of excitation and detection are mutually perpendicular. The interaction region between the atomic beam and laser light is typically inside a Fabry-Perot and the spectrum is recorded by changing the length of the interferometer. The details of the preparation of true two-level atoms involve the use of optical pumping and are given by Hartig *et al.* (1976).

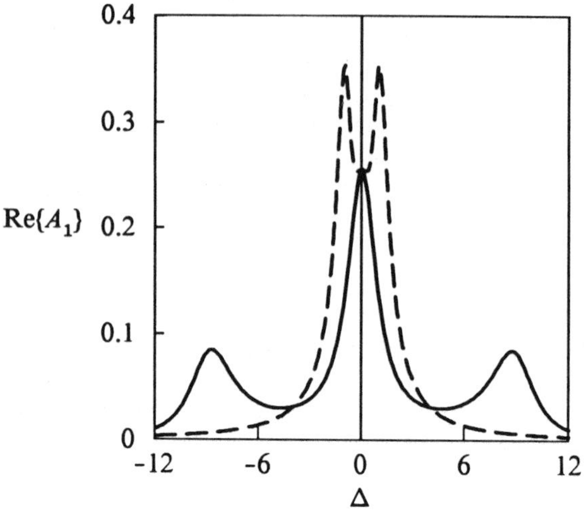

Fig. 15-2. Theoretical resonance fluorescence spectra corresponding to the experimental scheme of Fig. 15-1. The formula used is A_1 + c.c., where A_1 is given by Eq. (51), for a dimensionless pump intensity $I_2 = 40$ and the decay-rate choices $\Gamma = 2\gamma$ (solid line - pure radiative decay) and $\Gamma = 35\gamma$. Note that the Rabi frequencies $\mathscr{R}_0$ differ for these curves, but as Prob. 15-8 shows, coherent dips occur for a sufficiently long T_1.

Figure 15-2 shows resonance fluorescence spectra for large saturator intensity and two values of Γ/γ. For weak incident fields, the spectrum exhibits a single peak much narrower than the natural linewidth γ of the transition. As the laser power is increased, the spectrum splits into three peaks, consisting of a central peak centered at the laser frequency and two symmetrically placed sidebands displaced from the central peak by $\pm\mathscr{R}$, the generalized Rabi frequency. For a resonant pump wave, the central peak has a FWHM width 2γ and a height $(\Gamma+\gamma)/\gamma$ (=3 for pure radiative decay) times that of the sideband peaks, which each have the FWHM width $\Gamma+\gamma$.

The weak intensity limit is easy to understand. In this case, the atom is only weakly excited, and only the lowest order in perturbation theory is relevant, as symbolically illustrated in Fig. 15-3. Initially in the ground state b, the atom scatters only a single photon. By conservation of energy, the fluorescence light frequency ν_1 must equal the incident light frequency ν_2, and the spectrum is a delta-function

$$\mathcal{S}(\nu_1) = \mathcal{S}_0\delta(\nu_1-\nu_2)\ , \tag{1}$$

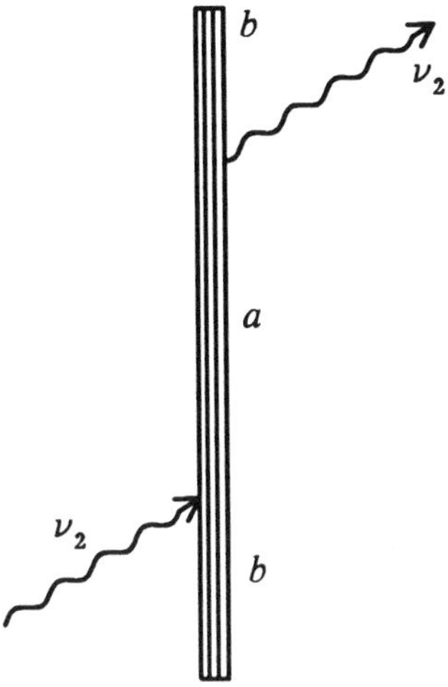

Fig. 15-3. Weak field resonance fluorescence.

which is just Rayleigh scattering. In practice, the incident laser always has a finite linewidth, which explains why the low intensity spectrum in Fig. 15-3 is not strictly a delta-function.

The splitting of the spectrum into a central peak and two sidebands can be understood intuitively by using the dressed atom picture of Sec. 13-1. We have seen that in the presence of a single-mode driving field, the eigenstates of the atom-field system are $|1n\rangle$ and $|2n\rangle$ of Eq. (13.16), which on resonance are symmetrical and antisymmetric superpositions of $|an\rangle$ and $|b\ n{+}1\rangle$. At resonance ($\nu_2 = \omega$), they have the eigenenergies

$$E_{1n} = \hbar(n + \tfrac{1}{2})\omega + \hbar g\sqrt{n+1}\ , \tag{2}$$

$$E_{2n} = \hbar(n + \tfrac{1}{2})\omega - \hbar g\sqrt{n+1}\ . \tag{3}$$

Four such levels are depicted in Fig. 15-4. From Eq. (13.20), the energy separation between $|1\ n\rangle$ and $|2\ n\rangle$ is equal to the Rabi frequency $\mathcal{R}_0 = g\sqrt{n+1}$, while the separation between $|1\ n\rangle$ and $|1\ n-1\rangle$ is the laser frequency ν_2. For strong fields we have $n+1 \simeq n$ and we can neglect the n-dependence in the Rabi frequency. In this limit we see that transitions at the three frequencies ν_2 and $\nu_2 \pm \mathcal{R}_0$ are possible (see Prob. 13-8). These correspond to the three peaks in the strong field limit of Fig. 15-2.

Since there are two possible transitions for ν_2 while only one each at $\nu_2+\mathcal{R}_0$ and $\nu_2-\mathcal{R}_0$, one would intuitively expect the integrated intensity

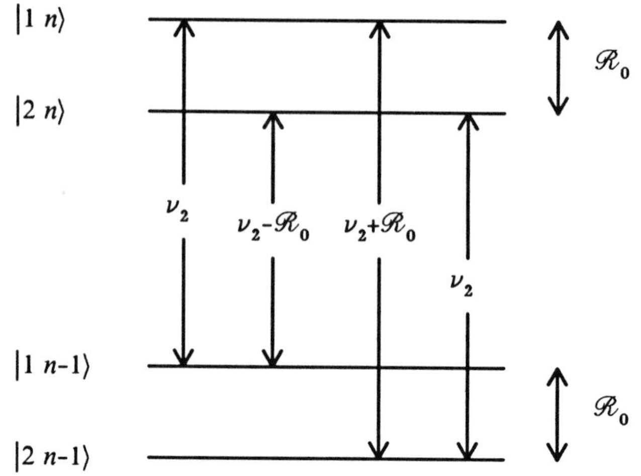

Fig. 15-4. Energy level diagram revealing three main frequencies in resonance fluorescence. See Fig. 13-1 for diagram showing variation with ω.

under the central peak to be twice that under a sideband, and this indeed turns out to be true. However the simple dressed-atom picture does not predict the widths of the various peaks, which requires a detailed quantum-mechanical treatment. This is the object of the next section.

The side peaks can also be understood intuitively in the bare-atom picture as the result of modulating the upper-level probability at the Rabi frequency. Whether in AM/FM radio wave transmission or in population pulsations, we know that modulators put sidebands on the wave they modulate. Resonance fluorescence is studied quantitatively from this point of view in Sec. 15-4.

15-2. Langevin Equations of Motion

Section 13-3 shows how the continuum of modes of the electromagnetic field in the vacuum state is responsible for spontaneous emission. Here the situation is somewhat more complicated, since the atom also interacts with a strong driving field. We treat this strong field classically and therefore decompose the Hamiltonian of the coupled atom-field system as

$$\mathcal{H} = \tfrac{1}{2}\hbar\omega\sigma_z + \hbar[\mathcal{V}_2\sigma_+ e^{-i\nu_2 t} + \text{adj.}] + \hbar\Sigma_s\Omega_s a_s{}^\dagger a_s + \hbar\Sigma_s(g_s a_s\sigma_+ + \text{adj.}), \quad (4)$$

where we have used Eqs. (13.48) and (3.109), $\mathcal{V}_2 = -\wp\mathcal{E}_2 U_2/2\hbar$, $\mathcal{E}_2$ is the complex classical pump field amplitude, and U_2 is the pump mode factor.

Under the assumption that we are dealing with an ergodic and stationary process, the Wiener-Khintchine theorem states that the spectrum $\mathcal{S}(\nu_1)$ is given by the two-time correlation function of the radiated field (Louisell 1973)

$$\mathcal{S}(\nu_1) \propto \lim_{T\to\infty} \frac{1}{T}\int_0^T dt \int_0^T dt' \, \langle E^-(t)E^+(t')\rangle \, e^{-i\nu_1(t-t')} , \quad (5)$$

where $E^+(t) = \Sigma_s a_s(t)$ is the positive frequency component of the electric field, as defined in Sec. 12-4. Using the stationarity condition, the correlation $\langle E^-(t)E^+(t')\rangle$ depends only on the time difference $\tau = t - t'$, so that we can reexpress Eq. (5) as

$$\mathcal{S}(\nu_1) \propto \lim_{T\to\infty} \frac{1}{T}\int_0^T dt \left[\int_0^t dt' + \int_t^T dt'\right] \langle E^-(t)E^+(t')\rangle \, e^{-i\nu_1(t-t')}$$

$$= \lim_{T\to\infty} \frac{1}{T}\int_0^T dt \left[\int_0^t d\tau \, \langle E^-(\tau)E^+(0)\rangle \, e^{-i\nu_1\tau} + \int_0^{T-t} d\tau \, \langle E^-(0)E^+(\tau)\rangle \, e^{i\nu_1\tau}\right]$$

or

$$\boxed{\mathcal{S}(\nu_1) = \beta\int_0^\infty d\tau \, \langle E^-(\tau)E^+(0)\rangle \, e^{-i\nu_1\tau} + \text{c.c.}} , \quad (6)$$

where we use the fact that $\langle E^-(0)E^+(\tau)\rangle = \langle E^-(\tau)E^+(0)\rangle^*$, since $\langle AB\rangle^* = \langle (BA)^\dagger\rangle = \langle B^\dagger A^\dagger\rangle$. In Eq. (6), we have extended the upper limit of the τ integrals to ∞ with negligible change provided the field operators are correlated only over a short time. In practice, the limit $T \to \infty$ is not reached, since the atoms interact with the laser beam for a finite transit time.

Equation (6) is the Fourier transform of $\langle E^-(\tau)E^+(0)\rangle$. To see that it is indeed a spectrum, that is, the absolute value square of the Fourier transform of the field, we use the ergodic hypothesis to replace the expectation value by a time average. This gives the Fourier transform of the convolution $\int dt E^-(t)E^+(t-\tau)$, which equals the desired product of Fourier transforms. We add a note of caution about using the Wiener-Khintchine theorem for time-dependent processes, since the resulting time-dependent spectrum can become negative. In such situations, one should use different definitions of the spectrum, such as e.g. the "physical spectrum" of Eberly and Wodkiewicz (1977).

Because the field has an infinite number of degrees of freedom, it might appear complicated to obtain an expression for $E^+(t)$. However we can take advantage of the result valid both classically and quantum-mechanically that the far field emitted by a dipole is proportional to this dipole, a result already expressed in (1.68) with the observation that $\ddot{\mathbf{x}} \simeq \omega^2 x$ for sufficiently slow decay. Omitting uninteresting constants and ignoring for simplicity the vectorial character of the field in the operator version of Eq. (1.68), we obtain the relation

$$E^+(\mathbf{r},t) \propto \frac{1}{r}\,\sigma_-(t-r/c)\ , \tag{7}$$

a very important result, since it teaches us that the knowledge of the dipole operator σ_- is all that is required to compute the fluorescence spectrum. The continuum of modes of the electromagnetic field in Eq. (4) plays the role of a bath for the two-level atom. Since the explicit dynamics of these modes is not required, they can be treated as in Sec. 14-3, yielding both decay and noise operator terms in the Heisenberg equations of motion for the spin operators σ_z, σ_+, and σ_-. The decay terms are precisely given by the Weisskopf-Wigner spontaneous emission rate, and after introducing the slowly varying operators

$$S_z = \tfrac{1}{2}\sigma_z\ , \tag{8a}$$

$$S_+ = \sigma_+ e^{-i\nu_2 t}\ , \tag{8b}$$

$$S_- = \sigma_- e^{i\nu_2 t}\ , \tag{8c}$$

we obtain readily the quantum mechanical Langevin-Bloch equations

$$\dot{S}_+ = -(\gamma - i\delta)S_+ - 2i\mathcal{V}_2^* S_z + F_+(t) \tag{9a}$$

$$\dot{S}_z = -\Gamma(S_z + \tfrac{1}{2}) + i\mathcal{V}_2^* S_- - i\mathcal{V}_2 S_+ + F_z(t) \tag{9b}$$

$$\dot{S}_- = -(\gamma + i\delta)S_- + 2i\mathcal{V}_2 S_z + F_-(t) \tag{9c}$$

where as usual $\delta = \omega - \nu_2$.

The noise operators $F_i(t)$ appearing in Eqs. (9) can be evaluated at least in principle, but Sec. 14-4 shows that their explicit form is not needed within the Markoff approximation and with the quantum regression theorem. This remarkable result eliminates the need for considerable algebra. All we need to know about the noise operators is that their mean value vanishes

$$\langle F_i(t)\rangle = 0 \,, \tag{10}$$

and that they are delta-correlated

$$\langle F_i^\dagger(t)F_j(t')\rangle = \langle 2D_{ij}\rangle\delta(t-t') \,, \tag{11}$$

where the diffusion coefficients $\langle 2D_{ij}\rangle$ give a measure of the strength of the correlated fluctuations. These strengths determine the decay constants γ and Γ in Eq. (9).

The semiclassical Bloch equations are recovered by taking the expectation value of Eqs. (9). With Eq. (10) this gives

$$\langle \dot{S}_+\rangle = -(\gamma - i\delta)\langle S_+\rangle - 2i\mathcal{V}_2^*\langle S_z\rangle \,, \tag{12a}$$

$$\langle \dot{S}_z\rangle = -\Gamma(\langle S_z\rangle + \tfrac{1}{2}) + i\mathcal{V}_2^*\langle S_-\rangle - i\mathcal{V}_2\langle S_+\rangle \,, \tag{12b}$$

$$\langle \dot{S}_-\rangle = -(\gamma + i\delta)\langle S_-\rangle + 2i\mathcal{V}_2\langle S_z\rangle \,. \tag{12c}$$

Note that these equations of motion are the same as those for the upper-to-ground-lower-level decay density matrix for ρ_{ba}, $D/2$, and ρ_{ab}, respectively, in Chap. 5. This is to be expected, since the strong field is treated classically. The steady-state solutions of these equations are

$$\langle S_+\rangle_s = -2i\mathcal{V}_2^*\langle S_z\rangle_s\mathscr{D}_2^* \,, \tag{13a}$$

$$\langle S_z\rangle_s = -\frac{1}{2}\,\frac{1}{1 + I_2\mathscr{L}_2} \,, \tag{13b}$$

$$\langle S_-\rangle_s = 2i\mathcal{V}_2\langle S_z\rangle_s\mathscr{D}_2 \,, \tag{13c}$$

where the dimensionless pump intensity I_2 is given by

$$I_2 = \frac{4|\mathcal{V}_2|^2}{\Gamma\gamma} \,, \tag{14}$$

and the complex Lorentzian denominator $\mathscr{D}_2$ is given by Eq. (3.8) and the Lorentzian $\mathscr{L}_2$ is given by Eq. (3.10). The steady-state values for the populations themselves are also useful

$$\langle S_a\rangle_s = \tfrac{1}{2} + \langle S_z\rangle_s = \frac{\tfrac{1}{2} I_2 \mathcal{L}_2}{1 + I_2\mathcal{L}_2} \tag{15a}$$

$$\langle S_b\rangle_s = \tfrac{1}{2} - \langle S_z\rangle_s = \frac{1}{2} + \frac{1}{2}\frac{1}{1 + I_2\mathcal{L}_2} \tag{15b}$$

15-3. Scattered Intensity and Spectrum

Chapter 12 shows that the counting rate I (or light intensity) measured on a photon detector is proportional to the single-time first-order correlation function of the field

$$I \propto \int_0^T dt\ \langle E^-(t)E^+(t)\rangle = \alpha \int_0^T dt\ \langle S_+(t)S_-(t)\rangle = \alpha T\langle S_a\rangle_s\ , \tag{16}$$

where we have used the field given by Eq. (7), introduced the proportionality constant α, neglected the trivial retardation factor, and used the relation

$$S_+(t)S_-(t) = |a\rangle\langle b|\,|b\rangle\langle a| = |a\rangle\langle a| \equiv S_a(t)\ . \tag{17}$$

We compute I by decomposing the spin operators as

$$\boxed{S_\pm(t) = \langle S_\pm(t)\rangle + \delta S_\pm(t)}\ , \tag{18}$$

where $\langle S_\pm(t)\rangle$ is a solution of the semiclassical equations (12) and $\delta S_\pm(t)$ the correction due to the effect of the noise operators $F(t)$. Substituting Eq. (18) into Eq. (16) yields

$$\begin{aligned} I &= \alpha \int_0^T dt\ |\langle S_+(t)\rangle|^2 + \alpha \int_0^T dt\ \langle \delta S_+(t)\delta S_-(t)\rangle \\ &\equiv I_{coh} + I_{inc}\ . \end{aligned} \tag{19}$$

The scattered intensity consists of two contributions: the first one, originating from the mean motion of the dipole driven by the laser field, is sometimes called the coherently scattered intensity I_{coh}, while the incoherent contribution I_{inc} is due to the fluctuations of the dipole motion produced by the vacuum field. Comparing Eqs. (16) and (19), we find that the incoherent intensity is given by

$$I_{inc} = \alpha T(\langle S_a\rangle - |\langle S_+\rangle|^2) . \tag{20}$$

Equations (19) and (20) give I_{coh} and I_{inc} in terms of average values of atomic operators. Using the steady-state values of Eqs. (13a) and (15), we have

$$I_{coh} = \alpha T|\langle S_+\rangle|^2 = \alpha T \frac{\frac{1}{2}I_2\mathscr{L}_2\Gamma/2\gamma}{(1 + I_2\mathscr{L}_2)^2} , \tag{21}$$

$$I_{inc} = \alpha T \frac{\frac{1}{2}I_2\mathscr{L}_2}{1 + I_2\mathscr{L}_2} - I_{coh} . \tag{22}$$

We note that the coherently scattered intensity goes to zero for strong driving fields, $I_2\mathscr{L}_2 >> 1$. This might appear surprising, as one would expect a semi-classical description of the problem to become better, the stronger the driving field, and hence that I_{coh} would dominate in this regime. However the coherent part is proportional to the squared magnitude of the steady-state dipole, which bleaches to zero for strong pump fields. In contrast, the incoherent part results from spontaneous emission from the upper-level, whose probability of occupation is approximately $\frac{1}{2}$ in strong fields. In the weak field limit with pure radiative decay, the atom remains essentially always in the ground state, with almost no spontaneous emission taking place. This is illustrated in Fig. 15-5, which shows

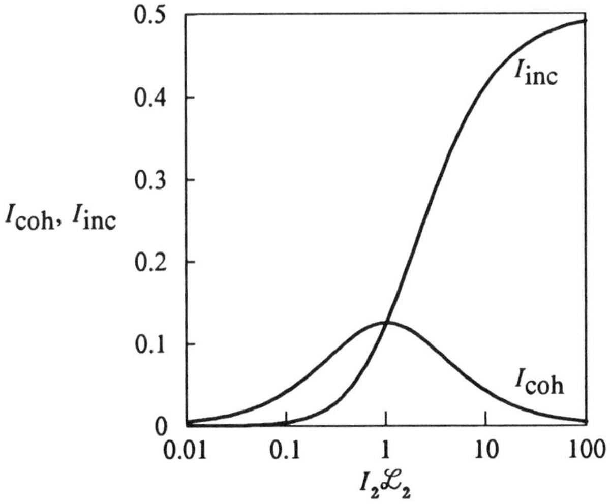

Fig. 15-5. Total coherent [Eq. (21) - solid line] and incoherent [Eq. (22) - dashed line] contributions to the scattered light intensity versus detuned pump intensity $I_2\mathscr{L}_2$ for pure radiative decay ($\Gamma = 2\gamma$) and $\alpha T = 1$.

the coherent and incoherent contributions to the scattered light intensity as a function of the detuned pump intensity $I_2\mathscr{L}_2$. For larger values of γ, I_{inc} has comparable or larger weak-field values than I_{coh}. In fact the two cross over at the intensity $I_2\mathscr{L}_2 = \Gamma/\gamma - 1$, e.g., $I_2\mathscr{L}_2 = 1$ for pure radiative decay and 0 for $\Gamma = \gamma$. For $\gamma > \Gamma$, I_{inc} exceeds I_{coh} for all nonzero intensities.

Spectrum

While the scattered intensity can be calculated without ever using the quantum noise operators, the situation is somewhat more complex for the scattered spectrum

$$\mathscr{S}(\nu_1) \propto \int_0^\infty d\tau\, \langle E^-(\tau)E^+(0)\rangle\, e^{-i\nu_1\tau} + \text{c.c.}$$

$$= \beta \int_0^\infty d\tau\, \langle S_+(\tau)S_-(0)\rangle\, e^{i(\nu_2-\nu_1)\tau} + \text{c.c.}\,, \tag{23}$$

where we have used Eqs. (6) through (8) and β is a constant of proportionality. As for the scattered intensity of Eq. (19), we can use Eq. (18) to decompose the light spectrum into a coherent and an incoherent part:

$$\begin{aligned}\mathscr{S}(\nu_1) &\equiv \mathscr{S}_{coh}(\nu_1) + \mathscr{S}_{inc}(\nu_1)\\ &= \beta|\langle S_+(t)\rangle|^2 \int_0^\infty d\tau\, e^{i(\nu_2-\nu_1)\tau} + \beta \int_0^\infty d\tau\, \langle \delta S_+(\tau)\delta S_-(0)\rangle\, e^{i(\nu_2-\nu_1)\tau}\\ &\quad + \text{c.c.},\end{aligned} \tag{24}$$

In steady state, the coherent part of the spectrum is readily found to be

$$\mathscr{S}_{coh}(\nu_1) = 2\pi\beta|\langle S_+\rangle_s|^2\delta(\nu_1-\nu_2) = \frac{\frac{1}{2}\pi\beta I_2\mathscr{L}_2\Gamma}{\gamma(1+I_2\mathscr{L}_2)^2}\delta(\nu_1-\nu_2)\,. \tag{25}$$

It consists simply of a delta-function peak centered at the laser frequency ν_2, and is known as the Rayleigh peak. This is precisely the result that would be expected from a simple energy conservation argument.

To evaluate the incoherent part, we need the two-time correlation function of the fluctuations $\delta S_i(t)$ of the atomic operators about their semiclassical mean values. The equations of motion for the fluctuations $\delta S_i(t)$ have the general form

$$\frac{d}{dt}\,\delta S_i(t) = \sum_j B_{ij}\;\delta S_j(t) + F_i(t)\;, \tag{26}$$

where $i = (+,z,-)$ and the matrix B is obtained by substituting Eq. (18) into Eq. (9) as

$$B = \begin{bmatrix} -\gamma + i\delta & -2i\mathcal{V}_2^{\,*} & 0 \\ -i\mathcal{V}_2 & -\Gamma & i\mathcal{V}_2^{\,*} \\ 0 & 2i\mathcal{V}_2 & -\gamma - i\delta \end{bmatrix} . \tag{27}$$

Equation (26) gives for the evolution of the two-time correlation functions

$$\frac{d}{d\tau}\langle\delta S_i(\tau)\delta S_-(0)\rangle = \langle(d\delta S_i(\tau)/d\tau)\;\delta S_-(0)\rangle$$

$$= \sum_j B_{ij}\langle\delta S_j(\tau)\delta S_-(0)\rangle + \langle F_i(\tau)\delta S_-(0)\rangle\;. \tag{28}$$

Since the noise operator $F_i(\tau)$ can not influence $\delta S_-(0)$ for $\tau > 0$, we find

$$\frac{d}{d\tau}\langle\delta S_i(\tau)\delta S_-(0)\rangle = \sum_j B_{ij}\langle\delta S_j(\tau)\delta S_-(0)\rangle \quad \text{for } \tau > 0\;. \tag{29}$$

This is a special case of the quantum regression theorem of Sec. 14-4. It shows that when the atom is removed from steady state, its subsequent evolution and damping are governed by precisely the same equation as the transient behavior of the mean dipole moment. The case for $\tau < 0$ need not be considered in detail, since the results are simply related to those for $\tau > 0$ by

$$\langle\delta S_i(\tau)\delta S_-(0)\rangle = \langle\delta S_+(0)\delta S_i{}^\dagger(\tau)\rangle^*\;, \tag{30}$$

which follows from $S_+ = S_-{}^\dagger$ (hence the c.c. in Eq. (6)).

One of the three equations of motion (29) is for the desired two-time correlation function $\langle\delta S_+(\tau)\delta S_-(0)\rangle$. Hence by solving these equations simultaneously, we find $\langle\delta S_+(\tau)\delta S_-(0)\rangle$, which we can then substitute into Eq. (24) to find the inelastic part of the resonance fluorescence spectrum. The easiest way to obtain these solutions is to note that the spectral quantity we need in Eq. (24) is actually the Laplace transform (of imaginary argument) of $\langle\delta S_+(\tau)\delta S_-(0)\rangle$. Hence we Laplace transform Eq. (29), thereby reducing the coupled differential equations into a set of coupled algebraic equations. This gives

$$(sI - B)\begin{bmatrix}\delta\mathcal{S}_{+-}(s)\\ \delta\mathcal{S}_{z-}(s)\\ \delta\mathcal{S}_{--}(s)\end{bmatrix} = \begin{bmatrix}\langle\delta S_+(0)\delta S_-(0)\rangle\\ \langle\delta S_z(0)\delta S_-(0)\rangle\\ \langle\delta S_-(0)\delta S_-(0)\rangle\end{bmatrix}, \tag{31}$$

where I is the identity matrix, the Laplace transform $\delta S_{ij}(s)$ is given by

$$\delta\mathcal{S}_{ij}(s) = \int_0^\infty d\tau e^{-s\tau}\langle\delta S_i(\tau)\delta S_j(0)\rangle , \tag{32}$$

and the matrix

$$sI - B = \begin{bmatrix}\gamma - i\delta + s & 2i\mathcal{V}_2^* & 0\\ i\mathcal{V}_2 & \Gamma + s & -i\mathcal{V}_2^*\\ 0 & -2i\mathcal{V}_2 & \gamma + i\delta + s\end{bmatrix}. \tag{33}$$

The initial conditions on the RHS of Eq. (31) are given by

$$\langle\delta S_i(0)\delta S_j(0)\rangle = \langle S_i(0)S_j(0)\rangle - \langle S_i(0)\rangle\langle S_j(0)\rangle , \tag{34}$$

with explicit values given in Prob. 15-1. Comparing Eq. (32) with Eq. (23), we see that the relevant Laplace transform variable s is $-i(\nu_2 - \nu_1) + \epsilon = -i\Delta + \epsilon$, where ϵ is an arbitrarily small positive real constant. Hence the incoherent spectrum $\mathcal{S}_{inc}$ is given by

$$\mathcal{S}_{inc} = \beta\delta\mathcal{S}_{+-}(-i\Delta) + \text{c.c.} \tag{35}$$

The complex correlation $\delta\mathcal{S}_{+-}(s)$ is, in turn, given according to Cramer's rule by the ratio of the determinant of the matrix in Eq. (33) with the first column replaced by the RHS of Eq. (31) to the determinant D of the matrix itself, which is given by

$$D = (\Gamma + s)(\gamma + s + i\delta)(\gamma + s - i\delta) + 4|\mathcal{V}_2|^2(\gamma + s). \tag{36}$$

The solution for $\delta\mathcal{S}_{+-}(-i\Delta)$ is given in Prob. 15-1.

Here we calculate the total spectrum of Eq. (23), directly using the $S_i(t)$ operators instead of the fluctuation operators $\delta S_i(t)$. This solution is mathematically simpler to obtain than Eq. (35) since only two of the corresponding initial conditions are nonzero. Furthermore it provides the spectral coefficients in the form used in the quantum theory of multiwave mixing in Secs. 15-4 and 16-2. In terms of the Laplace transforms

$$\mathcal{S}_{ij}(s) = \int_0^\infty d\tau e^{-s\tau} \langle S_i(\tau) S_j(0)\rangle \ , \tag{37}$$

the spectrum is given by

$$\mathcal{S}(\Delta) = \beta \mathcal{S}_{+-}(-i\Delta) + \text{c.c.} \tag{38}$$

Using the quantum regression theorem for the $S_i(t)$, we obtain

$$\frac{d}{d\tau} \langle S_i(\tau) S_-(0)\rangle = \sum_j B_{ij} \langle S_j(\tau) S_-(0)\rangle - \frac{\Gamma}{2} \langle S_-(0)\rangle \delta_{iz} \ , \tag{39}$$

where δ_{iz} is a Kronecker delta. The Laplace transform of these equations is

$$(sI - B) \begin{bmatrix} \mathcal{S}_{+-}(s) \\ \mathcal{S}_{z-}(s) \\ \mathcal{S}_{--}(s) \end{bmatrix} = \begin{bmatrix} \langle S_+(0) S_-(0)\rangle \\ \langle S_z(0) S_-(0)\rangle - \frac{1}{2}\Gamma s^{-1} \langle S_-(0)\rangle \\ \langle S_-(0) S_-(0)\rangle \end{bmatrix} . \tag{40}$$

Using Eq. (13) we find that the initial conditions on the RHS of Eq. (40) are given by

$$\begin{aligned} &\langle S_+(0) S_-(0)\rangle = \langle S_a\rangle \\ &\langle S_z(0) S_-(0)\rangle - \tfrac{1}{2}\Gamma s^{-1}\langle S_-(0)\rangle = -\tfrac{1}{2}\langle S_-\rangle(1 + \Gamma/s) \\ &\langle S_-(0) S_-(0)\rangle = 0 \ . \end{aligned} \tag{41}$$

Using Cramer's rule, we find

$$\mathcal{S}_{+-}(s) = \frac{\langle S_a\rangle[(\Gamma+s)(\gamma+i\delta+s) + 2|\mathcal{V}_2|^2] + i\mathcal{V}_2^{\,*}\langle S_-\rangle(1 + \Gamma/s)(\gamma+i\delta+s)}{D}$$

or

$$\mathcal{S}_{+-}(s) = \frac{I_2\mathcal{L}_2[(\Gamma+s)(\gamma+i\delta+s) + 2|\mathcal{V}_2|^2] + 2|\mathcal{V}_2|^2\mathcal{D}_2(1 + \Gamma/s)(\gamma+i\delta+s)}{2D(1 + I_2\mathcal{L}_2)} \tag{42}$$

Before writing this result in a more final form, we consider the limits of $s \to 0$ and of large Rabi flopping. Notice that the Γ/s term diverges for $s = 0$ ($i\Delta = 0$). In this limit we obtain

$$\mathcal{S}_{+-}(s{\to}0) \to \frac{\frac{1}{2}I_2\mathcal{L}_2}{\gamma^2\Gamma(1 + I_2\mathcal{L}_2)^2}\{\mathcal{L}_2[\Gamma(\gamma + i\delta) + 2|\mathcal{V}_2|^2] + \tfrac{1}{2}\Gamma\gamma(1 + \Gamma/s)\}\ ,$$

which gives the spectrum

$$\mathcal{S}(s{\to}0) \to \frac{\frac{1}{2}\beta I_2\mathcal{L}_2}{\gamma(1 + I_2\mathcal{L}_2)^2}\left[\mathcal{L}_2[2 + I_2] + 1 + \tfrac{1}{2}\Gamma\left(\frac{1}{s} + \text{c.c.}\right)\right]. \tag{43}$$

Noting that

$$\frac{1}{s} + \text{c.c.} = \lim_{\epsilon\to 0}\frac{1}{-i\Delta + \epsilon} + \text{c.c.} = \lim_{\epsilon\to 0}\frac{2\epsilon}{\Delta^2 + \epsilon^2} = 2\pi\delta(\Delta)\ , \tag{44}$$

where $\delta(\Delta)$ is a Dirac delta function, we see that the Γ/s term in Eq. (42) contains the coherent Rayleigh peak of Eq. (25).

To find the three-peaked spectrum of Fig. 15-2, we consider Eq. (42) for central pump tuning ($\delta = 0$) and for large Rabi flopping, i.e., $I_2 >> 1$ or equivalently, $4|\mathcal{V}_2|^2 >> \gamma\Gamma$. We also eliminate the coherent contribution proportional to Γ/s. We consider a running wave pump, for which $4|\mathcal{V}_2|^2 = \mathcal{R}_0^2$, where $\mathcal{R}_0$ is the pump Rabi flopping frequency $|\wp\mathcal{E}_2/\hbar|$. First consider the region around $s = -i\Delta \simeq 0$. Here D reduces to $\mathcal{R}_0^2(\gamma + s)$ and the numerator of Eq. (42) reduces to $\frac{1}{4}|\mathcal{R}_0|^2$ (aside from the δ-function considered above), so that

$$\mathcal{S}_{+-}(s \simeq 0)\Big|_{inc} \simeq \frac{1}{4(\gamma + s)}\ . \tag{45}$$

For $s \simeq \pm i\mathcal{R}_0$, i.e., in the vicinity of the sidepeaks, the numerator of Eq. (42) reduces to $-\frac{1}{4}\mathcal{R}_0^2$ and the denominator D reduces approximately to $s[\mathcal{R}_0^2 + s^2 + s(\gamma+\Gamma)]$. Substituting $s = -i\Delta$ in the ratio and noting that $\mathcal{R}_0^2 - \Delta^2 = (\mathcal{R}_0 + \Delta)(\mathcal{R}_0 - \Delta)$, we find

$$\mathcal{S}_{+-}(-i\Delta \simeq \pm\mathcal{R}_0) \simeq -\frac{\frac{1}{4}i\mathcal{R}_0^2}{\Delta[\mathcal{R}_0^2 - \Delta^2 - i\Delta(\gamma+\Gamma)]} \simeq \mp \frac{i}{8[\mathcal{R}_0 \pm \Delta \mp \frac{1}{2}i(\gamma+\Gamma)]} .$$

Substituting this along with Eq. (45) into Eq. (38), we find

$$\mathcal{S}(\Delta)\bigg|_{inc} \simeq \frac{\beta}{2\gamma}\left[\frac{\gamma^2}{\gamma^2+\Delta^2} + \frac{\gamma}{\Gamma+\gamma}\frac{\frac{1}{4}(\Gamma+\gamma)^2}{(\mathcal{R}_0 \pm \Delta)^2 + \frac{1}{4}(\Gamma+\gamma)^2}\right] . \tag{46}$$

The incoherent part of the scattered spectrum consists of three peaks, as expected from the phenomenological discussion of Sec. 15-1. The central peak, centered at $\nu_1 = \nu_2 = \omega$, has a width γ, while the sidebands centered about the frequencies $\nu_1 = \omega \pm \mathcal{R}_0$, have widths $\frac{1}{2}(\Gamma+\gamma)$. This is because the population decay constant Γ plays a negligible role for $\Delta \simeq 0$, while in the vicinity of the Rabi sidebands, it plays a role in the denominator D equal to the dipole decay rate γ. The population has a lifetime Γ, while the polarization decays at rate γ, and the mean of these two decay rates is $\frac{1}{2}(\Gamma + \gamma)$. This explains the 3:1 ratio between the heights of the central peak and sidebands for the case of pure radiative decay, for which $\Gamma = 2\gamma$. Since the argument of Sec. 15-1 predicts the ratio of central to side-peak integrated intensity to be 2:1, and the intensity is given by the product of peak × width, we must have

$$\frac{\text{central peak}}{\text{side peak}} = 2\,\frac{\text{side width}}{\text{central width}} = 2\,\frac{3\gamma/4}{\gamma/2} = 3 .$$

Figure 15-2 illustrates the resonance fluorescence spectrum in the strong field limit showing good agreement with the experimental results of Hartig *et al.* (1976) [see also Grove *et al.* (1977)].

15-4. Connection with Probe Absorption

The resonance fluorescence spectrum given by Eq. (24) or equivalently by Eq. (38) is closely related to the semiclassical absorption spectrum given by Eq. (8.17). Specifically both resonance fluorescence and probe stimulated emission are described by the same spectral function. The major difference is that in resonance fluorescence, the vacuum modes act like "one-way" probes, able to stimulate emission, but not absorption. In contrast, the semiclassical probe absorption coefficient is given by the difference between a stimulated absorption function and the emission function of resonance fluorescence. To demonstrate this important relationship and to lay a piece of the foundation for studying the generation of squeezed states in Sec. 16-2, we derive a master equation governing the evolution of the reduced density operator for the probe mode. We derive this equation using the second-order perturbation techniques of Sec. 14-1 in conjunction with the resonance-fluorescence spectral coefficients of Sec. 15-3. As before the driving field is treated to all orders, so that $S_+(\tau)$ and its correlations are determined to lowest order in $\mathcal{V}(\tau)$ by the Langevin-Bloch Eqs. (9). There is a small approximation involved in this step, since by removing a mode from the continuum and treating it explicitly as a probe field we change the number of modes in the reservoir by one. However removing a single term from the infinite sum in F_i is a negligible correction indeed. An important conceptual distinction between the probe and two-level atom master equations is that the probe mode is not directly coupled to the reservoir, but rather via the two-level atom. The response of the two-level atom to broad-band excitation has a finite bandwidth, and it passes the vacuum fluctuations as filtered by the dynamics of the Langevin-Bloch equations to the probe.

As in Sec. 15-3, we work in an interaction picture rotating at the pump frequency ν_2. The probe interaction energy $\mathcal{V}(\tau)$ is given by

$$\mathcal{V}(\tau) = \hbar g a_1 e^{i\Delta\tau} S_+(\tau) + \text{adj.}, \tag{47}$$

which we add to the total Hamiltonian of Eq. (4) transformed to the pump-frame interaction picture. To second-order in perturbation theory in $\mathcal{V}$, the reduced density matrix for the probe has a form similar to Eq. (14.21), except that the trace is now on both the vacuum modes and the two-level system. Again the first-order contribution vanishes when this partial trace is performed. The first term in the integrand of the second-order contribution is

$$\begin{aligned}\text{tr}_r\{\mathcal{V}(\tau')\mathcal{V}(\tau'')P_{sr}\} &= \hbar^2 g^2\, \text{tr}_{spin,r}\,\{[a_1 e^{i\Delta\tau'} S_+(\tau') + \text{adj.}]\\ &\qquad \times[a_1 e^{i\Delta\tau''} S_+(\tau'') + \text{adj.}]P_{sr}(t)\}\\ &= \hbar^2 g^2 a_1 a_1^\dagger \rho(t) e^{i\Delta(\tau'-\tau'')}\langle S_+(\tau')S_-(\tau'')\rangle\\ &\quad + \hbar^2 g^2 a_1^\dagger a_1 \rho(t) e^{-i\Delta(\tau'-\tau'')}\langle S_-(\tau')S_+(\tau'')\rangle\,,\end{aligned}$$

where we drop terms with $a_1 a_1$ and $a_1^\dagger a_1^\dagger$ since they don't conserve energy, and we have used the fact that at the initial time t, the system-reservoir density operator $P_{sr}(t)$ factors, i.e., $P_{sr}(t) = \rho(t)\rho_r(t)$. Note that the system consists now of both the two-level atom and the probe, and that the partial trace is over both reservoir and spins, since we are interested in the dynamics of the probe only. We consider times after which transients have died away and the spins are described by a stationary process, so we can write this expression in terms of time difference $T = \tau' - \tau''$. Similarly to Eq. (14.21) this gives the integral

$$-\frac{1}{\hbar^2\tau}\int_0^\tau d\tau'\int_0^{\tau'} d\tau''\langle\mathcal{V}(\tau')\mathcal{V}(\tau'')\rangle\rho(t) = -\frac{g^2}{\tau}\int_0^\tau d\tau'\int_0^{\tau'} dT[e^{i\Delta T}\langle S_+(T)S_-(0)\rangle a_1 a_1^\dagger$$

$$+ e^{-i\Delta T}\langle S_-(T)S_+(0)\rangle a_1^\dagger a_1]\rho(t)$$

$$= -A_1^* a_1 a_1^\dagger \rho(t) - B_1 a_1^\dagger a_1 \rho(t) , \qquad (48)$$

where

$$A_1^* = g^2\int_0^\infty dT\, e^{i\Delta T}\langle S_+(T)S_-(0)\rangle = g^2\mathcal{S}_{+-}(s) , \qquad (49)$$

$\mathcal{S}_{+-}(s)$ is given by Eqs. (37) and (42), and

$$B_1 = g^2\int_0^\infty dT\, e^{-i\Delta T}\langle S_-(T)S_+(0)\rangle = g^2\mathcal{S}_{-+}(s^*) , \qquad (50)$$

since to this order of perturbation the spin dynamics are given by the Langevin-Bloch equations (9). Dividing the denominator D of Eq. (36) into the numerator of Eq. (42), substituting the result into Eq. (49), and taking the complex conjugate, we find

$$\boxed{A_1 = \frac{g^2\mathcal{D}_1}{1 + I_2\mathcal{L}_2}\left[\frac{I_2\mathcal{L}_2}{2} - \frac{I_2\mathcal{F}\gamma}{2}\,\frac{\frac{1}{2}I_2\mathcal{L}_2\mathcal{D}_1 - \frac{1}{2}\mathcal{D}_2^*(1 + \Gamma/i\Delta)}{1 + I_2\mathcal{F}\frac{\gamma}{2}(\mathcal{D}_1 + \mathcal{D}_3^*)}\right]} , \qquad (51)$$

where the complex population-pulsation Lorentzian $\mathcal{F} = \Gamma/(\Gamma + i\Delta)$. Problem 15-2 shows that B_1 is given by

$$B_1 = \frac{g^2\mathcal{D}_1}{1 + I_2\mathcal{L}_2}\left[1 + \frac{I_2\mathcal{L}_2}{2} - \frac{I_2\mathcal{F}\gamma}{2}\,\frac{(1+\frac{1}{2}I_2\mathcal{L}_2)\mathcal{D}_1 + \frac{1}{2}\mathcal{D}_2^*(1-\Gamma/i\Delta)}{1 + I_2\mathcal{F}\frac{\gamma}{2}(\mathcal{D}_1 + \mathcal{D}_3^*)}\right] . \qquad (52)$$

Equation (51) looks similar to the absorption coefficient α_1 of Eq. (8.17), but unlike α_1, it does not bleach to zero as I_2 gets large. Indeed, it is the *difference* $B_1 - A_1$ that equals α_1! As we see shortly, A_1 can be interpreted as the emission part of α_1, while B_1 is the absorption part. Note that the resonance fluorescence spectrum for upper-to-ground-lower-level decay is given by $A_1 + A_1^*$, as seen by direct comparison with Eq. (46).

The second integral in the second-order contribution to the side-mode master equation is (compare with Eq. (14.21))

$$\frac{1}{\hbar^2\tau}\int_0^\tau d\tau' \int_0^{\tau'} d\tau'' \langle \mathcal{V}(\tau')P_{s-r}(t)\mathcal{V}(\tau'')\rangle$$

$$= \frac{g^2}{\tau}\int_0^\tau d\tau' \int_0^{\tau'} d\tau'' \,[e^{i\Delta(\tau'-\tau'')}\langle S_+(\tau')S_-(\tau'')\rangle a_1\rho(t)a_1^\dagger$$

$$+ e^{-i\Delta(\tau'-\tau'')}\langle S_-(\tau')S_+(\tau'')\rangle a_1^\dagger\rho(t)a_1]$$

$$= \frac{g^2}{\tau}\int_0^\tau d\tau' \int_0^{\tau'} dT \,[e^{i\Delta T}\langle S_-(0)S_+(T)\rangle a_1\rho(t)a_1^\dagger$$

$$+ e^{-i\Delta T}\langle S_+(0)S_-(T)\rangle a_1^\dagger\rho(t)a_1]$$

$$= B_1^* a_1\rho(t)a_1^\dagger + A_1 a_1^\dagger\rho(t)a_1 \,, \tag{53}$$

where used the fact that reservoir operators can be cyclically rotated inside the reservoir average, and the expectation values $\langle ...\rangle$ are on both the reservoir and the two-level atom.

Combining the second-order contributions of Eqs. (48) and (53) we find the sidemode master equation

$$\boxed{\dot\rho = -A_1(\rho a_1 a_1^\dagger - a_1^\dagger\rho a_1) - B_1(a_1^\dagger a_1\rho - a_1\rho a_1^\dagger) + \text{adj.}} \tag{54}$$

This equation has the same general form as Eq. (14.31) for a mode interacting with a reservoir of simple harmonic oscillators. However while the coefficients in Eq. (14.31) are real and depend only on the properties of the bath, those in Eq. (54) are complex functions of the beat frequency between the sidemode and pump frequencies. Furthermore, they account for the fact that the action of the reservoir on the side-mode is filtered by the two-level atom, hence they depend on the atomic relaxation times T_1 and T_2 and on the strength of the classical driving field.

Equation (54) can be used to calculate the equations of motion of various expectation values. In particular the average electric field amplitude $\mathcal{E}_1 = \mathcal{E}_\Omega \langle a_1 \rangle = \mathcal{E}_\Omega \mathrm{tr}\{a_1 \rho\}$ has the equation of motion

$$\dot{\mathcal{E}}_1 = (A_1 - B_1)\mathcal{E}_1 \,. \tag{55}$$

Comparing this with the Beer's law Eq. (5.3), we see that the sidemode absorption coefficient is given by

$$\alpha_1 = B_1 - A_1 \,, \tag{56}$$

which agrees with Eq. (8.17) derived from semiclassical theory. Note that for $\tau = 0$ in the integrand of Eq. (23), we have the average $\langle S_+(0)S_-(0)\rangle = \langle S_a(0)\rangle$. Hence the spectrum of Eq. (23), $\mathcal{S}_{+-}(s)$, and A_1 all provide measures of the fluctuations of the upper state and hence involve emission. Similarly $\mathcal{S}_{-+}(s^*)$ and B_1 provide measures of the fluctuations of the lower state and hence involve absorption. The difference between the two gives the semiclassical absorption coefficient α_1 of Eq. (8.17). Alternatively, the same interpretation is obtained by concentrating on field observables and noting that A_1 is the coefficient of the antinormally ordered side-mode operator $a_1 a_1^\dagger$, while B_1 is the coefficient of the normally ordered operator $a_1^\dagger a_1$.

The equation of motion for the average photon number $\langle n \rangle \equiv \mathrm{tr}\{a_1^\dagger a_1 \rho\}$ is given by

$$\frac{d}{dt}\langle n \rangle = A_1(\langle n \rangle + 1) - B_1 \langle n \rangle + \text{c.c.} \tag{57}$$

Note that here the characteristic 1 of quantized emission shows up associated with the A_1 coefficient. Unlike the intensity Beer's law of Eq. (5.5), Eq. (57) shows that the side-mode intensity $\langle n \rangle$ can build up (if $\mathrm{Re}\{A_1 - B_1\} > 0$) from resonance fluorescence. This provides the "spark" in side-mode buildup in a laser. By adding a cavity loss $\nu/2Q_1$ to B_1, Eq. (57) can be used to study the generation and amplification of resonance fluorescence in a cavity [see Holm, Sargent, and Stenholm (1987)]. From Eq. (14.31) we see that such a loss term results from including coupling to a reservoir of harmonic oscillators at zero temperature ($\bar{n} = 0$).

We see that the vacuum modes act as weak probe fields of the two-level response in the presence of a strong saturator wave. However they differ from a classical probe field in the very essential way that they only cause emission, that is, spontaneous emission. A classical probe field induces both emission *and* absorption, and hence the semiclassical absorption coefficient is the difference between these processes. As such it is a "net" absorption coefficient. Note that the $\Gamma/i\Delta$ term cancels out in the difference $B_1 - A_1$; a classical probe field is insensitive to Rayleigh (coherent) scattering.

By solving the coupled equations (40) by substitution instead of by Cramer's rule, we can see the role of population pulsations in this quantized field context. Specifically, the first and third components of Eq. (40) can be written in terms of $\mathcal{S}_{z-}$ as

$$\mathcal{S}_{+-} = \mathcal{D}_1^*[\langle S_a\rangle - 2i\mathcal{V}_2^*\mathcal{S}_{z-}]\,, \tag{58}$$

$$\mathcal{S}_{--} = 2i\mathcal{V}_2\mathcal{D}_3 S_{z_-}\,. \tag{59}$$

Substituting these equations into the $\mathcal{S}_{z-}$ equation, we find

$$\mathcal{S}_{z-} = -\frac{\frac{1}{2}(\langle S_-\rangle(1+\Gamma/s) + i\mathcal{V}_2\mathcal{D}_1^*\langle S_a\rangle)}{\Gamma + s + 2|\mathcal{V}_2|^2(\mathcal{D}_3 + \mathcal{D}_1^*)} \tag{60}$$

Evaluating this quantity at $s = -i\Delta$, taking the complex conjugate, using Eq. (14), and dividing by $\Gamma + i\Delta$, we find

$$\mathcal{S}_{z-}^* = \frac{i\mathcal{V}_2^*\mathcal{F}}{2\Gamma}\frac{\mathcal{D}_1 I_2\mathcal{L}_2 - \mathcal{D}_2^*(1+\Gamma/i\Delta)}{(1+I_2\mathcal{L}_2)(1+\frac{1}{2}I_2\mathcal{F}(\Delta)\gamma(\mathcal{D}_1 + \mathcal{D}_3^*))}\,. \tag{60}$$

Substituting this, in turn, into the complex conjugate of Eq. (58) we find A_1/g^2. As such we identify $\mathcal{S}_{z-}^*$ as a quantized population-pulsation term. The subscript $_-$ for $S_- = |b\rangle\langle a|$ identifies the initial state as being the upper state. A similar derivation for $\mathcal{S}_{z+}$ yields the corresponding population pulsation for the initial lower state. The difference $\mathcal{S}_{z-}^* - \mathcal{S}_{z+}$ is proportional to the semiclassical population pulsation d_{-1} of Eq. (8.16). The $\mathcal{S}_{z\pm}$ contain the Rayleigh scattering in their $\Gamma/i\Delta$ terms, while this scattering cancels out in d_{-1}.

Section 16-2 generalizes the equations of Secs. 15-3 and 15-4 to the case of two side-modes symmetrically placed about the strong driving field frequency. In this case, the master equation further includes two further quantum multiwave-mixing coefficients, $C_1 = -g^2\mathcal{S}_{--}(s^*)$ and $D_1 = -g^2[\mathcal{S}_{++}(s)]^*$. This master equation leads into a quantum theory of multiwave mixing, which builds a connection with the mode coupling of Chap. 9 and permits to study the generation of squeezed states in resonance fluorescence and in four-wave mixing.

15-5. Photon Antibunching

Further information and insight into the characteristics of resonance fluorescence light is obtained by investigating higher-order correlations of the scattered field. Experimentally this is done using two detectors to measure the joint probability of detecting a photoelectron at time $t = 0$ and a subsequent one at time $t = \tau$. The measured quantity is the second-order correlation function of Sec. 12-5

$$G^{(2)}(\tau) = \langle E^-(0)E^-(\tau)E^+(\tau)E^+(0)\rangle \propto \langle S_+(0)S_+(\tau)S_-(\tau)S_-(0)\rangle \ . \qquad (61)$$

Noting that $S_+(\tau)S_-(\tau) = S_a(\tau) = \frac{1}{2} + \frac{1}{2}S_z(\tau)$, we have

$$G^{(2)}(\tau) \propto \frac{1}{2}\langle S_a(0)\rangle + \langle S_+(0)S_z(\tau)S_-(0)\rangle \ . \qquad (62)$$

A derivation analogous to that of Eq. (29) yields, for $\tau > 0$,

$$\frac{d}{d\tau}\langle S_+(0)S_i(\tau)S_-(0)\rangle = \sum_j B_{ij}\langle S_+(0)S_j(\tau)S_-(0)\rangle - \frac{\Gamma}{2}\langle S_+(0)S_-(0)\rangle\delta_{iz} \ , \qquad (63)$$

where the matrix elements B_{ij} are given by Eq. (27), the $\Gamma\delta_{iz}$ term comes from the constant term in Eq. (12b), and we have used

$$\langle S_+(0)F_i(\tau)S_-(0)\rangle = 0 \quad \text{for } \tau > 0 \ . \qquad (64)$$

As for Eq. (29), the Laplace transform of Eq. (63) is of the form of Eq. (31), but here the only nonzero initial condition is

$$\langle S_+(0)S_z(0)S_-(0)\rangle = -\tfrac{1}{2}\langle S_a\rangle \ . \qquad (65)$$

Inserting Eq. (65) into (62) we find

$$G^{(2)}(0) = 0 \ . \qquad (66)$$

This result can be interpreted as follows: just after detection of a first photon, i.e., at $\tau \simeq 0$, the atom is certainly in the ground state. In order to be able to emit a second photon, the atom must first evolve back to the upper state, a process that takes an average time on the order of the Rabi frequency. Hence it cannot emit two photons in immediate succession. This phenomenon is called *photon antibunching*. We showed in Sec. 1-4 that it has no classical counterpart. The experimental verification of photon antibunching is a test of the validity of quantum electrodynamics.

Using Cramer's method as for Eq. (31), we find that the Laplace transform of $\langle S_+(0)S_z(\tau)S_-(0)\rangle$ is given by

$$\mathcal{S}_{+z-}(s) = -\frac{\langle S_a(0)\rangle(\gamma + i\delta + s)(\gamma - i\delta + s)}{2D} . \tag{67}$$

For a large centrally tuned pump field, we can neglect the elastic contribution and Eq. (67) reduces to

$$\begin{aligned}\mathcal{S}_{+z-}(s) &\simeq -\frac{\langle S_a\rangle(\gamma + s)}{2(s^2 + (\gamma+\Gamma)s + \mathcal{R}_0{}^2)} \\ &\simeq -\frac{\langle S_a\rangle}{4}\left[\frac{1}{s + \frac{1}{2}(\gamma+\Gamma) + i\mathcal{R}_0} + \frac{1}{s + \frac{1}{2}(\gamma+\Gamma) - i\mathcal{R}_0}\right] .\end{aligned}$$

Taking the inverse Laplace transform, we find

$$\langle S_+(0)S_z(\tau)S_-(0)\rangle \simeq -\tfrac{1}{2}\langle S_a(0)\rangle e^{-(\gamma+\Gamma)\tau/2}\cos(\mathcal{R}_0\tau) . \tag{68}$$

Finally substituting this result into Eq. (62), we obtain

$$\boxed{G^{(2)}(\tau) \propto \tfrac{1}{2}\langle S_a(0)\rangle[1 - e^{-(\gamma+\Gamma)\tau/2}\cos(\mathcal{R}_0\tau)]} . \tag{69}$$

To observe photon antibunching experimentally, it is crucial to detect the fluorescence light from a single atom (Kimble *et al.* (1978)). Early experiments with atomic beams required to work at exceedingly low densities, and even then, the Poissonian fluctuations inherent in the arrival times of the atoms in the interaction region lead to considerable difficulties. The advent of ion traps in which a single radiator can be isolated for considerable periods of time has recently removed this problem. Figure 15-6 illustrates the antibunching correlation of Eq. (69) normalized to 1 for $t \to \infty$.

15-6. Off-resonant Excitation

Although most studies of resonance fluorescence have been performed under resonant or near-resonant excitation, there is some advantage in considering strongly non-resonant lasers, $\nu_2 \neq \omega$. In this case, the study of intensity, spectrum, and intensity correlation function can be complemented by an investigation of the temporal correlations between the photons emitted in the two sidebands. Problem 15-8 shows that these sidebands are centered at the frequencies ν_1, $\nu_3 = \nu_2 \pm \mathcal{R}$, where the $\mathcal{R}$ is the

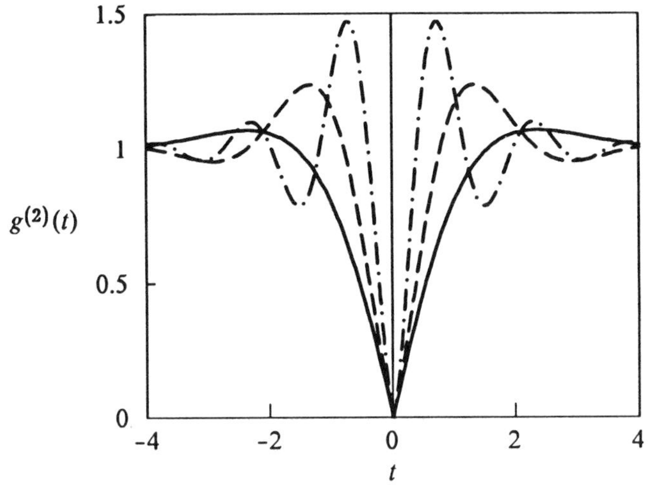

Fig. 15-6. Resonance fluorescence photon antibunching correlation $g^{(2)}(t)$ from Eq. (69) versus time for $\gamma = \Gamma = 1$ and $\mathcal{R}_0 = 1$ (solid), 2 (dashed), and 4 (dot-dashed). Values less than unity indicate nonclassical behavior.

generalized Rabi flopping frequency $[\delta^2 + |\mathcal{R}_0|^2]^{1/2}$. For large enough pump detuning $|\omega-\nu_2|$, it is possible to detect selectively the light emitted in the two sidebands by placing spectral filters between atom and detectors, as illustrated in Fig. 15-7. One can then measure the probability $p(\nu_1,t;\nu_3,t+\tau)$ of detecting a photon at frequency ν_1 at time t and a photon at frequency ν_3 at time $t + \tau$, where τ can be positive or negative.

Section 3-2 showed that for very large detunings, it is in general sufficient to treat the system using perturbation theory. To lowest order, the atom-field interaction can be pictorially represented by the diagram in Fig. 15-8. We have seen that this is the Rayleigh scattering contribution, which yields a delta-function spectrum. To obtain the off-resonance fluorescence spectrum, it is necessary to go one order higher in perturbation theory, with the absorption of two laser photons and the emission of two photons at frequencies ν_1 and ν_3. The relevant diagrams are shown in Fig. 15-9. The first diagram in this figure is a second-order correction to Rayleigh scattering. The dominant frequencies in Figs. 15-9*b* and 15-9*c* can be obtained either via the dressed-atom picture, as in Sec. 15-1, or by noting that the interaction is resonantly enhanced if one of the intermediate steps leads precisely to an atomic state. The only way to achieve this is shown in Fig. 15-9*b*, and yields $\nu_1 = \nu_2 - \Delta$ and $\nu_3 = \nu_2 + \Delta$. There are two diagrams yielding these frequencies, the second one (Fig. 15-9*c*) is not reso-

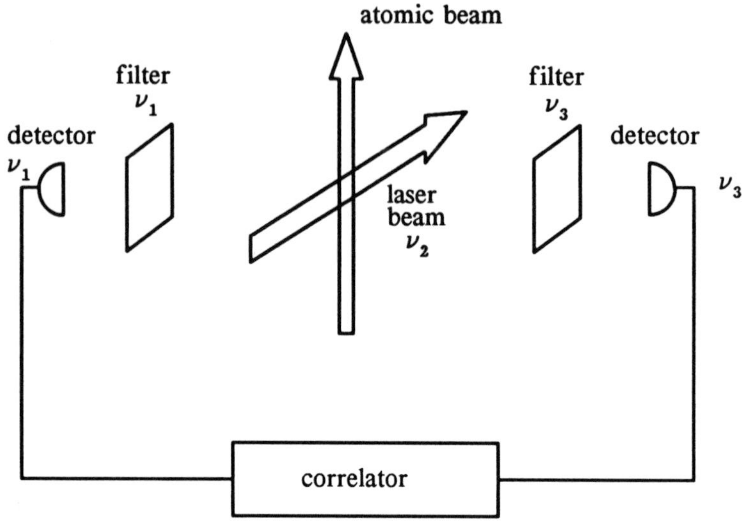

Fig. 15-7. Diagram allowing separate detection of the light in the two sidebands of resonance fluorescence.

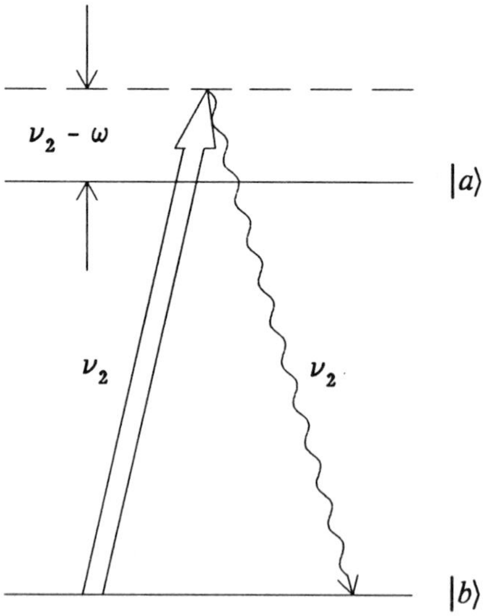

Fig. 15-8. Diagram showing lowest order (Rayleigh) scattering.

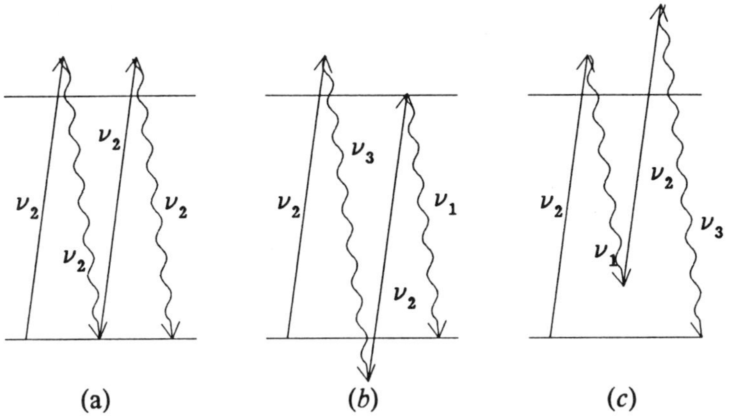

Fig. 15-9. (*a*) second-order correction to Rayleigh scattering. (*b*) four-photon process with an intermediate resonance. (*c*) four-photon process with no intermediate resonance.

nantly enhanced. Diagram 15-9*b* is thus privileged, and an event with a photon at frequency ν_1 occurring *after* an emission at ν_3 is more likely than the opposite. From this argument, one expects $p(\nu_1, t; \nu_3, t+\tau)$ to be strongly asymmetric, as illustrated in Fig. 15-10. If the incident laser fre-

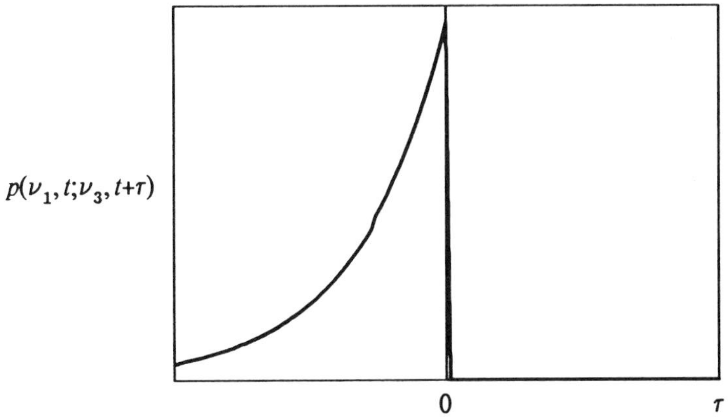

Fig. 15-10. The probability for a photon at frequency ν_1 to be emitted after a photon at ν_3 is much higher than the opposite.

quency ν_2 is larger than the atomic transition frequency ω, the scattered photon of higher frequency tends to be emitted first. The photon of lower energy is emitted first in the other case. This prediction was verified by Aspect *et al.* (1980). Note that in general perturbation theory is inadequate in the case of resonant excitation. One can however easily convince oneself that such an ordering of the scattered photons disappears in that case. The role of both sidebands is now perfectly symmetrical.

The ν_1-photon ν_3-photon correlation also leads to squeezing, but under slightly different conditions: δ is still taken to be large, but regions of significant resonance fluorescence are avoided, since the fluorescence swamps the squeezing. As discussed in Sec. 16-3, good squeezing is obtained when the magnitude of the coupled-mode fluorescence $C_1 + C_3$ given by Eq. (16.30) is maximized, while the resonance fluorescence $A_1 + A_1^*$ given by Eq. (51) is minimized. Since this squeezing occurs via multiwave mixing, phase matching can play an important role.

References

Aspect, A., G. Roger, S. Reynaud, J. Dalibard and C. Cohen-Tannoudji (1980), Phys. Rev. Lett. **45**, 617.

Carmichael, H. J. and D. F. Walls (1976), J. Phys. B9, 1199 gives the first prediction of photon antibunching in resonance fluorescence.

Cresser, J. D., J. Häger, G. Leuchs, M. Rateike, and H. Walther, Chap.3 in *Dissipative Systems in Quantum Optics*, Ed. by R. Bonifacio, Springer-Verlag, Berlin (1982) is a good review of resonance fluorescence including both experimental details and an outline of theoretical approaches.

Cohen-Tannoudji, C. (1977), in *Frontiers in Laser Spectroscopy*, Ed. by R. Balian, S. Haroche, and S. Liberman North-Holland, Amsterdam, Vol. I gives a theory of resonance fluorescence based on the Langevin-Bloch equations and stressing the dressed-atom interpretation.

Cohen-Tannoudji, C., J. Dupont-Roc, and G. Grynberg (1988), *Processus d'interaction entre photons et atomes*, InterEditions et Editions du CNRS, Paris discusses the Langevin-Bloch equations and the dressed atom picture in great detail.

Eberly, J. H. and Wodkiewicz (1977), J. Opt. Soc. Am. **67**, 1252 discuss definitions of the "physical spectrum".

Holm, D. A., M. Sargent III, and S. Stenholm (1987), J. Opt. Soc. Am. B2 1456.

Hartig, W., W. Rasmussen, R. Schieder, and H. Walther (1976), Z. Phys. A**278**, 205.

Kimble, H. J., M. Dagenais, and L. Mandel (1978), Phys. Rev. A**18**, 201 made the first antibunching measurements.

Knight, P. amd P. W. Milonni (1980), Phys. Rep. **66C**, 21.

Louisell, W. H. (1973), *Quantum Statistical Properties of Radiation*, Wiley, New York. The Wiener-Khintchine theorem is discussed in Appendix *I*.

Mollow, B. R. (1969), Phys. Rev. **188**, 1969. For an overview of resonance fluorescence with many references, see B. R. Mollow (1981), in *Progress in Optics XIX*, Ed. by E. Wolf, North-Holland, *p.* 1.

Sargent, M. III, M. O. Scully, and W. E. Lamb, Jr. (1977), *Laser Physics*, Addison-Wesley Publishing Co., Reading, MA. See Chap. 18 for two-level-atom model of a detector.

Shuda, F., C. R. Stroud, and M. Hercher (1974), J. Phys. B7, L198.

Problems

15-1. Show that the initial conditions in Eq. (31) are given by

$$\begin{aligned}\langle\delta S_+(0)\delta S_-(0)\rangle &= \langle S_a\rangle - |\langle S_+\rangle|^2\\ \langle\delta S_z(0)\delta S_-(0)\rangle &= -\langle S_a\rangle\langle S_-\rangle\\ \langle\delta S_-(0)\delta S_-(0)\rangle &= -\langle S_-\rangle^2 .\end{aligned}$$

Hence show that resonance fluorescence spectrum is given by Eq. (35) with

$$\begin{aligned}D\delta\mathcal{S}_{+-}(-i\Delta) = [\langle S_a\rangle - |\langle S_+\rangle|^2][(\Gamma-i\Delta)(\gamma+i\delta-i\Delta) + 2|\mathcal{V}_2|^2]\\ + 2i\mathcal{V}_2^*\langle S_-\rangle\langle S_a\rangle(\gamma + i\delta - i\Delta) - 2\mathcal{V}_2^{*2}\langle S_-\rangle^2 ,\end{aligned}$$

where the determinant D is given by Eq. (35).

15-2. Write the matrix equation corresponding to Eq. (40) that includes the average $\mathcal{S}_{-+}(s)$. Using Cramer's rule, solve for $\mathcal{S}_{-+}(s)$. Ans: B_1/g^2 with $s = i\Delta$, where B_1 is given by Eq. (69).

15-3. Show that the matrix of initial conditions, $\langle S_i(0)S_j(0)\rangle$, is given by

$$\begin{bmatrix} 0 & -\tfrac{1}{2}\langle S_+\rangle & \langle S_a\rangle \\ \tfrac{1}{2}\langle S_+\rangle & \tfrac{1}{4} & -\tfrac{1}{2}\langle S_-\rangle \\ \langle S_b\rangle & \tfrac{1}{2}\langle S_-\rangle & 0 \end{bmatrix}.$$

15-4. Using the initial-condition matrix of Prob. 15-5, derive the matrix of diffusion coefficients using Eq. (14.112) with $\mathbf{A} = -B$, where B is given by Eq. (27). Ans:

$$2D_{++} = i\mathcal{V}_2\langle S_+\rangle(\tfrac{1}{2} - \Gamma/s)$$

$$2D_{z+} = \tfrac{1}{2}\langle S_+\rangle(1 - \Gamma/s)(\Gamma + s + \gamma - i\delta) - i\mathcal{V}_2(\langle S_b\rangle - \tfrac{1}{4})$$

$$2D_{-+} = 2\gamma\langle S_b\rangle - i\mathcal{V}_2[\langle S_+\rangle(1 - \Gamma/s) - \tfrac{1}{2}\langle S_-\rangle]$$

$$2D_{+z} = -\tfrac{1}{2}\langle S_+\rangle(\Gamma + s + \gamma - i\delta) - 2i\mathcal{V}_2[\langle S_a\rangle - \tfrac{1}{4}]$$

$$2D_{zz} = i\mathcal{V}_2[\langle S_+\rangle(\tfrac{1}{2} - \Gamma/s) + \langle S_-\rangle(\tfrac{1}{2} + \Gamma/s)] + \tfrac{1}{2}(\Gamma + s)$$

$$2D_{z-} = \tfrac{1}{2}\langle S_-\rangle(\Gamma + s + \gamma + i\delta) + 2i\mathcal{V}_2(\langle S_b\rangle - \tfrac{1}{4})$$

$$2D_{-+} = 2\gamma\langle S_a\rangle - i\mathcal{V}_2[\langle S_-\rangle(1 + \Gamma/s) - \tfrac{1}{2}\langle S_+\rangle]$$

$$2D_{-z} = -\tfrac{1}{2}\langle S_-\rangle(1 + \Gamma/s)(\Gamma + s + \gamma + i\delta) + i\mathcal{V}_2(\langle S_a\rangle - \tfrac{1}{4})$$

$$2D_{--} = i\mathcal{V}_2\langle S_-\rangle(\tfrac{1}{2} + \Gamma/s)$$

15-5. Find the complete spectral matrix $\mathcal{S}_{ij}$ for resonance fluorescence by substituting the diffusion coefficient matrix of Prob. 15-4 into Eq. (14.114) with $\mathbf{A} = -B$, where B is given by Eq. (27). Spectral elements such as $\mathcal{S}_{--}$ and $\mathcal{S}_{++}$ are important in the quantum theory of multiwave mixing (see Sec. 16-2). (Lots of algebra!)

15-6. Show that for $\delta \neq 0$ and a generalized Rabi frequency $\mathcal{R} \gg \gamma$ and Γ, the resonance fluorescence sidebands are centered at approximately $\pm\mathcal{R}$.

15-7. Show for pure radiative decay ($\Gamma = 2\gamma$) but arbitrary values of δ and $\mathcal{R}$ that the resonance fluorescence spectrum is symmetrical. Hint: using Eq. (47), show that $A_3^* = A_1$. For this note that $i\Delta$, $\mathcal{F}$, the denominators, and the product $\mathcal{D}_1\mathcal{D}_3^*$ are unchanged by interchanging 1 and 3 and taking the complex conjugate. Put A_3^* over a common denominator, simplify, and note that the lone $\mathcal{D}_3^*$ is cancelled by a $1/\mathcal{D}_3^*$ provided $\Gamma = 2\gamma$.

15-8. Show for $T_1 \gg T_2$, i.e., $\Gamma \ll \gamma$, and a centrally tuned pump wave that the resonance fluorescence spectrum is given by

$$A_1 + A_1^* = \frac{g^2 I_2}{1 + I_2}\left[1 - \frac{\frac{1}{2}\Gamma I_2^2}{\Gamma^2(1+I_2)^2 + \Delta^2} + \frac{\pi}{2(1 + I_2)}\delta(\Delta)\right] .$$

Unlike the three-peaked spectrum, this short-T_2 formula features a power-broadened dip like the coherent dips in the absorption spectra of Sec. 8-2.

15-9. Using the initial condition in Eq. (65), show that the Laplace transform of $\langle S_+(0)S_z(\tau)S_-(0)\rangle$ is given by Eq. (67).

15-10. Given the Langevin equation for the slowly-varying annihilation operator

$$\dot{A} = -[\nu/2Q + i(\Omega - \nu) - \alpha]A(t) + F(t) ,$$

use the quantum regression theorem to find the spectrum defined by

$$\mathcal{S}(\omega) = \int_0^\infty dt\, e^{-i\omega t} \langle A^\dagger(t)A(0)\rangle + \text{c.c.}$$

Here $\nu/2Q$ is laser cavity decay rate, $\Omega - \nu$ is detuning of the mode oscillation frequency ν from the passive cavity resonance Ω, and α is the gain coefficient. Ans: see Sec. 17-3.

15-11. Describe the physics illustrated by the following fi, ure:

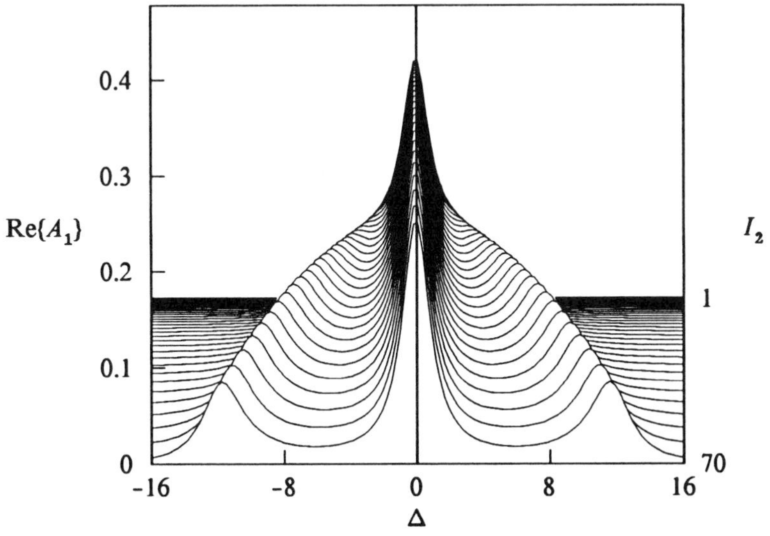

Chapter 16
SQUEEZED STATES OF LIGHT

The Heisenberg uncertainty principle $\Delta A \Delta B \geq \frac{1}{2} |\langle [A,B] \rangle|$ between the standard deviations of two arbitrary observables, $\Delta A = \langle (A - \langle A \rangle)^2 \rangle^{1/2}$ and similarly for ΔB, has a built-in degree of freedom: one can squeeze the standard deviation of one observable provided one "stretches" that for the conjugate observable. For example the position and momentum standard deviations obey the uncertainty relation

$$\Delta x \Delta p \geq \hbar/2 , \tag{1}$$

and we can squeeze Δx to an arbitrarily small value at the expense of accordingly increasing the standard deviation Δp. All quantum mechanics requires is that the *product* be bounded from below. As discussed in Sec. 12-1, the electric and magnetic fields from a pair of observables analogous to the position and momentum of a simple harmonic oscillator. Accordingly they obey a similar uncertainty relation

$$\Delta E \Delta B \geq (\text{constant})\hbar/2 . \tag{2}$$

In principle we can squeeze ΔE at the expense of stretching ΔB, or vice versa. Such squeezing of the electromagnetic field has recently been the object of considerable attention in view of potential application to high precision measurements. It offers the promise of achieving quantum noise reduction beyond the "standard shot noise limit" and might find applications in phase-sensitive detection schemes as required for the detection of gravitational radiation [see Meystre and Scully (1983)].

As an electromagnetic field mode oscillates, energy is transferred between E and B each quarter period. Hence to observe the squeezing in ΔE, we must somehow select its active quadratures from the general electromagnetic oscillation. In effect we need an optical frequency "chopper" that spins around opening up to show us a single quadrature. By varying the relative phase of the chopper and the light, we can look at any quadrature. In practice heterodyne detection provides such a selection process. It effectively multiplies the signal to be measured by a sine wave whose peaks correspond to openings in the chopper. By varying the relative phase of the signal and heterodyne waves, we can examine any quadrature of the signal. Generating the right single-mode radiation complete with a suitable heterodyne wave seems to be very difficult. Instead we can use four-wave mixing processes described in Sec. 2-4 and in Chap. 9 to create a pair of waves whose sum exhibits squeezing. The heterodyne wave is then derived directly from the pump wave.

Section 16-1 describes how squeezing in one quadrature results in stretching in the orthogonal quadrature for the simple harmonic oscillator. This treatment is a squeezed-state generalization of the Schrödinger oscillating packet that forms the basis for coherent states of light. This section also introduces the squeezing operator and shows how it turns the circular variance of the coherent state in the complex α plane into an ellipse. Section 16-2 extends the single-sidemode master equation of Sec. 15-4 to treat the two-sidemode cases found in three- and four-wave mixing. This theory quantizes the signal and conjugate waves, while leaving the pump wave classical. Section 16-3 applies this formalism to calculate the variances for squeezing via multiwave mixing. It gives some numerical illustrations for two-level media. Section 16-4 develops the theory of a squeezed "vacuum" and shows how an atom placed in such a vacuum has two dipole dephasing rates, a small one in the squeezed quadrature, and a correspondingly larger one in the stretched quadrature.

16-1. Squeezing the Coherent State

Section 12-4 shows how a displaced ground state of the simple harmonic oscillator of the correct width oscillates back and forth with unchanging width. However if we now squeeze this wave packet, it will spread for a quarter of a cycle, then return to the squeezed value at the half cycle, and so on as illustrated in Fig. 16-1. Looking at the mean and standard deviation of the electric field vector in the complex α plane, the coherent state appears as in Fig. 16-2*a*, while a squeezed state appears as in Fig. 16-2*b*.

Given a field described by the annihilation operator a, we form two hermitian conjugate operators giving its two quadratures as

$$d_1 = \frac{1}{2}(ae^{i\phi} + a^\dagger e^{-i\phi})\ , \quad d_2 = \frac{1}{2i}(ae^{i\phi} - a^\dagger e^{-i\phi})\ , \tag{3}$$

with $[d_1, d_2] = i/2$, so that $\Delta d_1 \Delta d_2 \geq 1/4$. These two operators correspond to position and momentum for the case of a mechanical oscillator. The variance $\Delta d_1{}^2$ is given by

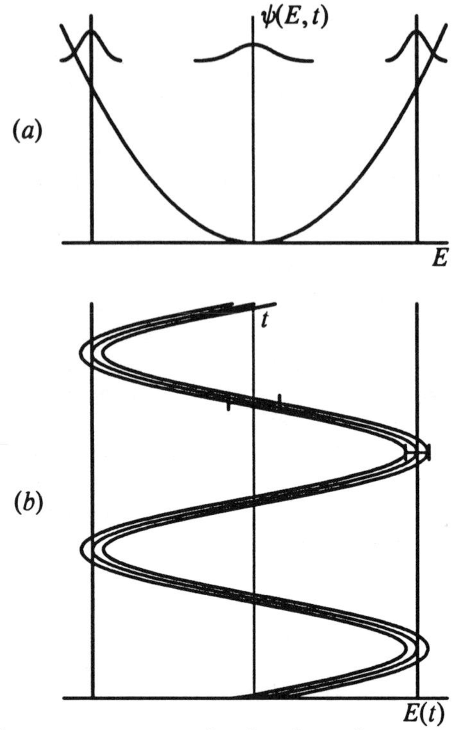

Fig. 16-1. (*a*) Wave packet oscillating in a simple harmonic oscillator potential well with an initially squeezed variance. (*b*) Path traced out by the oscillating packet as a function of time.

$$\Delta d_1^2 = \langle d_1^2\rangle - \langle d_1\rangle^2 . \tag{4}$$

Consider for a moment a quantum state such that the expectation value of the electric field is zero, $\langle a\rangle = \langle a^\dagger\rangle = \langle d_i\rangle = 0$. This reduces Δd_1^2 to

$$\Delta d_1^2 = \tfrac{1}{4}[\langle a^\dagger a\rangle + \langle aa^\dagger\rangle + (\langle a^2\rangle e^{2i\phi} + \text{c.c.})] = \tfrac{1}{4} + \tfrac{1}{2}\langle a^\dagger a\rangle + \tfrac{1}{2}Re\{\langle aa\rangle e^{2i\phi}\} . \tag{5}$$

For a given set of expectation values, the minimum variance is given by the ϕ that yields $\langle aa\rangle e^{2i\phi}$ + c.c. = $-2|\langle aa\rangle|$, that is

$$\Delta d_1^2 = \tfrac{1}{4} + \tfrac{1}{2}\langle a^\dagger a\rangle - \tfrac{1}{2}|\langle aa\rangle| . \tag{6}$$

The conjugate variance Δd_2^2 is given by

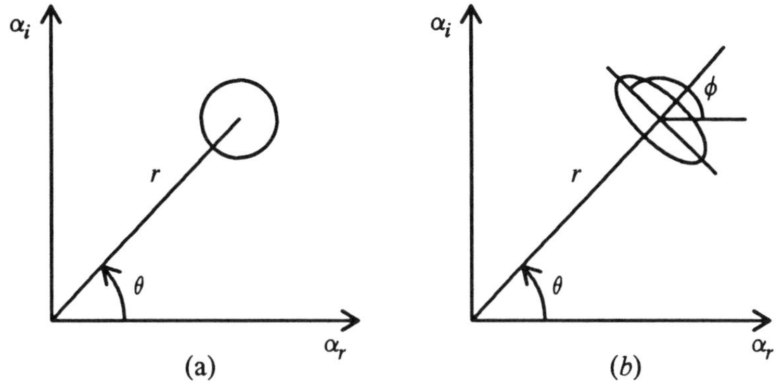

Fig. 16-2. (*a*) Coherent state amplitude vector $\alpha = re^{i\theta}$ and its variance. (*b*) Squeezed state amplitude vector $\alpha = r'e^{i\theta}$ and variance squeezed by the operator $S(\zeta)$ of Eq. (9) at an angle ϕ with respect to the real axis.

$$\Delta d_1{}^2 = \tfrac{1}{4} + \tfrac{1}{2}\langle a^\dagger a\rangle + \tfrac{1}{2}|\langle aa\rangle| \,, \tag{7}$$

which is greater than or equal to Eq. (6). These equations give the uncertainty principle

$$\Delta d_1 \Delta d_2 = \frac{1}{4}\sqrt{[1 + 2\langle a^\dagger a\rangle - 2|\langle aa\rangle|][1 + 2\langle a^\dagger a\rangle + 2|\langle aa\rangle|]} \geq \frac{1}{4}\,. \tag{8}$$

Squeezing occurs for d_1 if $\Delta d_1{}^2$ becomes smaller than $\frac{1}{4}$, i.e., squeezed below the minimum uncertainty product value. This does not violate the uncertainty principle, since d_2 has a correspondingly increased variance.

In the present example it is the $|\langle aa\rangle|$ term that leads to squeezing. A way to obtain such squeezing formally is to "squeeze" the state vector with the squeeze operator

$$S(\zeta) = e^{\zeta a^{\dagger 2} - \zeta^* a^2}\,. \tag{9}$$

This operator converts the circular variance of a coherent state (Fig. 16-2*a*) into a rotated ellipse as illustrated in Fig. 16-2*b*. To show this, we calculate the standard deviations Δd_i in the state $S(\zeta)|\alpha\rangle$. To do this, we need the expectation values of a, $a^\dagger$, a^2, and $a^{\dagger 2}$ in this state. These expectation values involve the operator products $S^\dagger(\varsigma)aS(\varsigma)$ and $S^\dagger(\varsigma)a^\dagger S(\varsigma)$. We can express these products in terms of simple powers of a and $a^\dagger$ by using the operator identity of Eq. (3.151), where here we take $e^{-B} = S$ and note that $S^\dagger(\varsigma) = S^{-1}(\varsigma)$, i.e., $S(\varsigma)$ is a unitary operator. In working with the operator S, it is convenient to write ς in polar coordinates as

$$\varsigma = \tfrac{1}{2}re^{-2i\phi} . \tag{10}$$

Using the relations (3.149) and (3.150), we have that

$$[B, ae^{i\phi}] = [\varsigma^* a^2 - \varsigma a^{\dagger 2}, ae^{i\phi}] = \varsigma e^{i\phi}[a, a^{\dagger 2}] = ra^\dagger e^{-i\phi} , \tag{11}$$

where we include $e^{i\phi}$ since it simplifies the derivation. The adjoint of Eq. (11) is

$$[B, a^\dagger e^{-i\phi}] = rae^{i\phi} . \tag{12}$$

Using these commutators repeatedly in Eq. (3.151), we obtain the series

$$\begin{aligned} S^\dagger(\varsigma)ae^{i\phi}S(\varsigma) &= ae^{i\phi} + ra^\dagger e^{-i\phi} + \frac{r^2}{2!}ae^{i\phi} + \frac{r^3}{3!}a^\dagger e^{-i\phi} + \ldots \\ &= ae^{i\phi}\cosh r + a^\dagger e^{-i\phi}\sinh r , \end{aligned} \tag{13}$$

which has the adjoint

$$S^\dagger(\varsigma)a^\dagger e^{-i\phi}S(\varsigma) = a^\dagger e^{-i\phi}\cosh r + ae^{i\phi}\sinh r . \tag{14}$$

The corresponding squeezed versions of the Hermitian operators d_i given by Eq. (3) are

$$S^\dagger(\varsigma)d_1S(\varsigma) = \tfrac{1}{2}d_1[\cosh r + \sinh r] = d_1e^r , \tag{15}$$

$$S^\dagger(\varsigma)d_2S(\varsigma) = d_2e^{-r} . \tag{16}$$

Furthermore

$$S^\dagger(\varsigma)d_1^2S(\varsigma) = S^\dagger(\varsigma)d_1S(\varsigma)S^\dagger(\varsigma)d_1S(\varsigma) = d_1^2e^{2r} , \tag{17}$$

$$S^\dagger(\varsigma)d_2^2S(\varsigma) = d_2^2e^{-2r} . \tag{18}$$

Hence the unitary transformation in Eqs. (15) through (18) has the effect of squeezing and stretching the operators d_1 and d_2.

These equations provide all the pieces to calculate the standard deviations Δd_1 and Δd_2 in the squeezed state $S(\varsigma)|\alpha\rangle$. With Eqs. (15) and (16), we have

$$\langle\alpha|S^\dagger(\varsigma)d_1S(\varsigma)|\alpha\rangle = e^r\langle\alpha|d_1|\alpha\rangle = \tfrac{1}{2}e^r(\alpha e^{i\phi} + \alpha^* e^{-i\phi}) , \tag{19}$$

$$\langle\alpha|S^\dagger(\varsigma)d_2S(\varsigma)|\alpha\rangle = \tfrac{1}{2}ie^r(\alpha e^{i\phi} - \alpha^* e^{-i\phi}) , \tag{20}$$

while Eqs. (17) and (18) give

$$\begin{aligned}\langle\alpha|S^\dagger(\varsigma)d_1{}^2S(\varsigma)|\alpha\rangle &= \tfrac{1}{4}e^{2r}\langle\alpha|a^2e^{2i\phi} + a^{\dagger 2}e^{-2i\phi} + 2a^\dagger a + 1|\alpha\rangle \\ &= \tfrac{1}{4}(\alpha^2e^{2i\phi} + \alpha^{*2}e^{-2i\phi} + 2\alpha^*\alpha + 1) .\end{aligned} \tag{21}$$

Combining these results with Eqs. (19) and (20) gives the standard deviation

$$\boxed{\Delta d_1 = \tfrac{1}{2}e^r} . \tag{22}$$

Similarly we find the standard deviation

$$\boxed{\Delta d_2 = \tfrac{1}{2}e^{-r}} . \tag{23}$$

Hence we find that the standard deviation of the field quadrature at the angle ϕ with respect to the real and imaginary α axes are stretched, and that of the field quadrature at the angle $\phi + \frac{1}{2}\pi$ is squeezed. The angle ϕ is determined by the squeeze parameter ς of Eqs. (9) and (10). The angle θ that the phasor α makes with respect to its real and imaginary axes is in general independent of ϕ. The state with $\phi = \theta$ is called a phase squeezed state and the state with $\phi = \theta+\frac{1}{2}\pi$ is called an amplitude squeezed state. Equations (22) and (23) reveal that the general squeezed state $S(\varsigma)|\alpha\rangle$ is a minimum uncertainty state, or *squeezed coherent state*, since $\Delta d_1\Delta d_2 = \frac{1}{4}$, independent of r and ϕ. More generally, squeezed states exist that yield variances less than the average minimum uncertainty for one quadrature, but whose uncertainty product exceeds the minimum uncertainty value of $\frac{1}{4}$. As Probs. 16-2 and 16-3 note, both the magnitude and the photon numbers of the squeezed states increase with squeezing. In particular, the squeezed vacuum has a nonzero photon number.

Note that the squeeze operator $S(\varsigma)$ involves two-photon processes like a^2. It resembles the evolution operator $U(t) = \exp(-i\mathcal{H}t/\hbar)$ associated to a two-photon Hamiltonian of the form

$$\mathcal{H} = \hbar\Omega a^\dagger a + i\hbar\Lambda a^{\dagger 2}e^{-2i\Omega t} - i\hbar\Lambda^* a^2e^{2i\Omega t} , \tag{24}$$

where the coupling strength Λ is proportional to ς. This suggests that two-photon interactions might be a good way to generate squeezing. Equation (24) gives the Heisenberg equation of motion

$$\dot{a} = -i\Omega a + 2\Lambda a^\dagger e^{-2i\Omega t} , \tag{25}$$

i.e., an equation that couples a to $a^\dagger$. A nondegenerate version of such a two-photon process is four-wave mixing, for which the annihilation operator a_1 for one mode is coupled to the creation operator $a_3^\dagger$ of the conjugate mode. This involves processes such as the absorption of two pump photons and the corresponding emission of a signal/conjugate photon pair. This mechanism led to the first observation of squeezed states of light [by Slusher *et al.* (1985)]. Section 16-2 gives the quantum theory of four-wave mixing, which Sec. 16-3 applies to the generation of squeezed states of light. Section 16-4 discusses a related problem in which the vacuum modes are squeezed, which can dramatically change the way a two-level dipole dephases.

16-2. Two-Sidemode Master Equation

This section shows how one strong classical wave and one or two weak quantal side waves (see Fig. 2-1) interact in a nonlinear two-level medium. The derivation allows the waves to propagate in arbitrary directions, and can be used to treat noise in weak-signal phase conjugation, build-up from quantum noise of sidemodes in lasers and optical bistability, resonance fluorescence, Rayleigh scattering, and to treat the effects of stimulated emission and phase conjugation on resonance fluorescence as might be found in cavity configurations. In particular, it allows us to generate squeezed states of light as described in the next section.

To find a master equation for two sidemodes symmetrically detuned with respect to the frequency ν_2 of the classical field we extend the single-sidemode discussion of Sec. 15-4 by using the two-mode interaction energy $\mathcal{V}(\tau)$

$$\mathcal{V}(\tau) = \hbar g[a_1 e^{i\Delta\tau} + a_3 e^{-i\Delta\tau}]S_+(\tau) + \text{adj.}, \tag{26}$$

where $S_+(\tau)$ and its correlations are determined by the Langevin-Bloch Eqs. (15.9). We derive the coarse-grained time rate of change of the sidemode density operator $\rho(t)$ by using a derivation that parallels exactly that of Sec. 15-4. Again, we use an equation similar to Eq. (14.21), except that the trace is over the reservoir and the two-level system. The terms involving a_1 and $a_1^\dagger$ alone are the same as those in Sec. 15-4. Similarly the terms involving a_3 and $a_3^\dagger$ alone are given by the corresponding a_1 terms with $_1$ replaced by $_3$. But in addition there are now terms involving the products $a_1 a_3$ and $a_1^\dagger a_3^\dagger$, which conserve energy and couple the two modes. In particular similarly to the derivation of Eq. (15.48), we find that the integrand of the first second-order contribution is

$$\begin{aligned}\langle \mathcal{V}(\tau')\mathcal{V}(\tau'')\rangle &= g^2\langle\, [(a_1 e^{i\Delta\tau'} + a_3 e^{-i\Delta\tau'})S_+(\tau') + \text{adj.}] \\ &\qquad \times [(a_1 e^{i\Delta\tau''} + a_3 e^{-i\Delta\tau''})S_+(\tau'') + \text{adj.}]\rangle \\ &= \hbar^2 g^2 \Big[a_1 a_1^\dagger e^{i\Delta(\tau'-\tau'')}\langle S_+(\tau')S_-(\tau'')\rangle + a_1^\dagger a_1 e^{-i\Delta(\tau'-\tau'')}\langle S_-(\tau')S_+(\tau'')\rangle \\ &\quad + a_1 a_3 e^{i\Delta(\tau'-\tau'')}\langle S_+(\tau')S_+(\tau'')\rangle + a_1^\dagger a_3^\dagger e^{-i\Delta(\tau'-\tau'')}\langle S_-(\tau')S_-(\tau'')\rangle\Big]\,\rho(t) \\ &\quad + (\text{same with } 1 \longleftrightarrow 3)\ ,\end{aligned}$$

where we have used the fact that at the initial time t, the system-reservoir density operator $P_{s\text{-}r}(t)$ factors, i.e., $P_{s\text{-}r}(t) = \rho(t)\rho_r(t)$, and the expectation values involve traces over both the reservoir and the spin-1/2 particles. After transients have died away the spins are described by a stationary process and we can write this expression in terms of time difference $T = \tau' - \tau''$. This gives an integral similar to that in Eq. (14.21)

$$-\frac{1}{\hbar^2\tau}\int_0^\tau d\tau' \int_0^{\tau'} d\tau''\ \langle \mathcal{V}(\tau')\mathcal{V}(\tau'')\rangle$$

$$\begin{aligned}&= -\frac{g^2}{\tau}\int_0^\tau d\tau' \int_0^{\tau'} dT \Big[a_1 a_1^\dagger e^{i\Delta T}\langle S_+(T)S_-(0)\rangle + a_1^\dagger a_1 e^{-i\Delta T}\langle S_-(T)S_+(0)\rangle \\ &\qquad + a_1 a_3 e^{i\Delta T}\langle S_+(T)S_+(0)\rangle + a_1^\dagger a_3^\dagger e^{-i\Delta T}\langle S_-(T)S_-(0)\rangle\Big]\,\rho(t) \\ &\qquad + (\text{same with } 1 \longleftrightarrow 3) \\ &= -[A_1^* a_1 a_1^\dagger + B_1 a_1^\dagger a_1 - C_1 a_1^\dagger a_3^\dagger - D_1^* a_1 a_3]\rho(t) \\ &\qquad + (\text{same with } 1 \longleftrightarrow 3)\ ,\end{aligned} \tag{27}$$

where A_1^* and B_1 are given by Eq. (15.49) and (15.50), respectively, and the new three-mode mixing coefficients C_1 and D_1^* are given by

$$C_1 = -g^2 \int_0^\infty dT\, e^{-i\Delta T} \langle S_-(T) S_-(0)\rangle = -g^2 \mathscr{S}_{--}(s^*)\,, \tag{28}$$

$$D_1^* = -g^2 \int_0^\infty dT\, e^{i\Delta T} \langle S_+(T) S_+(0)\rangle = -g^2 \mathscr{S}_{++}(s)\,, \tag{29}$$

and the Laplace transforms $\mathscr{S}_{ij}(s)$ are defined by Eq. (15.37). For the two-level medium, these transforms can be evaluated using the Bloch Langevin Eqs. (15.12) with the quantum regression theorem as in Sec. 15-3. Problem 16-4 shows that

$$C_1 = -\frac{g^2 \mathscr{D}_1}{1 + I_2 \mathscr{L}_2} \frac{I_2 \mathscr{F} \gamma}{2} \frac{\tfrac{1}{2} I_2 \mathscr{L}_2 \mathscr{D}_3^* - \tfrac{1}{2}\mathscr{D}_2(1 + \Gamma/i\Delta)}{1 + I_2 \mathscr{F} \frac{\gamma}{2}(\mathscr{D}_1 + \mathscr{D}_3^*)}\,, \tag{30}$$

and Prob. 16-5 shows that D_1 is given by

$$D_1 = -\frac{g^2 \mathscr{D}_1}{1 + I_2 \mathscr{L}_2} \frac{I_2 \mathscr{F} \gamma}{2} \frac{(1 + \tfrac{1}{2} I_2 \mathscr{L}_2)\mathscr{D}_3^* + \tfrac{1}{2}\mathscr{D}_2(1 - \Gamma/i\Delta)}{1 + I_2 \mathscr{F} \frac{\gamma}{2}(\mathscr{D}_1 + \mathscr{D}_3^*)}\,. \tag{31}$$

Problem 16-6 shows that the second second-order integral in Eq. (14.21) is

$$\frac{1}{\hbar^2 \tau} \int_0^\tau d\tau' \int_0^{\tau'} d\tau'' \langle \mathcal{V}(\tau') P_{s-r}(t) \mathcal{V}(\tau'')\rangle = B_1^* a_1 \rho a_1^\dagger + A_1 a_1^\dagger \rho a_1$$

$$- C_1^* a_1 \rho a_3 - D_1 a_1^\dagger \rho a_3^\dagger + (\text{same with } 1 \longleftrightarrow 3)\,. \tag{32}$$

Substituting the second-order contributions of Eqs. (27) and (32) into the generalization of Eq. (14.21) to the present problem and using the adjoint to write terms without complex conjugates, we find the dual-sidemode master equation of Sargent, Holm, and Zubairy (1985)

$$\begin{aligned} \dot\rho = &- A_1(\rho a_1 a_1^\dagger - a_1^\dagger \rho a_1) - B_1(a_1^\dagger a_1 \rho - a_1 \rho a_1^\dagger) \\ &+ D_1(\rho a_3^\dagger a_1^\dagger - a_1^\dagger \rho a_3^\dagger) + C_1(a_1^\dagger a_3^\dagger \rho - a_3^\dagger \rho a_1^\dagger) \\ &+ (\text{same with } 1 \to 3) + \text{adjoint}\,. \end{aligned} \tag{33}$$

Equation (33) yields the correct semiclassical coupled-mode equations for the mode amplitudes $\mathcal{E}_1 = \mathcal{E}_\Omega\langle a_1\rangle$ and $\mathcal{E}_3^* = \mathcal{E}_\Omega\langle a_3^\dagger\rangle$, where $\mathcal{E}_\Omega$ is the electric field "per photon." Specifically, we find

$$\dot{\mathcal{E}}_1 = \mathcal{E}_\Omega\langle a_1\dot{\rho}\rangle = (A_1-B_1-\nu/2Q_1)\mathcal{E}_1 + (C_1-D_1)\mathcal{E}_3^* , \tag{34}$$

$$\dot{\mathcal{E}}_3^* = \mathcal{E}_\Omega\langle a_3^\dagger\dot{\rho}\rangle = (A_3^*-B_3^*-\nu/2Q_3)\mathcal{E}_3^* + (C_3^*-D_3^*)\mathcal{E}_1 , \tag{35}$$

where we include phenomenological cavity loss terms. These equations can be used for detuned ($\nu_2 \neq \omega$) instability studies or in phase conjugation. In semiclassical phase conjugation, C_1-D_1 is the coupling coefficient χ_1, with the value (from Eqs. (28) and (31))

$$\chi_1 = \frac{g^2 2T_1\mathcal{D}_1U_1^*U_3^*\mathcal{V}_2^2}{1+I_2\mathcal{L}_2}\;\frac{\mathcal{F}(\Delta)(\mathcal{D}_2+\mathcal{D}_3^*)}{1+I_2\mathcal{F}\frac{\gamma}{2}(\mathcal{D}_1+\mathcal{D}_3^*)} . \tag{36}$$

This agrees with the semiclassical value of Eq. (9.21) if we assume perfect phase matching.

A somewhat cavalier way to find the Langevin equations corresponding to the master equation (33) is to simply remove the average values brackets from and adding noise operators to Eqs. (34) and (35). This gives

$$\dot{a}_1 = (A_1-B_1-\nu/2Q_1)a_1 + (C_1-D_1)a_3^\dagger + F_1 , \tag{37}$$

$$\dot{a}_3^\dagger = (A_3^*-B_3^*-\nu/2Q_3)a_3^\dagger + (C_3^*-D_3^*)a_1 + F_3^\dagger . \tag{38}$$

More properly, we can obtain this result from the Einstein relations as outlined in Chap. 14. Rigorous stochastic equation techniques are discussed by Gardiner (1980). In the next section these equations are used with the quantum regression theorem to derive the spectrum of squeezing.

Furthermore Eq. (33) yields the coupled photon-number rate equations

$$\frac{d}{dt}\langle n_1\rangle = (A_1-B_1-\nu/Q_1)\langle n_1\rangle + (C_1-D_1)\langle a_1^\dagger a_3^\dagger\rangle + A_1 + \text{c.c.} , \tag{39}$$

$$\frac{d}{dt}\langle a_1^\dagger a_3^\dagger\rangle = (A_1-B_1-\nu/Q_1)\langle a_1^\dagger a_3^\dagger\rangle + (C_1-D_1)\langle n_1\rangle + C_1 + (1 \longleftrightarrow 3) , \tag{40}$$

in extension of Eq. (15.57). The choice ν_2=ω yields A_1=A_3^*, B_1=B_3^*, etc. $\langle a_1^\dagger a_3^\dagger\rangle$ is the quantum version of Lamb's combination tone, responsible for three-mode mode locking [compare with E_1 times Eq. (9.48) of Sargent, Scully, and Lamb (1977)]. To lowest nonzero order in a_2, it results from the four-wave mixing process $a_1^\dagger a_3^\dagger a_2^2$, in which two pump ($\nu_2$) photons are annihilated and both a ν_1 and a ν_3 photon are created.

As the following section shows, the *coupled-mode fluorescence* sum $C_1 + C_3$ is the source term for the generation of squeezed states by quantum multiwave mixing. The C_1 and D_1 coefficients originate from correlations the complex dipole develops with itself in the course of time due to multiwave mixing phenomena. Specifically, the C_1 correlation occurs because of the absorption of two pump (ν_2) photons and the emission of a ν_1 photon and a ν_3 photon. The D_1 correlation occurs because of the absorption of a ν_1 photon and a ν_3 photon and the emission of two ν_2 photons. Similarly to the semiclassical absorption coefficient, the semiclassical coupling coefficient is the difference between the absorption and emission of sidemode photons, while the coupled-mode fluorescence ($C_1 + C_3$) results only from the emission of sidemode photons. $C_1 + C_3$ gives a frequency-domain measure of the amplitude correlations developed between vacuum modes symmetrically placed on either side of the pump frequency. Section 15-6 discusses related intensity correlations in the time domain.

16-3. Two-Mode Squeezing

We consider the generation of squeezed light via four-wave mixing in a cavity. Figure 16-3 depicts a simplified diagram of the experimental configuration of Slusher *et al.* (1985). The pump laser at frequency ν_2 is reflected off a mirror to form a standing wave through which the atomic

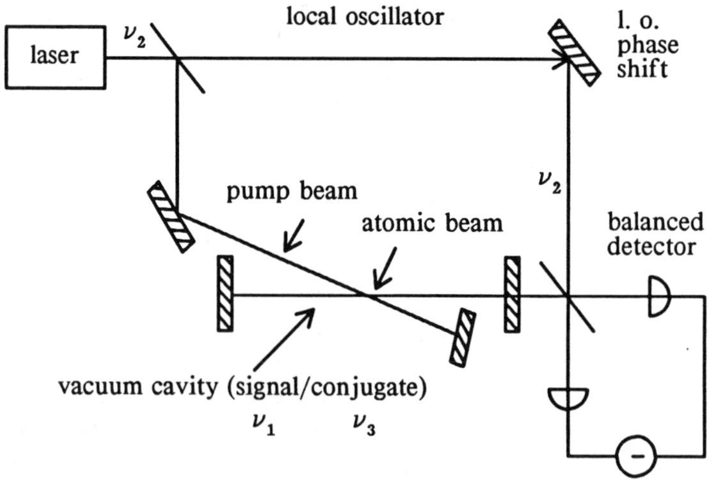

Fig. 16-3. Cavity four-wave mixing experiment used to observe squeezed states.

beam is passed. The vacuum cavity is servo-locked to the sidemode frequencies ν_1 and ν_3 and encloses the interaction region. Quantum noise builds up in the cavity that, due to the four-wave mixing process, develops correlations between the sidemodes. Squeezing is observed in a linear combination of the sideband amplitudes a_1 and a_3. To detect the squeezing, a balanced homodyne detection scheme is used wherein the sidemode fields exiting the cavity are mixed with a local oscillator whose phase is at an angle θ with respect to the pump field ν_2. This homodyne detection permits the direct measurement of the variance for any relative phase shift θ.

The total complex amplitude operator of the squeezed field is given by

$$d = 2^{-1/2}(a_1 e^{i\phi} + a_3^\dagger e^{-i\phi}) \,. \tag{41}$$

From this operator we define two Hermitian operators d_1 and d_2 as

$$d_1 = \frac{1}{2}(d + d^\dagger) \text{ and } d_2 = \frac{1}{2i}(d - d^\dagger) \,, \tag{42}$$

with

$$[d_1, d_2] = i/2 \,, \tag{43}$$

where we have made use of the commutation relations $[a_1, a_1^\dagger] = [a_3, a_3^\dagger] = 1$. With Eqs. (41) and (42) and noting that $\langle a_1 \rangle = \langle a_3 \rangle = \langle a_1^\dagger a_3 \rangle = 0$ if the sidebands arise from the vacuum [see Eqs. (15) and (16)], we find the variance $\Delta d_1^2 = \langle d_1^2 \rangle - \langle d_1 \rangle^2$ as

$$\Delta d_1^2 = \frac{1}{4} + \frac{1}{4}[\langle a_1^\dagger a_1 \rangle + \langle a_3^\dagger a_3 \rangle + (\langle a_1 a_3 \rangle e^{2i\phi} + \text{c.c.})] \,, \tag{44}$$

With the commutator (43), the Heisenberg uncertainty relation gives $\Delta d_1 \Delta d_2 \geq 1/4$ so that squeezing occurs whenever Δd_1^2 drops below 1/4. The expectation values $\langle a_1^\dagger a_1 \rangle$ and $\langle a_3^\dagger a_3 \rangle$ are the average number of photons in modes 1 and 3 respectively, and consequently are never negative. Thus for squeezing to occur the quantity in parentheses in Eq. (44) must be negative. Maximum squeezing occurs when $\langle a_1 a_3 \rangle e^{2i\phi} = -|\langle a_1 a_3 \rangle|$, that is

$$\boxed{\Delta d_1^2 = \frac{1}{4} + \frac{1}{4}[\langle a_1^\dagger a_1 \rangle + \langle a_3^\dagger a_3 \rangle - 2|\langle a_1 a_3 \rangle|]} \,. \tag{45}$$

This equation shows that squeezing occurs if $2|\langle a_1 a_3 \rangle| > \langle a_1^\dagger a_1 \rangle + \langle a_3^\dagger a_3 \rangle$. Physically this means that the sideband fields are more correlated with each other than with themselves, and thus we seek a range of parameters to enhance this coupling. Equation (45) is a nondegenerate version of Eq. (6).

The operators in the correlation functions of Eq. (45) are evaluated at a single time $t = 0$ long enough for the system to have reached steady state. Hence $\langle a_1{}^\dagger a_1\rangle$ is the steady-state mean number of photons in mode 1, etc. The identity

$$\langle a_i{}^\dagger(0)a_j(0)\rangle = \int d\omega \int d\tau\, e^{-i\omega\tau}\, \langle a_i{}^\dagger(\tau)a_j(0)\rangle \tag{46}$$

shows that one-time correlation functions can be interpreted as the sum over frequency of spectral components $\int \exp(-i\omega\tau)\ \langle a_i{}^\dagger(\tau)a_j(0)\rangle$. This suggests that a narrow-band detector might measure peak values larger than these average values (45). In a cavity, the *squeezing spectrum* is obtained by spectral analysis of $\Delta d_1{}^2$ at the optimum output frequency detuning $\delta = \nu_3 - \nu_2$, i.e. at the detuning that minimizes $\Delta d_1{}^2$. To predict this value, we Fourier transform the cavity fields to find their fluctuation spectra to get

$$\Delta d_1{}^2(\delta) = \frac{1}{4} + \frac{1}{4}(\mathcal{S}_{12} + \mathcal{S}_{34} - 2|\mathcal{S}_{13}|)\ . \tag{47}$$

If the narrow-band detector is outside the cavity, it is necessary to calculate the corresponding spectral density outside the cavity. This inside-outside transfer problem has been studied in detail by Collett and Gardiner (1984). They find that to relate the spectra inside and outside the cavity we must multiply $\mathcal{S}_{ij}$ by the density of states $\mathcal{D}(\Omega_2)$ outside the cavity. Furthermore, the amount passed is proportional to the square of the coupling constant g between a field mode inside the cavity and those outside. Hence we expect that the spectral densities outside the cavity are given by

$$\mathcal{S}_{out}(\delta) = 2\pi|g|^2\, \mathcal{D}(\Omega_2)\, \mathcal{S}(\delta)\ . \tag{48}$$

As shown in Sec. 14-1, the damping constant of a simple harmonic oscillator coupled to a bath of simple harmonic oscillators with density of states $\mathcal{D}(\Omega_2)$ is given by $2\pi|g|^2\mathcal{D}(\Omega_2)$. In the present case, this damping constant is the cavity linewidth ν/Q, so that

$$\mathcal{S}_{out}(\delta) = \frac{\nu}{Q}\,\mathcal{S}(\delta)\ . \tag{49}$$

With Eq. (47) this gives the maximum-squeezing variance

$$\Delta d_1{}^2(\delta)\Big|_{out} = \frac{1}{4} + \frac{1}{4}\frac{\nu}{Q}(\mathcal{S}_{12} + \mathcal{S}_{34} - 2|\mathcal{S}_{13}|)\ , \tag{50}$$

where the sidemode Fourier transforms $\mathcal{S}_{ij}$ are the matrix elements of Eq. (14.102) in the basis $\boldsymbol{\alpha}^T = (a_1, a_1^\dagger, a_3, a_3^\dagger)$. These transforms are given by the general Eq. (14.108), in which the Langevin matrix

$$\mathbf{A} = \begin{bmatrix} \alpha_1 & 0 & 0 & -\chi_1 \\ 0 & \alpha_1^* & -\chi_1^* & 0 \\ 0 & -\chi_3^* & \alpha_3 & 0 \\ -\chi_3 & 0 & 0 & \alpha_3^* \end{bmatrix} . \tag{51}$$

and the diffusion matrix

$$2\mathbf{D} = \begin{bmatrix} 0 & A_1+A_1^* & C_1+C_3 & 0 \\ A_1+A_1^* & 0 & 0 & C_1^*+C_3^* \\ C_1+C_3 & 0 & 0 & A_3+A_3^* \\ 0 & C_1^*+C_3^* & A_3+A_3^* & 0 \end{bmatrix} . \tag{52}$$

Carrying out the inversions and matrix products in Eq. (14.108), Reid and Walls (1986) and Holm and Sargent (1987) found

$$D\mathcal{S}_{12} = (\alpha_3-i\delta)(\alpha_3^*+i\delta)A_1 + |\chi_1|^2A_3 + (\alpha_3^*+i\delta)\chi_1^*(C_1+C_3) + \text{c.c.} \,, \tag{53}$$

$$\begin{aligned} D\mathcal{S}_{13} = {} & (\alpha_3^*+i\delta)\chi_3(A_1+A_1^*) + (\alpha_1^*-i\delta)\chi_1(A_3+A_3^*) \\ & + (\alpha_1^*-i\delta)(\alpha_3^*+i\delta)(C_1+C_3) + \chi_1\chi_3(C_1+C_3) \,, \end{aligned} \tag{54}$$

$$\mathcal{S}_{34} = \mathcal{S}_{12} \text{ (same with }_1\text{ and }_3\text{ interchanged)} \,, \tag{55}$$

where $D = |(\alpha_1+i\delta)(\alpha_3^*+i\delta) - \chi_1\chi_3^*|^2$. Note that these expressions depend on the specific mixing medium only through the coefficients A_n, B_n C_n, and D_n, which are, in turn, given by Laplace transforms of two-time spin correlation functions. As such they are quite general and apply for example to two-level atoms, semiconductors, and both of these in a squeezed vacuum (see Sec. 16-4). Figure 16-4 shows an example of the minimum variances predicted by Eq. (50) for two-level media. Substantially larger squeezing can be obtained with an optical parametric amplifier [see Wu *et al.* (1986)]. The squeezing observed in Fig. 16-4 results from 1) choosing the pump-cavity detuning δ to minimize the denominator D of Eqs. (53) - (55), and 2) detuning the pump from the atomic resonance sufficiently to cause the coefficients A_n and C_n to be nearly imaginary. This causes the resonance fluorescence given by $\text{Re}\{A_n\}$ to nearly vanish, while the magnitude of the coupled-mode fluorescence $C_1 + C_3$ remains substantial, causing the latter to be primarily responsible for the squeezing.

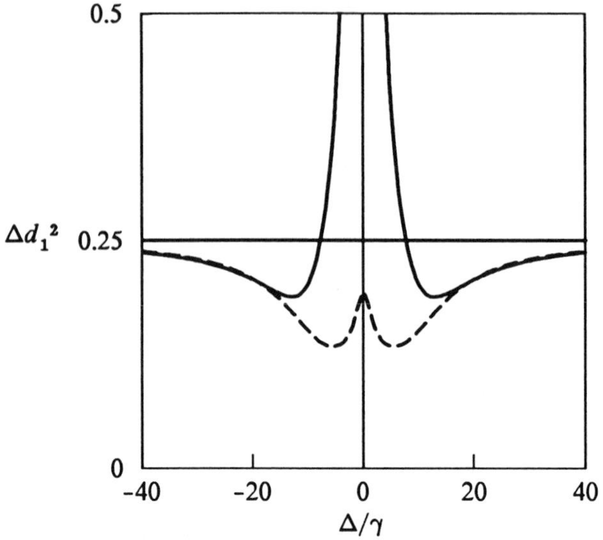

Fig. 16-4. Squeezed variance Δd_1^2 *vs* Δ/γ for $\Gamma = 2\gamma$, $I_2 = 40$, $\delta_2 = 2, 12$, and C (cooperativity) = 10.

16-4. Squeezed Vacuum

This section develops the master equation for a two-level system in a "squeezed vacuum" reservoir. As noted by Gardiner (1986), such a reservoir leads to two dipole decay constants (T_2's), a small T_2 for the dipole component in the squeezed quadrature and a correspondingly larger T_2 for the dipole component in the stretched quadrature. The derivation of the master equation proceeds as in Sec. 14-1, except that the aa and $a^\dagger a^\dagger$ terms in Eq. (14.24) must now be retained: The master Eq. (14.31) neglects these terms under the assumption that the noise-operator reservoir average $\langle F(\tau')F(\tau'')\rangle_r$ and its complex conjugate vanish. This in turn results from the assumption that $\langle b_k b_{k'}\rangle_r$ and $\langle b^\dagger{}_k b^\dagger{}_{k'}\rangle_r$ vanish, where b_k is the reservoir annihilation operator at the frequency ω_k. But if multiwave mixing is used to squeeze the vacuum, that is, to correlate conjugate pairs b_k and b_{-k}, where ω_k and ω_{-k} are placed symmetrically about the system operator frequency Ω, then just as $\langle a_1 a_3\rangle$ is nonzero in Secs. 16-2 and 16-3, the average $\langle b_k b_{-k}\rangle_r$ may be nonzero for a range of ω_k's about the system frequency Ω. Accordingly we set

$$\langle b_k b_{k'}\rangle_r = \langle b_k b_{-k}\rangle_r \, \delta_{k',-k},$$

and use Eq. (14.23) and the logic leading to Eq. (14.27) to find

$$\int_0^\tau d\tau' \int_0^{\tau'} d\tau'' \langle F(\tau')F(\tau'')\rangle_r$$

$$= -\int_0^\tau d\tau' \int_0^{\tau'} d\tau'' \sum_{kk'} g_k g_{k'} \langle b_k b_{k'}\rangle_r \, e^{i(\Omega-\omega_k)\tau'} e^{i(\Omega-\omega_{k'})\tau''}$$

$$= -\int_0^\tau d\tau' \sum_k g_k g_{-k} \langle b_k b_{-k}\rangle_r \int_0^{\tau'} dT e^{i(\Omega-\omega_k)T} = -\tfrac{1}{2}\gamma \bar{m}\, \tau \,, \tag{58}$$

where γ is the Weisskopf-Wigner decay constant of Eq. (14.28) and $\bar{m}$ is the complex squeezing number

$$\bar{m} = \langle b(\Omega)^2\rangle_r \frac{g(\Omega)^2}{|g(\Omega)|^2} \,. \tag{59}$$

The $g(\Omega)^2$ factor introduces spatial variations typical in multiwave mixing and causes significant phase mismatch for vacuum modes not parallel to a running-wave pump mode. In principle a standing-wave pump mode can squeeze vacuum modes for all directions, since as discussed in Secs. 2-4 and 9-2 such four-wave mixing is nearly perfectly phase matched.

The magnitude of the complex squeezing number $\bar{m}$ is limited by

$$|\bar{m}|^2 \le \bar{n}(\bar{n} + 1) \,. \tag{60}$$

To see this, we note that for any real r and θ

$$0 \le \langle (re^{i\theta}b + b^\dagger)^\dagger (re^{i\theta}b + b^\dagger)\rangle = r^2\langle b^\dagger b\rangle + \langle bb^\dagger\rangle + r[e^{i\theta}\langle b^2\rangle + \text{c.c.}]$$

$$= r^2\bar{n} + \bar{n} + 1 + r[e^{i\theta}\bar{m} + \text{c.c.}] \,. \tag{61}$$

Similarly to the derivation of Eq. (6), the minimum of the RHS with respect to θ is given by $e^{i\theta}\bar{m} = -|\bar{m}|$, which yields

$$0 \le r^2\bar{n} + \bar{n} + 1 - 2r|\bar{m}| \,.$$

The minimum of this RHS with respect to r gives $r = |\bar{m}|/\bar{n}$, which, in turn, gives Eq. (60). This inequality implies that to squeeze the vacuum, we have to increase $\bar{n}$, i.e., the energy in the reservoir. The limit of infinite squeezing requires the deposition of an infinite amount of energy in the bath. From this point of view, the expression "squeezed vacuum" is certainly somewhat of a misnomer.

Inserting Eq. (58) and its complex conjugate into Eq. (14.24), we find the squeezed-reservoir master equation

$$\boxed{\begin{aligned}\dot{\rho}(t) = &- \frac{\gamma}{2}(\bar{n}+1)\,[a^\dagger a\rho(t) - a\rho(t)a^\dagger] - \frac{\gamma}{2}\bar{n}\,[\rho(t)aa^\dagger - a^\dagger\rho(t)a] \\ &+ \frac{\gamma}{2}\bar{m}\,[\rho(t)a^\dagger a^\dagger - a^\dagger\rho(t)a^\dagger] + \frac{\gamma}{2}\bar{m}\,[a^\dagger a^\dagger\rho(t) - a^\dagger\rho(t)a^\dagger] + \text{adj.}\end{aligned}} \quad (62)$$

in generalization of Eq. (14.31). Note that this single-mode master equation has the same general form as the dual-sidemode master Eq. (35), except that both the C_1 and D_1 coefficients are replaced by the squeezed bath parameter $\frac{1}{2}\gamma\bar{m}$.

Atomic Damping by a Squeezed Vacuum

As noted for Eq. (14.39), the two-level master equation is given by the simple harmonic oscillator master equation with the annihilation operator a replaced by the spin-flip operator σ_- and $a^\dagger$ replaced by σ_+. Replacing γ by Γ to describe the upper-to-ground-lower-level model of Chap. 5, we find

$$\begin{aligned}\dot{\rho}(t) = &- \frac{\Gamma}{2}(\bar{n}+1)\,[\sigma_+\sigma_-\rho(t) - \sigma_-\rho(t)\sigma_+] - \frac{\Gamma}{2}\bar{n}\,[\rho(t)\sigma_-\sigma_+ - \sigma_+\rho(t)\sigma_-] \\ &+ \frac{\Gamma}{2}\bar{m}\,[\rho(t)\sigma_+\sigma_+ - \sigma_+\rho(t)\sigma_+] + \frac{\Gamma}{2}\bar{m}\,[\sigma_+\sigma_+\rho(t) - \sigma_+\rho(t)\sigma_+] + \text{adj.}\ ,\end{aligned} \quad (63)$$

which is valid provided the bandwidth of the squeezed vacuum is large compared to Γ. This equation is set in a frame rotating at the atomic frequency ω. Since the spin-flip operators have the same time dependence in this picture, we have $\sigma_+\sigma_+ = \sigma_-\sigma_- = 0$, $\sigma_+\sigma_- = \sigma_a$, and $\sigma_-\sigma_+ = \sigma_b$. This simplifies Eq. (63) to

$$\begin{aligned}\dot{\rho}(t) = &- \frac{\Gamma}{2}(\bar{n}+1)\,[\sigma_a\rho(t) - \sigma_b\rho_{aa}(t)] - \frac{\Gamma}{2}\bar{n}\,[\rho(t)\sigma_b - \sigma_a\rho_{bb}(t)] \\ &- \Gamma\bar{m}\,\sigma_+\rho_{ba}(t) + \text{adj.}\end{aligned} \quad (64)$$

The equation of motion for the dipole matrix element ρ_{ab} is given by $\langle a|\dot{\rho}|b\rangle$, that is, in the absence of other interactions by

$$\dot{\rho}_{ab} = -\Gamma(\bar{n} + \tfrac{1}{2})\rho_{ab} - \Gamma\bar{m}\rho_{ba} \,. \tag{65}$$

The complex conjugate of this equation is

$$\dot{\rho}_{ba} = -\Gamma(\bar{n} + \tfrac{1}{2})\rho_{ba} - \Gamma\bar{m}^*\rho_{ab} \,. \tag{66}$$

Similarly the probability equations of motion are given by

$$\dot{\rho}_{aa} = -\dot{\rho}_{bb} = -\Gamma(\bar{n} + 1)\rho_{aa} + \Gamma\bar{n}\rho_{bb} \,,$$

which gives

$$\dot{\rho}_{aa} - \dot{\rho}_{bb} = -\Gamma(2\bar{n} + 1)(\rho_{aa} - \rho_{bb}) - \Gamma \,. \tag{67}$$

In the limit of a zero temperature reservoir ($\bar{n} = \bar{m} = 0$, no squeezing), these equations give standard pure radiative decay ($\gamma = 1/T_2 = \frac{1}{2}\Gamma$). However for a nonzero-temperature squeezed reservoir, the dipole decay constant splits into two constants, one each for the quadratures of minimum and maximum variance. To see this, we write the corresponding Bloch-vector components as

$$U = \rho_{ab}e^{i\phi} + \text{c.c.}, \tag{68}$$

$$V = i\rho_{ab}e^{i\phi} + \text{c.c.}, \tag{69}$$

where $\bar{m} = |\bar{m}|e^{-2i\phi}$ and the difference component W is given as usual by $\rho_{aa} - \rho_{bb}$. Using Eqs. (65) and (66), we find

$$\dot{U} = -\Gamma(\bar{n} + |\bar{m}| + \tfrac{1}{2})U = -\gamma_u U \,, \tag{70}$$

$$\dot{V} = -\Gamma(\bar{n} - |\bar{m}| + \tfrac{1}{2})V = -\gamma_v V \,. \tag{71}$$

According to Eq. (60), the maximum squeezing is given by

$$|\bar{m}| = \sqrt{\bar{n}(\bar{n} + 1)} \simeq \bar{n} + \tfrac{1}{2} - 1/8\bar{n} \,. \tag{72}$$

Hence for sufficiently large $\bar{n}$, the decay constant γ_v can be squeezed arbitrarily close to zero. This gives a T_2 that is arbitrarily large compared to the corresponding T_1, which violates the inequality $T_2 \le 2T_1$ valid for unsqueezed reservoirs. While γ_v is squeezed to a small value, γ_u and γ_w both become large. Figure 16-5 illustrates how pump/probe spectra in a squeezed bath can be much sharper than in an unsqueezed (normal) vacuum. Note that since the UV phase is chosen here by the phase of the squeezed variance, the full Bloch equations contain real and imaginary parts of the electric-dipole interaction energy. In general semiclassical calculations involving such squeezed vacua are more easily carried out using ρ_{ab} rather than U and V [see, for example, An and Sargent (1989)].

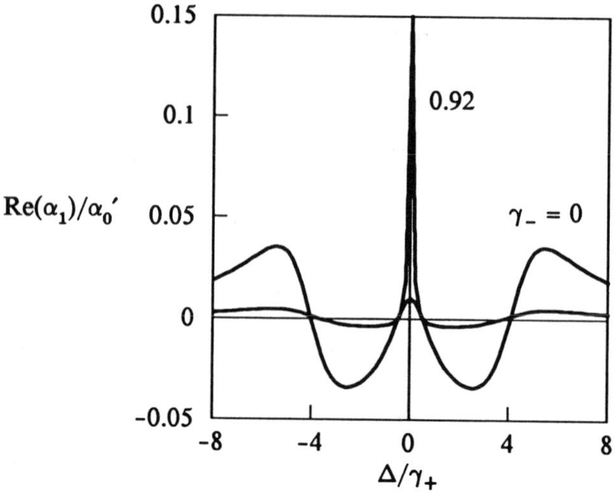

Fig. 16-5. Pump probe spectra in an unsqueezed vacuum from Eq. (8.17) and in a squeezed vacuum with $\gamma_- = .92\gamma_+$.

References

An, S. and M. Sargent III (1989), Phys. Rev. A**40**, 7039.

Cohen-Tannoudji, C., J. Dupont-Roc, and G. Grynberg (1988), *Processus d'interaction entre photons et atomes*, InterEditions et Editions du CNRS, Paris give a clear tutorial discussion of squeezing. English edition to be published by John Wiley, New York.

Collett M. J. and C. W. Gardiner (1984), Phys. Rev. A**30**.

Gardiner C. (1980), *Handbook of Stochastic Methods*, Springer, Berlin.

Gardiner, C. W. (1986), Phys. Rev. Lett. **56**, 1917 gives the original derivation of the master equation describing an atom in a squeezed bath.

Holm, D. A. and M. Sargent III (1987), A**35**, 2150.

Meystre, P. and M. O. Scully (1983), Eds. *Quantum Optics, Experimental Gravitation and Measurement Theory*, Plenum, New York. These proceedings contain reviews on the potential applications of squeezed states in gravitational wave detection and "quantum nondemolition" measurements.

Reid, M. D. and D. F. Walls (1986), Phys. Rev. **A34** 4929.

Sargent, M. III, D. A. Holm, M. S. Zubairy (1985), Phys. Rev. **A31**, 3112.

Slusher, R. E., L. W. Hollberg, B. Yurke, J. C. Mertz, J. F. Valley (1985), Phys. Rev. Lett. **55**, 2409 reports the first observation of squeezed light.

Walls, D. F. (1983), Nature **306**, 141 is a tutorial review of squeezed states.

Wu, L. A., H. J. Kimble, J. L. Hall, and H. Wu (1986), Phys. Rev. Lett. **57**, 2520.

Problems

16-1. Calculate the variance of d_1 of Eq. (3) in a) a number state, *b*) a coherent state. Calculate the variance of the number operator in *c*) a number state, *d*) a coherent state.

16-2. Calculate $|\langle\alpha|S^\dagger(\varsigma)ae^{i\phi}S(\varsigma)|\alpha\rangle|^2$ using Eq. (13). Show that your result exceeds the unsqueezed value of $\alpha\alpha^*$.

16-3. Calculate the expectation value of the photon number operator $a^\dagger a$ in the squeezed state $S(\varsigma)|\alpha\rangle$. Note that the photon number increases with increased squeezing and that the squeezed vacuum has a nonzero photon number.

16-4. Using Cramer's rule, solve Eq. (15.40) for $\mathcal{S}_{--}(s)$. Ans: $-C_1/g^2$, where C_1 is given by Eq. (30).

16-5. Solve the matrix equation of Prob. 15-2 for $\mathcal{S}_{++}(s)$. Ans: $-D_1^*/g^2$, where D_1 is given by Eq. (31).

16-6. Show that the second second-order integral in Eq. (14.21) is given by Eq. (32). Hint: the derivation is an extension of that for Eq. (15.53).

16-7. Show using the generalized Einstein relation (14.36) and the quantum coupled-mode equations (39) and (40) that the multiwave-mixing diffusion matrix is given by Eq. (52).

16-8. Using the master Eq. (62), calculate the equations of motion for $\langle a\rangle$ and $\langle a^\dagger a\rangle$ in a squeezed vacuum.

16-9. Write the Bloch equations for a squeezed vacuum including the contributions for the electric-dipole interaction energy. Hint: add appropriate terms to Eqs. (65) and (66), and use Eqs. (68) and (69).

16-10. Write the Langevin Bloch equations in a squeezed vacuum corresponding to Eqs. (15.9). Ans: [here $\gamma_\pm = \frac{1}{2}(\gamma_u \pm \gamma_v)$]

$$\dot{S}_+ = -(\gamma_+ - i\delta)S_+ - \gamma_- S_- - 2i\mathcal{V}_2^* S_z + F_+(t) \,, \tag{73a}$$

$$\dot{S}_z = -\Gamma(S_z + \tfrac{1}{2}) + i\mathcal{V}_2^* S_- - i\mathcal{V}_2 S_+ + F_z(t) \,, \tag{73b}$$

$$\dot{S}_- = -(\gamma_+ + i\delta)S_- - \gamma_- S_+ + 2i\mathcal{V}_2 S_z + F_-(t) \,, \tag{73c}$$

16-11. Applying the two-time spin correlation method of Secs. 15-3, 15-4, and 16-2 to Eq. (73), solve for the multiwave mixing coefficients A_1, B_1, C_1, and D_1 in a squeezed vacuum. Ans: see An and Sargent (1989).

16-12. Describe the physics illustrated by the following figure (I_2 is varied from 1 to 30, and $\Gamma = \gamma = 1$, i.e, *not* pure radiative decay):

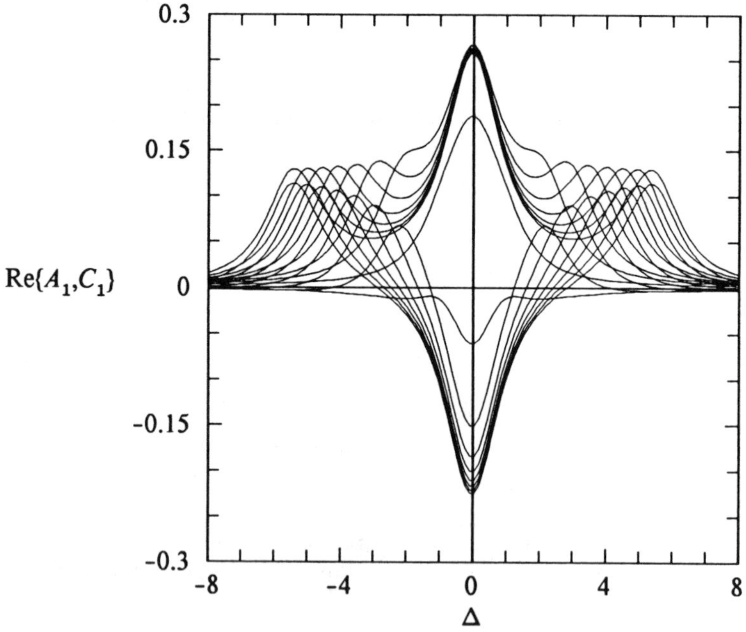

Chapter 17
QUANTUM THEORY OF A LASER

This is the last chapter of this book, and the third that applies the quantum theory of the interaction of radiation with matter developed in Chaps. 12 - 14. So far, our treatment of quantized field-matter interactions has concentrated mostly on situations where the quantized field acts as a noise source: under these conditions the nonlinear atomic response originates from a *classical* driving field, while the effects of the quantized modes are treated linearly. A notable exception was the Jaynes-Cummings model, where the Hamiltonian can be diagonalized exactly and hence the quantized mode is treated to all orders.

In general, it is an extremely complex task to handle fully quantized atom-field problems to all orders. This is because the Heisenberg operator equations of motion lead to an infinite hierarchy of differential equations for the moments of the operators, and it is usually not clear how to truncate such hierarchies. Oftentimes, a useful strategy is to handle the build-up of fields from noise quantum mechanically until a time when the excitation becomes macroscopic and one can somehow make a transition to a classical description. Such strategies have been useful, for example, in the theory of superfluorescence, the collective emission of radiation from an ensemble of atoms [see Haroche and Raimond (1985)].

This chapter treats the quantum theory of the laser, a device where the quantum field can become strong enough to saturate the atomic medium and linearization procedures are not sufficient. The semiclassical theory of a laser (Chap. 6) predicts that if a laser mode has zero amplitude, it will continue to have zero amplitude. Some kind of noise must be added to start the modes oscillating. Physically the source of such noise is spontaneous emission. Spontaneous emission also ultimately prevents the width of the laser line from vanishing. As compared to the semiclassical theory, the quantum theory of the laser permits to predict numerous new properties such as its photon statistics, its linewidth, and the build-up from spontaneous emission of one and two modes. We begin in Sect. 17-1 by presenting the theory of a micromaser. While resonance fluorescence and squeezing involve *linear* quantum mechanical fluctuations, which do not depend on the photon number explicitly, the micromaser problems can involve nonlinear quantum effects involving what might be called the granular, i.e., n-dependent, nature of the field. Unfortunately only relatively simple cases can be treated analytically and some of the most interesting effects require substantial numerical computation.

The single mode laser problem has much in common with the micromaser, and we show how that theory can be manipulated to describe these more conventional lasers in Sect. 17-2. The laser involves an interaction of quantized field modes with atoms and various media as depicted in Fig.

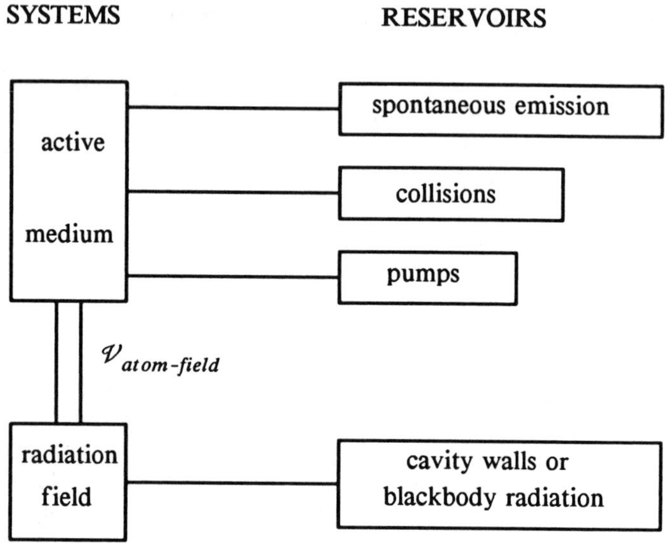

Fig. 17-1. Block diagram showing atom and field systems interacting with their reservoirs.

17-1. Hence we use an extension of the system-reservoir formalism of Chap. 14 to derive a master equation for the laser mode density matrix.

We describe the steady-state laser photon statistics using variations of the Scully-Lamb (1967) theory. The statistics are similar to those for the micromaser subject to Poisson injection times, and comparing these two situations help shed light on the origin of properties of laser light. The laser linewidth is derived by applying the quantum regression theorem to the laser field Langevin equation.

The laser linewidth is discussed in Sect. 17-3. It is due almost completely to phase fluctuations, which unlike the amplitude fluctuations are not restrained by a steady-state operating point. However by changing the saturation, amplitude fluctuations can induce index changes that, in turn, yield phase shifts that contribute significantly to the linewidth.

The onset of multimode operation discussed semiclassically in Chap. 10 is given by the answer to the question, can more than one mode oscillate? This leads to sidemode equations of motion, which also describe resonance fluorescence, that is, the emission spectrum of an atom in the presence of a strong single-mode field. Section 17-4 outlines a quantized sidemode theory based on the quantum Langevin method of Secs. 15-4 and 16-2, here including a pump term to obtain gain. The sidemodes are assumed to be sufficiently weak that they cannot, by themselves, saturate the response of the medium.

To simplify the discussion, we assume in this chapter that the lower laser state is the ground state of the atom, such as in a ruby laser. This is also the level configuration typically used in resonance fluorescence. The corresponding single-mode atom-field levels are depicted in Fig. 17-2. A more general derivation including the present scheme as well as systems with two excited states has been given by Sargent, Holm, and Zubairy (1985). Their approach takes place in the Schrödinger picture, revealing the interplay of atom-field density matrix elements.

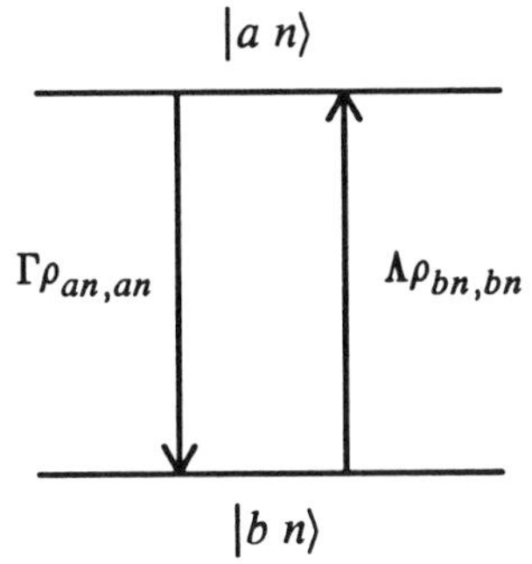

Fig. 17-2. Atom-field level scheme for a quantum theory of the laser. The upper atomic level decays to a ground-state lower level with decay constant Γ. To obtain laser action, a pump process is included to pump ground state atoms to the upper laser level. This model is appropriate for lasers like the ruby laser.

17-1. The Micromaser

A number of interesting atom-field interactions can be studied with very highly excited atoms known as Rydberg atoms. These hydrogen-like atoms have special properties such as long wavelength and large dipole moments as discussed by Gallas *et al.* (1985) that allow the experimental investigation of such simple quantum mechanical models as the Jaynes-Cummings problem of Secs. 13-1 and 13-2, and the micromaser discussed in this section.

We consider a single-mode resonator into which excited two-level atoms are injected at a rate low enough that at most one atom at a time is inside the resonator. There, they interact with the cavity mode, which is taken to be in the microwave region, hence the name maser. The low density beam is required to avoid collective atomic radiative effects such as superradiance. In addition, we assume that the atom-field interaction time τ is much shorter than the cavity damping time Q/ν, so that the relaxation of the resonator field mode can be ignored while an atom is inside the cavity. The strategy to describe the maser is then straightforward (Filipowicz *et al*, 1986): while an atom flies through the cavity, the coupled field-atom system is described by the Jaynes-Cummings Hamiltonian, and during the intervals between successive atoms the evolution of the field is governed by the master equation of a harmonic oscillator damped by a thermal bath, as discussed in Section 14-1.

The Jaynes-Cummings Hamiltonian is given by Eq. (13.10) with Ω replaced by ν to match the laser notational convention,

$$\mathcal{H} = \tfrac{1}{2}\hbar\omega\sigma_z + \hbar\nu a^\dagger a + \hbar(g\sigma_+ a + \sigma_- a^\dagger) \ . \tag{1}$$

We have seen in Sec. 13-2 that this model is exactly solvable, and the effect of the time evolution operator $U(t) = \exp(-i\mathcal{H}t/\hbar)$ on an arbitrary initial state is known.

At time t_i, the ith atom enters the cavity containing the field described by the reduced density operator $\rho(t_i)$. At this time, the density operator $\rho_{a\text{-}f}(t_i)$ of the combined atom-field system is simply the outer product of $\rho(t_i)$ and the initial atomic density operator $\rho_a(t_i)$. After the interaction time τ the atom exits the resonator, and leaves the field in the state described by the reduced density operator return map

$$\rho(t_i+\tau) = \mathrm{tr}_{atom}\{U(\tau)\rho_{a\text{-}f}(t_i)U^{-1}(\tau)\} \equiv F(\tau)\rho(t_i) \ , \tag{2}$$

where tr_{atom} stands for the trace over the atomic variables. This equation defines the operator $F(\tau)$, which we use to simplify the notation.

In the interval between $t_i + \tau$ and the time t_{i+1} at which the next atom is injected, the field evolves at rate ν/Q toward a thermal steady state with a temperature-dependent mean photon number $\bar{n}$, its evolution given by the master equation of a damped harmonic oscillator given by Eq. (14.31) with γ replaced by ν/Q

$$\dot{\rho} \equiv L\rho = -\frac{\nu}{2Q}(\bar{n}+1)[a^\dagger a\rho(t) - a\rho(t)a^\dagger] - \frac{\nu}{2Q}\bar{n}[\rho(t)aa^\dagger - a^\dagger\rho(t)a] + \text{adj} . \quad (3)$$

Here the evolution operator L is called the Liouvillian operator. At the time t_{i+1} the field density matrix is given by

$$\rho(t_{i+1}) = \exp(Lt_p)F(\tau)\rho(t_i) , \quad (4)$$

where $t_p = t_{i+1} - t_i - \tau \simeq t_{i+1} - t_i$ is the time interval between atom i leaving the resonator and atom i+1 entering it. The problem remains to find how the L and F operators change the field density operator $\rho(t_i)$.

Suppose that the field density matrix is initially diagonal in the energy representation and that an atom in the upper level is injected inside the resonator. The reduced density operator of the field then remains diagonal in time, that is, $\langle n|\rho|n'\rangle = p_n\delta_{n,n'}$ and we can restrict our considerations to the diagonal elements p_n. The initial atom-field density operator is

$$\rho_{a\text{-}f}(t_i) = |a\rangle\langle a| \otimes \sum_n p_n(t_i)\,|n\rangle\langle n| . \quad (5)$$

In the course of time the Hamiltonian couples $|an\rangle$ and $|b\; n+1\rangle$. Tracing over the atomic states, we find the field reduced density operator

$$\rho(t_i+\tau) = \sum_n p_n(t_i)\,[|C_{an}(\tau)|^2|n\rangle\langle n| + |C_{bn+1}(\tau)|^2|n+1\rangle\langle n+1|] . \quad (6)$$

Identifying the diagonal element p_n, we find

$$p_n(t_i+\tau) = [1 - \mathcal{C}_{n+1}(\tau)]p_n(t_i) + \mathcal{C}_n(\tau)\,p_{n-1}(t_i) , \quad (7)$$

where using Eq. (13.26), we find

$$\mathcal{C}_n(\tau) = \frac{4ng^2}{(\omega-\nu)^2 + 4ng^2}\sin^2[\tfrac{1}{2}\sqrt{(\omega-\nu)^2 + 4ng^2}\,\tau] . \quad (8)$$

Similarly for atoms injected in the lower level, we have

$$p_n(t_i+\tau) = [1 - \mathcal{C}_n(\tau)]p_n(t_i) + \mathcal{C}_{n+1}(\tau)\,p_{n+1}(t_i) . \quad (9)$$

The diagonality of the field is preserved during its decay, so that the master equation (3) can be restricted to its diagonal elements as for (14.35)

$$\dot{p}_n = -\frac{\nu}{Q}(\bar{n} + 1)[np_n - (n+1)p_{n+1}] - \frac{\nu}{Q}\bar{n}\,[(n+1)p_n - np_{n-1}]\,. \tag{10}$$

Under these conditions, successive iterations of Eqs. (7) or (9) eventually yield a diagonal steady-state field density matrix ρ_{st} which is the solution of this equation with $\rho(t_{i+1}) = \rho(t_i)$. Note that this is not a "true" steady-state, but rather a steady state of the quantum mechanical return maps (7) and (9). Physically it corresponds to a situation where the same field repeats at the precise instants when successive atoms exit the cavity.

For simplicity we assume that the atoms enter the cavity according to a Poisson process with mean spacing $1/R$ between events, where R is the atomic flux. We want to average over the random times t_p between events. Since $\rho(t_i)$ in Eq. (4) depends only on earlier time intervals, it is statistically independent of the current $\exp(Lt_p)$, and we can factor the average of Eq. (4) as

$$\begin{aligned}\langle\rho(t_{i+1})\rangle &= \langle\exp(Lt_p)\rangle F(\tau)\langle\rho(t_i)\rangle \\ &= R\int_0^\infty dt_p \exp[-(R - L)t_p]F(\tau)\langle\rho(t_i)\rangle \\ &= \frac{R}{R - L}F(\tau)\langle\rho(t_i)\rangle\,. \end{aligned}\tag{11}$$

Here we have averaged the damping operator $\exp(Lt_p)$ over an exponential distribution of the intervals between atoms with the average rate of injection R.

With the injection of a succession of atoms inside the cavity, the stochastic average of the field density matrix evolves toward a steady state $\bar{\rho}$ satisfying the relation

$$R[1 - F(\tau)]\,\bar{\rho} = L\bar{\rho}\,. \tag{12}$$

The matrix element $\langle n|F(\tau)\bar{\rho}|n\rangle$ is given by Eqs. (7) and (9) for atoms injected in the upper and lower levels, respectively. The damping contribution $\langle n|L\bar{\rho}|n\rangle$ is given by Eq. (10). Substituting these contributions into the nth diagonal element of Eq. (12), we find a three-term recursion relation for the occupation numbers $\bar{p}_n \equiv \langle n|\bar{\rho}|n\rangle$ similar to Eq. (14.35) in steady state. We can express this relation in the form

$$S_n = S_{n+1}\,, \tag{13}$$

where

$$S_n \equiv [n\mathcal{A}_n + \bar{n}n\nu/Q]\bar{p}_{n-1} - [n\mathcal{B}_n + (\bar{n}+1)n\nu/Q]\bar{p}_n \ . \tag{14}$$

Here the coefficients $\mathcal{A}_n$ and $\mathcal{B}_n$ are given by

$$\mathcal{A}_n = \frac{4R_a g^2}{(\omega-\nu)^2 + 4ng^2} \sin^2[\tfrac{1}{2}\sqrt{(\omega-\nu)^2 + 4ng^2}\tau] \ , \tag{15}$$

$$\mathcal{B}_n = \frac{4R_b g^2}{(\omega-\nu)^2 + 4ng^2} \sin^2[\tfrac{1}{2}\sqrt{(\omega-\nu)^2 + 4ng^2}\tau] \ , \tag{16}$$

and R_α is the rate at which atoms are injected into level α (= a or b). In steady state S_1 must vanish, which with Eq. (13) implies that $S_n = 0$ for all n. Equation (14) then gives the ratio of successive occupation numbers

$$\bar{p}_n = \frac{\bar{n}\nu/Q + \mathcal{A}_n}{(\bar{n}+1)\nu/Q + \mathcal{B}_n} \bar{p}_{n-1} \ ,$$

that is,

$$\boxed{\bar{p}_n = \bar{p}_0 \prod_{k=1}^{n} \frac{\bar{n}\nu/Q + \mathcal{A}_k}{(\bar{n}+1)\nu/Q + \mathcal{B}_k}} \ . \tag{17}$$

The normalization condition $\Sigma_n \, \bar{p}_n = 1$ determines $\bar{p}_0$. Equation (17) is the central result of this section. The $\sin x/x$ character of $\mathcal{A}_n$ and $\mathcal{B}_n$ causes the micromaser to exhibit a number of features that are absent in the conventional lasers discussed in Sec. 17-2.

Features of the photon statistics

Since the intracavity field always remains diagonal, the photon statistics Eq. (17) contain all information about the statistical properties of the steady-state field reached by the micromaser. Figure 17-3 shows the normalized average photon number

$$\mathcal{n} \equiv \langle n \rangle / N_a = \Sigma_n n\bar{p}_n / N_a \tag{18}$$

in which $N_a = R_a Q/\nu$ as a function of the dimensionless pump parameter Θ

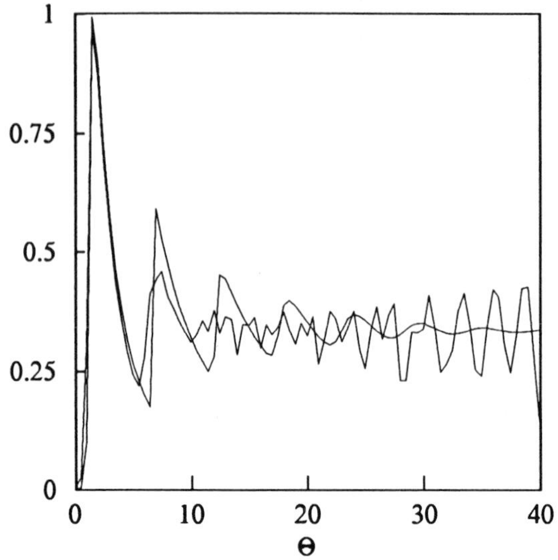

Fig. 17-3. Normalized average number of photons $\mathscr{n}$ from Eq. (18) *vs* the pump parameter Θ of Eq. (19) with N_a = 20. and 200., and N_b = 0. The number of thermal photons is $\bar{n}$ = .1.

$$\Theta = \tfrac{1}{2}\sqrt{N_a}\,g\tau \ . \tag{19}$$

We show later on that Θ plays the role of a pump parameter for the micromaser.

The two curves correspond to N_a = 20 and 200 with N_b = 0, and the number of thermal photons is $\bar{n}$ = 0.1. A common feature to all cases is that $\mathscr{n}$ is nearly zero for small Θ, but a finite $\mathscr{n}$ (and $\langle n\rangle$) emerges at the threshold value $\Theta = 1$. For Θ increasing past this point, $\mathscr{n}$ first grows rapidly, but then decreases to reach a minimum at about $\Theta \simeq 2\pi$, where the field abruptly jumps to a higher intensity. This general behavior recurs roughly at integer multiples of 2π, but becomes less pronounced for increasing Θ. Finally, a stationary regime with $\mathscr{n}$ nearly independent of Θ is reached. Outside the time scale of Fig. 17-3 there is additional structure reminiscent of the Jaynes-Cummings revivals.

The number and in particular the sharpness of the features in the photon number depend on N_a. At the onset of the field around $\Theta = 1$ the function $\mathscr{n}(\Theta)$ essentially does not depend on N_a if $N_a \gg 1$, but the subsequent transitions become sharper for increasing N_a. In the limit $N_a \to \infty$, this hints at an interpretation of the first transition in terms of a continuous phase transition, while the others are similar to first-order phase transitions [see Guzman *et al.* (1989)].

It is possible to give a simple interpretation of the first transition (threshold) of the micromaser in terms of a gain/loss argument similar to that in Chap. 6. In the spirit of a "rate equation" analysis, one would expect the average number of photons in the cavity mode to be governed by an equation of the form

$$\frac{d}{dt}\langle n\rangle = \gamma N_a \sin^2(\tfrac{1}{2} g\sqrt{\langle n+1\rangle}\tau) - \gamma\langle n\rangle \ . \tag{20}$$

The first term in Eq. (20) is the gain due to the change in atomic inversion as deduced from the Rabi oscillations formula, where we have used Eq. (15) and the "+1" accounts for spontaneous emission into the resonator mode, while the second term describes cavity losses (here $N_b = 0$). The possible mean photon numbers $\langle n\rangle$ are approximately given by the stable stationary solutions of Eq. (20). For $\Theta << 1$ the only solution for the field is $\langle n\rangle \simeq \Theta^2 << 1$. The maser threshold occurs when the linearized (stimulated) gain for $\langle n\rangle \simeq 0$ compensates the cavity losses:

$$\gamma N_a \left.\frac{d}{d\langle n\rangle} \sin^2(g\sqrt{\langle n\rangle}\ \tau/2)\right|_{\langle n\rangle=0} \simeq \gamma N_a\ (g\tau)^2/4 = \gamma \ , \tag{21}$$

which reduces precisely to the threshold value $\Theta = 1$ obtained from the exact photon statistics Eq. (17). This justifies interpreting Θ as the pump parameter of the micromaser.

Figure 17-4 shows the normalized standard deviation

$$\sigma \equiv \frac{(\langle n^2\rangle - \langle n\rangle^2)^{1/2}}{\langle n\rangle^{1/2}} \tag{22}$$

of the photon distribution as a function of Θ for $N_a = 200$ and $\bar{n} = 0.1$. Above the threshold $\Theta = 1$ the photon statistics is first strongly super-Poissonian, with $\sigma \simeq 4$. (Poissonian photon statistics would yield $\sigma = 1$.) Further super-Poissonian peaks occur at the positions of the subsequent transitions. In the remaining intervals of Θ, σ is typically of the order of 0.5, a signature of the sub-Poissonian nature of the field. These predictions have recently been verified by Rempe *et al.* (1990).

Figure 17-5 shows the photon statistics (17). For $\Theta = 3\pi$, clean subPoissonian photon statistics emerge, whereas for 15π the distribution has three peaks. Far above threshold, the micromaser does not tend toward the Poissonian photon statistics typical of conventional single-mode, homogeneously broadened lasers, described in Sec. 17-2.

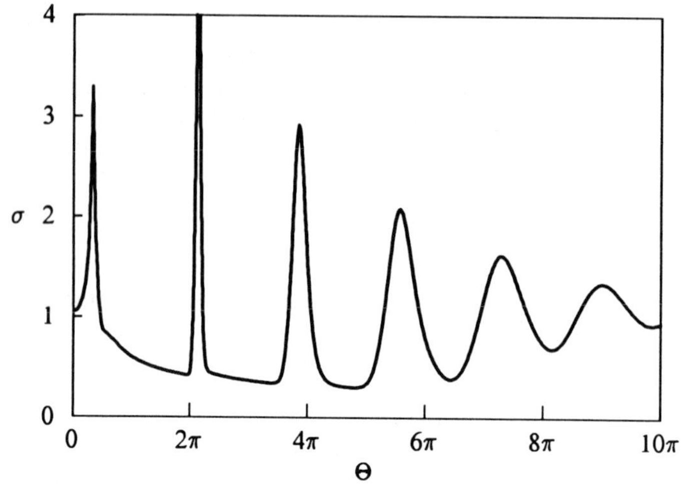

Fig. 17-4. Normalized deviation σ of Eq. (22) *vs* Θ.

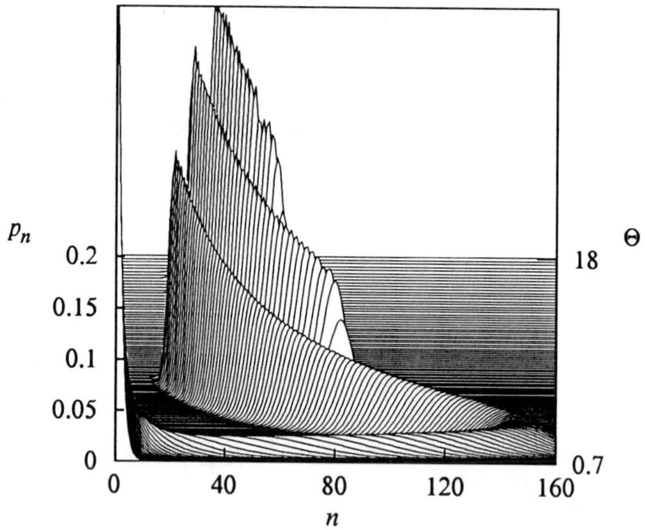

Fig. 17-5. Steady-state photon statistics $\bar{p}_n$ for $N_a = 150$, $\bar{n} = 0.1$, and values of the pump parameter Θ varying logarithmically from .7 to 18.

The temperature dependence of the micromaser steady state illustrates particularly simply the difference between the effects of quantum and thermal noise. To see this, note that the zeros of Eq. (8) imply the existence of number states $|n_q\rangle$ that cause successive atoms to experience $2q\pi$ pulses, where q is an integer during their interaction time τ. For example, on resonance the zeros are given by $g(n_q+1)^{1/2}\tau = 2q\pi$. These *trapping states*, which prohibit the growth of the cavity field past them, result from the coherent nature of the atom-field interaction. Competing with their action is dissipation, which leads to an incoherent transfer of population between the cavity mode levels. According to Eq. (10), for finite $\bar{n}$, this dissipation transfers population both upward ($n \rightarrow n + 1$) and downward ($n + 1 \rightarrow n$). Hence thermal fluctuations allow the micromaser to jump past the trapping states and rapidly wash out their effects. In the limit $\bar{n} \rightarrow 0$ ($T \rightarrow 0$), however, Eq. (10) reduces to

$$\dot{p}_n = -\frac{\nu}{Q}[np_n - (n+1)p_{n+1}] , \tag{23}$$

so that dissipation only causes downward transitions. In contrast to thermal fluctuations, vacuum fluctuations alone do not permit the growth of the maser past the trapping states. In this limit we can expect remnants of these states to appear in the steady-state properties of the maser.

Figure 17-6 shows the normalized steady-state mean photon number $\nu = \langle n\rangle/N_a$ given by Eq. (18) for $N_a = 50$ with $\bar{n} = 10^{-7}$, as a function of the micromaser pump parameter $\Theta = \frac{1}{2}N_a{}^{1/2}\,\kappa t_{int}$. The "resonances" in the $\nu(\Theta)$ curve are easily interpreted in terms of the trapping condition which becomes, in terms of the parameters N_a and Θ,

$$\frac{N_a}{\Theta^2} = \frac{n_q + 1}{q^2\pi^2} . \tag{24}$$

For fixed N_a, the successive resonances in Fig. 17-6 correspond to values of Θ such that decreasing Fock states $|n\rangle$ become trapping states with $q = 1$.

We anticipate the results of Sec. 17-2 to emphasize that the photon statistics of the microscopic maser exhibits features alien to ordinary masers and lasers: The field is typically strongly "nonclassical", where by classical we mean a field with a positive-definite $P(\alpha)$ distribution. It has no particular tendency, even far above threshold, of being Poissonian, and extra phase transitions take place when the pump parameter is increased. Section 17-2 shows that these differences originate in the fact that the micromaser possesses *less* stochasticity and noise than macroscopic masers and lasers for which the atom-field interaction is terminated by exponential atomic decays rather than a transit time. As a result the coherence of the quantum-mechanical light-matter interaction is lost or averaged over in conventional lasers, and the purely quantum-mechanical effects appearing in micromasers are largely lost.

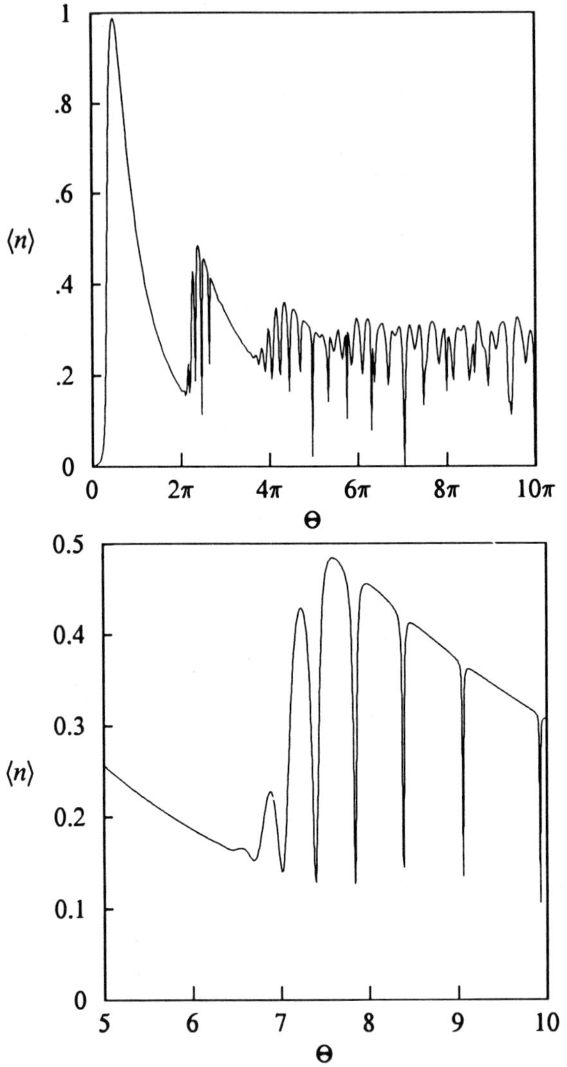

Fig. 17-6. Normalized mean photon number $\ell = \langle n \rangle / N_a$ as a function of the pump parameter Θ for $\bar{n} = 10^{-7}$.

17-2. Single Mode Laser Master Equation

In the micromaser it was essential to assume that at most one atom at a time is present inside the resonator. If that condition is not fulfilled, one must in principle take into explicit account the cooperative dynamics of the atoms. This is because in the high-Q microwave cavities used in these experiments the radiation wavelength is comparable to the dimensions of the cavity, and hence all atoms "see" a common electromagnetic field. Hence such systems are well suited for studying the collective dynamics of atoms interacting with the electromagnetic field, such as superfluorescence and superradiance (see Haroche *et al.* (1985)). In conventional lasers, these effects are neglected on the ground that in the optical regime the atoms move more or less randomly over distances large compared to the wavelength of the field and the collective effects are averaged out. Even at reasonably large densities, it is a good approximation to assume that the atoms respond independently.

To obtain an equation of motion for the photon-number probability for a field interacting with atoms we can convert the atomic contributions of Eqs. (7) and (9) to a coarse-grained time derivative similar to that in Eq. (14.20) by the equation

$$\dot{p}_n = \frac{p_n(t_i+\tau) - p_n(t_i)}{\tau}, \tag{25}$$

where τ is the cavity transit time. Adding the electric-dipole contributions of Eqs. (7) and (9) and the cavity-loss contributions of Eq. (10), we have the photon-number equation of motion

$$\boxed{\begin{aligned}\dot{p}_n = &-(n+1)\left[\mathcal{A}_{n+1} + \frac{\nu}{Q}\bar{n}\right]p_n + (n+1)\left[\mathcal{B}_{n+1} + \frac{\nu}{Q}(\bar{n}+1)\right]p_{n+1}\\ &+ n\left[\mathcal{A}_n + \frac{\nu}{Q}\bar{n}\right]p_{n-1} - n\left[\mathcal{B}_n + \frac{\nu}{Q}(\bar{n}+1)\right]p_n\,,\end{aligned}} \tag{26}$$

where the coefficients $\mathcal{A}_n$ and $\mathcal{B}_n$ are given by Eq. (15). Equation (26) has the general form of Eq. (14.35) and describes photon number probability flows similar to those in Fig. 14-1. With the coefficients of Eq. (15), Eq. (26) describes a maser or laser for which the transit time through the cavity is short compared to the atomic decay times. Since Eq. (26) is based on a coarse-grained derivative, it is valid provided the change in p_n during the transit time is sufficiently small and that the atoms act independently, that is, not collectively. We have seen how to solve this kind of equation in steady state in Sec. 14-1. As we see in Sec. 17-3, the corresponding steady-state photon statistics are given by Eq. (17).

In typical lasers, atoms are excited to their initial states within the cavity and they decay due to spontaneous emission and collision processes. The decay times are often short compared to the times over which the field mode amplitudes vary appreciably, so that we may still be able to use a coarse-grained derivative. The contribution of the atoms is then given by averaging over their lifetimes. For simplicity, we suppose that only one lifetime is needed, which we call γ^{-1}, and we average the coefficients $\mathcal{A}_n$ and $\mathcal{B}_n$ in Eq. (15) using the function $\gamma\int d\tau e^{-\gamma\tau}$. In this average we extend the upper limit to ∞, since that adds a negligible amount and simplifies the result. This gives

$$\mathcal{A}_n = \frac{R_a R}{2(1 + nR)}, \tag{27}$$

$$\mathcal{B}_n = \frac{R_b R}{2(1 + nR)}, \tag{28}$$

where the dimensionless rate constant

$$R = 4|g/\gamma|^2 \mathcal{L}(\omega-\nu), \tag{29}$$

and the dimensionless Lorentzian

$$\mathcal{L}(\omega-\nu) = \frac{\gamma^2}{\gamma^2 + (\omega-\nu)^2}. \tag{30}$$

Note that Eqs. (27) and (28) no longer have the zeroes given by the $\sin^2$ factors in Eqs. (15) and (16). This effect of the "granular" nature of the field is eliminated by the averaging process.

Before analyzing Eq. (26), it is instructive to recalculate it by solving the atom-field dynamics directly using a density matrix. This allows us to readily treat two-level systems with upper-to-ground-lower-level decay and with a pump from the ground to the excited state (see Fig. 17-2). The results have the correct form to describe still more complicated relaxation and pumping schemes. This technique can also be generalized to treat multimode phenomena such as resonance fluorescence, sidemode buildup, and quantum four-wave mixing. As a fringe benefit, it also shows how the single-mode semiclassical density-matrix equations of motion follow from fully quantal equations.

In the interaction picture at the frequency ν, the total Hamiltonian describing the laser system of Fig. 17-1 is

$$\mathcal{H} = \hbar(\omega-\nu)\sigma_z + [\hbar g a \sigma_+ + \text{adjoint}] + \text{pump/decay terms}\ . \qquad (31)$$

We wish to find the equation of motion for the photon-number probability $p_n \equiv \rho_{nn}$, where as before ρ is the reduced density matrix for the field mode. This probability is given by the trace over the atoms of the atom-field density matrix

$$p_n \equiv \rho_{nn} = \rho_{an,an} + \rho_{bn,bn}\ . \qquad (32)$$

Hence to find $\dot{p}_n$, we need $\dot{\rho}_{an,an}$ and $\dot{\rho}_{bn,bn}$. In the interaction picture at frequency ν with the pump/decay scheme of Fig. 17-2, the atom-field density matrix elements have the equations of motion

$$\dot{\rho}_{an,an} = -\Gamma\rho_{an,an} + \Lambda\rho_{bn,bn} - [i\mathcal{V}_{an,bn+1}\rho_{bn+1,an} + \text{c.c.}]\ , \qquad (33)$$

$$\dot{\rho}_{bn+1,bn+1} = \Gamma\rho_{an+1,an+1} - \Lambda\rho_{bn+1,bn+1} + [i\mathcal{V}_{an,bn+1}\rho_{bn+1,an} + \text{c.c.}]\ , \qquad (34)$$

$$\dot{\rho}_{an,bn+1} = -[\gamma + i(\omega-\nu)]\rho_{an,bn+1} + i\mathcal{V}_{an,bn+1}(\rho_{an,an} - \rho_{bn+1,bn+1})\ , \qquad (35)$$

where γ is the dipole decay constant ($\equiv 1/T_2$), and $\mathcal{V}_{an,bn+1} = g\sqrt{n+1}$. Note that while Eqs. (33) through (35) look semiclassical, they describe quantized transitions between quantized atom-field levels. The decay and pump terms result from tracing the complete density matrix over the corresponding reservoirs. Hence Eqs. (33) through (35) describe a reduced density matrix. Taking the time rate of change of the trace (32) and using Eqs. (33) and (34), we have the equation of motion for the probability of having n photons

$$\begin{aligned}\dot{p}_n &= \dot{\rho}_{an,an} + \dot{\rho}_{bn,bn} \\ &= -i\mathcal{V}_{an,bn+1}\rho_{bn+1,an} + i\mathcal{V}_{an-1,bn}\rho_{bn,an-1} + \text{c.c.}\end{aligned} \qquad (36)$$

This reduces the problem to finding the dipole density-matrix element $\rho_{bn+1,an}$. The derivation is almost the same as for the semiclassical case in Sec. 5-1. Although derived for the simple level scheme of Fig. 17-2, Eq. (36) is valid for two-level interactions with arbitrary decay and pumping contributions, since these contributions inevitably cancel out in the trace (32). Using the trace in Eq. (32) to eliminate $\Lambda\rho_{bn,bn}$ in Eq. (33) and $\Gamma\rho_{an+1,an+1}$ in Eq. (34), we find

$$\dot{\rho}_{an,an} = \Lambda p_n - (\Gamma+\Lambda)\rho_{an,an} - [i\mathcal{V}_{an,bn+1}\rho_{bn+1,an} + \text{c.c.}] \qquad (37)$$

$$\dot{\rho}_{bn+1,bn+1} = \Gamma p_{n+1} - (\Gamma+\Lambda)\rho_{bn+1,bn+1} + [i\mathcal{V}_{an,bn+1}\rho_{bn+1,an} + \text{c.c.}] \,. \quad (38)$$

As in the semiclassical theory of Chap. 5, we suppose that the atoms react quickly compared to variations in the field amplitude. Hence we can solve Eqs. (35), (37), and (38) in steady state by setting the time rates of change equal to zero [note that more generally these atom-field equations include the cavity damping terms that lead to the ν/Q terms in Eq. (52)]. The steady-state solution to the dipole Eq. (35) is

$$\rho_{an,bn+1} = i\mathcal{V}_{an,bn+1}\mathcal{D}(\omega-\nu)[\rho_{an,an} - \rho_{bn+1,bn+1}] \,, \quad (39)$$

where the complex Lorentzian denominator

$$\mathcal{D}(\omega-\nu) = 1/(\gamma + i(\omega-\nu)) \,. \quad (40)$$

Similarly subtracting Eq. (38) from (37) and solving in steady state, we find the probability difference

$$\rho_{an,an}-\rho_{bn+1,bn+1} = N_a p_n - N_b p_{n+1} - 2T_1[i\mathcal{V}_{an,bn+1}\rho_{bn+1,an} + \text{c.c.}], \quad (41)$$

where the unsaturated probabilities of being in the upper and lower levels are given by

$$N_a = \Lambda T_1 \,, \; N_b = \Gamma T_1 \,, \quad (42)$$

the dimensionless rate constant

$$R = 4|g|^2 T_1 T_2 \mathcal{L}(\omega-\nu) \,, \quad (43)$$

and the probability-difference decay time

$$T_1 = 1/(\Gamma + \Lambda) \,. \quad (44)$$

Substituting the complex conjugate of Eq. (39) into Eq. (41), we find

$$\rho_{an,an} - \rho_{bn+1,bn+1} = \frac{N_a p_n - N_b p_{n+1}}{1 + (n+1)R} \,. \quad (45)$$

Substituting this into Eq. (39), we obtain the complex electric-dipole term

$$\rho_{an,bn+1} = i\mathcal{V}_{an,bn+1}\mathcal{D}(\omega-\nu)\left[\frac{N_a p_n - N_b p_{n+1}}{1 + (n+1)R}\right] . \quad (46)$$

Substituting this, in turn, into Eq. (36) and including the cavity loss contributions, we once again find Eq. (26), this time with $\mathcal{A}_n$ and $\mathcal{B}_n$ given by Eqs. (27) and (28), in which R_α is given by

$$R_\alpha = N_\alpha / T_1 \,. \tag{47}$$

In fact with appropriate definitions of T_1, N_a, and N_b, Eqs. (26) through (28), and (47) describe two-level systems with arbitrary relaxation and pumping schemes. For two excited laser levels, the N_b term comes from excitation to the lower laser level.

Reduction to Semiclassical Equations of Motion

We can recover the semiclassical density-matrix equations of motion used in Chap. 5 directly from the Eqs. (33) through (35). To this end on the RHS of these equations, we *factor* the atom-field density matrix as

$$\begin{bmatrix} \rho_{an,am} & \rho_{an,bm} \\ \rho_{bn,am} & \rho_{bn,bm} \end{bmatrix} = \rho_{nm} \begin{bmatrix} \rho_{aa} & \rho_{ab} \\ \rho_{ba} & \rho_{bb} \end{bmatrix}, \tag{48}$$

Substituting this into Eq. (33) and tracing over n, we find the semiclassical equation of motion (5.32)

$$\dot{\rho}_{aa} = -\Gamma \rho_{aa} + \Lambda \rho_{bb} - [i\mathcal{V}_{ab}\rho_{ba} + \text{c.c.}] \,, \tag{5.32}$$

where $\mathcal{V}_{ab} = g\langle a \rangle$, which is in an interaction picture rotating at the frequency ν. $\langle a \rangle$ is the expectation value for the mode annihilation operator. Classically it corresponds to $-\frac{1}{2}\wp\mathcal{E}/g$. Similarly we find $\dot{\rho}_{bb} = -\dot{\rho}_{aa}$ and

$$\dot{\rho}_{ab} = -[\gamma + i(\omega - \nu)]\rho_{ab} + i\mathcal{V}_{ab}[\rho_{aa} - \rho_{bb}] \,, \tag{50}$$

which is an interaction-picture version of Eq. (5.9). The steady-state solutions to these equations are given simply by setting Eqs. (5.32) and (50) equal to zero, instead of using an integrating factor as in Eq. (5.9). The results are the same as for Eqs. (5.9) and (5.32). We see that a fundamental approximation used to obtain the semiclassical equations of motion is to *factorize* the atom-field density matrix. This implies neglecting the correlations that develop between these systems over time, and hints at the fact that even more important than strong fields, the concept of time scales is essential to the validity of the semiclassical approximation. Indeed, we have seen that the Jaynes-Cummings model of Chap. 13 never reaches a semiclassical limit, regardless of the field intensity. This is because no mechanism exists to destroy the correlations built by the quantum dynamics between the atom and the field mode. In the laser problem, an important consequence of the factorization Ansatz is that it eliminates the terms that can lead to a nonzero laser linewidth.

17-3. Laser Photon Statistics and Linewidth

In this section, we solve the master equation Eq. (26) in steady-state to determine the laser photon statistics, and then use this same equation to find the equation of motion for the average number of photons, $\langle n\rangle$. The calculations are simplified by noting that Eq. (26) can be written as

$$\dot{p}_n = -(n+1)\mathcal{A}_{n+1}{}'p_n + (n+1)\mathcal{B}_{n+1}{}'p_{n+1} + n\mathcal{A}_n{}'p_{n-1} - n\mathcal{B}_n{}'p_n \, , \tag{51}$$

where

$$\mathcal{A}_n{}' = \mathcal{A}_n + \bar{n}\nu/Q \quad \text{and} \quad \mathcal{B}_n{}' = \mathcal{B}_n + (\bar{n}+1)\nu/Q \, . \tag{52}$$

We solve the set of photon-number probabilities in steady state ($\dot{p}_n$=0) using the principle of detailed balance discussed in Sec. 14-1. This states that for a steady state to occur, the flow of probability from p_n to p_{n-1} has to equal the flow from p_{n-1} to p_n for all values of n. Specifically we set

$$\mathcal{B}_n{}'p_n = \mathcal{A}_n{}'p_{n-1} \, , \tag{53}$$

which is the same as setting $S_n = 0$ in Eq. (14). Similarly solving for p_n by iteration, we recover Eq. (17)

$$\boxed{p_n = p_0 \prod_{k=1}^{n} \frac{\bar{n}\nu/Q + \mathcal{A}_k}{(\bar{n}+1)\nu/Q + \mathcal{B}_k}} \, . \tag{17}$$

Here p_0 is a normalization factor that can be determined by setting $\Sigma_n\, p_n = 1$. For optical frequencies, we can usually set the average thermal photon number $\bar{n} = 0$. Further using $\mathcal{A}_n$ and $\mathcal{B}_n$ given by Eqs. (27) and (28) with Eq. (47), we find

$$p_n = p_0 \prod_{k=1}^{n} \frac{N_a}{N_b + 2T_1(R^{-1} + k)\nu/Q} \, . \tag{54}$$

It is instructive to examine this photon distribution way below, at, and way above the laser threshold. Way below threshold, $N_a \ll N_b + 2T_1\nu/QR$. Hence $p_1 \ll p_0$ and in general $p_n \ll p_{n-1}$. This implies that the photon numbers of interest are sufficiently small that the k in the denominator of Eq. (54) can be dropped. This gives the approximate distribution

$$p_n = p_0 x^n = (1-x)x^n \, , \tag{55}$$

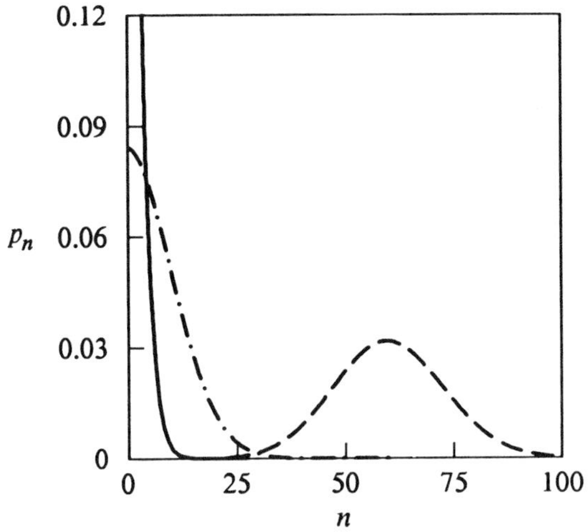

Fig. 17-7. Graph of steady-state solution of Eq. (54) for excitation below (solid line), at (dot-dashed line), and above (dashed line) threshold. N_b was taken to be zero.

where $x = N_a/(N_b + 2T_1\nu/QR)$. This is the exponential decay formula of a thermal distribution and is illustrated by the solid-line curve in Fig. 17-7.

At threshold, $N_a = N_b + 2T_1\nu/QR$. The k in the denominator of Eq. (54) then causes $p_n < p_{n-1}$ for all n. For small n, the multiplicative factors in Eq. (54) are nearly unity, so the function of n gradually starts to fall off. For larger values of n, the $+k$ term begins to dominate leading to a rapid decay as shown by the dot-dashed line in Fig. 17-7. Way above threshold, $N_a >> N_b + 2T_1\nu/QR$. Hence for small n, $p_n > p_{n+1}$, and the distribution increases as a function of n. For the n yielding $N_a = N_b + 2(R^{-1} + n)T_1\nu/Q$, the distribution tops out and then starts to decrease as shown by the dashed line in Fig. 17-7.

We can write the distribution (54) in terms of a factorial function as

$$\boxed{p_n = \frac{p_0}{(n+C)!}\left[\frac{N_a}{2T_1(\nu/Q)}\right]^{n+C}}, \tag{56}$$

where $C = R^{-1} + N_b/2T_1(\nu/Q)$. Using the identity $n = n + C - C$, we find that the average number $\langle n \rangle$ given by Eq. (56) is given by

$$n_{ss} = \sum_n n p_n = p_0 \sum_n \frac{n + C - C}{(n+C)!} \left[\frac{N_a}{2T_1(\nu/Q)}\right]^{n+C} = \frac{N_a - N_b}{2T_1\nu/Q} - \frac{1-p_0}{R} \,. \quad (57)$$

Far above threshold, the values of n's around n_{ss} are much larger than C. Hence in this limit we can write Eq. (56) as

$$p_n = \frac{e^{-\langle n\rangle} \langle n\rangle^n}{n!} \,, \quad (58)$$

that is, a Poisson distribution. Thus a single-mode laser way above threshold radiates a state with the photon statistics of a coherent state. However, this radiation is not a pure state: Due to the finite laser linewidth, the phase diffuses away, yielding a "phase-diffused" coherent state, i.e., one with unknown phase.

Taking the time rate of change of the average photon number

$$\langle n(t)\rangle = \Sigma_n \, n \, p_n(t) \,, \quad (59)$$

we find

$$\begin{aligned}\frac{d}{dt}\langle n\rangle &= \sum_n n \, \dot{p}_n = -\sum_n [\mathcal{A}_{n+1}'(n^2 + n) - \mathcal{B}_n'(n^2 - n) \\ &\qquad - \mathcal{A}_{n+1}'(n^2 + 2n + 1) + \mathcal{B}_n' n^2] \, p_n \\ &= \Sigma_n (\mathcal{A}_{n+1}' - \mathcal{B}_n') n p_n + \Sigma_n \mathcal{A}_{n+1}' p_n \\ &\simeq (\mathcal{A} - \mathcal{B} - \nu/Q)\langle n\rangle + \mathcal{A} + \bar{n}\nu/Q \,, \end{aligned} \quad (60)$$

where in the last equality we approximate $\mathcal{A}_{n+1}$ and $\mathcal{B}_n$ by their semiclassical values (exact for a coherent state)

$$\mathcal{A} = \frac{RN_a}{2T_1(1+\langle n\rangle R)} \text{ and } \mathcal{B} = \frac{RN_b}{2T_1(1+\langle n\rangle R)} \,. \quad (61)$$

This approximation is good if the average number of photons is peaked around $\langle n\rangle$, which we have seen to occur for a laser above threshold, or if $\langle n\rangle R \ll 1$, which is true for a laser below threshold. For very large $\langle n\rangle$ the $\mathcal{A} + \bar{n}\nu/Q$ factor in Eq. (60) can be neglected. The steady-state value is then determined by setting $\mathcal{A} - \mathcal{B} = \nu/Q$, which gives Eq. (57).

The classical dimensionless laser intensity is given by

$$I_n = \langle n\rangle R \,. \quad (62)$$

Hence Eq. (60) is just a quantum intensity equation of motion. Unlike the semiclassical equation of motion (6.20), Eq. (60) builds up even if $\langle n\rangle$ is initially 0. This is due to the presence of the $\mathcal{A} + \bar{n}\nu/Q$ term, which is the sum of spontaneous emission and nonzero-temperature cavity contributions. Substituting Eqs. (44) and (61) through (62) into Eq. (60), we recover the semiclassical equation (6.20) plus the spontaneous emission coefficient $\mathcal{A} + \bar{n}\nu/Q$. The saturation factor $\mathcal{S}(I_n)$ found is given by the unidirectional formula (6.16).

The steady-state solution of Eq. (60) gives the alternate formula for the steady state photon number

$$n_{ss} = \frac{N_a - N_b}{4T_1\kappa} - \frac{1}{2R} \pm \sqrt{\left[\frac{N_a - N_b}{4T_1\kappa} - \frac{1}{2R}\right]^2 + \frac{N_a}{2T_1\kappa}} \ . \qquad (63)$$

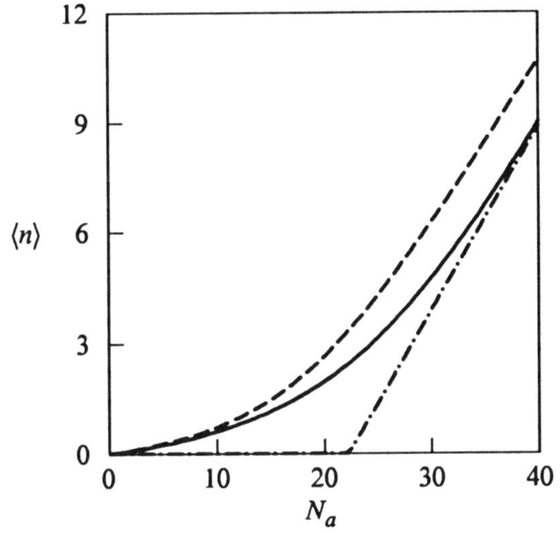

Fig. 17-8. Steady-state photon number n_{ss} given by the + value of Eq. (63) (dashed curve), the semiclassical value (dot-dashed line) given by Eq. (57) with $1\text{-}p_0 \simeq 1$, and the quantum value (solid line) given by Eq. (59) with the p_n of Eq. (54).

Figure 17-8 compares the + value of this formula to the semiclassical value given by Eq. (57) with $1\text{-}p_0 \simeq 1$, and to the quantum value given by Eq. (59) with p_n of Eq. (54). We see that both quantum values predict a substantial average photon number at threshold, while the classical value starts up from zero at threshold. Equation (63) overestimates n_{ss}, due to the second term in the radicand, but this term allows the formula to give more accurate values near threshold. For larger photon numbers, this difference is smaller.

Laser Linewidth

From the Wiener-Khintchine theorem, the laser linewidth is given by the width of the real part of the complex spectrum

$$\mathcal{S}(s) = \int_0^\infty dt\, e^{-st} \langle a^\dagger(t)a(0)\rangle \,. \tag{64}$$

As for resonance fluorescence in Chap. 15, we find this spectrum by applying the quantum regression theorem to the Langevin equation for a quantized field mode interacting with a medium described by a gain operator $\alpha(t)$. This equation is

$$\boxed{\dot{a}(t) = [\alpha(t) - \upsilon]a(t) + f(t)}\,, \tag{65}$$

where $a(t)$ and $f(t)$ are slowly-varying annihilation and noise operators, respectively, the complex cavity-loss coefficient

$$\upsilon = \frac{\nu}{2Q} + i(\Omega - \nu)\,, \tag{66}$$

Ω is the cavity resonance frequency, and ν is the self-consistent field frequency. Equation (65) is essentially the semiclassical Eq. (6.5) written as an operator equation, which includes the noise operator $f(t)$. In general the complex gain $\alpha(t)$ is an operator that is saturated by the number operator $a^\dagger(t)a(t)$. In other respects, we allow α to be an arbitrary operator, so that Eq. (65) applies to lasers in general, including those with two-level and semiconductor media.

To obtain the traditional laser linewidth most quickly, we consider the case for which the complex gain α is real. Its value is then completely determined by solving the photon number equation of motion (60) written as

$$\frac{d}{dt}\langle n\rangle = 2\langle \alpha_r - \upsilon_r\rangle\langle n\rangle + \mathcal{A} + \bar{n}\frac{\nu}{Q} \tag{67}$$

in steady state. This gives the saturated gain

$$\langle \alpha_r\rangle = \upsilon_r - \frac{\mathcal{A} + \bar{n}\nu/Q}{2n_{ss}}\,, \tag{68}$$

which is slightly less than the cavity loss due to the reservoir fluctuations. Substituting Eqs. (68) and (65) into Eq. (64), we find the spectrum

$$\mathcal{S}(s) = \frac{n_{ss}}{s + (\mathcal{A} + \bar{n}\nu/Q)/2n_{ss}}, \tag{69}$$

which gives a FWHM laser linewidth of

$$\Delta\nu = \frac{\mathcal{A} + \bar{n}\nu/Q}{n_{ss}}. \tag{70}$$

At threshold and with atoms injected only into the upper state, $\mathcal{A}$ equals ν/Q and at optical frequencies $\bar{n} \simeq 0$, which gives the value $\Delta\nu \simeq \nu/Qn_{ss}$ often seen in the literature.

More generally, we suppose that the saturated gain α is complex. It is convenient to write Eq. (65) as

$$\dot{a}(t) = [\alpha_r(t) - v_r][1 + i\beta(t)]a(t) + f(t), \tag{71}$$

where $\alpha_r(t) - v_r = \mathrm{Re}\{\alpha - v\}$ and

$$\beta(t) = \frac{\alpha_i(t) - v_i}{\alpha_r(t) - v_r}. \tag{72}$$

In principle Eq. (71) is sufficiently general to determine the laser linewidth subject only to the approximation that the medium follows the field fluctuations adiabatically.

A general calculation of the laser linewidth is quite lengthy and even then not entirely satisfying. Indeed, it is not at all a trivial task to find the operator $\alpha(t)$ in general! Consequently we restrict our discussion to two simple semi-heuristic derivations that seem to contain the most important physical features. First we suppose that we can split the complex gain term in Eq. (71) into two parts, one for which α is independent of the fluctuations (pure phase fluctuations) and the other for which the fluctuation dependence is given by a first-order Taylor series so that

$$\dot{a}(t) = \langle\alpha_r(t) - v_r\rangle[1 + i\langle\beta(t)\rangle]a(t) + \frac{\partial\alpha}{\partial n}\delta n a(t) + f(t), \tag{73}$$

For small number fluctuations Δn, we approximate β by

$$\beta = \frac{\langle\alpha_i\rangle - v_i + \Delta\alpha_i\Delta n}{\langle\alpha_r\rangle - v_r + \Delta\alpha_r\Delta n} \simeq \frac{\Delta\alpha_i}{\Delta\alpha_r}, \tag{74}$$

where $\Delta\alpha_r$ and $\Delta\alpha_i$ are the semiclassical real and imaginary changes in α brought about by a linear change in the saturation. For example from Eq. (5.27), a homogeneously broadened two-level medium has the value

$$\beta = -(\omega - \nu)/\gamma. \tag{75}$$

Substituting Eq. (68) into Eq. (73) gives

$$\dot{a}(t) = -\frac{(\mathcal{A} + \bar{n}\nu/Q)(1 + i\beta)}{2n_{ss}}a(t) + \frac{\partial\alpha}{\partial n}\delta n a(t) + f(t)\ , \tag{76}$$

Inserting this into Eq. (64), neglecting the δn term for the moment, and using the quantum regression theorem, we readily find

$$\mathcal{S}(s) = \frac{n_{ss}}{s + (\mathcal{A} + \bar{n}\nu/Q)(1 + i\beta)/2n_{ss}}\ , \tag{77}$$

which gives the laser linewidth of Eq. (70). In this approximation, the β contribution amounts to a tiny frequency shift.

In reaching Eq. (77), we explicitly neglected contributions resulting from fluctuation-induced saturation changes represented by the δn term in Eq. (76). This approximation amounts to including fluctuation components that change the field phase directly, but neglecting those in the field intensity. Most notably in semiconductor lasers, the intensity fluctuations cause a larger contribution to the laser linewidth by changing the saturation, which in turn momentarily modifies both the index and the gain. During the time that such a fluctuation relaxes back to zero, the index change causes the instantaneous field phasor to rotate at a slightly different frequency, thereby giving a random phase shift. This frequency change is analogous to that of the atomic dipole induced by a van der Waal's collision (see Sec. 4-2), which leads to a broadening of the natural linewidth (decrease in the dipole decay time T_2), except that here the intensity-induced frequency fluctuations enhance the laser linewidth. Although according to Eq. (76) intensity fluctuations can lead directly to a small increase in the linewidth, the coupling of the intensity fluctuations to the phase via index changes is potentially much larger.

To estimate the laser linewidth including this intensity/phase coupling, we use an essentially classical argument similar to that of Henry (1982). In Fig. 17-9, we see how the instantaneous field phasor is shifted by a fluctuation of length 1, which corresponds to spontaneous emission. In terms of this figure, the reduction of the linewidth by the $1/n_{ss}$ factor can be understood by noting that fluctuations of constant average magnitude add to a field phasor amplitude proportional to $\sqrt{n_{ss}}$. Hence the ability of the fluctuations to rotate the field phasor is inversely proportional to $\sqrt{n_{ss}}$. This corresponds to a linewidth inversely proportional to n_{ss}.

The linewidth is due primarily to phase fluctuations, so that we approximate the annihilation operator as

$$a(t) = a_0 e^{i\phi(t)}\ , \tag{78}$$

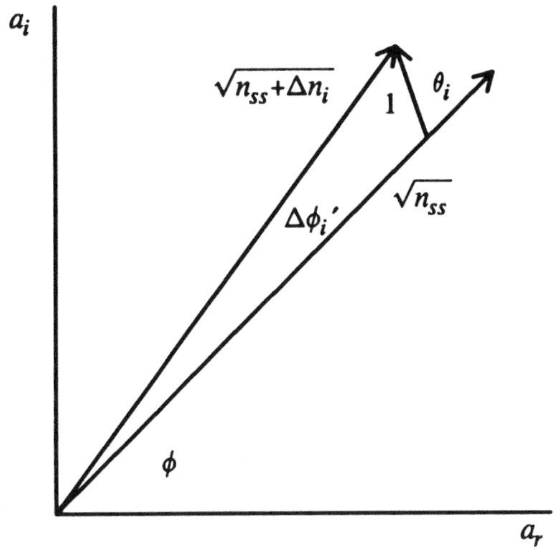

Fig. 17-9. Diagram showing how the instantaneous field phasor is changed by a field fluctuation. Fluctuation components perpendicular to the phasor cause no change in saturation and hence no index change, while those parallel to the phasor induce a change in saturation and hence in index.

where a_0 is the steady-state field expectation value. A phase fluctuation $\Delta\phi_i$ shifts $a(t)$ by the multiplicative factor $\exp(i\Delta\phi_i)$. Similarly to the T_2 discussion of Sec. 4-2, an average of $a(t)$ over the Gaussian random phase shift $\Delta\phi_i$ is given by

$$a(t) = a_0\exp[i\phi(0)]\langle\exp(i\Delta\phi_i)\rangle_{av} = a_0\exp[i\phi(0)]\exp[i\langle\Delta\phi_i\rangle_{av} - \langle(\Delta\phi_i)^2\rangle_{av}] ,$$

where $_{av}$ indicates an average over the field fluctuations. If these phase fluctuations occur at the sum of the spontaneous-emission and thermal-fluctuation rates, $\mathcal{A} + \bar{n}\nu/Q$, this gives

$$a(t) = a_0\exp[i\phi(0)]\exp\{(\mathcal{A} + \bar{n}\nu/Q)t[i\langle\Delta\phi_i\rangle_{av} - \langle(\Delta\phi_i)^2\rangle_{av}]\} . \tag{79}$$

The phase shift $\Delta\phi_i$ caused by the ith fluctuation is given by

$$\Delta\phi_i = \Delta\phi_i' + \Delta\phi_i'' , \tag{80}$$

where $\Delta\phi_i{}'$ is the direct contribution

$$\Delta\phi_i{}' = \frac{\sin\theta_i}{\sqrt{n_{ss}}} , \tag{81}$$

and $\Delta\phi_i{}''$ is contribution resulting from index changes caused by intensity fluctuations. Applying the law of cosines to the triangle in Fig. 17-9, we find that the intensity component of the ith fluctuation is given by

$$\Delta n_i = 1 + 2\sqrt{n_{ss}}\cos\theta_i . \tag{82}$$

To relate this change to $\Delta\phi_i{}''$, we write Eq. (6.9) in terms of the imaginary part of the complex gain coefficient as

$$\nu + \dot{\phi} = \Omega + \alpha_i , \tag{83}$$

We suppose that the oscillation frequency ν is pulled an amount given by α_i in the absence of a fluctuation. Then $\dot{\phi}$ is the value due to a change $\Delta\alpha_i$

$$\dot{\phi} = \Delta\alpha_i = \beta(\alpha_r - \upsilon_r) , \tag{84}$$

where we continue to approximate β by the semiclassical value of Eq. (74). Comparing Eq. (84) with the photon-number equation of motion (67) (without the quantum source terms and the averages), we find that

$$\frac{d\phi}{dt} = \frac{\beta}{2n_{ss}}\frac{dn}{dt} .$$

Integrating both sides of this equation over time and using Eq. (82), we find the saturation-coupled random phase shift

$$\Delta\phi_i{}'' = \frac{\beta}{2n_{ss}}\Delta n_i = \frac{\beta}{2n_{ss}}(1 + 2\sqrt{n_{ss}}\cos\theta_i) . \tag{85}$$

The first term in this expression gives a value of $\langle\Delta\phi_i\rangle_{av}$ that leads to the tiny frequency shift in Eq. (77). The total second-order phase shift due to the ith fluctuation is given by

$$\langle(\Delta\phi_i)^2\rangle = \frac{\langle(\sin\theta_i + \beta\cos\theta_i)^2\rangle}{n_{ss}} .$$

Since the angle θ_i is random, the cross terms average to zero leaving

$$\langle(\Delta\phi_i)^2\rangle = \frac{1+\beta^2}{2n_{ss}}. \tag{86}$$

Substituting this value into Eq. (79) and the result into Eq. (64), we find the complex spectrum

$$\mathcal{S}(s) = \frac{n_{ss}}{s + \dfrac{(\mathcal{A} + \bar{n}\nu/Q)[1 + i\beta + \beta^2]}{2n_{ss}}}, \tag{87}$$

which gives a FWHM linewidth of

$$\boxed{\Delta\nu = \frac{(\mathcal{A} + \bar{n}\nu/Q)(1+\beta^2)}{n_{ss}}}. \tag{88}$$

The $1 + \beta^2$ contribution is called the *linewidth enhancement factor*. While Eq. (75) shows that this factor is typically very small for homogeneously broadened two-level media, it can be important for semiconductor lasers, provided that they operate in a region with significant mode pulling. Note that this theory uses plane waves; the full story for semiconductor diode lasers should include transverse field variations and other effects such as injection-current fluctuations.

17-4. Quantized Sidemode Buildup

To consider the buildup of a sidemode in the presence of a strong classical laser mode and to determine the resonance fluorescence spectrum, we generalize the quantized probe absorption discussion of Sec. 15-4 to include a pump to transform the atoms into a gain medium. The equation of motion (15.9b) for the population-difference operator is given by

$$\dot{S}_z = -(\Gamma + \Lambda)S_z - \tfrac{1}{2}(\Gamma - \Lambda) + i\mathcal{V}_2^{*}S_- - i\mathcal{V}_2S_+ + F_z(t), \tag{89}$$

Hence Γ is replaced by $\Gamma + \Lambda$ in the B matrix of Eqs. (15.27) and (15.33), and Γ is replaced by $\Gamma - \Lambda$ in Eqs. (15.40) and (15.41). This then yields the sidemode coefficients

$$A_1 = \frac{g^2\mathscr{D}_1}{1+I_2\mathscr{L}_2}\left[\frac{I_2\mathscr{L}_2}{2}+N_a - I_2\frac{\gamma}{2}\mathscr{F}\frac{\mathscr{D}_1\left[\frac{I_2\mathscr{L}_2}{2}+N_a\right] + \mathscr{D}_2^*(N_a-N_b)\left[1+\frac{\Lambda-\Gamma}{i\Delta}\right]}{1 + I_2\mathscr{F}\frac{\gamma}{2}(\mathscr{D}_1+\mathscr{D}_3^*)}\right] \tag{90}$$

$$B_1 = \frac{g^2\mathscr{D}_1}{1+I_2\mathscr{L}_2}\left[\frac{I_2\mathscr{L}_2}{2}+N_b - I_2\frac{\gamma}{2}\mathscr{F}\frac{\mathscr{D}_1\left[\frac{I_2\mathscr{L}_2}{2}+N_b\right] - \mathscr{D}_2^*(N_a-N_b)\left[1-\frac{\Lambda-\Gamma}{i\Delta}\right]}{1 + I_2\mathscr{F}\frac{\gamma}{2}(\mathscr{D}_1+\mathscr{D}_3^*)}\right] \tag{91}$$

where $\mathscr{D}_n$ is given by Eq. (8.8), $\mathscr{L}_2$ by Eq. (8.10), $I_2 = |\wp\mathscr{E}_2/\hbar|^2 T_1 T_2$, and $\mathscr{F}(\Delta)$ is given by

$$\mathscr{F}(\Delta) = \frac{1}{1 + iT_1\Delta}, \tag{92}$$

T_1 is defined in Eq. (44), and N_a and N_b are given by Eq. (42). Including the cavity loss ν/Q_1 for mode 1, we generalize Eq. (15.54) to the sidemode master equation

$$\dot{\rho} = -A_1(\rho a_1 a_1^\dagger - a_1^\dagger \rho a_1) - (B_1 - \nu/2Q)(a_1^\dagger a_1 \rho - a_1 \rho a_1^\dagger) + \text{adj.} \tag{93}$$

The A_1 and B_1 coefficients can also be derived in the Schrödinger picture using the level scheme of Fig. 17-9. While this approach requires more algebra, it reveals how the atom-field quantum levels are mixed by the multimode interactions (see Prob. 17-12).

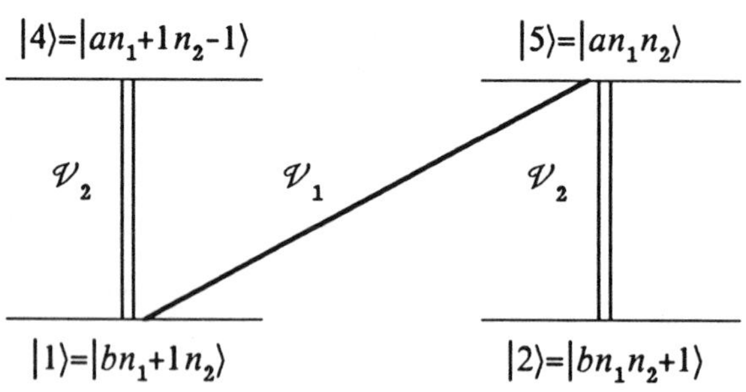

Fig. 17-9. Atom-field level scheme for two-wave mixing.

We are primarily interested in the build-up of the sidemode 1, which can be described by the average photon number $\langle n_1 \rangle = \Sigma_{n_1} n_1 p_{n_1}$. With Eq. (93) we readily find the equation of motion

$$\frac{d}{dt}\langle n_1 \rangle = (A_1 - B_1 - \nu/2Q_1)\langle n_1 \rangle + A_1 + \text{c.c.}, \tag{94}$$

which resembles the single-mode Eq. (60). A combination of the net gain $(A_1 - B_1 - \nu/2Q_1)$ and resonance-fluorescence (A_1 + c.c.) spectra tells us which sidemode frequencies might build up in a laser cavity. As discussed in Chap. 10 and 16, typically two sidemodes collaborate in building up in the presence of a central mode. We have seen how these more complicated mixing problems provide ways to generate squeezed states of light.

References

Filipowicz, P., J. Javanainen, and P. Meystre (1986), Phys. Rev. A**34**, 3086 gives the original theory of the micromaser.

Gallas, J. A. C., G. Leuchs, H. Walther, and H. Figger (1985), in *Advances in Atomic and Molecular Physics*, D. Bates, B. Bederson, Eds., vol 20Academic Press, Orlando, FA, *p*. 413. A detailed review of Rydberg atoms from the point of view of quantum optics.

Guzman, A. M., P. Meystre, and E. M. Wright (1989), Phys. Rev. A**40**, 2471. This paper also gives a semiclassical description of the micromaser.

Haken, H. (1970), *Laser Theory*, in *Encyclopedia of Physics*, **XXV/2c**, Ed. by S. Flügge, Springer-Verlag, Heidelberg. The quantum theory of the laser was developed independently by three schools around H. Haken, M. Lax and W. E. Lamb, Jr. This book is an encyclopedic review of work by the Haken school through 1969.

Haroche, S.(1984), in *New Trends in Atomic Physics*, Les Houches, Session XXXVIII, Ed. by G. Grynberg and R. Stora, Elsevier Science Pubs. B. *V*.

Haroche, S. and J. M. Raimond (1985), *Advances in Atomic and Molecular Physics*, Vol. 20, Academic Press, New York, *p*. 350. A review of applications of Rydberg atoms in quantum optics as of 1984, and has a nice section on superfluorescence.

Haroche, S. and D. Kleppner (1989), Phys. Today (Jan) gives a tutorial discussion of "cavity quantum electrodynamics", including micromasers, enhanced and inhibited spontaneous emission, etc...

Henry, C. H. (1982), IEEE J. Quant. Electronics **QE-18**, 259.

Lax, M. (1968), in *Brandeis University Summer Institute Lectures (1966)*, Vol. II, ed. by M. Chretien, E. P. Gross, and S. Deser, Gordon and Breach, New York. See also M. Lax (1966), in *Physics of Quantum Electronics*, Ed. by P. L. Kelly, B. Lax, P. E. Tannenwald (McGraw-Hill, New York), p. 795 ("Quantum Noise V"); M. Lax (1967), Phys. Rev. **157**, 213 ("QNX").

Meschede, D., H. Walther, and G. Muller (1984), Phys. Rev. Lett. **54**, 551 performed the first micromaser experiment.

Rempe, G., F. Schmidt-Kaler, and H. Walther (1990), Phys. Rev. Lett. **64**, 2483 verified the subpoissonian nature of micromaser radiation.

Sargent, M. III, D. A. Holm, M. S. Zubairy (1985), Phys. Rev. A**31**, 3112.

Sargent, M. III, M. O. Scully, and W. E. Lamb, Jr. (1977), *Laser Physics*, Addison-Wesley Publishing Co., Reading, MA. This book gives a quantum theory of the laser following the Lamb school. See especially Chap. 17.

Scully, M. O. and W. E. Lamb, Jr. (1967), Phys. Rev. **159**, 208.

Problems

17-1. Consider the Jaynes-Cummings model

$$\mathcal{H} = \tfrac{1}{2}\hbar\omega\sigma_z + \hbar\omega a^\dagger a + \hbar g(a^\dagger\sigma_- + a\sigma_+).$$

Show that $a^\dagger a + \frac{1}{2}\sigma_z$ is a constant of motion. Determine the mean field intensity as a function of time. Assume that the field is initially in a coherent state $|\alpha\rangle$ and the atom in the upper state $p_a(0) = 1$, so that

$$p_a(t) = \exp(-|\alpha|^2)\sum_n \frac{|\alpha|^{2n}}{n!}\cos^2(g\sqrt{n+1}\,t).$$

17-2. Show for atoms injected in the lower level that $p_n(t_i+\tau)$ is given by Eq. (9).

17-3 Calculate p_n of Eq. (17) for an inhomogeneously broadened medium. Hint: see Sec. 11-1 on optical nutation.

17-4. Calculate p_n of Eq. (17) for a sufficiently weak interaction that first-order perturbation theory is valid. Evaluate the product explicitly and identify the kind of distribution you find.

17-5. Show that an average over exponential decay in Eq. (17) yields the $\mathcal{A}_n$ and $\mathcal{B}_n$ coefficients of Eqs. (27) and (28).

17-6. Calculate the off-diagonal elements $\rho_{nm}(t_i+\tau)$ corresponding to the diagonal elements given by Eqs. (7) and (9). Assume an arbitrary initial field value $\rho_{nm}(t_i)$. What happens if $\rho_{nm}(t_i) = 0$ for $n \neq m$?

17-7. Expand the $\mathcal{A}_n$ and $\mathcal{B}_n$ coefficients of Eqs. (27) and (28) to fourth order in g (second order in R) to find the fourth-order equation of motion for $\dot{p}_n$. Find the equation of motion for the average photon number $\langle n(t)\rangle$.

17-8. Using the fourth-order values of $\mathcal{A}_n$ and $\mathcal{B}_n$ found in Prob. 17-7, show that the steady-state photon statistics are given by

$$p_n = p_0 \prod_{k=1}^{n} \frac{\bar{n}\nu/Q + \frac{1}{2}R_a R(1 - kR)}{(\bar{n}+1)\nu/Q + \frac{1}{2}R_b R(1 - kR)} . \tag{95}$$

17-9. For a zero temperature cavity reservoir and a vanishing lower-level pump rate, Eq. (95) reduces to

$$p_n = p_0 \prod_{k=1}^{n} \frac{A - kB}{\nu/Q} , \tag{96}$$

where $A = \frac{1}{2}R_a R$ and $B = RA$. Using this distribution, show that

$$\langle n\rangle = \frac{A - \nu/Q}{B} - 1 . \tag{97}$$

Hint: write n in terms of $[A - B(n + 1)]Q/\nu$.

17-10. Show that the mean-square deviation is given by $\langle n^2\rangle - \langle n\rangle^2 = \nu/QB$. Hint: write n^2 in terms of $[A - B(n + 1)][A - B(n + 2)](Q/\nu)^2$ and use (97).

17-11. Calculate $\dot{\rho}_{nm}$ by generalizing the analysis of Eq. (36).

17-12. Write the equations of motion for the density matrix elements for the level scheme in Fig. 17-9. These matrix elements correspond to the semiclassical Fourier components of the polarization and populations of Sec. 8-1 as follows: ρ_{51} to the sidemode polarization component p_1, ρ_{24} to p_3^*, ρ_{52} and ρ_{41} to the pump-mode polarization p_2, ρ_{54} to the population pulsation component n_{a1}, and ρ_{21} to n_{b1}. A general derivation based on this approach is given by Sargent, Holm, and Zubairy (1985).

INDEX